The Accidental CIO

The Accidental CIO

A Lean and Agile Playbook for IT Leaders

Scott Millett

WILEY

Published by John Wiley & Sons, Inc., Hoboken, New Jersey.
Published simultaneously in Canada and the United Kingdom.

ISBNs: 9781119612094 (paperback), 9781119612193 (ePDF), 9781119612100 (ePub)

Library of Congress Control Number: 2022941855

Cover image: © Nora Carol Photography/Getty Images
Cover design: Wiley

SKY10069873_031824

For my wife. What you do makes a difference.

About the Author

Scott Millett is the CIO for Iglu.com. During his 20+ years in IT, he has held roles from analyst programmer to software development manager, and enterprise architect to CIO. He was awarded the ASP.NET MVP in 2010 and 2011. He is also the author of *Patterns, Principles, and Practices of Domain-Driven Design*, Wrox; (20 April 2015), *Professional ASP.NET Design Patterns*, Wrox; (10 Sept. 2010), and *Professional Enterprise .NET*, Wrox; (2 Oct. 2009). If you would like to contact Scott, feel free to write to him at scott@elbandit.co.uk, give him a tweet @ScottMillett, or become friends via www.linkedin.com/in/scottmillett.

Acknowledgments

It took a long time to write this book, perhaps too long. This book represents my knowledge gained over the years, not just from the theories from industry experts or other authors but also from the many people I have had the pleasure of working with and learning from. While I don't have the time, or the space, to list everyone that contributed to helping me understand how to be an effective CIO, I would like to acknowledge at least a few people to whom I am forever thankful for giving me their time and sharing their wisdom.

The Wiley Team: Thanks to Jim Minatel, Tom Dinse, Pete Gaughan, Archana Pragash, and all at Wiley for keeping faith in me and helping get this book finished.

For the inspiration and contribution to the field of IT and strategic thinking: Simon Wardley.

For my colleagues for supporting me on my journey and teaching more than you will ever know: Niall Gray, Pete Beams, Dan Clay, Phil Banyard, Renato Oliveira, Nancy Emery, Matt Compton, Dean Thomas, Dave McDermid, Nick Tune, Seif Attar, Richard Limburn, Marcin Frycz, Tobias Randall, Reece Campbell, Ian Whillians, David Gooch, Kirsteen Fox, Dave Mills, Ed Burke, Siobhan Commins, Lorna Vincent, Richard Downs. . . . Oh and (Ski) Simon McIntyre!

For strategic clarity, vision, and humour: Glen Gill, Rob Brennan, David ODonnell, and Dan Brown.

For the board meetings on the seafront: Nicholas Pink.

For the influential ghosts of Christmas past: Mark Dermody, Aidan Fitzpatrick, Stefan Barden, Donal Linehan, Andreas Panteli, Steve Sharpe, Bob Mulumudi, Steve Mills, Dan Glyde, Julian Heritage, Isabel Mack, and Dean Maskell.

For helping me understand what happiness is: My wife, for your kindness and consideration for others. For always putting the we before the me. To my dearest Bertie and Primrose, thank you for being perfectly imperfect. Good luck as you begin your journey on changing from children into young adults.

Contents at a Glance

Contents

Foreword

In the seminal "Manager's Theories about the Process of Innovation" (*Journal of Management Studies*, March 2002), Graeme Salaman and John Storey elegantly unravel the innovation paradox pervading contemporary organizations. They argue that the key to enduring success lies in the meticulous balance between survival today, which requires the efficient exploitation of existing competencies (marked by virtues of coherence, coordination, and stability), and survival tomorrow, which requires exploration and the dissolution of these very same virtues.

This intricate dance of dichotomies is further complicated by the axiom that today's breakthroughs morph into tomorrow's utilities. In an ever-evolving marketplace, the territories once charted in the name of exploration swiftly transition into arenas of exploitation. Furthermore, as innovations evolve into commodities, they lay the groundwork for venturing into new, adjacent domains, thereby perpetuating a cycle of renewal. In history, the standardization of simple mechanical hardware such as the nut and bolt enabled complex machinery. More recently, the transformation of bespoke computing into ubiquitous cloud services has catalyzed the emergence of Big Data and AI. This process underscores the relentless progression of technology and enterprise.

This dynamic interplay between the poles of exploration and exploitation underlines a perpetual motion, challenging the conventional understanding of management. Nothing stands still; everything is in motion; everything has a context. There are no simple solutions.

Into this turbulent landscape, organizations try to navigate and distinguish themselves with novel offerings while striving for operational excellence in commoditized domains. The constant motion and evolution from novelty to utility demands contextual agility in leadership and strategy. What works in one domain does not work in the other or even in the transition between the two. Amidst this maelstrom, leaders are tasked with decisions of monumental consequence, shaping the strategic direction, governance, talent development, and clarity with which their organizations communicate and survive both today and tomorrow. Missteps can precipitate misalignments with severe repercussions.

While many frameworks offer navigational aids in this environment, their applicability varies significantly across different contexts. The challenge of leadership lies in discerning the appropriate approach guided by the unique economic and technological contours of one's own organization. To do this, you have to understand the landscape that the organization operates within.

Dedicated to demystifying this landscape, *The Accidental CIO* comprehensively explores frameworks, methodologies, and techniques used by contemporary leadership and places them in context. From the strategic precision of the Hoshin Kanri process to the agility embedded in the OODA loop, from the creativity of design thinking to the adaptability of flexible planning horizons, this book weaves a rich tapestry of insights. It integrates strategic, architectural, and operational perspectives, offering a lucid understanding of the terrain leaders must navigate.

For executives championing innovation while managing commodity operations, fostering consensus, and driving strategic collaboration across all levels of the organization, *The Accidental CIO* emerges as an indispensable guide. In an era where adaptability and strategic foresight are paramount, this book is not just a resource but a beacon for the forward-thinking leader.

—Simon Wardley

Introduction

If there is no struggle, there is no progress.

—*Frederick Douglass*

Be yourself. Everyone else is taken.

—*Oscar Wilde*

How do you get to Carnegie Hall? - Practice, practice, practice

—*Anonymous*

We think too much and feel too little. More than machinery, we need humanity; more than cleverness, we need kindness and gentleness. Without these qualities, life will be violent and all will be lost.

—*Charlie Chaplin*

It was my first day on my new job, and it hit me like a bucket of ice-cold water. It wasn't that I had imposter syndrome. I just didn't know what to do.

Back in January of 2015 I received an offer from `Iglu.com`, an online travel agent for the cruise and ski markets, to become its first IT director. I wanted a change from `Wiggle.co.uk`, where I had been the first full time developer working along the founder, development manager, and most recently an enterprise architect. The new role was a step up, but I was confident in my technical ability. After all, it was only another form of e-commerce, and I was comfortable with that. But I wasn't prepared for a role as the most senior leader in IT and one that was part of the exec group—one that not only needed to lead and inspire an IT department, ensuring day-to-day reliable operational running, but one that had a pivotal role in contributing to the organization's digital transformation. It was a role that required me to be a business leader as well as a technical leader. I realized I had a lot to learn. Over the following years I studied and grew professionally. I learned both the theory and how to put it into practice. I had begun my journey to become a strategic CIO.

This book is the codification of all the knowledge I acquired, a playbook that I hope will be useful on your journey as you transition to a CIO or an IT leadership role. My context, like yours, is unique; the challenges you experience will differ from what I faced. However, if like me you have found yourself in an IT leadership position where you were unsure on your next move, then this book will provide you the guidance to help your orientation as you navigate the trials and tribulations of a life as a CIO.

Why Should You Care? The CIO Challenge

Becoming a CIO is a hugely rewarding role and one that is critical to nearly all modern businesses. Because of their unique position in the organization, CIOs understand the constraints and opportunities of the business as well as having knowledge of how to mitigate or capitalize on them. This makes them best placed to take a more active role in digital transformation projects, moving beyond implementing new technologies to spearheading organizational transformation and driving business value. However, it's still a relatively new role that's not very well understood by the rest of the business and the board; it is stressful and rapidly evolving. All of this is occurring within an environment of accelerated transformation, emerging technology, constant disruptions, rising customer expectations, against a backdrop of huge sociopolitical challenges. In short, a lot is expected from IT leaders and CIOs in this most turbulent of times.

The evolving expectations of the CIO to lead, disrupt and transform, run, mitigate risk, consolidate, and grow can appear to be contradictory based on the archetypes of IT leaders we have come to know. These contradictions form what Martha Heller calls the CIO Paradox as detailed in her book *The CIO Paradox: Battling the Contradictions of IT Leadership (Routledge 2012)*:

- The "Innovator's Dilemma" paradox refers to the conflict of having to manage the balance between the requirement to stay operationally stable, secure, reliable, and compliant with the need to innovate, experiment, and take chances.
- The "Business-IT Alignment" paradox alludes to the need for CIOs to be technical experts as well as understand the business to ensure strategic alignment and coordination. This is difficult due to the complexity and rapid evolution of technology, the speed of business needs, and the time it takes to deliver technology.
- The "Digital Literacy" paradox relates to the challenges that CIOs face caused by other executives' ignorance of the consequences of technology choices and how it affects the company.
- The "Influence" paradox refers to the difficulty of acquiring authority to make decisions inside the business when they are often considered a service provider or an overhead.

- The "Blame" paradox speaks to the difficulty of CIOs accepting accountability for the outcomes of technological projects when they don't have complete control over decision-making. This is chiefly caused by asking CIOs to deliver defined scope or output rather than outcomes.

This set of conflicting forces is deeply embedded in the operating and mental models of organizations that have been formed within contexts that are no longer relevant today. The problem is that the purpose the system (the IT operating model) was designed for has changed. The old system is based on archetypes of CIOs that are mutually exclusive in that either they specialize in running efficient operations (the service provider, order taking, stable, secure, process-oriented) or they are focused on innovation (the disruptor, adaptable, innovative, lightweight governance, and fast). CIOs don't need to be innovators *or* operational; they need to focus on innovation *and* operational stability. They need to manage digital transformation, digital optimization, and operation efficiency. A good CIO can make or break a company. However, boards hire or promote technologists. What they need are strategic leaders who specialize in technology.

Taking Action: Becoming a Strategic Leader

Great CIOs are sought after; they are partners, cocreators, consultants, and advisors. They are business leaders first, ones that just happen to be accountable for the technology within an organization. They report to the CEO, have a seat at the top table, contribute, and sometimes lead an organization to digital transformation and strategic success. They achieve this by balancing and adapting to meet the variety of challenges they face. These are the core behaviors that they exhibit:

- **Coauthor strategy.**
  Great CIOs are not order takers. They coauthor strategy and focus on what matters, namely generating enterprise value. They can achieve this because of their deep knowledge of the business and operating models. They understand the context that the business operates in and the material factors that can affect the organization. They know what the business needs to do to be successful—where it will play and how it will win. They interpret where technology contributes and where it can lead.
- **Focus on outcomes over output.**
  It doesn't matter what you do if it doesn't make an impact. Great CIOs bridge the gap between business impact and technology output by focusing on what

business outcomes are required for success, how technology can be used to achieve them, and how best to organize teams to execute them.

- **Structure teams for intrinsic motivation.**
  Great CIOs know that they are only as good as their team. Great CIOs excel in recruiting, developing, and retaining talent. They do this by ensuring teams are motivated to solve complex problems and deliver value. They achieve this by designing an operating model to support people's need for purpose, autonomy, and mastery.
- **Focus on being agile, not doing agile.**
  Great CIOs know that to bridge the paradoxes, they need to be adaptable. There is no single way of doing something. Agile is appropriate for some problems, whereas big upfront design is suitable for others. Failure is expected when exploring uncharted problem spaces but not when working in well-known and understood areas. Sometimes it's best to buy and sometimes it makes sense to build. Great CIOs adapt their methods and team dynamics depending on context.
- **Manage the flow of work, not people.**
  Great CIOs work on the system, not in it. They leverage their power to remove impediments and inspire teams with an aligned vision, using their strategic, social, and relationship skills to influence and lead in change and innovation. They manage the flow of work; they lead the people.

CIOs that show aptitude in these areas will have an impact greater than any other exec on business success. However, to get there you will need the right attitude. You will need to embrace a growth mindset. You need to continuously learn, adapt, and develop. Picking up this book is your first step on that journey.

What Will You Learn?

I am going to show you how I became a strategic CIO. I'll walk you through, step-by-step, how to create an IT strategy and a tactical plan to execute it. I will show you how to design an operating model to deliver results. You'll discover how to create an IT organization that is empowered and focused on solving customer problems and generating enterprise value. You'll learn to adapt your methods depending on the context of the problem you are facing. You'll understand the science behind what motivates teams and how to change behavior. You'll be taught how to think like a business leader and focus on impactful business outcomes rather than IT output alone.

This book is organized into three parts, as illustrated in Figure I.1.

- Part 1: "A New System of Work"
  I explain the underlying factors that require us to change the system, the philosophies that we need to embrace for a new way of working and thinking, and the science of how we can change the system and inspire our teams.

- Part 2: "Designing an Adaptive Operating Model"
  We examine each component of the operating model, from ways of working to governance, leadership, talent, and organizational structure. I'll show you how each component relates to the others, and how they can adapt to the problem context they are addressing.
- Part 3: "Strategy to Execution"
  Where I show you how you to understand your business at a deeper level so that you can interpret business needs and define an IT strategy that will contribute to business success. Then how to deploy and execute that strategy, ensuring alignment across the organization at both a tactical and operational level.

Each chapter revolves around a central argument that we need a balanced and adaptive way of working across all of IT to manage the paradoxes and the extremes of being a CIO.

Feel free to read this book from beginning to end, or if you wish, dip into any chapter that is of interest. I do suggest, however, that you read Part 1, "A New System of Work" first. Part 1 lays the groundwork and context for the philosophies that underpin much of Part 2, "Designing an Adaptive Operating Model," and Part 3, "Strategy to Execution."

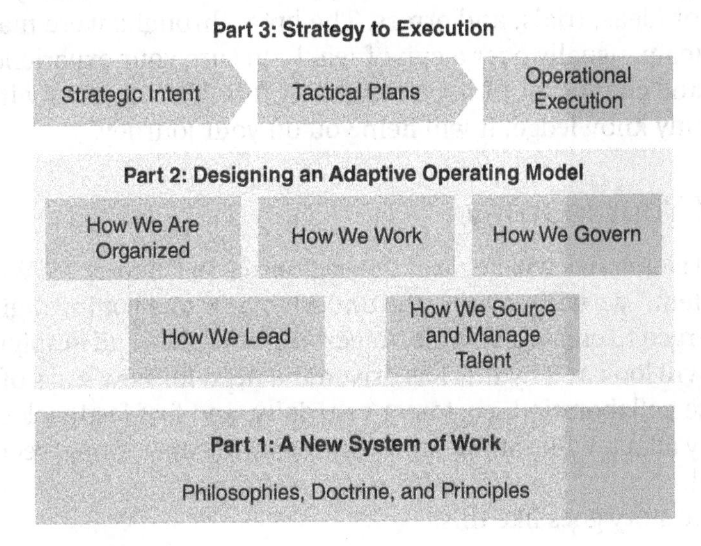

Figure I.1: How this book is organized

Why Should You Listen to Me?

This book is based on more than 20 years of experience in IT, from an operational level as engineer, to a tactical level as a development manager and solution architect, to most recently at a strategic level as an enterprise architect and CIO. I have worked for large multinational companies, startups, and high-growth organizations. I have worked with many brilliant CIOs, CTOs, IT directors, enterprise architects, CEOs, founders, and experts across many fields in IT and the wider business. When I stepped up to IT director, I looked for a book that would support me to become an IT strategic leader, one that would help me design the IT organization to address the paradoxes and challenges covered earlier, one that would show me how to effectively build and execute strategy. But there was no book. No manual. No holistic view of how to build a system for success. I had to learn the hard way. I attended many CIO/CTO events, I read anything I could get my hands on, I watched all the videos, I listened to the podcasts, I read blog articles, I spoke to other IT leaders, my old bosses, and new friends. I learned through trial and error. Many errors. Over time I was able to put the puzzle together piece by piece.

I wrote this book because I wanted to codify my knowledge and help refine my own understanding of how to be an effective CIO. I wanted to demystify the problem space and help others who will be going through a similar experience. This book shows the end of my journey, or rather the end of the beginning. The format of this book, the diagrams and content, will appear structured and polished. My journey to get here wasn't as black-and-white as the text in the book. It was never as neat and tidy, or as lined up as I might like it to have been. My journey was full of messy white boards, notebooks full of ideas, trials, and errors. The breakthroughs were made collaboratively with my team, usually over a cup of tea. I am sure your experience will be just as challenging and chaotic, but I hope ultimately fun. My intention with this book is that by sharing my knowledge, it will help you on your journey.

The Start of Your Journey

As with all good stories, we will begin at the beginning. In Chapter 1, "Why We Need to Change the System," we will examine the underlying factors behind digital disruption that have converged to create a volatile, uncertain, complex, and ambiguous business landscape. We will look at why this has driven the need for new ways of working: the need to embrace collaboration, customer centricity, and fast feedback cycles and the ability to quickly adapt, while at the same time ensuring operational security, stability, and cost control.

The rest of the story goes like this:

- **Chapter 2: Philosophies for a New System.** Where I introduce you to lean, design thinking, agile, and Wardley mapping. These underlying philosophies will be used to shape your strategic thinking.

- **Chapter 3: How to Change the System.** Where I explain systems thinking as a way of making sense of complexity and how mastery, purpose, and autonomy can instill intrinsic motivation in teams.
- **Chapter 4: The Anatomy of an Operating Model.** Where I examine the various components of an operating model that work together to determine "how things are done."
- **Chapter 5: How We Are Organized.** Where I delve into the how and why of structuring a department and the development teams within it.
- **Chapter 6: How We Work.** Where I cover the many approaches to understanding and solving problems and how to choose the most appropriate method.
- **Chapter 7: How We Govern.** Where I discuss how the various elements of governance can adapt and complement the ways of working depending on unique problem context.
- **Chapter 8: How We Source and Manage Talent.** Where I detail perhaps one of the most important responsibilities of a CIO succeeding. How we can attract, develop, and retain talent.
- **Chapter 9: How We Lead.** Where I introduce you to the notion of servant leadership and how, by support and leading rather than managing teams, you can get the best out of them.
- **Chapter 10: Understanding Your Business.** Where I enable you to understand your business at a deeper level.
- **Chapter 11: IT Strategic Contribution.** Where I show you how to interpret business need and create an IT strategy.
- **Chapter 12: Tactical Planning: Deploying Strategy.** Where I explain how to deploy strategy and define the tactical initiatives that will bridge between strategic intent and operational action.
- **Chapter 13: Operational Planning: Execution, Learning, and Adapting.** Where I examine operational execution and how to review and adapt strategic, tactical, and operational planning based on feedback.

A New System of Work

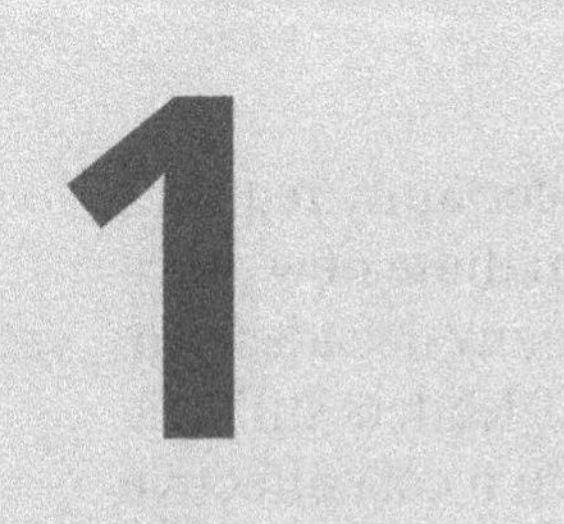

Why We Need to Change The System

It is not the most intellectual of the species that survives; it is not the strongest that survives; but the species that survives is the one that is able best to adapt and adjust to the changing environment in which it finds itself.

—Leon C. Megginson on Darwin's On the Origin of Species

It is not necessary to change. Survival is not mandatory.

—W. Edwards Deming

The techniques that worked so extraordinarily well when applied to sustaining technologies, however, clearly failed badly when applied to markets or applications that did not yet exist.

Clayton M. Christensen, The Innovator's Dilemma: When New Technologies —Cause Great Firms to Fail

The strategic role of IT has increased greatly in the last number of decades due to the impact of digital technologies at both a business and a social level. The old model of IT as a support center or an order taker is outdated in today's complex digital business world, where pace, adaptability, creativity, innovation and collaboration are fundamental to succeed in both existing and new business endeavors. This requires a new IT operating model. One that can contribute to digital exploration and the discovery of new opportunities but at the same time can maximize and exploit the performance of an incumbent business model.

This chapter begins with an examination of the impact that new technology, rising customer expectations, and an adaptive competitive set have had on the business landscape and why it is now often characterized as a volatile, uncertain, complex, and ambiguous environment. This context explains the complex and unpredictable challenges the business, and therefore IT leaders, will face. This uncertainty requires a change in how we operate, as the characteristics and approaches to these new problems are very different to the traditional ways of working formed many decades ago. But at the same time, we must understand what has not changed. IT leaders still face the same problems of supporting an organization's need to scale in a cost effective and secure manner.

To understand how best to manage challenges in support of these two extreme ends of business need, and everything in between, we will leverage the Cynefin decision framework. The framework will aid your situational awareness, helping you to categorize problems based on observable characteristics. Through this categorization of problems, you will learn of the most appropriate methods of approaching them, which will give you a greater chance of succeeding in solving them.

The bottom line is that we need an adaptive IT operating model that is fit for purpose for today's complex and volatile business environment, one that can innovate and cocreate but at the same time provide scalable and efficient solutions to exploit the organization's current model and maximize value.

The Age of Digital Disruption

Software is eating the world, it is the rise of the knowledge worker, and according to Davos World Economic Forum, we are in the Fourth Industrial Revolution. Call it what you will, but the speed at which technology has impacted our lives, from both a business and social context, has focused the need for a more strategic IT leadership. This digital disruption is important to understand as it is the fundamental reason IT leaders, and businesses, need to change the way they lead and make decisions. The factors causing this disruptive business landscape are faster and cheaper technology, evolving customer expectations, adaptive organizations, and new business models.

Disruptive Technology

Access to cheap, pay as you go, infinite cloud computing power and storage, means low cash flow startups that can't afford large capital investments are no longer prohibited from launching into, and disrupting, established markets with speed. Managed cloud services for databases, infrastructure, and machine learning allow companies to focus on value-add activities such as innovation and experimentation by building upon a rising platform of leading-edge capabilities rather than focusing on keeping the lights on. The adoption of mobile technologies, voice-activated assistants, and connected homes has meant that technology is penetrating every aspect of our lives, and businesses have been keen to capitalize on being able to reach customers 24/7.

As technology became cheaper and more powerful, so did data storage and the tools to analyze it. This, coupled with the explosion in the amount of data that was available on customer behaviors due to the always-on connected devices and IOT, enabled companies to start creating powerful and relevant experiences for end users. Digital companies are now able to make better choices on how to evolve their products and services by analyzing the mass of information gained from customer behavior. Data visualization and analytics platforms are easier to use than ever before to gain deeper insights and understanding; furthermore, these platforms are made directly available to the teams that run business departments and make decisions, vastly reducing the lead time from insight to action.

The Rise of Customer Expectations and Influence

Perhaps the single biggest impact from the advances in technology and new value propositions is that of the shift in expectations and the influence of customers on today's business environment. Customers have high expectations for an immediate response to service requests and the fulfillment of purchases 24 hours a day and seven days a week. Personal time by end users is increasingly spent in the cloud on highly polished user experiences with high levels of customer service. This results in customers, both internal and external, having much lower tolerances and higher expectations, when it comes to user experience than before.

Consumer influence is now greater than ever due to the wide adoption of social media platforms, which enable customer networks to have a direct impact on brands. Because of the power of customers, there has been a shift to how organizations are selling and positioning themselves. Customer-first and customer-centered strategies are now the norm due to the value customers place on service and product experience. This is heavily influenced by the large tech companies such as Amazon and Google.

Adaptive Organizations

Companies that have thrived in the digital era have been those that have been able to adapt their ways of working. Highly collaborative, customer focused businesses who learn and adapt at pace, are far better positioned to remove constraints and exploit opportunities. Traditional methods of having a strategy with a fixed three-year detailed plan is no longer as useful as they once were due to the rapid pace at which competition and customer expectations are moving. Companies that have embraced and pushed down a learn-and-adapt feedback cycle to employees versus a command-and-control mentality are finding that they can innovate far more effectively. Pushing down accountability and autonomy to highly skilled and talented employees that are close to the problem, and aligned to the company strategic need and vision, is proving to be an effective way of working. Empowering workers to analyze data to identify patterns and trends to make more informed choices and better decisions when determining what to do to achieve goals is reaping rewards. Embracing the reality of the sometimes-chaotic rate of change and unpredictability in the business context rather than trying to control it is really the only strategy.

Innovation has proved to be ever more essential for businesses to adapt and transform. Because of the art of the possible that new digital technologies afford, organizations should challenge long-held assumptions around business model propositions, test hypotheses through fast and inexpensive experimentation, and be comfortable with making mistakes along the way to learn. Cloud-based technology has made it much faster to test ideas without committing to large upfront costs. To avoid being disrupted, progressive businesses are disrupting their own business models. This ruthless focus on innovation and self-disruption through continuous learning, all enabled by a culture that values experimenting and not being afraid to fail, is now a fundamental capability that is needed to succeed.

New Business Models

The convergence of technology, shifting customer expectations, and the rise of highly adaptive organizations has led to a tremendous amount of business model change, not only through the digitalization of existing value propositions but also through the creation of new business models. This evolution has been powered largely by a move from products to services and platforms. Products are increasingly moving from an ownership model to an access model, and customers are valuing experiences that evolve daily over waiting for the next version of a product. We no longer buy music; we rent it. We don't own films; we stream them. We buy smart devices connected to the Internet along with subscriptions to services. Free models, subscription, on-demand, and freemium are just some of the new ways to sell products as services that have helped to disrupt the incumbent business models.

In addition to services, there has been an explosion of platforms, perhaps the most notable being Amazon's marketplace, Apples app store, Etsy, Facebook, Uber, and Airbnb. These platforms rely on partners to drive value, with their value proposition being the connection between partner and consumer on a global, secure and highly polished user experience platform.

Operating in a Volatile, Uncertain, Complex, and Ambiguous Business Environment

In today's digital world, the size of a company is no longer an advantage, and neither is a history of dominance in a market. Businesses that have sat on their laurels are ready for disruption, and not just from known competitors but also by tech-savvy startups and entrepreneurs that have identified untapped customer needs in the market. Access to data and analysis on customer behavior and the commoditization of technology capabilities has leveled the playing field and enabled new competitors to enter a market at pace. These new entrants, who hold the customer central to the business model, can move at pace, adapt, and embrace new ways of working to quickly take market share from the slow-to-react incumbents. All these changes are happening on a global scale due to the ease of the delivery of new digital services. The pace, widespread scale of changes, and inability to predict the next industry disruption has made for a chaotic and unprecedented business environment.

VUCA is an acronym that originated at the US Army War College to describe the volatility, uncertainty, complexity, and ambiguity of the world after the Cold War and geopolitical instability and to reflect on what military operations are increasingly faced with, especially with the rise in extremism and terrorism. However, since the turn of the century, the term has been adopted by business leaders to describe the pace of unpredictable change and evolution in the business environment.

VUCA applied to the business environment can be thought of as follows:

Volatility There are no longer any untouchable business models; no industry is exempt from disruption. Change is impossible to predict, but it is now the norm—the market is volatile.

Uncertainty There are no longer any certainties; investments and big bets based on ideas and plans that take years to develop are extremely risky.

Complexity Cause and effect are not linear in a global market that has high customer expectations of digital services and products. It is difficult to understand what will succeed and what changes (your own and competitors) will have undesired effects.

Ambiguity Interpreting meaning, analyzing trends and spotting patterns is extremely difficult in an ever-moving business environment.

As new business models are created, while others evolve beyond recognition and some become irrelevant, we find that three-year plans are too rigid and result in slow reactions to competitors and changing customer expectations. Companies without the skills and the ability to sense and adapt to changing markets will struggle in the VUCA environment and will cease to be relevant to customers.

IT is now central to all businesses, and the role of IT leaders is more demanding than ever in an increasingly VUCA business world. IT leaders must learn to embrace this new reality and adjust operating models accordingly. They must place a focus on collaborative learning and adapting to customer needs at pace to manage initiatives in increasingly hard to predict and complex problem domains. Without the ability to adapt ways of working and other operating model components based on the problems they face, IT leaders run the risk of being the very blockers to their enterprise's digital future.

Leading IT in a Complex and Adaptive World

As previously discussed, the business environment is more volatile than ever due to technology advances, greater access to data, higher customer expectations, the speed of business model evolution, and competitors challenging from across the globe. In this new and more challenging environment, IT leaders no longer only face simple problems that can be solved through best practice and by following a plan. Instead, we are seeing more complex and complicated problems and opportunities that require a different type of approach to solutions.

IT leaders need to accept the uncertainty of working in complex and unpredictable problem domains and avoid behaviors that at best give the illusion of control and predictability. You can certainly plan in complex adaptive problem domains but you need to accept that your plans are likely to change as you receive feedback on your actions. The unescapable fact is that the way we make sense and understand a complex system is by interacting with it, not by analyzing it. This is achieved through experimentation of ideas and hypotheses. Leaders need to manage the expectations of the enterprise and be comfortable with the fact that because teams need to validate ideas and hypotheses there will be failures. These failures should be viewed as learnings, and through these learnings a deeper understanding of the problem domain and its underlying system and patterns will emerge which will result in the discover of actions that will result in the desired outcomes.

However, not all businesses are startups, and not all problems are within the complex and unpredictable domain. To fund digital transformation and exploration, there is a need to exploit and maximize the profits of existing business models. While these problems are still challenging, they are based on proven value propositions and therefore more knowable. They need a different approach to how we might tackle something new with more unknowable unknowns.

What this requires is an IT operating model that can adapt to consider the context of the problem domain and apply the most appropriate method of dealing with it. Operating in high levels of uncertainty requires new behaviors and ways of working—and not only for IT but for the entire enterprise. However, we should not apply these same ways to problems that are more straightforward and predictable. We must apply a fit-for-purpose way of working that's dependent on context. By explicitly understanding the context of a problem domain, IT leaders can apply an appropriate and relevant solution strategy. There is no such thing as one size fits all and no magic frameworks. IT leaders will need to create an operating model that can manage the various problem contexts that are in play in modern businesses. This requires a high level of situational awareness and the ability to make the most appropriate decisions on which methods you use for each problem you face.

Decision-Making with the Cynefin Framework

The Cynefin framework, shown in Figure 1.1, is the work of David Snowdon and his research network Cognitive Edge. The framework helps leaders to understand what context they are operating in and therefore how best to make decisions and lead efforts within.

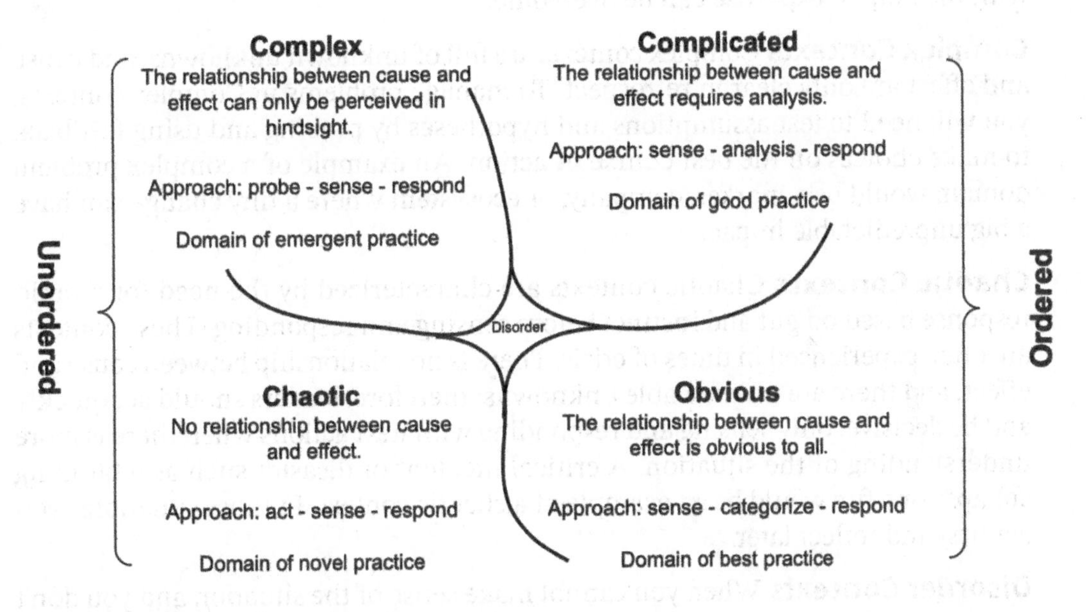

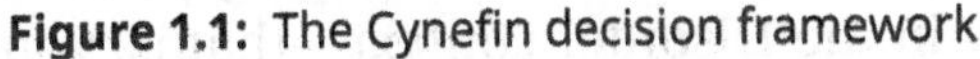
Figure 1.1: The Cynefin decision framework

The framework is composed of five sections that help leaders understand the context of the problem domain and how best to approach solving them:

Obvious Contexts Leaders should follow best practice solutions in the context that is defined of known knowns and avoid reinventing the wheel. Problems in this context are linear in nature and have a clear cause and effect that present an obvious solution approach. Leaders should be able to recognize the problem domain through sensing and categorizing and then select an appropriate response to deal with it. Problems that are typically handled by a service desk will likely be in the obvious context, where there is a set procedure to identify and resolve. Problems in this context are great candidates for automation.

Complicated Contexts Complicated contexts are made up of known unknowns, and as with an Obvious context, there is a correlation between cause and effect, but it requires analysis. In this context, the framework recommends to sense and analyze the data before responding with a solution. Leveraging expert knowledge to analyze the data is a good practice to quickly getting to a suitable solution. An example of a complicated problem domain is that of building an extension on a house. It is a difficult task with challenges but one with known unknowns that with the help of expertise can be overcome.

Complex Contexts Complex contexts are full of unknown unknowns, and cause and effect are only clear in retrospect. To manage problems in Complex contexts, you will need to test assumptions and hypotheses by probing and using feedback to make choices on the best course of action. An example of a complex problem domain would be a market, company, or ecosystem where a tiny change can have a big unpredictable impact.

Chaotic Contexts Chaotic contexts are characterized by the need for a rapid response based on gut and instinct before sensing and responding. These contexts are often experienced in times of crisis. There is no relationship between cause and effect, and there are unknowable unknowns; therefore, leaders should act quickly and be decisive, only sensing and responding with next actions when there is more understanding of the situation. A critical incident or disaster such as a building collapse or a fire would be an example of a chaotic context. In those situations, you act first and reflect later.

Disorder Contexts When you cannot make sense of the situation and you don't know the best way forward, you are likely in the Disorder context—that is, you are unaware of which context you are in. The best course of action is to distill the problem into subproblems and understand the relative context of each in order to move forward. The Disorder context highlights one of the underlying principles of the Cynefin framework—the importance of understanding the context of the problem domain you find yourself in. With an understanding of context, you can adapt your problem-solving skills accordingly.

IT Needs to Operate in Both the Ordered and Unordered Problem Spaces

The obvious and complicated domains in the Cynefin framework relate to problems in ordered systems. We can predict cause and effect in these problem domains with relative confidence. Software development in ordered problem domains should be straightforward, in that you can determine what to do; it may be challenging, but you have a good understanding of the way forward. Ordered systems are good candidates for off-the-shelf software as they tend to be systems of record, highly evolved or commodity capabilities in areas such as HR or finance (assuming that these are not your business's areas of differentiation). You can also apply traditional project methodologies and perform upfront planning by leveraging best practice or analysis model patterns when the problem domain is obvious and well understood.

However, due to the need for more creative and innovative business solutions brought on by the factors that are causing digital disruption, IT leaders are increasingly finding themselves tackling problems within the unordered spaces. Trying to obtain predictability for problems in these contexts is impossible – in that you don't know what you don't know. Instead, leaders should allow teams to propose a hypothesis and then run experiments to discover, learn, and adapt their plans. This may be uncomfortable, but it is nevertheless the reality. Trying to lay out a detailed road map and plan beyond the shortest of timescales will not be worth the paper it's written on and will ultimately be a waste of time. Leaders should empower teams to deliver in small chunks and deliver often, allowing for fast feedback and the ability to correct their course, rather than trying to predict the future with detailed and fixed plans.

IT leaders will face problems that land in both the ordered and unordered space as businesses need to innovate and explore at the same time they are exploiting incumbent business models. As shown in Figure 1.2, this means that the operating model of IT needs to be able to support and contribute to both of these objectives. Therefore, we need an adaptive mode of operating, one that complements the reality of the complex and challenging environment within which businesses exist, one that is comfortable in managing problems in both problem spaces.

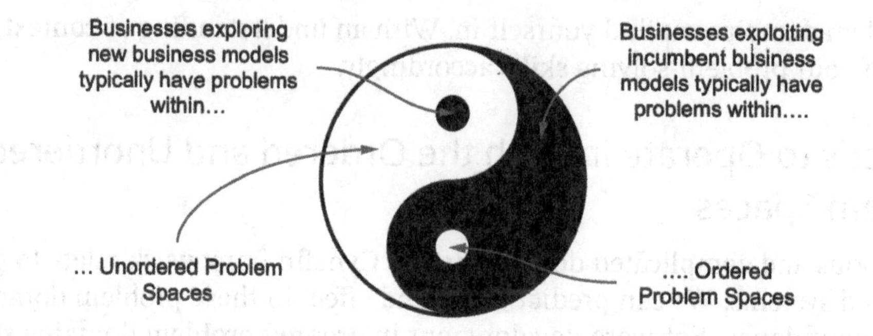
Businesses exploring new business models typically have problems within....
Businesses exploiting incumbent business models typically have problems within....
... Unordered Problem Spaces
.... Ordered Problem Spaces

Figure 1.2: IT needs to operate in both the ordered and unordered problem spaces to support business exploration and exploitation.

Summary

This chapter examined the underlying factors behind digital disruption, namely technology and data, customer expectations, adaptive organizations and new business models. These have converged to create a volatile, uncertain, complex, and ambiguous business landscape, requiring a new way of working that encompasses collaboration, customer centricity, fast feedback cycles, and the ability to quickly adapt. And as IT is more central than ever to business strategy in this digital environment, these complex and adaptive business problems are IT leaders' problems.

However, not all problems are complex. It is essential that IT leaders can coauthor a business exploring new digital opportunities, but at the same time they need to be able to support organizations exploiting incumbent and mature business models to maximize value. These different problems require very different approaches.

To help understand the different ways IT leaders will need to adapt their thinking and ways of working, we looked at the Cynefin decision framework, which shows the different characteristics and the approaches required to solve the problem contexts that exist in modern organizations. The conclusion we can distill from this is that IT needs to move to a new adaptive and balanced operating model that can manage problems in both the ordered and unordered spaces. In Chapter 2, "Philosophies For A New System", we will look at the underlying set of principles and values that will guide us in designing an operating model that will contribute to both business exploration and exploitation.

Philosophies for a New System

Lean thinking defines value as providing benefit to the customer; anything else is waste.

—Eric Ries

The Toyota style is not to create results by working hard. It is a system that says there is no limit to people's creativity. People don't go to Toyota to "work" they go there to "think."

—Taiichi Ohno

I think it's very important to have a feedback loop, where you're constantly thinking about what you've done and how you could be doing it better. I think that's the single best piece of advice—constantly think about how you could be doing things better and questioning yourself.

—Elon Musk

How do we approach solving problems in the exploration of new business models and exploitation of incumbent ones? How do we manage development and delivery in the unordered problem space? And how do we gain situational awareness so that we can apply the most suitable method for problem discovery, solution approach, and delivery execution for each specific problem context?

To answer these questions this book will explore a new adaptable operating model for IT, and in this chapter you will learn about the fundamental principles behind four philosophies that will form the foundations of that operating model:

Design Thinking A creative, nonlinear and iterative approach for complex problem solving based on empathizing with customer needs.

Lean Thinking A human-centered philosophy focused on increasing the flow of customer value, achieved through continuous improvement and the elimination of waste.

Agile A development philosophy that embraces uncertainty and unpredictability in order to manage the complexity of change.

Wardley Mapping A strategic thinking approach that takes situational awareness into account when making strategic decisions.

These four philosophies will run as constant themes throughout this book and will enable you to change your mental model of what an IT department is capable of achieving. Therefore this chapter is essential reading as the operating model that will be introduced in Part 2, "Designing An Adaptive Operating Model," has its structure firmly based around these belief systems. Between Lean thinking and Design thinking's approaches to problem discovery, Agile's approach to managing complexity and the strategic insights that Wardley Mapping can give you will have the foundations in place to create an adaptive operating model that can manage problems for both exploration and exploitation.

Philosophies vs. Methodologies

It's important to understand the relationship between methodologies and philosophies as often ceremonies such as a daily stand up or using a specific tool can be misinterpreted as being agile. Figure 2.1 shows how the lean, agile, design thinking and Wardley Mapping philosophies influence methodologies and how these methodologies in turn influence tools. It is critical to have a good understanding of the fundamental meaning of what it is to be lean and work in an agile manner, as it is often easy to overlook the aims and values, and therefore lose the benefits of these philosophies when we focus only on the methodology. Part 1, "A New System of Work," of this book focuses on the philosophies, whereas Parts 2 & 3, "Strategy To Execution," show the methods and frameworks to put these philosophies into practice.

Discovering Value Using Design Thinking

Design thinking is an iterative and non-linear approach to problem solving, or rather understanding what problem is best to solve. It is useful in exploring complex or "wicked" problem domains characterized by unpredictability and a high level of unknownable unknowns. Design thinking was a way to teach engineers to approach problems in a creative and open-minded way like designers do when faced with complex challenges. At its core it is a human-centric way of working, with empathic research playing a key part to determine what solutions customers really need. The underlying principle is that by focusing on what is best for the customer rather than just what is best for a business you will discover a better product or service. Design thinking is a solution-based framework, as opposed to a problem-focused approach, this small difference has a big impact on how you approach solving a problem:

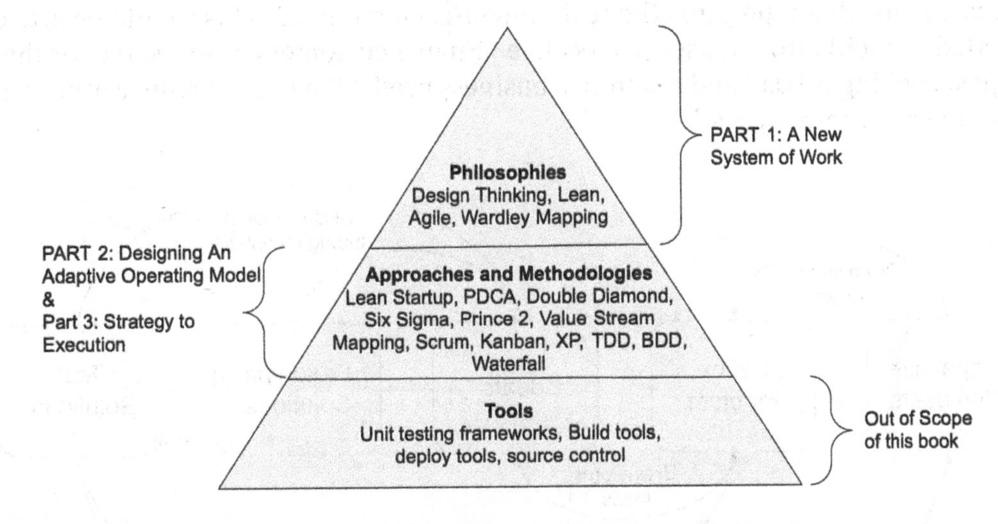

Figure 2.1: The difference between philosophies, methodologies, and tools

- Problem-Focused is a structured approach to finding solutions to immediate problems. It follows a scientific method using both logic and analysis to identify the root cause of a problem and then find the best solution.
- Solution-Focused encourages creative problem solving. It questions the problem itself and the assumptions behind it. The focus is on generating ideas to find a solution to a real need rather than fixing immediate problems. There is a need to think "outside the box," in order to discover alternative ways to identify innovative solutions.

There are Five Stages of Design Thinking, as shown in Figure 2.2; however, these stages are not necessarily sequential, and they can also be run in parallel and be repeated. The approach is iterative in nature with later stages influencing earlier stages by confirming or challenging assumptions, or by the discovery of new ideas.

Empathize When looking to explore a new opportunity for a product or service the first thing you should establish is what is the human need behind it? To discover the underlying need you have to understand people, empathise with them and gather deep insights into what motivates them and the real problems that they are trying to solve. By understanding customers' needs on an emotional and psychological level you can suspend all assumptions and biases you may have about what you think they need, you will then find it easier to "think outside of the box" and challenge the status quo. Empathy is fundamental in order to produce creative and innovative solutions.

Define Once you have gathered information through research and observations, analyzed it, and gained insight, you can state and define the core set of problem statements that represents the real human-centered need. This should be articulated explicitly from a user perspective "Enable customers to self-serve for their post booking needs." and not from a business need: "We need to reduce overheads in the call centre by 5%."

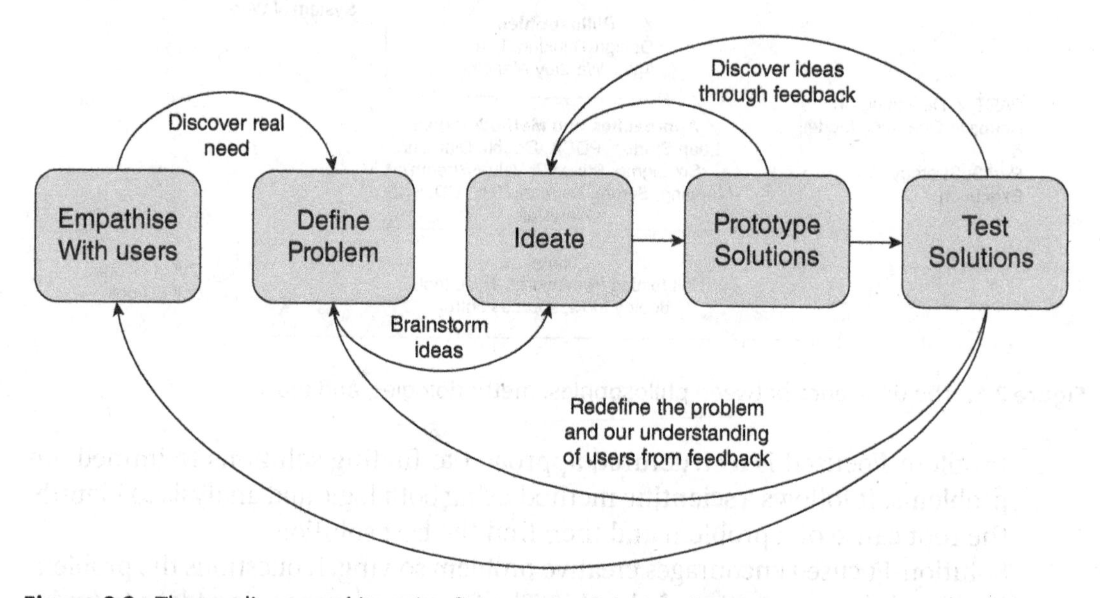

Figure 2.2: The nonlinear and iterative five stages of design thinking

Ideate With a fundamental knowledge of user needs and a clear problem statement you can now begin to generate ideas to solve the problem. Design thinkers will look at the problem from alternative viewpoints and challenge assumptions in order to generate innovative solutions.

Prototype Stage four is to experiment by creating cheap prototypes to discover the best solution for each problem based on the ideas in stage 2. Each prototype is tested within a closed group either by the team itself, with colleagues or a small group of real customers to determine suitability. Having a prototype allows us to understand how real users will experience and interact with the end product or service. Based on this feedback we can either move to the next stage, modify the prototype or reject it, and go back to the ideas stage or even the problem definition itself. By the end of the Prototype stage, the team will have a better idea of the problem and an idea of a solution that they can test with a wider set of real end users.

Test Stage five is about testing the solution based on all of the successful prototype ideas from stage 4 with a larger set of users. Typically we will discover much more about our potential solution, the problem itself, and users' behavior at this stage. Therefore it is highly likely that we will make changes and improvements, iterating through several or even all of these stages before we arrive at a product or service that meets the needs of users.

Eliminating Waste with Lean

Lean is a philosophy that is built upon purpose, people, and process, the focus of which is to create the flow of value through eliminating wasteful activities. In order to achieve this, a culture of continuous learning coupled with trust and respect in employees is needed to ensure that those closest to the value stream are empowered to make decisions to improve flow. Lean, in its essence, is a different way of thinking and acting for an enterprise, the benefits of which are increased competitiveness, innovation, and motivation. Lean looks at problems and failures as opportunities to remove waste and learn more about the value stream. Lean encourages a servant leadership style that embraces reality, where leaders and managers take on the role of coaches for employees, helping to push down accountability and enable autonomy through trust.

The term *lean* was coined in a book entitled *The Machine That Changed the World* (Scribner, 1990) by James Womack, Daniel Jones, and Daniel Roos in 1990 that described what lean production was, which itself was based on an exploration of lean production as presented in an article by John Krafcik on productivity levels between different car manufacturers—notably Toyota. Toyotas way of working has been codified in the Toyota Production System (TPS) and The Toyota Way which represent the foundational principles that lean is based on.

Lean Production: The Toyota Production System and The Toyota Way

The Toyota Production System (TPS) is a set of principles adopted by the Toyota company during the rebuilding effort in Japan after World War II. A disciplined process of practices, principles and cultural ways of working focused on the elimination of wasteful activities that added no value to the production system were developed to deal with shortages of material and money. Although Toyota was itself influenced by its ways of working, TPS is regarded as the origin of lean thinking practices. Figure 2.3 shows the main concepts that make up the TPS.

The two pillars of the TPS are Jidoka and Just-in-Time:

Just-in-Time To enable quick responses to change, reduce waste, and deliver value quickly, apply Just-in-Time manufacturing to ensure we only process what is needed when it is needed. Apply a pull rather than a push system for continuous flow and the reduction of waste.

Jidoka To prevent passing defects down the line from equipment malfunctions or other quality problems, Jidoka refers to the practice of production being stopped automatically by machines that can recognize problems in quality or manually by humans when there is a defect. By stopping production immediately, we can fix the immediate problem before investigating and solving the root cause.

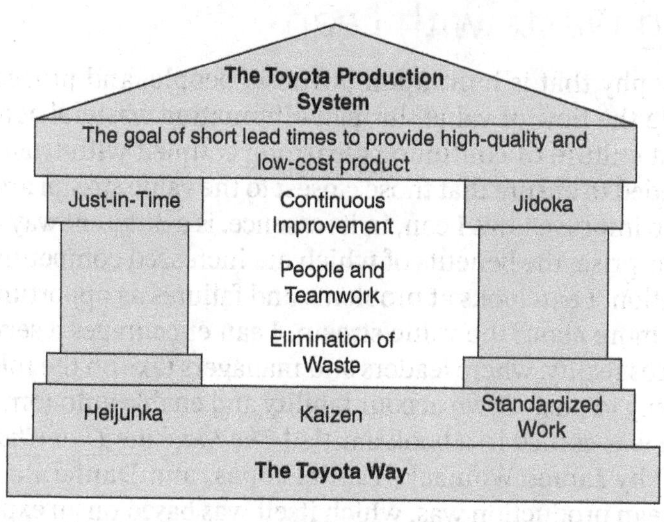

Figure 2.3: The Toyota House—the main concepts that make up the Toyota Production System

In turn the foundations of the philosophy are as follows:

Kaizen This is an approach for continuous improvement that focuses on constant, small, and incremental changes. Many small improvements are easier to implement than large, big-bang change initiatives. The culture is to empower employees to take ownership and accountability for their own process improvement, which builds intrinsic motivation.

Heijunka This refers to the leveling of production with unpredictable customer demand patterns, thus eliminating manufacturing output waste.

Standardized Work This refers to the outcome of baselining and documenting best practices in order to amplify learning. Continuous improvement raises the baseline for all as the improvement becomes the new standardized way of working.

As can be seen in Figure 2.3, underpinning the TPS house is the "Toyota Way": this was the management approach that Toyota formulated. Acting on the principles within the TPS would be impossible without management embracing a deeper attitude change and a mental shift toward the central values of the Toyota Way and instilling these in the workforce. In his book *The Toyota Way* (McGraw Hill, 2020), Dr. Jeffrey Liker interprets this philosophy into 14 principles organised into four sections.

1. Long-term philosophy.
   1.1. Base your management decisions on a long-term philosophy, even at the expense of short-term financial goals.
2. The right process will produce the right results.
   2.1. Create a continuous process flow to bring problems to the surface.
   2.2. Use "pull" systems to avoid overproduction.
   2.3. Level out the workload (work like the tortoise, not the hare).
   2.4. Build a culture of stopping to fix problems, to get quality right the first time.
   2.5. Standardized tasks and processes are the foundation for continuous improvement and employee empowerment.
   2.6. Use visual controls to avoid hiding problems.
   2.7. Use only reliable, thoroughly tested technology that serves your people and process.
3. Add value to the organization by developing your people.
   3.1. Grow leaders who thoroughly understand the work, live the philosophy, and teach it to others.
   3.2. Develop exceptional people and teams who follow your company's philosophy.
   3.3. Respect your extended network of partners and suppliers by challenging them and helping them improve.
4. Continuously solving root problems drives organizational learning.
   4.1. Go and see for yourself to thoroughly understand the situation.
   4.2. Make decisions slowly by consensus, thoroughly considering all options; implement decisions rapidly.
   4.3. Become a learning organization through relentless reflection and continuous improvement.

Lean Enterprise

Inspired by the Toyota production system along with Just-in-Time practices, James Womack and Daniel Jones developed the principles of *Lean Production* to codify how Toyota achieved superior productivity and quality through the flow of value and elimination of waste. This system of techniques and activities of lean production focuses on the elimination of all non-value-adding activities and waste from the business, and can be applied to services' operations as well as manufacturing. In their book *Lean Thinking* (Productivity Press), Womack and Jones, extended the principles of lean beyond production through to the optimisation of value delivery across the entire enterprise. Where it differs from lean production is that you can't simply focus on incrementally improving efficiency over time; you need to adapt the value proposition to ensure you are delivering real value to customers and their needs as the market changes or risk becoming irrelevant. After all, the ultimate waste is providing the wrong value proposition in an optimised manner.

The aim of Lean Thinking is to understand or discover the value you bring to customers and eliminate waste in delivering it to them. There are similarities with design thinking in terms of value discovery and thinking of value from the customer's position, but the core concept of lean thinking focuses on what changes need to be made to what we already do to retain existing customers and deliver to them a product or service of the highest quality. This is achieved by solving immediate problems for customers, hence, problem-focused. The focus is not to start from a blank page and invent, but to start from what is already there and improve it. This is in contrast with design thinking, which is about inventing and exploring completely new value propositions. However we will see in Chapter 6, "How We Work," that the concept of lean has been applied to design thinking via Eric Rees' lean startup framework to optimise the process of exploration.

The philosophy of lean thinking is built upon a continuous feedback cycle that minimizes waste while maximizing customer value. It is based on five principles as shown in Figure 2.4.

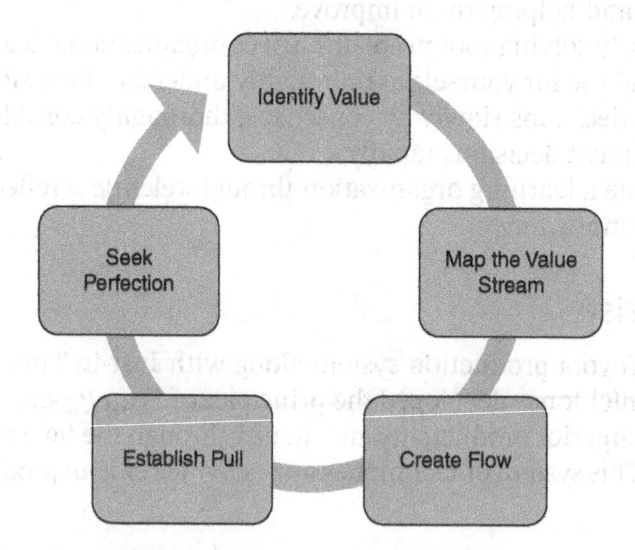

Figure 2.4: The five-step thought process proposed by Womack and Jones in 1996 to guide managers through a lean transformation

Identify Value Lean is based on maximizing the delivery of value. Therefore, it is important to begin with the discovery of what value is worth delivering—in other words, what need are you fulfilling for your customer? You need to start with a vision or a hypothesis on what will be valuable for customers. The vision is your purpose; the practices and people will enable the delivery of that value.

Map the Value Streams Once you have a firm understanding of what value you need to deliver, you can identify the steps to deliver that value. This is known as *mapping the value stream*. The value stream is composed of each process that leads toward producing the goal of creating value, whether that goal is a physical product turning from raw material into a finished production or a digital service moving from idea to code in production. Mapping the value stream enables you to understand the constraints in producing value.

Create Flow The third step of lean thinking is to ensure that there is flow between the steps in the value stream. It is about analyzing each step and finding ways to eliminate waste while increasing knowledge and efficiency. The guiding principle is how can we reduce lead time in the production of value.

Establish Pull Step four is about looking at the value stream in reverse and understanding the customer's point of view and needs. After optimizing the flow, there is no point in delivering value that the customer doesn't need at this time. In manufacturing, this is about avoiding a build-up of inventory and preferring just-in-time delivery. In software, this could be limiting features to those that are critical to a customer rather than including all the bells and whistles.

Seek Perfection The last step in lean thinking is about the continuous and constant focus on change and identifying improvement, not only in the delivery through the value stream but from the value of the proposition to the customer as well.

Achieving Flow with the Theory Of Constraints

How do we ensure all effort is focused on making the greatest impact to the flow of value? The theory of constraints (TOC) is a management philosophy to help achieve the flow of value introduced by Eli Goldratt in his 1984 best-selling book *The Goal*. Much like a value stream TOC regards any business as a system that is composed of a series of connected processes that form a chain, transforming a customer request into customer value. However, one part of the chain, the weakest link, acts as a constraint upon the entire system. Goldratt goes on to state that any improvement made to a system should be focused on the limiting factor—the weak link. Focusing effort anywhere else will not have an effect on the performance of the system, in fact any improvements made anywhere besides the bottleneck are an illusion. This will ensure that any output of effort results in maximum outcome to the business goal. In very simple terms, identify the bottleneck that restricts the production of business value and remove it. Used with value stream mapping presented earlier, TOC is a powerful tool.

The simplest way to demonstrate the TOC is by using a manufacturing example. Consider Figure 2.5: the system requires an output of 100 units a day; each process is shown with how much inventory it can manage in a day. You can see that process B and D are both below 100. Process B defines the constraint of the system, in that it can only produce 50 units a day. Therefore any effort to increase the capacity of anything other than process B has no impact on the overall system. If we eliminate the constraint of process B, and change the capacity to 100, then the constraint of the system moves to process D. This is a very simplistic example, but highlights the fact that any effort to improve anything other than the system constraint, while having a local improvement, fails to have a system-level improvement.

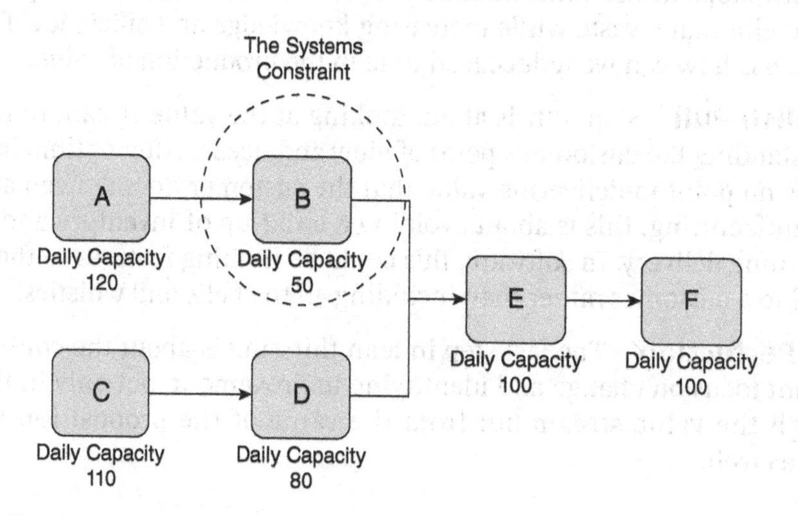

Figure 2.5: The Theory of Constraints

Whilst TOC is typically shown in a manufacturing context, the method is applicable to knowledge work and many other business domains such as healthcare and teaching, where often the constraint is technology, people or process. *The Phoenix Project* by Gene Kim is a very good book written in a novel format similar to Eli Goldratt's *The Goal*, but focuses on the application of TOC to an IT rather than a manufacturing context.

The Five Focusing Steps, also referred to as the process of ongoing improvement, is a methodology for identifying and eliminating constraints, as shown in Figure 2.6.

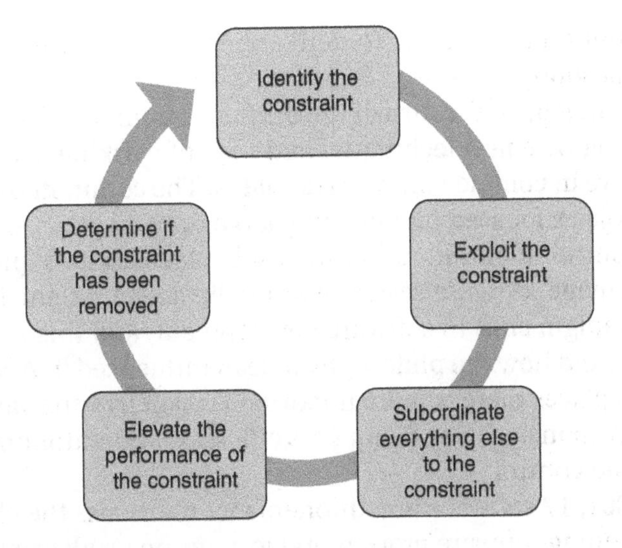

Figure 2.6: The Five Focusing Steps

1. **Identify the constraint.**
   Understand the key issue in the value stream that is causing a bottleneck or preventing your business generating value.
2. **Determine how to eliminate the constraint.**
   Focus all efforts at removing the constraint as fast as possible.
3. **Subordinate everything else to the constraint.**
   Don't be distracted by any other problems that are not contributing to the constraint. Any other effort elsewhere is wasted effort. The constraint is the priority to resolve.
4. **Remove the constraint.**
   Apply the solution to the constraint in order to remove it.
5. **Determine if the constraint has been removed.**
   Sometimes you can push a problem down or up stream. If this is the case simply repeat the steps for the new constraint.

Managing Complexity in Software Development with Agile

Whether you are solving opportunities discovered using the principles of design thinking or removing obstacles identified through lean practices, often you will be faced with complex problems. These complex problems will need to be tackled without knowledge upfront on a solution. To deliver here we can utilise the agile software development philosophy.

In the years leading up to the turn of the century, a number of software practitioners were experimenting with new techniques and ways of working in order to improve the delivery of software in complex problem domains. The common frustration they held was with how projects focused on following large upfront plans, extensive documentation, and the control of output rather than solving customers' problems.

The need to manage software design in unpredictable problem domains has necessitated a new paradigm shift in software development, and this is how the agile philosophy was born, and how the philosophy of lean influenced it. And coupled with the principles of lean places there is more importance on delivering value over delivering to a plan, being responsive over being cost efficient, and autonomy and motivation over command and control.

In February 2001, 17 software practitioners met to discuss the challenges with the current way of running software projects and to agree on an alternative way to develop software. What they came up with was a manifesto for agile software development made up of four values along with 12 principles known as the Manifesto for Agile Software Development.

The Manifesto for Agile Software Development

The Agile Manifesto states:

We are uncovering better ways of developing software by doing it and helping others do it. Through this work we have come to value:

- *Individuals and interactions over processes and tools*
- *Working software over comprehensive documentation*
- *Customer collaboration over contract negotiation*
- *Responding to change over following a plan*

That is, while there is value in the items on the right, we value the items on the left more.

These values were followed by 12 principles, which solidified a few weeks later after the initial Snowbird session.

1. Our highest priority is to satisfy the customer through early and continuous delivery of valuable software.
2. Welcome changing requirements, even late in development. Agile processes harness change for the customer's competitive advantage.
3. Deliver working software frequently, from a couple of weeks to a couple of months, with a preference to the shorter timescale.

4. Business people and developers must work together daily throughout the project.
5. Build projects around motivated individuals. Give them the environment and support they need, and trust them to get the job done.
6. The most efficient and effective method of conveying information to and within a development team is face-to-face conversation.
7. Working software is the primary measure of progress.
8. Agile processes promote sustainable development. The sponsors, developers, and users should be able to maintain a constant pace indefinitely.
9. Continuous attention to technical excellence and good design enhances agility.
10. Simplicity—the art of maximizing the amount of work not done—is essential.
11. The best architectures, requirements, and designs emerge from self-organizing teams.
12. At regular intervals, the team reflects on how to become more effective, then tunes and adjusts its behavior accordingly.

The agile philosophy embraces the reality of working in complex contexts where problems are unpredictable and nonlinear by focusing less on detailed upfront planning and estimation and more on close collaboration and fast feedback and adapting cycles. Designs, requirements, and architecture emerge as a team's and the customer's knowledge of the problem space increases. Teams develop iteratively and incrementally in order to deliver something of value into the hands of a customer as soon as possible. Agile culture is one of close participation of customers in order to deliver valuable outcomes over just output.

The Values of the Agile Manifesto

When reading the four value statements, you should understand that the preference is to focus on the concepts on the left-hand side of the statement rather than them being alternative or removing the need for the concepts to the right. The values of the Agile Manifesto are as follows:

Value 1: Individuals and Interactions over Processes and Tools Project methodologies, detailed Gantt charts, and frameworks do not deliver projects—people do. While agile recognizes that frameworks are not inherently bad, we should trust people to make judgements rather than sticking to a rigid process or a prescribed methodology or way of working. Agile focuses on increasing intrinsic motivation in order to be more responsive to change. This is achieved through greater autonomy and the empowerment of self-organizing teams so that they can adopt processes and tools that fit their needs.

Value 2: Working Software over Comprehensive Documentation There is a myth that following agile values and principles means that you need not create documentation. This is of course not true. In traditional project methodology, time was spent and delays incurred on generating detailed technical drawings based on assumptions that often were proved incorrect, and thus all the upfront work was wasted. Agile instead promotes an emergent design and just-in-time architecture, in that we should delay decisions to the last responsible moment. This enables us to have maximum knowledge before making choices and reduces waste.

Value 3: Customer Collaboration over Contract Negotiation Trying to determine upfront all of the requirements for complex problems is impossible; therefore, instead of spending time on negotiating scope, which will lead to feature bloat as customers will be wary of leaving anything out, you should instead promote collaborating on the highest priority needs to the customer and delay decisions on everything else until required. Collaboration and engagement between IT people who understand the art of the possible and customers who can articulate need and constraints allow opportunities to innovate and deliver valuable business impact and outcome.

Value 4: Responding to Change over Following a Plan There is another myth that following agile values and principles means that you don't need to follow a plan. As we have seen, problems in a complex context are unpredictable, and constant feedback and adaptation are more valuable than following a plan that was created when knowledge of the problem was at its lowest. In fact, agile is about always planning and replanning by challenging assumptions, continuous learning, and discovery.

The Principles of the Agile Manifesto

The following list expands on the 12 principles of agile:

1. Our highest priority is to satisfy the customer through early and continuous delivery of valuable software. Writing software for complex problems is unpredictable. Therefore, we have much more chance of satisfying customers' needs by getting working code into production and getting feedback to determine if it is valuable or not.

2. Welcome changing requirements, even late in development. Agile processes harness change for the customer's competitive advantage. Change in an unpredictable complex problem domain is the norm and should be embraced. It is a fallacy that people know what they want; requirements will emerge, and unless the problem and solution is clear and obvious, we should treat features as a hypothesis or an assumption. We should plan but know how to change, adapt, and replan when we discover more about the problem.

3. Deliver working software frequently, from a couple of weeks to a couple of months, with a preference to the shorter timescale. The quicker we can get working software out into production, the faster we can begin to get feedback from it and reduce risk of an incorrect assumption or hypothesis.

4. Business people and developers must work together daily throughout the project. Collaboration is essential between department experts and IT experts. As we have seen with digital disruption, it is no longer about IT and the business; IT is the business. All business projects are IT projects, or perhaps all IT projects are business projects. IT is so critical to business survival that without close collaboration and sharing vision, constraints, and opportunities, value will not be produced.

5. Build projects around motivated individuals. Give them the environment and support they need and trust them to get the job done. Respect people, give them a problem to solve and the autonomy and trust to find a solution rather than handing them a set of requirements. This will enable accountability and motivate a team that is closest to a problem.

6. The most efficient and effective method of conveying information to and within a development team is face-to-face conversation. Complex systems will have many moving parts and many teams; after all, no man is an island. People relationships are as important as code relationships; therefore, refactor your personal relationships as they are key to delivering effective solutions. Focus on co-located teams that have safe and open communications channels.

7. Working software is the primary measure of progress. While working software is a measure of progress, the essence behind this principle is that solving customers' problems and meeting needs is the true measure of progress. Working software could be considered great output, but if it does not represent a good outcome for the customer, can we regard this as progress? Working software does give us the opportunity to learn because even if we have not achieved the desired outcome, we may have the experience of what doesn't work.

8. Agile processes promote sustainable development. The sponsors, developers, and users should be able to maintain a constant pace indefinitely. Sustainable development is about working smarter, not harder, or faster. Momentum is extremely motivating. Teams working at a steady pace who are not burnt out and have the time to learn and improve are likely to reach good outcomes.

9. Continuous attention to technical excellence and good design enhances agility. By providing support for teams to master their technical skills we improve the ability of software to be able to adapt to changing business needs.

10. Simplicity—the art of maximizing the amount of work not done—is essential. Keep solutions simple and focused and strive for boring plain code. Teams will often fall into the trap of overcomplicating a problem. Keeping a solution simple does not mean opting for the quick and dirty; it's about avoiding mess and unnecessary complexity. Often "good" is good enough. Often, we can solve a problem without writing code through changing a poorly thought-out business process or leaving complex edge cases for humans rather than investing in overcomplex software.

11. The best architectures, requirements, and designs emerge from self-organizing teams. Design emerges as teams learn more about a problem over time. Requirements emerge due to close collaboration and a focus on the real customer need. Trust a team, empower them, and give them accountability to solve a problem.

12. At regular intervals, the team reflects on how to become more effective, then tunes and adjusts its behavior accordingly. Continuous improvement and feedback on performance and the need to change is the key to being agile rather than just doing agile.

Strategic Decision-Making Using Wardley Mapping

As leaders, an important part of our role is to make decisions that have a significant impact on the organisation. To help us make these strategic decisions we can utilise the research of Simon Wardley in the form of his mapping philosophy. Wardley Mapping is a visual strategic thinking framework for designing and evolving strategies. It is heavily influenced by military history and in particular Sun Tzu's Book, *The Art of War*. From this research Simon distilled that the process of making strategic decisions is based on five factors:

- **Purpose:** The scope of what you are trying to do, the problem you are trying to solve, the core principle that prompts you to do this work, your North Star, your why. Everything you do is for this, everything should be anchored to the purpose.
- **Landscape:** A map of the situation represented as the value chain of needs (e.g., capabilities), necessary to deliver on the purpose, arranged according to their dependent relationships and evolution.

- **Climate:** The external forces, such as competitor actions, customer changes, or technology evolution, acting on the landscape that can have a material impact.
- **Doctrine:** The basic universal principles, or standard ways of operating that are applicable to all industries regardless of context.
- **Leadership:** The informed decisions on the appropriate strategic actions you are able to make given your Purpose, and informed by the Landscape, the Climate, and your Doctrine.

As shown in Figure 2.7, the process of strategy is not a linear process but an iterative cycle. Strategy will need to adapt to change. This can be caused from a change in climate affecting your purpose, the landscape or feedback from actions. Therefore, the philosophy is based around a strategy cycle. We will look at Wardley Maps in more detail in Chapter 12, "Tactical Planning: Deploying Strategy."

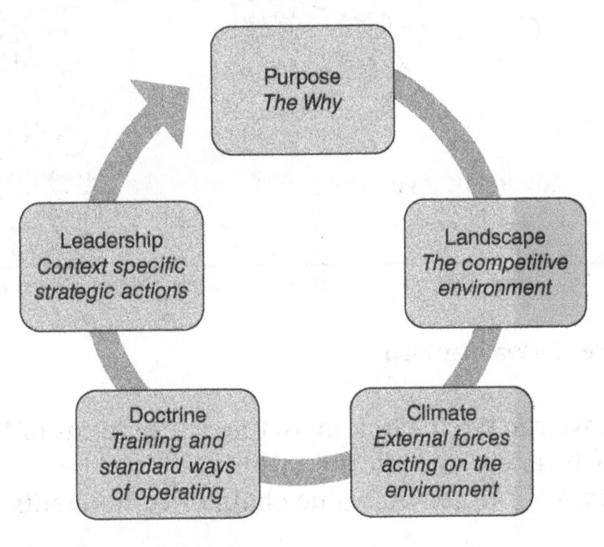

Figure 2.7: Simon Wardleys Strategy Cycle

Maps

In order to adequately implement The Strategy Cycle you need a map. A map to capture purpose, the landscape, climatic patterns, and decisions. Wardley Maps will be used extensively in Parts 2 and 3 of this book to help visualise the impact of doctrine so it is important to cover it at a high level now. The map, as shown in Figure 2.8, is a map of the structure of an organisation or a service including all the components in the value chain, required to serve a customer's needs. Space and position in a Wardley Map have meaning. A Wardley Map has:

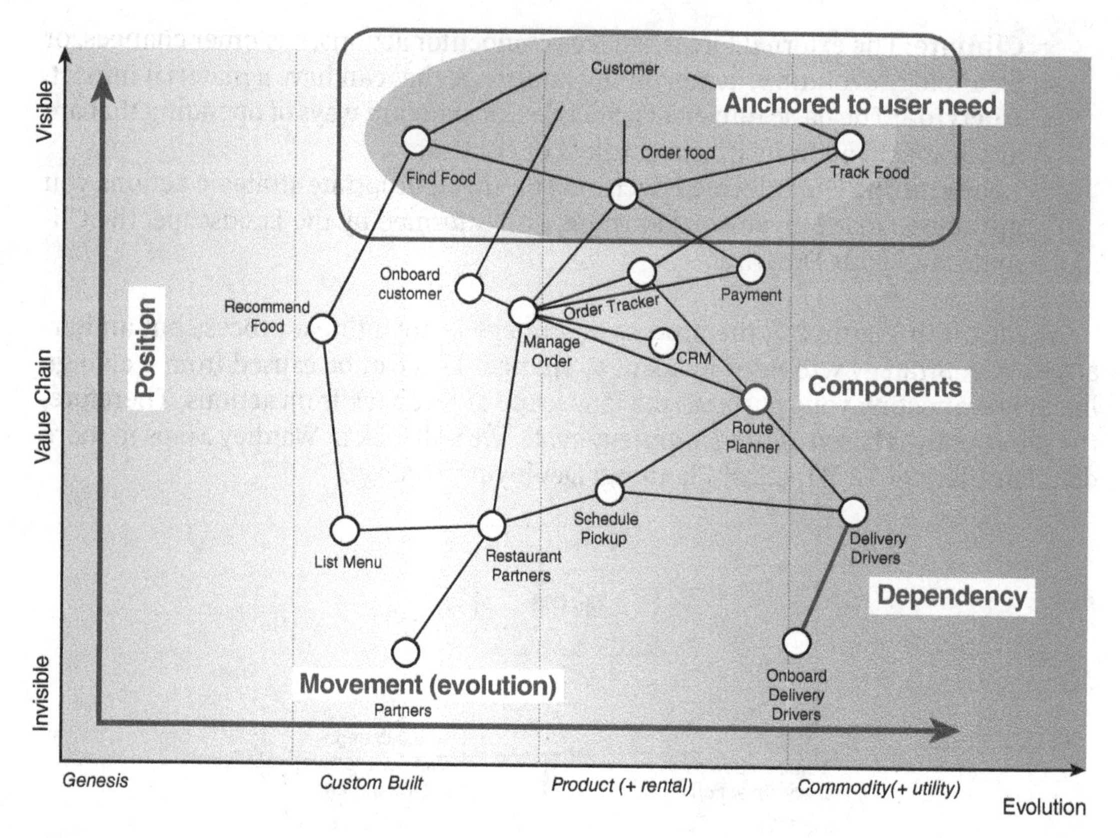

Figure 2.8: An example of a Wardley Map

- **An Anchor:** We can see the link to user needs. In terms of IT strategy we can go back to the business strategic objective.
- **Components:** We can see the value chain of components needed to fulfil a user need.
- **Position:** We can see how close the components are to a user, thus how visible they are.
- **Movement:** Components are constantly evolving, over time moving from custom-made to commodity.
- **Dependency:** We can see what components rely on each other.

To make a map we begin by understanding the user and her needs (the anchor). We then list the components needed to satisfy that user's needs. These components are then ordered by evolution. This visualisation of component, dependency, evolution, visibility to user all anchored to a user's needs aids strategic conversations and critical thinking through situational awareness and shared assumptions, which in turn makes it easier to challenge and make informed strategic choices on which actions to take.

Doctrine

Although Doctrine is part of the strategy cycle, each of the other parts are influenced by it. Doctrine is a codification of useful beliefs, techniques or principles that give guidance on how to act. However, just as components evolve so will doctrine as we discover new and improved ways of working. Simons doctrine principles will influence the operating model in Part 2 as well as how we execute strategy in Part 3, "Strategy to Execution."

The list of doctrinal principles is broken down into the categories of Communication, Development, Operation, Learning, Leading, and Structure. Furthermore these principles are then sequenced into four consecutive phases—Phase I: Stop Self-Destructive Behaviour (Table 2.1) , Phase II: Become More Context Aware (Table 2.2), Phase III: Better for Less (Table 2.3), and Phase IV: Continuously Evolve (Table 2.4). Before we think of the strategic plays we will make, it is important to assess the state of an organization's doctrinal fitness in order to understand if it has the capability to carry out strategic actions. Improving your doctrine improves the way you operate and therefore your chances of successfully executing strategy. The thinking behind the phases is to focus on fundamental principles before expanding to high order principles—walking before you run.

Table 2.1: Phase I: Stop self-destructive behaviour

Communication	Use a common language (necessary for collaboration).	Challenge assumptions (speak up and question).	Focus on high situational awareness (Understand what is being considered).	
Development	Know your users (e.g. customers, shareholders, regulators, staff).	Use appropriate methods (e.g., agile vs, lean vs, six sigma).	Focus on user needs.	Remove bias and duplication.
Learning	Use a systematic mechanism of learning bias towards data.			
Operations	Think small (know the details).			

Table 2.2: Phase II: become more context aware

Communication	Be transparent (A bias towards open).			
Development	Focus on the outcome, not a contract (e.g., Worth-based development).	Use appropriate tools (e.g., mapping, financial models).	Be pragmatic (It doesn't matter if the cat is black or white, as long as it catches mice).	Use standards where appropriate. Think Fast, Inexpensive, Restrained, and Elegant (FIRE).
Leading	Strategy is iterative, not linear fast reactive cycles.	Move fast. An imperfect plan executed today is better than a perfect plan executed tomorrow.		
Learning	A bias towards action (learn by playing the game).			
Operations	Manage inertia (e.g., existing practice, political capital, previous investment).	Effectiveness over efficiency.	Manage failure.	
Structure	Think aptitude and attitude.	Think small (as in teams).	Distribute power and decision-making.	

Table 2.3: Phase III: Better for less

Leading	Be the owner (Take responsibility). Think big (Inspire others, provide direction).	Strategy is complex (There will be uncertainty).	Commit to the direction, be adaptive along the path (crossing the river by feeling the stones).	Be humble (Listen, be selfless, have fortitude).
Learning	A bias towards the new (Be curious, take appropriate risks).			

Operations	Do better with less (Continual improvement).	Set exceptional standards (Great is just not good enough).	Optimise flow; remove bottlenecks.
Structure	Provide purpose, mastery & autonomy.	Seek the best.	

To illustrate how applying doctrine and visualising on a map can help steer strategic action, let's take the principle 'Use Appropriate Methods' as an example. Consider Table 2.5, this shows the various characteristics of the evolution of a component, which can help us determine the most appropriate method to use. Figure 2.9 shows a map with this doctrine applied informed by the characteristics of components to reveal how to approach problem solving, the most appropriate solutions and how to deliver.

On the far left side of the map is genesis. Components here are unique, and require innovation to create or discover. In this area of the map there is a strong hint that you will need to approach the problem using design thinking . Design thinking will help with exploration and revealing which "experience" is right for your customers. Because you are in uncharted territory you will highly likely need to build here. As you don't know what you are building you'll almost certainly take an agile approach to reduce the cost of change typically with an in-house team.

Toward the middle of the map components move from Genesis into Custom and Product as they evolve. Our approach here moves away from exploration into exploitation. We focus on eliminating waste and optimising the component. Therefore we leverage lean and take a problem focused approach to problem solving. In terms of solutions, we may build, but increasingly we will leverage off-the-shelf systems. Our approach may still require agile methods if our problems are complex, but we may be able to use other more traditional approaches if we are utilising existing technologies.

At the other extreme of the map, on the right we are dealing with capabilities that are highly evolved and are a commodity, so there is still a strong hint to take a lean approach and eliminate waste. However we can take this to the extreme and outsource a solution altogether and use a plan-driven well-trodden approach to delivery.

Table 2.4: Phase IV: Continuously evolve

Leading	Exploit the landscape.	There is no core (Everything is transient).
Learning	Listen to your ecosystems (acts as future sensing engines).	
Structure	Design for constant evolution.	There is no one culture (e.g., Pioneers, Settlers, Town Planners).

Table 2.5: The various stages of evolution on a Wardley Map

Questions to help understand a component's evolutionary stage:

- How evolved is each component? Is it common? It is generic or unique to us?
- How does the rest of the market view it? Can you buy it or rent it?

Genesis	Custom	Product	Commodity
Innovative, unique, you need to create or discover it. Others don't have this.	Exists, others use it but needs to be customized to suit your needs. Can be built from existing technologies but customized for your needs.	You can rent it without customization. Available as COTs with minor configuration.	Ubiquitous and standardised such a power. Candidate to outsource as not a capability that will make a difference if you build it yourself.

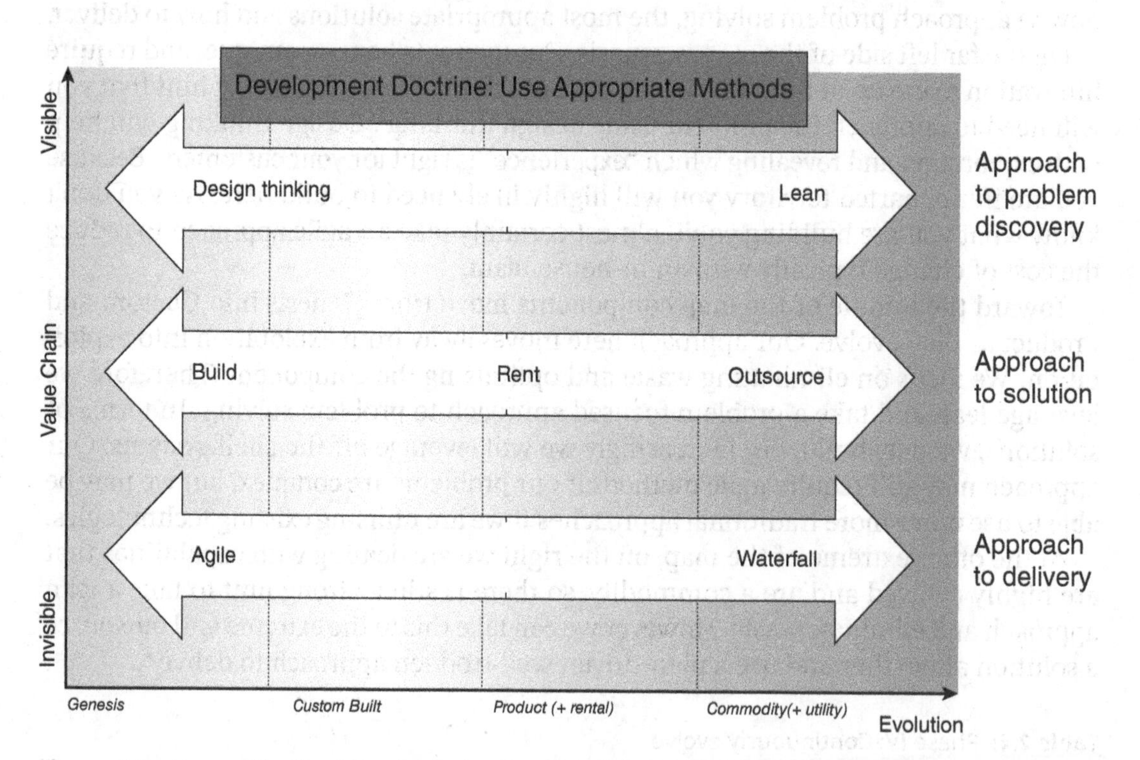

Figure 2.9: How a Wardley Map can help visualise the appropriate method to use

Summary

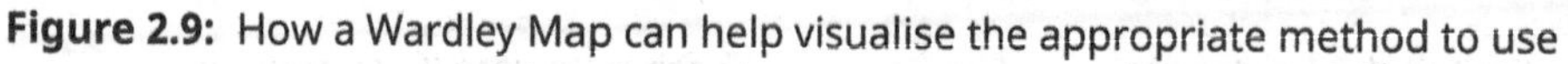

To meet the needs of modern businesses that are exploring new opportunities as well as exploiting existing ones we need a mode of operating that can adapt to address the differences in these fundamentally different problem domains. This chapter introduced

four ways of thinking that will form the belief system of a new IT operating model that will define approaches to structure, governance, ways of working, and leadership.

To begin we looked at two philosophies for approaching problem discovery. Namely Design thinking and Lean thinking. These philosophies share many similarities, in particular, the focus on delivering value to customers and an iterative approach to problem solving. But the underlying paradigms are radically different as highlighted in Table 2.6. By understanding the key differences between the two, we can understand which approach is useful based on your specific context; in addition, it is not an either or choice; often you will find that complex problems will require both approaches to be used together.

Table 2.6: The difference between lean thinking and design thinking

	Design Thinking	Lean Thinking
Purpose	Exploration. To discover new markets, opportunities and products.	Exploitation. The optimisation of an existing value proposition and all of the activities that enable it through the solving of immediate problems for customers and the reduction of waste.
Approach	Understand an unmet customer need via empathy. Define the underlying problem. Ideate to solve the problem. Prototype to test the idea. Test the Prototype with customers. Iterate.	Be clear on the value you are trying to deliver. Map the value stream of activities that deliver the value. Remove waste in the flow of the value stream. Control the flow and limit work in progress using a pull system. Iterate until all waste has been removed.
Way of thinking	Divergent—Solution First.	Convergent—Problem First.
Useful for	Defining solutions to wicked or complex problems that are ill-defined or unknown.	Continuous improvement and optimization through eliminating waste.
Measure of success	Finding a market fit.	Improving efficiency and the flow of value.

Regardless of where problems come from we are increasingly finding that they are complex in nature. To approach delivering a solution within uncertain and complex problem domains we looked at agile. Agile is a collection of values and principles that have inspired a number of practices for effectively dealing with software in complex and complicated problem domains that promotes collaboration and regular customer feedback and participation designed for dealing with uncertainty. Agile culture accepts that software design is difficult and that things change at a fast rate, so instead of fighting against it, change is welcomed. Scope is not limited, and requirements will change as more is discovered about what works and what does not. Agile promotes a process of iterative and incremental delivery of value, largely due to the reality that large upfront planning is not effective in anything other than an obvious or complicated problem.

Lastly we looked at Wardley Mapping. A strategic thinking approach that takes situational awareness into designing and evolving our strategic actions. This is key to understand where to use the design thinking and lean methods and where an approach using agile methods is appropriate. A key factor of the Wardley Mapping strategy cycle is doctrine—the universal principles applicable to any industry, that will be embedded into the operating model and how we form and execute strategy, which we will cover in Parts 2 and 3 of this book.

The ideas behind these four new ways of thinking overlap greatly and share a common set of values and principles:

- **A Focus on Purpose:** A focus on the bigger picture and root causes rather than symptoms in order to discover underlying customers' needs and deliver meaningful change.
- **The Delivery Of Value:** The relentless experimentation and learning in pursuit of delivering value over following a plan.
- **A Process of improvement:** An understanding of all the processes that are involved in the production of the value and identify opportunities to improve the flow by removing waste.
- **People-Centric Ways of working:** A focus on high-trust, collaborative, and motivated teams with clarity on purpose, who share empathy for a customer and their needs.
- **Iterative process with constant Feedback:** An iterative process with a focus on fast and amplified feedback loops to ensure constant adaptive planning.

Chapter 3, "How To Change the System," will show you how to embed these philosophies into the mental models of you and your teams so that their way of working is underpinned by these new ways of thinking.

How to Change the System

Attempts to change an organizations culture is a folly, it always fails. People behavior (culture) is a product of the system; when you change the system peoples' behavior changes.

—John Seddon

Everyone thinks about changing the world, but no one thinks about changing himself.

—Leo Tolstoy

When the thinking changes, the organization changes, and vice versa.

—Gerald M Weinberg

As introduced in Chapter 1, "Why We Need to Change The System," the problems that IT leaders are having to tackle are increasingly defined within a Complex context. In order to drive business outcomes and impacts, we need an IT organization made up of people who can work with unscripted collaboration, are autonomous but aligned with what is important, are focused on continuous improvement, and are able to analyze feedback from what they have delivered, learn what is and is not effective, and adapt plans as necessary—in other words, embrace the values and principles of the agile and lean philosophies. However, this new way of operating represents a fundamental shift not only for IT departments but also for the rest of the enterprise. The underlying values and ideas of a plan-based, command-and-control IT management paradigm have been ingrained into leaders' and managers' mental models for years as "best practice." It is not simply a case of following an agile framework or a set of practices to change this; instead, it is a total shift in leaders' underlying thinking on how their organization's work needs to happen, how teams are structured, who is accountable for results, how they are measured and who can make decisions. To put it bluntly a change in each and every component in the organisation's operating model is needed.

This chapter begins by introducing why simply copying the ways of working and the practices of successful agile companies won't make a long-term difference and you will ultimately fail to achieve the outcomes you were expecting. It's not just a case of doing agile, you must be agile. Being agile is about (a) changing your mental models and the values and beliefs of your teams and (b) building quality and adaptability into the technical architecture. For the former we will look at leveraging systems thinking; for the latter, the practices of eXtreme Programming and DevOps. Systems thinking helps us to think of the IT department, and the rest of the enterprise, as a complex adaptive system. The actions of the people who work within it are directly influenced by working patterns and structures of the system, which are fundamentally based upon mental models and visions of how work should be executed. Therefore, to transform our operating model, we must change the rules and constraints that people work within. To do this we must first change the way leaders lead, and how they lead is influenced by their own mental models, values, and beliefs, and the way they think.

But what values should leaders hold in their belief system? We will look at the traditional command-and-control style of management and its shortcomings in today's world. The beyond budgeting movement offers leadership and management principles that at first may seem a poor way to manage due to how we have been taught to lead and manage in the past, but they offer a way to intrinsically motivate people to achieve better outcomes and greater impacts as well as accept the ever-changing and uncertain business context. Finally, you will be introduced to the three factors that drive employee intrinsic motivation, required for working in complex problem domains, namely the combination of purpose, autonomy, and mastery.

Being Agile vs. Doing Agile

Over the years the concept of "agility" in agile practice has unfortunately lost a lot of its meaning and understanding due to the overvaluing of methodologies and frameworks, the underestimating of technical practices, and the commercialization of quick-fix agile consultants and certifications. Organizations have either enforced agile practices upon their workforce by sending teams on crash courses without leadership attendance, and most importantly without leadership change, or enforced strict adherence to a particular methodology such as scrum with full governance and reporting via tools such as Jira. Alternatively, small teams of developers have tried to introduce Agile practices in isolation and without the support of leadership and the rest of the organization. In both cases there is merely a focus on the rituals and practices and a lack of attention on appreciating the objectives and goals that the practices are designed to support.

This obsession on activities associated with agile rather than the goals will not lead to any meaningful change to the delivery of value. This in turn leads organizations to conclude that agile was overhyped and no better than what they had been doing before.

If we take it back to first principles, we should remind ourselves that being agile is about being able to adapt quickly based on new information and changes to the context that you are working within. Following agile practices and methodologies can assist, but if you are not fully engaged with the principles, or you do not understand them, then you are merely laying new frameworks over traditional ways of working, which will not deliver an improved operation or produce the outcomes and impacts that the business needs. It is important too that as well as management practices, the underlying architecture needs to be subtle enough to change or to be quickly replaced. A focus on technical practices must be made in conjunction with leadership and management changes. From a leadership perspective, agility is a mental state that must be adopted. It's a completely different frame of mind that requires not only a change of process but also one of culture. From a technical point of view, agility is the ability for code to adapt to changes in context without a loss in quality, and therefore the technical practices as well as the management practices need to work in parallel to become more agile. IT leaders must avoid being fixated on ceremony but instead embrace the underlying management values and technical practices that promote empowering teams to continuously improve, practice short cycle times for delivery, gather fast feedback, respond and adapt to new information, and focus on the delivery of value for customers. To realign with the fundamental principles of agile, we must first understand why adopting only the practices and rituals, the things we can easily see, won't have any meaningful effect on delivering customer impacts.

Why Only Adopting the Practices of Agile Won't Work

Spotify is a company that has been the poster child for agile organizations ever since Henrik Kniberg documented the squads and tribes operating model in 2014. However, the success that Spotify has had with its organization model has often been imitated but without the same results. The reason is due to the other organizations' taking the practices they can see but not taking on the principles and values that have evolved the organization to look like it does. W. Edwards Deming, who was a leading management thinker in the field of quality, said, "People copy examples and then they wonder what is the trouble. They look at examples and without theory they learn nothing." To emulate the success of organizations such as Spotify, we need to change not only practices but how the enterprise thinks and tackles problems. We need to fundamentally change how we work or, put another way, how the organization, or system, works. Spotify doesn't even follow Spotify's model—Henrik Kniberg's published article on it was a moment-in-time view of a company in a state of continuous evolution and improvement. Copying another organization's design without changing your own mental models and belief systems is known as cargo culting. As shown in Figure 3.1, cargo culting is a case of doing agile, adopting only the practices, and ignoring cultural change, rather than being agile, where values are changed in how work is tackled in a complex and adaptive business environment.

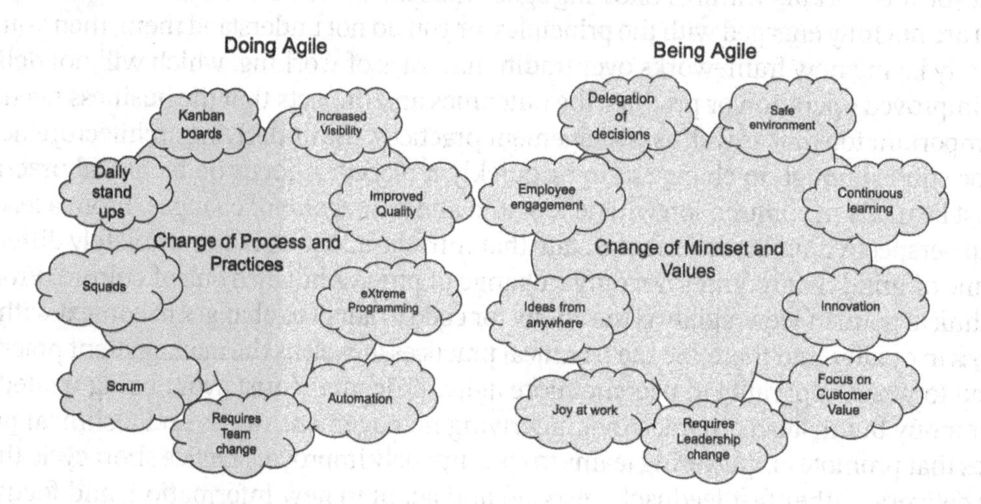

Figure 3.1: The difference of doing agile and being agile

Adopting the practices of agile without embracing the core values and technical practices is merely paying lip service to the philosophy and is akin to showing that there is a lack of trust between leaders and the people doing the work. The result is that you work in sprints, have daily standups, and use sticky notes, but leadership makes all the decisions and team members must adhere to their plan. This fosters a lack of trust, which is a fundamental problem when working in complex problem domains—as those with deep knowledge who are closest to the problem need to be empowered to change direction and resolve issues themselves without having to navigate through an overly bureaucratic governance process. The key point is that agility is not the adoption of a process or a set of practices or a framework; agility can't be achieved by sending development teams on a scrum training course for a few days. The central core of an organization must change, and agility is ultimately a cultural change—from both a technical and leadership perspective. Just like lean, it changes the relationships between people, changing from a command-and-control paradigm to an organizational design that removes unnecessary hierarchies and siloed departmental thinking, giving teams responsibility underlined with empowerment and trust to get the job done. In this environment, people are aligned and understand the overall objectives of an organization and intrinsically collaborate to deliver meaningful outcomes to contribute to achieving them.

To truly become agile, we must look beyond the adoption of agile frameworks and methodologies and fundamentally change how the organization operates—the way the work works—and the best way to change the organization is to understand it as a system and change the core values and beliefs of that system.

Use Systems Thinking to Change Behavior

Systems thinking is one of the most powerful management tools that can be employed to address and transform an organization to embrace an agile and lean way of working. In a nutshell, systems theory states that the structure of a system, along with the environment the system exists within, will determine the behavior of the system. A system is defined as a group of interdependent and interactive parts that form a whole, all working for a common purpose. Systems thinking appreciates that these parts influence one another. Understanding this allows us to reach deeper insights into the underlying structures that influence behavior within a system, which in turn allows us to make more impactful and lasting changes. Systems thinking takes a holistic approach to problem solving, emphasizing the need to look at a problem as a whole rather than distilling it down and analyzing its components in isolation, which can lead to siloed and localized thinking. This avoids the trap of jumping to the first "obvious" solution that may address a symptom but can miss the underling root cause altogether.

The Fundamentals of Systems Thinking

We perceive system behavior through events—these are the things we can see. However, a naive approach is to look at events only and think of systems in a linear fashion, with events being the result of cause and effect; it's also naive to think that to change the behavior of the system, we need only to react to the direct cause of events. In ordered systems, this may be a valid way of thinking, and we can reason in this way. However, complex problem domains are unordered systems where there is no clear and direct relationship between cause and effect. The parts of these complex adaptive systems affect each other in unpredictable ways. Therefore, reacting to events will not resolve the root cause of the problem. Complex problem domains are made of many complex relationships and interrelationships, so it's important to understand the patterns and structures beneath the events that are at work in the systems. Change them and you will have a lasting impact; this is the essence of systems thinking.

As shown in Figure 3.2, events are the results of patterns of behavior. These patterns are formed based on systemic structures in place, all of which are influenced by mental models, beliefs, and values that underpin the system. Events (What happened?), patterns (What are the trends over time?), systemic structures (What forces are in play that can influence the patterns?), and mental models (What belief systems are held to support the structure?) are the four concepts that we can use to reason with a system.

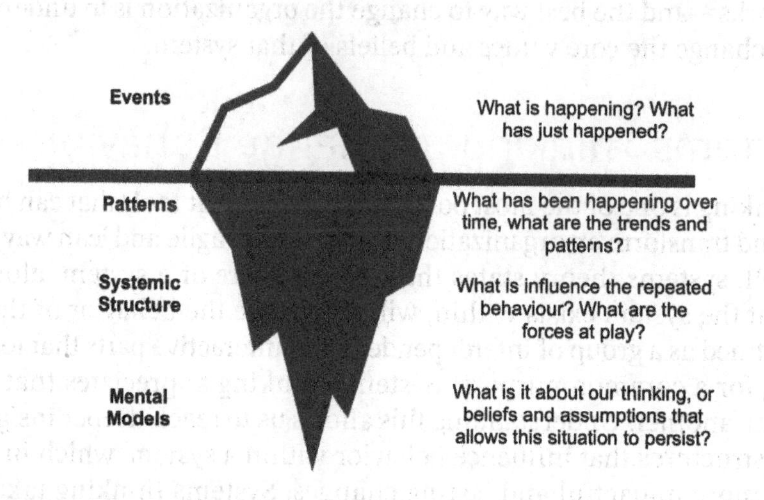
Events
What is happening? What has just happened?
Patterns
What has been happening over time, what are the trends and patterns?
Systemic Structure
What is influence the repeated behaviour? What are the forces at play?
Mental Models
What is it about our thinking, or beliefs and assumptions that allows this situation to persist?

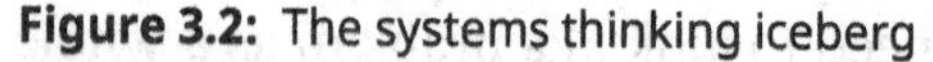
Figure 3.2: The systems thinking iceberg

Complex systems don't have a linear cause and effect; you need to look beyond events and patterns to understand the structures that are enabling the patterns and the mental models that inform structures. Events are easy to grasp and understand in complex contexts, but often these events are symptoms of a bigger event or series of events at play. We can solve the immediate problems, or perceived problems, but if the fundamental underlying structure that caused that event, the symptom, is not addressed, then the problem will persist. Along with structural change, we need to make a mindset change. Leading transformation requires deep fundamental changes to how we perceive work.

To help illustrate the concept of systems thinking, consider the following example, which shows how a belief system can contribute to poor health and a degradation in performance.

Event	I caught a cold.
Patterns	I have been going to sleep later and later. I often get colds.
Systemic Structures	I am working later at work and not eating healthy meals.
Mental Models	I believe that asking for help is a sign of weakness.

If we want to make changes, we can react to events, such as buying medicine for a cold, which will fix the symptom for a time but not the root cause. However, if we want to make impactful and lasting changes, we need to change the way we think, as shown in Figure 3.3. In systems thinking, this is referred to as our mental models. The ingrained values, beliefs, culture, and experiences shape our mental models and that influences how we act, tackle, and reason about problems and how we discover and deliver solutions. To change the system and have any meaningful impact on change, we have to change our mental models.

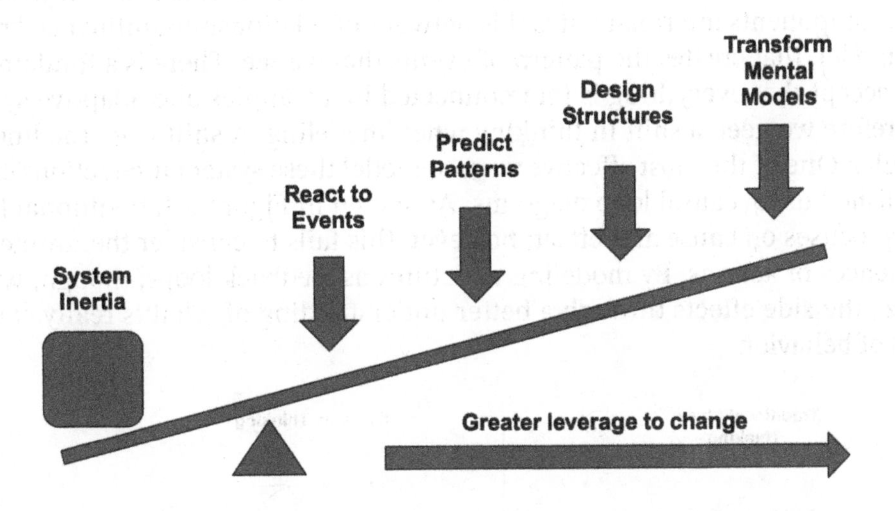

Figure 3.3: Where we have the greatest leverage to make change

Consider how a change in the mental model of the person suffering from colds and poor performance changes their system structure and their pattern of behavior, which results in an improved performance and less ill health. While this is a simplistic example, it is used to illustrated how a change in belief system has a far-reaching effect.

Mental Models	I have the safety to be honest with my leader.
Systemic Structures	I get help on difficult tasks and finished them more quickly, giving me back time to cook a healthy meal.
Patterns	I can go to bed early and I feel healthy.
Event	I am achieving more at work, and I have fewer colds.

Understanding the concepts behind systems thinking enables both leaders and their teams to understand the true impact of their actions and the impact they can cause. This knowledge and ability to see the long-lasting impact of our actions enables leaders to manage their organizations more effectively and allows teams to manage complex problems in a smarter way.

Tools for Exploring and Understanding Systems

To understand a system, we need to model its structure. Structure is the way that the system's components interrelate. It is this network of relationships, influenced by the mental model, that creates the pattern of events that we see. There is a fundamental need to accept that everything is interconnected in a complex and adaptive system, and therefore we need a shift in thinking when modeling. A shift from the linear to the circular. One of the most effective ways to model these system interactions can be accomplished using causal loop diagrams. As shown in Figure 3.4, traditional linear thinking focuses on cause and effect; however, this fails to consider the unintended consequences of actions. By modeling structures as feedback loops, though, we can minimize the side effects through a better understanding of what is really causing patterns of behavior.

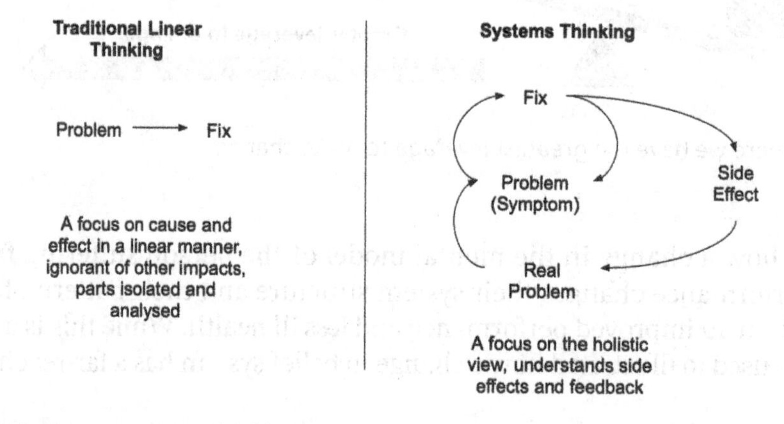

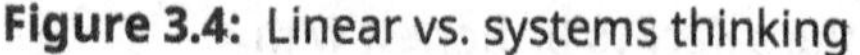
Figure 3.4: Linear vs. systems thinking

As an example, consider Figure 3.5, where we have a mental model that is based on the premise that all failures are bad. The impact to our structure is that we have a system that criticizes and assigns blame to failure in the hope that it will improve performance. However, this leads to a lack of psychological safety, which leads to fear—fear of speaking up, raising concerns, or taking risks on new and innovative ideas. This in turn leads to poor decision making on how to effectively solve problems, which itself leads to failures and again to blame, which completes this reinforcing negative feedback loop. The patterns we see from this structure are a decrease in productivity and low employee morale. The events are that we are missing out on opportunities, not removing constraints, and employees are leaving the business.

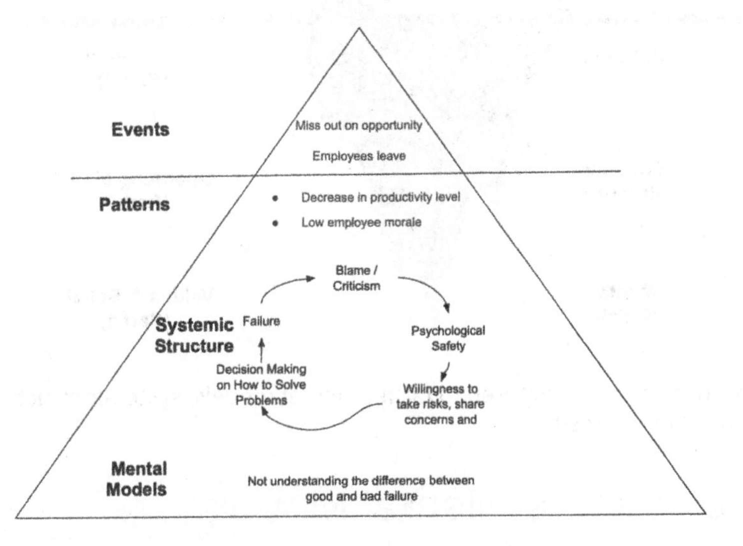

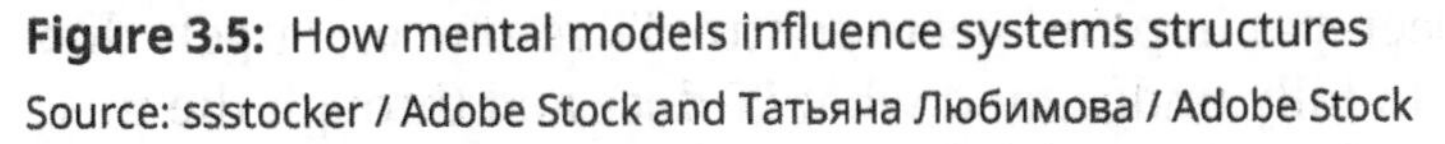

Figure 3.5: How mental models influence systems structures

Source: ssstocker / Adobe Stock and Татьяна Любимова / Adobe Stock

Applying Systems Thinking to Organizational Change

If we apply systems thinking to an organization, we find that it is the operating model that forms the structure. The operating model defines the ways of working, decision making, team organization, and various other components. It is this structure that has a big impact on the effectiveness of how we perform as an IT department. This book will show you how to transform the dynamics of how IT operates using the principles of systems thinking to ensure we can support a business's need to both explore and exploit. As shown in Figure 3.6, the three chapters in Part 1, "A New System of Work," cover the philosophies, values, and principles that will form your and your team's mental model and belief system. Part 2, "Designing an Adaptive Operating Model," changes the systemic structures, in this case the operating model (the ways of working, structure, and governance), to be more adaptable. Part 3, "Strategy to Execution," will show how the structure supports an IT organization that is able to contribute to business success, from strategy to planning and execution, resulting in a pattern of delivering positive business outcomes and long-lasting impactful change.

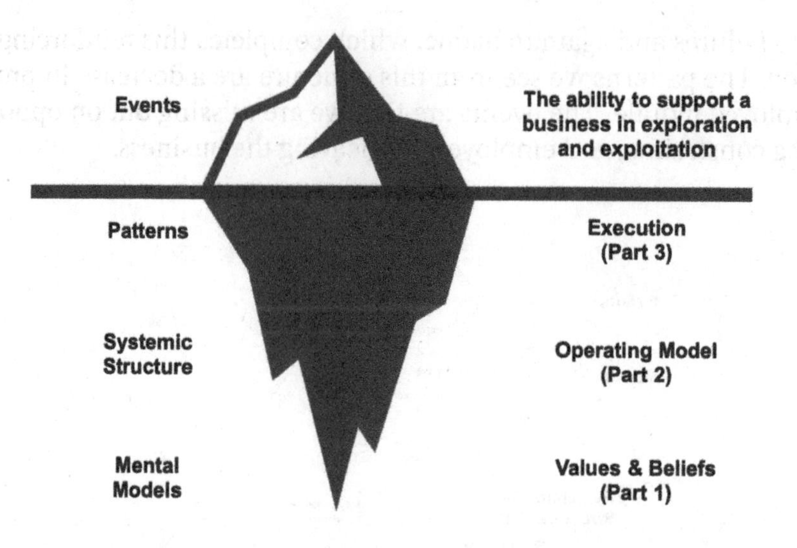

Figure 3.6: How this book is structured to target mental models, systemic structures, and patterns to produce better result

Changing Leaders' Mental Models

IT leaders have historically been put in positions to control IT resources, largely due to other execs not being able to fully understand what IT does and through the misunderstanding that it was a predictable production process rather than an emergent design process. Because of this, there has been a heavy focus on IT leaders to command and control, to meet project plan forecasts, to commit to estimates and remove all risk up front, to demonstrate their expertise. IT leaders were rewarded for operational performance, cost savings, and control.

However, as IT has moved beyond a service provider and is now deeply integrated into the core of what makes a business work, IT must adapt, innovate, and move at the same pace as the business. Outdated ways of control are unable to react fast enough to changes in the business context and to capitalize on opportunities or remove constraints. IT leaders need to focus more on enabling a highly collaborative workforce to adapt, pivot, and respond in the new reality of the chaotic business environment rather than simply to act as project delivery managers. To enable the enterprise, which is so reliant on technology, to succeed, we need to move beyond thinking of ourselves as the guardians of IT cost and control and move to a culture of influence, empowerment, and trust focused on profit and loss and customer value rather than only cost control. The bottom line, as W. Edwards Deming puts it, "Everyone is already doing their best; the problems are with the system . . . only management can change the system." But to change the system, we need to change how leaders and management think.

Systems Thinking vs. Linear Thinking

To be effective at making change in complex adaptive organizations, we need to change the way we think. Table 3.1 shows the shift in thinking required for leaders. To get the improved outcomes that agile and lean ways of working can bring, you need to be fully committed to that way of working; your working practices, patterns, and structure and your core values need to change. You can't half do agile if you truly aspire to become an effective and influential IT leader.

Table 3.1: The difference between linear and systems thinking

Linear Thinking	Systems Thinking
Blame someone for a failure.	Understand the influences that led to a failure. Identify and change the behaviors and the patterns in order to prevent them in the future.
Focus on fixing symptoms.	Focus on underlying patterns, structures, and values that cause symptoms.
Focus on a single events.	Look at patterns and reoccurring behaviors that cause events.
Concerned with output.	Concerned with the process.
Care about what is said.	Care about how things are said and what is not said and why.
Divide and break problems into silos.	Understand problems as a whole.
Try to control chaos within complex adaptive systems.	Look for patterns and how to leverage influence within complex adaptive systems.
Believe organizations can be predictable.	Understand that complex organizations are inherently unpredictable and chaotic.
Direct relationship between cause and effect.	Effects are caused by underlying patterns and behaviors.
People are the problem.	The system is the problem.

The Problems with Command-and-Control Leadership and Management

Command-and-control management theory is based around maintaining control of people and process. It puts a strong emphasis on hierarchies, rigid rules, strictly defined specialized roles, and a siloed workforce. The unwritten premise behind command and control is that leaders know best; they have clarity on the future and believe the context won't change. They have a plan on how to get there that they don't want to deviate from as that will be viewed as failure. They need to control the behavior of employees so that they adhere to their plans because they don't trust them.

Command-and-control leadership styles were designed to enforce standardization, compliance, and cost efficiency. Henry Ford famously said, "Nothing is particularly hard if you break it down into small jobs." In an ordered system, one with problems that are routine and algorithmic, this is certainly true, but it is not true of complex and unordered systems. The roots of the command-and-control leadership style were created during the Industrial Revolution by Frederic W. Taylor's scientific management and Max Weber's bureaucratic management (known as "Taylorism") and were focused on production, where cost efficiency was valued far greater than workforce engagement and adaptability. "Taylorism" was a popular management method when organizations were solely production oriented—one that streamlines the production process for its product or service. Modern businesses have evolved to become more consumer driven—influenced by the actions and needs of consumers. This is a more complex and adaptable environment to the one that Frederic W. Taylor's scientific management techniques were designed for, one that requires a leadership paradigm that promotes trust, autonomy, responsiveness, unscripted collaboration, experimentation, adaptability, and innovation. Command-and-control management practices are not effective for complex adaptive businesses that require an engaged and skilled knowledge workers. At best they can only offer the illusion of control.

The HIPPOs (highest paid person's opinion) can no longer make all the decisions. Instead, they must set strategic direction and give context, the what and the why, but allow a skilled workforce to solve the how. We can no longer separate those making decisions about the work from those closest to, and doing, the work. Unless leaders lose the bad habits of command and control and adopt the values and principles of lean and agile philosophies when managing and leading teams, there is a real danger that leaders themselves, the architects of the system, will be the greatest obstacle when trying to deliver outcomes that have a positive impact on the business. Leaders need a new set of principles to guide them on how to lead and manage in the new world. For this we can look to the beyond budgeting movement.

The Beyond Budgeting Movement

Moving beyond the command-and-control management paradigm of the past to a leadership style that accepts the futility of trying to predict the future, embraces uncertainty, and empowers its workforce through trust to adapt to ever-changing contexts is the philosophy of the beyond budgeting movement. The beyond budgeting movement was created by Jeremy Hope and Robin Fraser in 1997 to look for a better governance model that would enable organizations to react to a faster rate of change. Initially their work focused on the process of developing a better alternative to the traditional way of setting budgets, but quickly they realized that the entire leadership and management models needed to change to reflect the faster pace of business change and volatility. They documented their findings in the book *Beyond Budgeting* (Harvard Business Review Press, 2003). In it they describe an organizational design with principles that are built around autonomous teams aligned on a common purpose with the responsibility to make their own decisions on when to pivot and adapt, with support from leadership, to deliver business outcomes. In addition, they describe a set of management processes that support the organizational design by removing overly controlling and bureaucratic governance processes in favor of a more emergent management model built around trust and relative performance measures. Beyond budgeting represents the need to change leadership thinking to adapt to the new reality of a global 24/7 business environment that is heavily reliant on technology and where innovation, collaboration, and the ability to pivot quickly is much more valuable than adherence to rigid upfront plans.

The beyond budgeting leadership model is based around 12 principles, split into two themes: decentralized leadership to instill intrinsic motivation and an emergent and adaptive set of management processes. These are shown in Table 3.2.

Table 3.2: Beyond budgeting's leadership and management principles

Leadership Principles	Management Principles
1. **Purpose:** Engage and inspire people around bold and noble causes, not around short-term financial targets.	7. **Rhythm:** Organize management process dynamically around business rhythms and events, not around the calendar year only.
2. **Values:** Govern through shared values and sound judgment, not through detailed rules and regulations.	8. **Targets:** Set directional, ambitious, and relative goals; avoid fixed and cascaded targets.
3. **Transparency:** Make information open for self-regulation, innovation, learning and control; don't restrict it.	9. **Plans and forecasts:** Make planning and forecasting lean and unbiased processes, not rigid and political exercises.
4. **Organization:** Cultivate a strong sense of belonging and organize around accountable teams; avoid hierarchical control and bureaucracy.	10. **Resource allocation:** Foster a cost-conscious mindset and make resources available as needed. not through detailed annual budget allocations.
5. **Autonomy:** Trust people with freedom to act; don't punish everyone if someone should abuse it.	11. **Performance evaluation:** Evaluate performance holistically and with peer feedback for learning and development, not based on measurement only and not for rewards only.
6. **Customers:** Connect everyone's work with customers' needs; avoid conflict of interests.	12. **Rewards:** Reward shared success against competition, not against fixed performance contracts.

Leadership Principles to Instill Intrinsic Motivation

The leadership principles empower people by instilling intrinsic motivation—behavior that is driven by internal rewards. There is a strong emphasis on trust and purpose, autonomy, and decentralization, along with unscripted collaboration to drive outcomes and customer impacts, which is vital in complex problem domains and organizations. We cannot afford to separate those who make decisions about the work from those who are closest to the work; doing so drastically slows down throughput and leads to decisions that are made by those less informed of the problem at hand. To adapt faster and improve lead times for business impacts, we must give the responsibility of decision-making to the people who are doing the work. This requires moving away from the illusion of control to a position of trust underlined though clear alignment to strategic objectives. Leaders still need to retain governance at a steering level, but day-to-day decisions need to be made the responsibility of the team who are focused on the problem at hand. The leadership principles are underpinned on trust, which is an essential ingredient for delivering at pace.

Emergent and Adaptive Management Processes

The six management processes support the six leadership principles. They embrace a more emergent way of managing that takes reality into account by managing relative to the context you are in rather than against fixed targets and plans made 12 months ago. Trying to predict the future is impossible in a complex and adaptive organization; therefore, setting targets and budgets way into the future to control behavior is not effective. Instead, giving relative targets based on the current context and competitors is far more effective. Setting targets that do not reflect current reality can lead to people gaming the system to hit a number even if it is to the detriment of the business. Emergent management focuses on a more holistic view of people and business performance that is beyond rigid and fixed targets and budgets.

Instilling Drive through Purpose, Mastery, and Autonomy

In his book *Drive: The Surprising Truth About What Motivates Us* (Canongate Books, 2010), Daniel Pink explains that the methods that worked to motivate people before the advent of digital business do not support the type of work we have today. As you read earlier in this chapter, Taylorism and the scientific method that promoted command and control, specialism and siloed working used the carrot and stick approach, extrinsic rewards, to ensure the work was done. This was successful as the work was routine and algorithmic and required no creative thinking. Pink refers to this as Type X behavior.

Somewhat unintuitively, however, Pink shows that extrinsic rewards do not work well for nonroutine and heuristic work that requires critical thinking and high levels of unscripted collaboration. In fact, extrinsic rewards have the opposite effect and can negatively impact work. This type of work requires a workforce that is motivated by internal rewards, intrinsic motivation, where the work itself is a joy and therefore the reward. This is what Pink refers to as Type I behavior. Table 3.3 shows the differences between the two different ways to motivate and where they are effective.

Table 3-3: The difference between Type X and Type I behavior

	Type X Behavior	Type I Behavior
Applicable to work that is	Algorithmic Work that follows a set of established formulas. Routine in nature can be boring but necessary.	Heuristic Complex tasks that have no defined algorithm or routine manner of being accomplished. Experimentation, sensing, and adapting are required to devise a solution.
Motivation	External	Internal
Requirement	Outsourced to machines	Required for complex and new problem domains
Promotes a	Fixed mindset	Growth mindset
Reward	***If-then*** If you do the work, you will get the reward.	***Now-that*** Primary reward is the satisfaction of the activity. However, now that the work is complete, rewards can be offered but never as an incentive.
How to motivate	Carrot and the stick. Reward for completing the work, punished for failure.	1. The freedom and autonomy to choose the best path to complete the work 2. The challenge and opportunity to learn and master "just outside my comfort zone" 3. Understanding the higher purpose behind the activity

Type I behavior and intrinsic motivation is achieved by people having purpose, mastery, and autonomy:

Purpose People need clarity on how their work relates to the strategic direction of the company and the role they play in contributing to problem solving rather than simply solution execution. Without this sense of purpose, without understanding the *why* behind the *what*, without understanding that they are part of something bigger than themselves, we cannot expect people to be intrinsically motivated.

Mastery Type X behavior is defined for compliance, whereas Type I behavior encourages engagement. Engagement is a key factor required for learning and bettering yourself. Pink describes mastery as being defined by three rules. The first is that it is a mindset and that people need to understand that their ability is infinite. Second, mastery requires effort; it is challenging to improve and requires dedication. Third, mastery is itself infinite; you can always learn more, and you will never finish.

Autonomy People have a desire to be self-directed and the masters of what they do, how they do it, and when they do it. Therefore, give responsibility and trust people to make the right decisions and choose the path they believe will lead to a solution. Autonomy increases engagement, which leads to high motivation.

However, as Pink points out, you do need to ensure people are compensated well enough as to "take the issue of money off the table." When compensation is no longer a problem, that is, you are paying the market rate, then the focus on optimizing for joy of working will cultivate the environment that will foster motivation. The joy needs to be the activity, the pursuit, not the result. Motivation is a product of the system, and the design of the system is the responsibility of the leaders. Set up the right conditions and structure to clarify purpose, support mastery, and enable autonomy and your teams will be motivated and work unscripted and collaborate on driving business and customer outcomes.

Summary

In this chapter we focused on why and how the values and principles of the lean and agile philosophies can be embedded in your operating model. We started with a look at the difference between being agile vs. doing agile. Simply copying the ceremonies and practices of agile ways of working without changing the underlying value system or focusing on trust and people will not make a meaningful difference. It is important to understand that taking an agile or lean approach to IT isn't about copying the practices of successful companies or their models or by following a particular framework. It's about understanding the deep cultural change that is at the heart of both these philosophies. It's about changing how work fundamentally happens.

With an understanding that agile isn't something you do, that it's more of a mindset and set of practices on how to approach problems and solutions, we needed to understand the best way to change the values, visions, and mental models of people in order to change the system and motivate people to adopt new ways of working; simply changing the project methodology or using a new set of tools won't make a meaningful difference. To achieve this, we looked at systems thinking, a philosophy for managing complex problems. The essence of systems thinking is that a system will only improve by optimizing the whole instead of focusing on individual parts.

Furthermore, to change a system, you must look beyond the events to the underlying patterns, structures, and mental models.

At the heart of a system is its mental model: to be agile we need to change the values and beliefs of the organization. To change the system, we first need to change the mental models of leaders from command and control, designed to manage in conditions like those of the Industrial Age of mass production, to a more inclusive and agile style for today's Digital Age, which is heavily built around knowledge workers and trust. Command-and-control leadership works well in obvious problem contexts with little variation and when there isn't a need for skilled, intrinsically motivated knowledge workers. Therefore, it's not suited to the complex problems we are increasingly finding ourselves confronted with. The beyond budgeting movement presents 12 principles for a more inclusive leadership and management style that focuses on people, purpose, and autonomy. These are three factors that Daniel Pink points out that are essential to promote intrinsic motivation, which is critical to working in complex problem domains.

II Designing An Adaptive Operating Model

The Anatomy of an Operating Model

People work in the system that management created.

The role of management is to change the system rather than badgering individuals to do better.

A bad system will beat a good person every time.

—*Dr. W. Edwards Deming*

To change performance, we need to change system design. To do that requires a change in our thinking, change in our assumptions and beliefs.

—*Hermanni Hyytiala*

If the strategy is *what* we will do, the operating model describes *how* we will organize to do it. The operating model defines "how we work around here." In its simplest terms, the operating model describes the choices on how people, process, and technology will be used to deliver the strategic business outcomes. The operating model communicates where accountability lies; how decisions are made; the ways of working; and how we will invest, govern, and measure performance. The operating model is our system's structure, the purpose of which is to deliver business benefits by producing customer value. Changes made at the operating model level will have an influence on the patterns of behavior within IT, leading to the improved delivery of positive business outcomes. As illustrated in Figure 4.1, the operating model is the link between a company's strategy and the IT organization's ability to execute it. An effective operating model will ensure the successful execution of strategy. In Part 3, "Strategy to Execution," we will look in detail at how strategy shapes the design of the operating model.

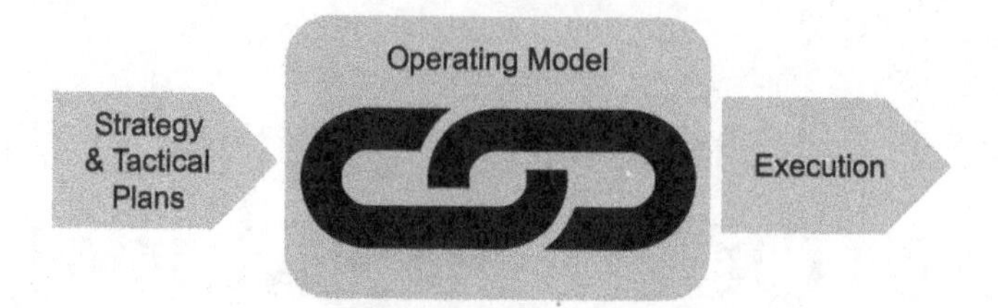

Figure 4.1: The operating model is the link between strategy and execution.

The Anatomy of an Operating Model

The operating model is made up of five components:

- **How we are organized.** The organizational component defines the teams, reporting lines, and boundaries of responsibilities and accountability. The structure of an organization or a department is the most visual and perhaps often what people think of when they hear the term *operating model.*
- **How we govern.** Governance refers to how we manage and prioritize demand, how we fund, how we measure, how we review performance, and where decisions rights lay.
- **How we work.** The how we work component defines the set of processes that dictate how IT will approach work in terms of both discovery and delivery. These are the methodologies and frameworks influenced by the philosophies covered in Chapter 2 "Philosophies for A New System."
- **How we source and manage talent.** The talent component defines how we will source, retain, and develop people that are required to deliver the IT strategy.
- **How we lead.** The leadership component defines the role leaders and managers will play in the IT department.

As Dr. Russell Ackoff said, "A system is never the sum of its parts. It is the product of the interactions of its parts." As shown in Figure 4.2, the five components that form the operating model are interconnected; they act as a system. Therefore, if there is a change in one component, there needs to be a change in the others to align all parts of the operating model to ensure the system effectively fulfills its purpose. If we change how we work, then we need to ensure we change how we govern and how we lead; this will also have an impact on the talent we need. Each component of the IT operating model will need to be optimized to support a holistic agile and adaptable method of operation.

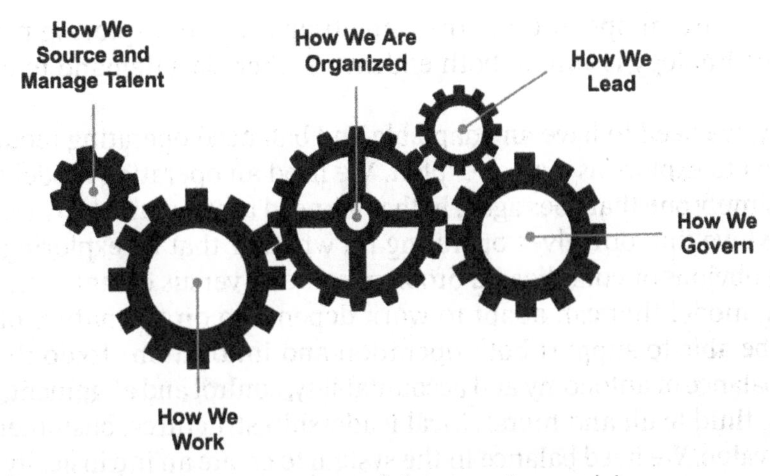

Figure 4.2: The interrelated five components of the operating model

The Themes of an Adaptable Operating Model

IT has traditionally been organized around the paradigm of an order taker or a service provider, with the operating model based on linear planning and centered around control and compliance. Performance was measured on output and conformance to budgets and plans, with teams siloed by function within a rigid structural hierarchy. Work was funded and prioritized based around detailed financial business cases that focus on operational efficiency and margin maximization for shareholders. Teams were formed to deliver discrete projects, which were typically handed off to a BAU (Business As Usual) or service team once complete. This mode of operating was suited to supporting the exploitation of a low complex and stable business model with a focus on operational effectiveness and efficiency.

However, as you have read in Chapter 1, "Why We Need to Change the System," the world has moved on. Due to the disruptive trends of heightened customer expectations, the rapidly changing ways customers want to interact, the scarcity of talent, and the advances in technology, this model of IT operation is no longer effective. Incumbent business models and their value streams need to incrementally innovate to improve operations and meet customers' experience expectations. Organizations need to extend their offering to more market segments or geographies or new channels making use of new technology advances. In addition, businesses need to explore new and radical business models to avoid being disrupted by fast-moving competitors enabled by new technology that lowers the barrier to entry in established markets. The new reality is that businesses are having to be ambidextrous, needing to constantly innovate on exploiting the current operation at the same time they are exploring new business models. As technology is core to both endeavors, there is a demand to change how IT operates.

Ultimately, we need to have an adaptable and balanced operating model that supports the need to explore as well as exploit. We need an operating model that is agile rather than simply one that does agile, in that we need to structure the ways of working to the context we find ourselves operating in, whether that be exploring or exploiting, tackling obvious or complicated problem contexts versus complex ones. We need an operating model that can adapt to work depending on the nature of the work. We need to be able to support both operation and innovation. To do this, we need balance—a balance of autonomy and accountability, control and alignment, innovation and stability, fluid team and hierarchical leadership structures, customer value and shareholder value. We need balance in the system to create an intrinsically motivated, engaged, and inspired workforce.

To manage this paradox our operating model will be based on four themes, shown in Figure 4.3, inspired by the values and principles of Design Thinking, Lean and Agile.

- A focus on impacts and outcomes over output.
- Structured for intrinsic motivation.
- A focus on being agile, not doing agile.
- Manage the flow of work, lead the people.

A Focus on Impacts and Outcomes over Output

Beyond running an efficient enterprise operation, IT leaders need to ensure efforts are focused primarily on impactful business outcomes rather than just activities that produce output. As shown in Figure 4.4, an outcome is a change in organizational or customer behavior that leads to a positive business impact. An outcome links high-level business goals to actionable work for teams. Outputs, the programs and projects, can generate outcomes that can lead to positive business impacts. Delivering successful outputs, even if on budget, on time, and to spec, has little value if they don't deliver a favorable business outcome. Therefore, we need to obsess about the problem or outcome rather than being hung up on a particular solution.

IT leaders can no longer sit back and wait to act only when a clear need has been articulated by the rest of the business. The very notion of a need or requirement in complex problem domains is a fallacy. What the enterprise really needs are outcomes to create business impacts; exactly how this is achieved in a complex domain is uncertain. Do not seek requirements but look for the behaviors that will lead to business success and the outcomes that will support them. Delegate the job of achieving those outcomes to cross-skilled teams who will explore hypotheses and tactics for delivering the outcomes. IT leaders should focus on aligning effort and investment around discovering and driving business outcomes rather than solely on the management of delivering predescribed output.

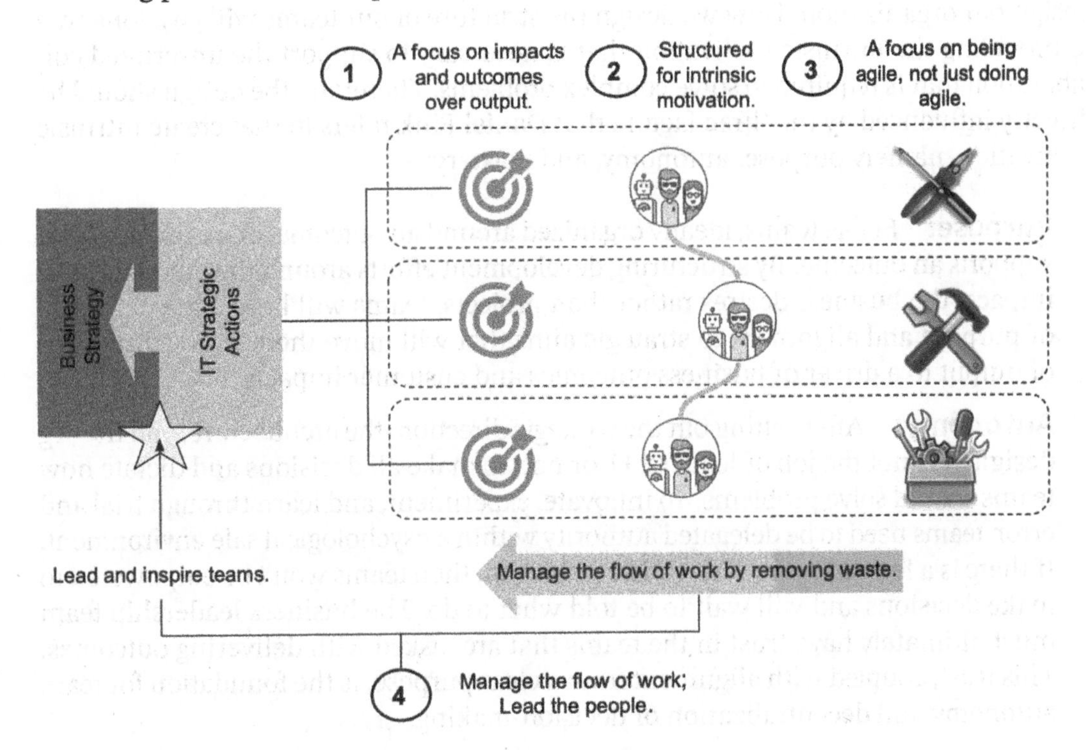

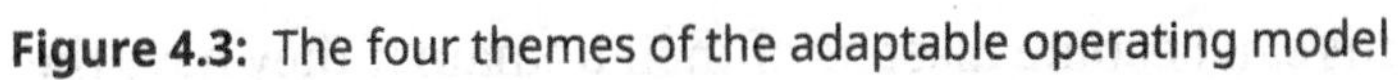

Figure 4.3: The four themes of the adaptable operating model

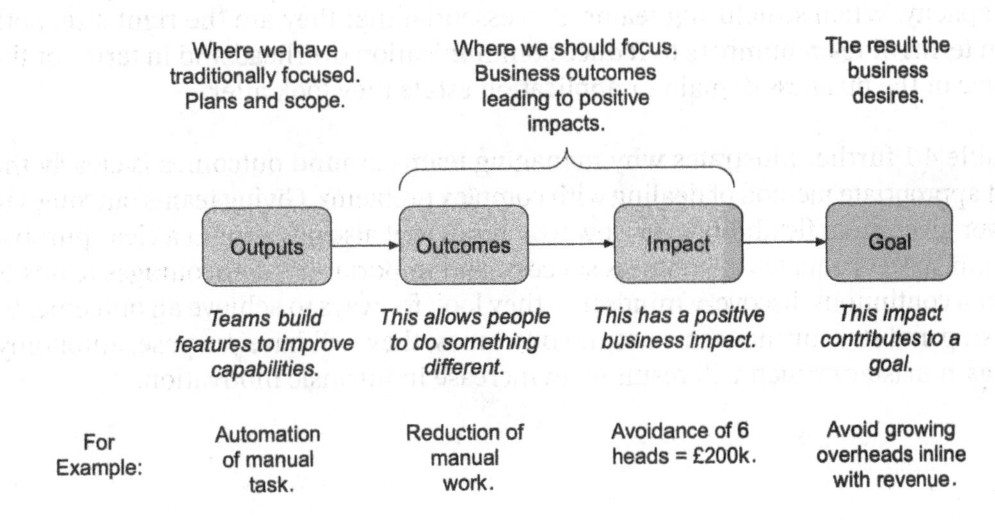

Figure 4.4: The relationship of output to outcomes and impact

Structured for Intrinsic Motivation

Stable and empowered teams should be the fundamental building block of how we design our organization. How we design the structure of our teams will go a long way to providing the intrinsic motivation that is necessary to support the unscripted collaboration that is required to solve complex problems. Therefore, the design should be directly influenced by the three factors that Daniel Pink refers to that create intrinsic motivation, namely purpose, autonomy, and mastery:

Purpose: Stable teams, ideally organized around an outcome, or a capability that supports an outcome. By structuring development efforts around the outcomes and impacts the business desires rather than projects, teams will have a greater sense of purpose and alignment to strategic aims that will move them from a producer of output to a driver of business outcomes and customer impacts.

Autonomy: After setting out the strategic direction, the architecture, and the org design, it is not the job of leaders, IT or not, to make all decisions and dictate how teams should solve problems. To innovate, experiment, and learn through trial and error, teams need to be delegated authority within a psychological safe environment. If there is a fear that mistakes will be punished, then teams won't be empowered to make decisions and will wait to be told what to do. The business leadership team must ultimately have trust in the teams that are tasked with delivering outcomes. This trust, coupled with alignment from a clear purpose, is the foundation for team autonomy and decentralization of decision-making.

Mastery: Long-lived teams can become masters of their domain if they have the capacity. When structuring teams, it is essential that they are the right size, both in terms of team numbers to reduce communication overhead and in terms of the size of the business domain or application estate they look after.

Table 4.1 further illustrates why managing teams around outcomes is clearly the most appropriate method of dealing with complex problems. Giving teams outcomes to deliver gives them flexibility over how to achieve it; it also gives them a clear purpose as team success equates to business success, and importantly, it encourages teams to adopt a continuous discovery mindset as they look for ways to achieve an outcome. By focusing and structuring teams around outcomes, they will have purpose, autonomy, and gain mastery which will result in an increase in intrinsic motivation.

Table 4.1: Managing by output, outcome, and goal

Managing by Output	Managing by Outcome	Managing by Goal
Manage teams by giving them detailed upfront requirements and asking them to deliver to the agreed scope. As our work is increasingly in the complex problem space, which is characterized by unpredictability and uncertainty, managing by output does not guarantee success and discourages teams from proactively seeking value.	Manage teams by asking them to change customer or business behavior to achieve a specific outcome that will result in a positive business outcome. Teams have the autonomy, within meaningful constraints, to determine the best solution to achieve the outcome. Teams are clear that success is measured by progress on the outcome.	Manage teams by giving them high-level goals to hit. For example, asking teams to increase profits offers no meaningful constraints on which direction to focus on.

A Focus on Being Agile, Not Just Doing Agile

There isn't one way to do IT; there are multiple ways of working, leading, and governing. There are even different types of people that are suited to one way of operating over another. Each combination has its own strengths and is appropriate within a given context. An agile operating model will not rigidly follow one way of operating, but it will adapt to the nature of a problem. As illustrated on the Wardley Map in Figure 4.5, there are multiple operating models for teams within the overall operating model depending on the nature of the problem you are facing. The talent attributes and attitudes are characteristics of what Simon Wardley refers to as Explorers, Villagers, and Town Planners, which we will look at in detail in Chapter 8, "How We Source And Manage Talent."

Teams Dealing with Unique and Uncharted Problems:

- Ways of working: Design thinking and Agile practices work well here.
- Governance: In areas of exploration the value is in reducing uncertainty rather than measuring ROI, where rapid learning and fast decision cycles are important to discovering opportunities.
- Leadership: Task teams with discovering market opportunities. Here we have good failure. As we experiment, we understand what doesn't work and that helps us evolve how we tackle a problem and our understanding of it.
- Talent: Entrepreneur and innovative creative types, happy to experiment and fail often in the quest to learn what works and what makes customers happy.

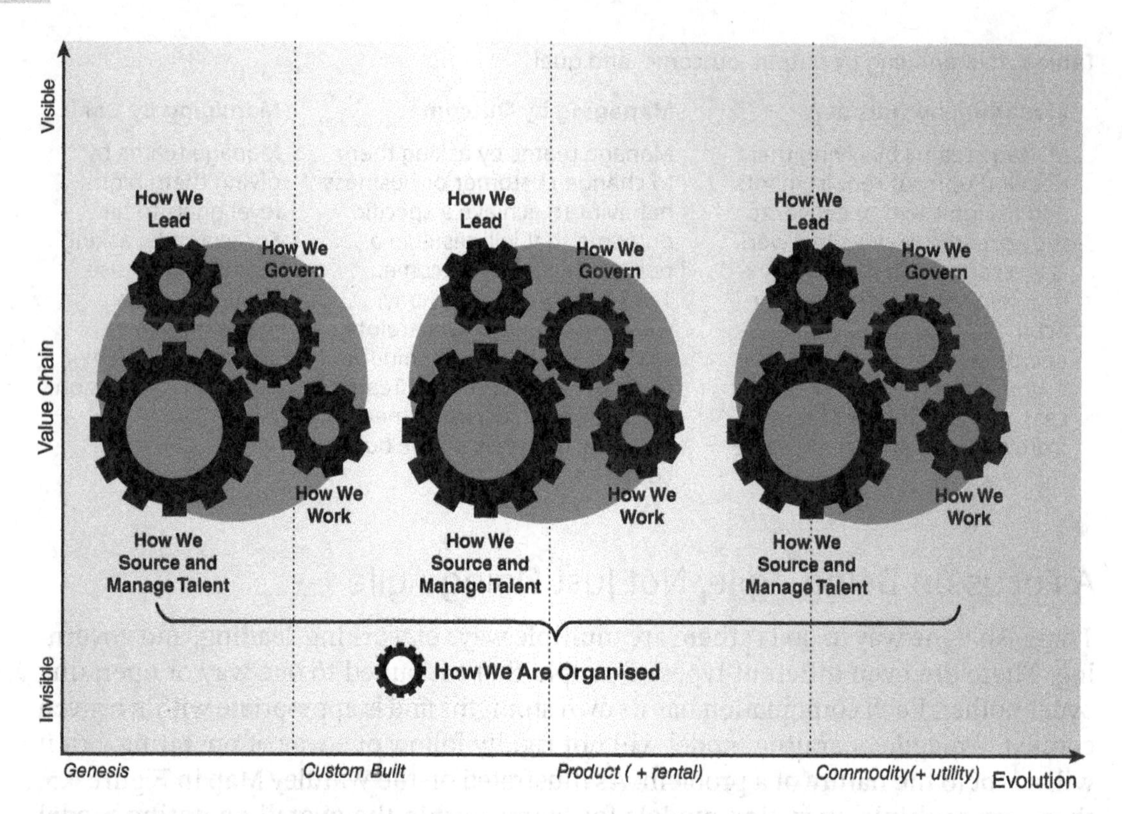

Figure 4.5: Multiple operating models are required to manage the spectrum of work.

Teams Dealing with Reducing Waste in Known Problems:

- **Ways of working:** Lean and agile practices work well here, where we need to remove waste and focus on operational efficiency.
- **Governance:** Teams focus on producing outcomes that improve operational metrics such as revenue growth and profitability.
- **Leadership:** Direct teams on the desired business outcomes.
- **Talent:** People that like improving well understood but still complicated problems, with a focus on scale and efficiency.

Teams Dealing with Problems That Are Highly Evolved:

- **Ways of working:** Waterfall and upfront planning work well here.
- **Governance:** Business cases with a clear return on investment; project governance is conforming to a plan or budget. Focus is on reducing deviation from well-defined processes.

- **Leadership:** Be more prescriptive and direct teams on output required. Have less tolerance for operational, aka bad, failure through poor execution of a solution in a well-understood problem area.
- **Talent:** People that comfortable with a focus on operational efficiency and standardization.

Manage the Flow of Work; Lead the People

Rear Admiral Grace Hopper famously said, "You cannot manage men into battle. You manage things; you lead people." The things we should manage in the operating model are the impediments to the flow of value. In other words, leaders need to work on the system, not in the system. We need to support the team's ability to deliver by focusing management efforts on the waste outside of teams and outside of their control. We need to focus on the entire "idea to impact" delivery value stream. Outside of the delivery step of the value stream, waste manifests itself as unnecessary bureaucracy, a lack of decision rights, too much work in progress, a lack of a well-thought-through buffer of work for teams to pull from, and the waste caused by work waiting in queues due to dependencies. The greatest impact management can have on the flow of work is not by man managing but from the removal of obstacles that stand in a team's way of delivering value.

While we don't want to manage people, we do want to lead them. Leadership, at all levels, is vital to inspire and motivate teams when dealing with complex problems within an ever increasingly chaotic and challenging world. Leading is not about hand-holding or controlling; it's about articulating a future vision and inspiring those that you lead. It's about clear and consistent communication on the strategic direction, especially when it changes. It is creating an aligned and shared desire to collaborate and achieve that vision, empowering them through trust, giving them the support and encouragement they need to successfully achieve the shared vision. Both lean and agile have respect for people at the center of their philosophies. Empathy and respect for people in your teams and departments should be at the center of yours.

Summary

The operating model defines how we are set up to do the work required to execute the strategy. As you will see in Part 3, the operating model is shaped directly from the strategic actions that IT will contribute to business success. The operating model is composed of five components: how we work, how we govern, how we are organized, how we source and manage talent, and how we lead.

There are four themes, inspired by the values of lean and agile, that will run through an agile operating model:

- A focus on outcomes over output.

  It doesn't matter what you do if it doesn't make an impact. IT teams should predominantly be measured on business outcomes and not by project outputs. IT should avoid the build trap by delivering meaningful business impacts and not just features.
- Structured for intrinsic motivation.

  Teams require a level of motivation when dealing with complexity and uncertainty, where there is a need to explore and discover solutions. This is only achieved through purpose, autonomy, and mastery.
- A focus on being agile, not doing agile.

  We need to apply the most appropriate methods depending on context. There will be mini operating models within teams depending on the type of problem they are dealing with that reside in the overarching operating model.
- Manage the flow of work, not people.

  IT leaders must manage the system, not people, by removing impediments from the flow of work. IT leaders must lead by inspiring and supporting teams to achieve business outcomes and goals.

We will explore each of the operating model components over the next five chapters.

How We Are Organized

Any organization that designs a system . . . is constrained to produce a design whose structure is a copy of the organization's communication structure.

—*Melvin Conway, 1968*

We don't hire smart people to tell them what to do. We hire smart people so they can tell us what to do.

—*Steve Jobs*

Companies need hierarchies to run the business, and a network structures to adapt and change the business.

—*J. P. Kotter*

It won't surprise you to learn that there isn't a single perfect design to organize the IT department that will suit all contexts. What is likely to be universally common is the need for a structure that can support a business's ability to both explore and exploit, to manage complexity, to drive innovation while maximizing efficiencies and simplifying processes. In this chapter, we will look at how employing traditional command-and-control hierarchies coupled with the recent emergence of holacratic networked teams can provide a balanced structure that combines flexibility with control, alignment with autonomy, and purpose with ownership. These are all required to manage the cross section of challenges and opportunities that a modern business will present.

We will take a detailed look at network teams, known as product teams, and how they are a move away from organizing people around short-term projects to persistent teams designed around key technical, social, and business boundaries that require long-term investment. We will look at how to define these team boundaries, the benefits of them, and why, contrary to what the name may suggest, the principles behind product teams are applicable to all teams in the IT organization.

Organizational Structure

The organizational structure of a department is the most visible component of its operating model, and the one that is likely to have the biggest impact on the shape of the technical landscape and the execution of work. Therefore, structure should be deliberately designed to support the strategic contribution that IT will make to business success, which we will cover in depth in Part 3, "Strategy to Execution." In other words, as Naomi Stanford, a leading author on organizational design, says, there should be a compelling reason to change the structural design of teams and "Part of a decision to design rests on making a very strong, strategic, widely accepted business case for it—based on the operating context. If there is no business case for design or redesign, it is not going to work." In addition to the strategic needs, we should also take into consideration social, technical, and business seams when designing team boundaries to ensure alignment and collaboration, team health, and an increase in the flow of value. The organizational design will have a material effect on how people function and their relationships to other areas of the business, but if the other components of the operating model don't complement the changes, there will be limited positive business impact. We will explore both how we work and how we govern the need to adapt to changes in how teams are designed in the following two chapters.

To understand how best to organize teams, I will compare two extreme approaches: hierarchical and holacracy, and show how employing a mixture of both, rather than choosing an either-or approach, will support a structure that can enable business performance in exploring new business models and opportunities as well as exploiting and maximizing the incumbent model.

Hierarchy

At its most extreme, a hierarchy is a system based around the concept of command and control that separates thinking from doing. The world of hierarchy is vertical, with power flowing from the top down. As shown in Figure 5.1, prescriptive orders are given by leaders, feedback is passed up from the workers via management, and decisions are then made at the top and passed down. The purpose of the hierarchy is to ensure control and predictable operational results. Thinkers at the top, doers at the bottom.

The popular flavors of hierarchical organizational design are functional, divisional, and matrix.

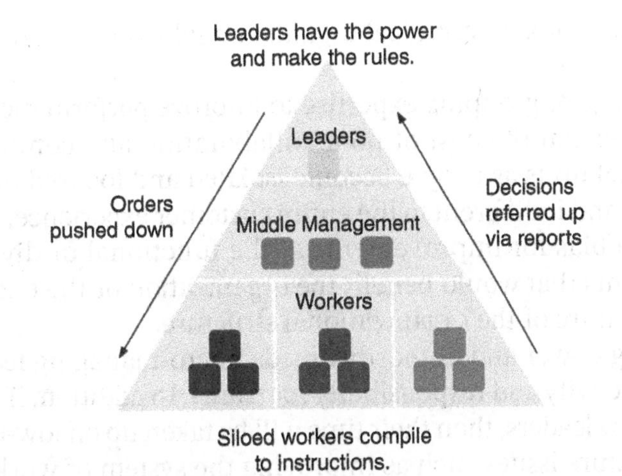

Figure 5.1: A hierarchy and the system of command and control

- In a functional structure, an enterprise groups people based on roles or tasks, such as marketing, commercial, operations, or IT.
- In a divisional structure, the enterprise is organized so that each division, or business unit, has the necessary capabilities (people and process for a given product line or location).
- In a matrix structure, some individuals can report to multiple leaders. For example, an individual may report to a functional leader but also have a dotted line reporting to a project manager.

The advantages of a hierarchical organizational structure are as follows:

- There are clear lines, and a clear direction, of communication. Individuals are clear on who they report to, where to get direction, and who their spokesman is. There is a high-level of coordination within a function due to the command-and-control nature of leadership.
- The boundaries of accountability are clear between the various functions or divisions within an organization. Authority and power are based on an individual's management level. This clarity in the chain of command is particularly useful when there is a need for a quick decision as it is clear who has the authority to give an order.
- There are efficiency gains to be had by organizing people by role or task. Cost savings can be realized by sharing a central pool of people all with the same or similar skill sets. Specialists can learn and train together, ensuring improved expertise.

The disadvantages of a hierarchical organizational structure are as follows:

- The advantage of grouping expertise to improve performance at a department level can come at the cost of poor collaboration and communication across departmental units as people become isolated and focused on their part of the value chain and indifferent to the entire customer experience. This tunnel vision can act as a bias for improvements at the functional or divisional level over improvements that would benefit the organization or the end customer due to the siloed nature of the organizational structure.
- Centralizing power and authority breeds micro-managing leaders leading to a lack of proactivity and responsibility for teams. In addition, if all decisions need to be made by leaders, then their time will be taken up on low-level details rather than big-picture issues such as improving the system of work. Often decisions are best made, and are quicker to be made, by those closer to the problem. If there is a need to wait for sign-off, then this is a major blocker for teams trying to be responsive to rapid change.
- For larger organizations with many levels of hierarchy, making decisions and communicating can become slow and bureaucratic. Decisions need to go up and back down the value chain, and the larger the organization, the longer it takes. This slowing down of decision-making reduces the ability for teams to pivot and adapt based on feedback from customers.

Holacracy

At its extreme, a holacracy is a management strategy and an organizational structure where the power to make important decisions is not held solely by those paid the highest but instead is distributed throughout an organization to people at all levels. The fundamental concept behind holacracy is self-organization. Given a common goal, employees should organize themselves into small teams and determine the best use of their time to achieve that goal. The structure encourages distributed decision-making to those closest to the problem without the need to pass through layers of bureaucracy to authorize pursuing an idea. Along with empowerment to make decisions, trust, safety, and transparency are core principles of a holacratic way of operating. A loose form of hierarchy exists, as shown in Figure 5.2, but this light structure is just to ensure boundaries of responsibility are clear and that leaders can articulate the organization's goals and direction. Contrast this to the traditional hierarchy, where thinking and doing are separated in that those at the top decide what to do and how to do it and hand off to those at the bottom to do it in a very prescriptive manner. In a holacratic organizational design, these roles are merged. Thinkers and doers work as a team guided by the strategic objectives of the business.

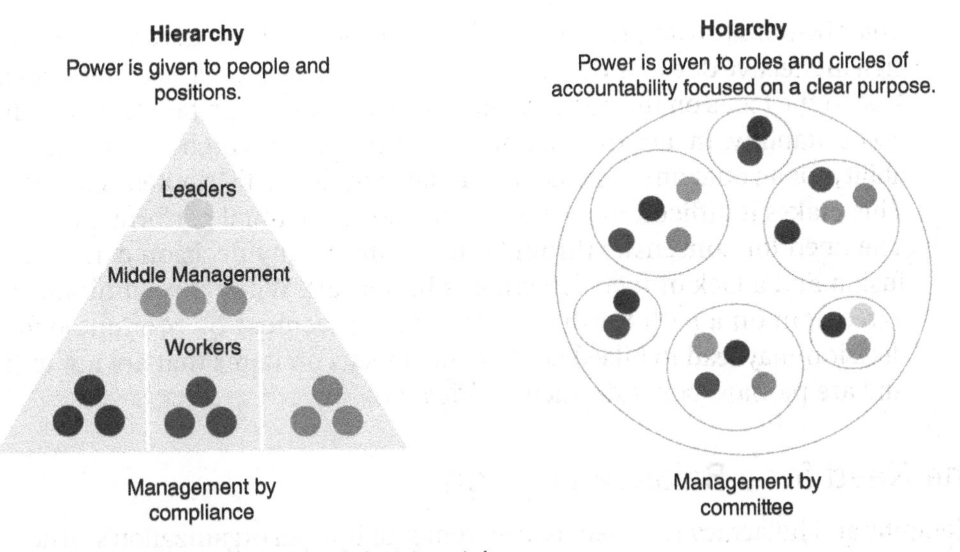

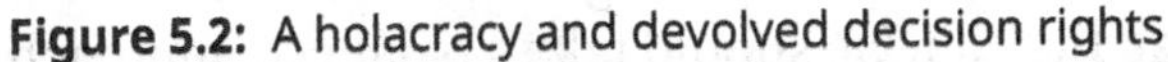
Figure 5.2: A holacracy and devolved decision rights

The advantages of a holacratic organizational structure are as follows:

- Everyone in the organization has a line of sight to what is important and works directly toward a common goal. How best to achieve that goal is everybody's responsibility. This differs from a more hierarchical structure where people at the bottom take commands from people at the top. A holacratic structure results in a more purpose-driven environment where autonomy and empowerment grow, which is very good for discovering innovative solutions.
- Decisions can be made quickly and easily by those closest to the customer due to the absence of a bureaucratic management requiring every decision to be approved by someone further up the chain. This results in every employee being empowered to deliver exceptional customer service through the ability to make decisions quickly and to feel a sense of ownership for problems and solutions.
- Everyone in a holacratic organization knows what is expected of them and the direction of the business due to the open and transparent objectives and information that are shared. The organization's focus on a shared set of outcomes and common goals prevents local optimization over system-wide benefits.

The disadvantages of a holacratic organizational structure are as follows:

- The holacratic organizational design is not familiar to many; it's a radical change and therefore can be a big jump for many organizations. It is difficult to implement since it is as much of a mindset and cultural change as it is an organizational structural change. Often this is overlooked, rendering any redesign ineffective due to the same entrenched management and hierarchies at play.

- Due to the focus on the performance of a team, there can be a lack of individual accountability. In a traditional hierarchy, it is clear who has ultimate accountability for an outcome, whereas in a holacracy, this is distributed across the team. This makes it difficult to assess and manage individual performance.
- The need for consensus throughout a team on any decision can lead to confusion and a lack of focus on what is important. While it is admirable to have agreement on a path for the whole team, agreement by committee for every decision may lead to bikeshedding and a focus on items that are not important and are perhaps better decided by a leader.

The Need for a Balanced Design

Hierarchy and holacracy represent two extremes of how an organization's structure can be designed, but neither of these is ideal. Instead, we should take a balanced approach, opting for a mixture of the two or a "flatter" organizational structure as shown in Figure 5.3. This structure retains the essence of a hierarchy to give direction, coordinate across teams, and provide guardrails while utilizing the benefits of a holacracy by focusing on the team as the primary unit of an organization and distributing decisions, rights, and giving employees autonomy to innovate. This blend of both a hierarchy and a holacracy takes out the unnecessary layers of bureaucracy but still leaves a framework in place for coordination, control, and communication. In truth, depending on the problem context you are dealing with, you may favor one structure more than the other. For example, you may favor a more hierarchical structure for simple problem contexts, such as when working with third parties to integrate an off-the-shelf system and a more holacratic team structure for more complex and unknowable problem contexts that require the team to innovate and adapt to feedback quickly.

Figure 5.4 gives an example of how a balanced structure may look in reality. There are networked teams of people based around business domains, be they customer journeys, value streams, or business capabilities. There are hierarchies of people for cross-cutting concerns such as enterprise architecture, project management, security, governance, service, and infrastructure.

A hierarchy brings clarity in decision-making, leadership, and accountability and is therefore not inherently bad. However, we need just enough hierarchy for direction setting, coordination, and guidance, while avoiding unnecessary bureaucracy. Too much of a hierarchy or command-and-control structure will stifle innovation and customer-centric approaches, which will prevent an organization from being successful in managing complex problem domains. A hierarchy is critical to effectively deal with communication and coordination as well as size and complexity as the network teams grow and are distilled into coherent smaller teams. The greater the number of teams, the greater the need for a hierarchy to control at a macro level; therefore, hierarchy sits above a holacratic set of networked teams as an organization grows.

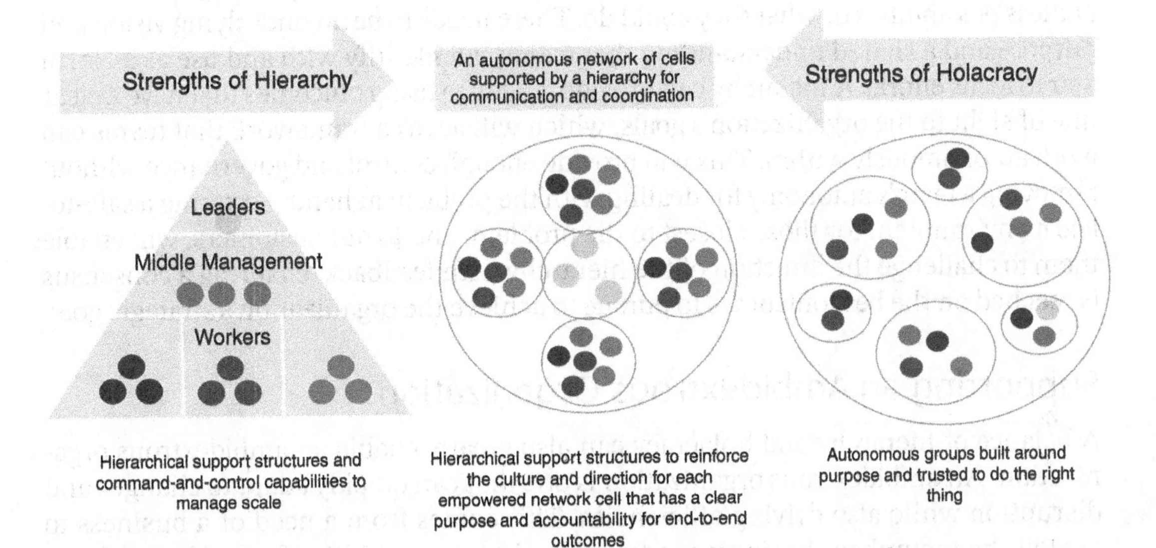

Figure 5.3: An organizational structure that balances a holacracy and hierarchy

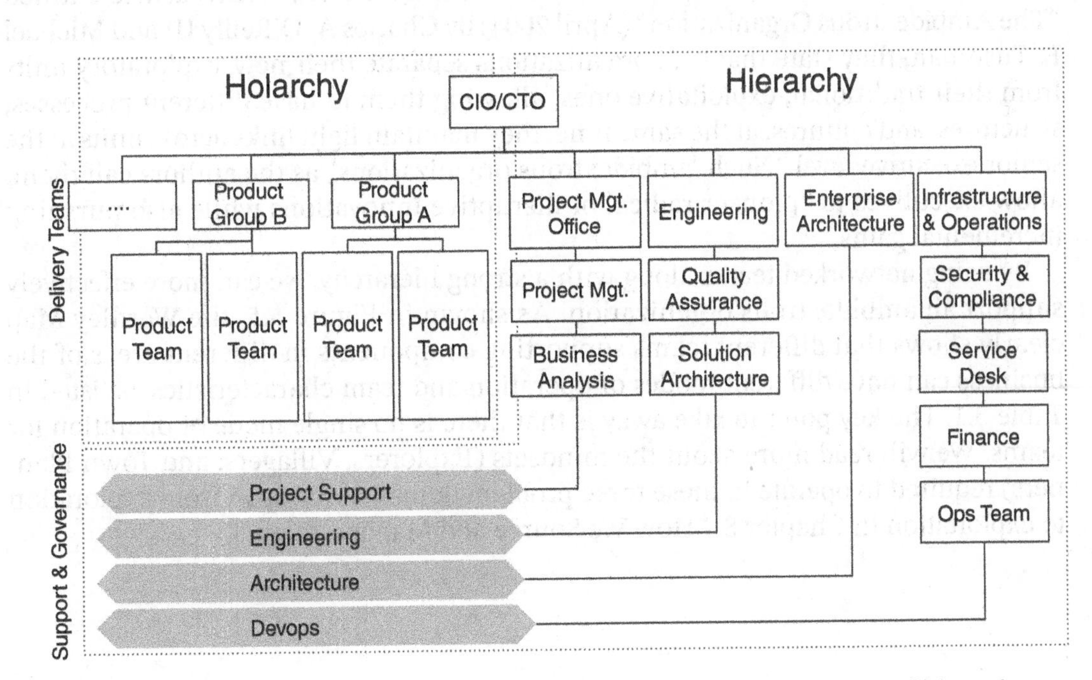

Figure 5.4: An example of an organizational structure that balances holacracy and hierarchy

There is also nothing wrong with seeking a more holacratic design; however, this doesn't mean ignoring the advantages of a hierarchy and just letting teams do what they want. Often people need constraints in the form of guardrails or guiding principles as well as shared outcomes and visions to avoid becoming paralyzed with the endless possibilities of what they could do. There needs to be an underlying vision and purpose and a shared understanding that teams can identify with and use as a North Star to focus efforts. A hierarchy can provide guidance using outcomes that have a clear line of sight to the organization's goals, which will act as a framework that teams can work autonomously within. This will provide enough control and governance without removing a team's autonomy for dealing with the problem at hand. Fostering a safe-to-learn environment for those closest to the problem, and to our customers, will enable them to challenge the direction of the hierarchy with feedback, ensuring a consensus is reached on the best outcomes to pursue to achieve the organization's strategic goal.

Supporting an Ambidextrous Organization

A balance of hierarchy and holacracy can also greatly enable an ambidextrous organization. An ambidextrous organization is one that can quickly adapt to changes and disruption while also driving efficiencies. This comes from a need of a business to exploit the incumbent business model and existing capabilities for profit and focus on exploring new models and new opportunities for growth to ensure they do not become obsolete due to inertia. In a *Harvard Business Review* (HBR) article entitled "The Ambidextrous Organization" (April 2004) by Charles A. O'Reilly III and Michael L. Tushman, they state that ". . . organizations separate their new, exploratory units from their traditional, exploitative ones, allowing them to have different processes, structures, and cultures; at the same time, they maintain tight links across units at the senior executive level." Such 'ambidextrous organizations', as the authors call them, allow executives to "pioneer radical or disruptive innovations while also pursuing incremental gains."

By using networked teams along with a strong hierarchy, we can more effectively support an ambidextrous organization. As shown in Figure 5.5, the Wardley Map clearly shows that different teams supporting components in different areas of the business can have different modes of operation and team characteristics as listed in Table 5.1. The key point to take away is that there is no single mode of operation for teams. We will read more about the mindsets (Explorers, Villagers, and Town Planners) required to operate in these three problem domains that span from exploration to exploitation in Chapter 8, "How We Source and Manage Talent."

Table 5.1: Different modes of operating to support an ambidextrous organization

	Exploratory	Transitional	Exploitative
Team Characteristic	The mode of operation that can manage uncertainty while exploring new growth opportunities.	The mode of operation focused on components of the business model that are evolving from the exploration toward the exploit.	The mode of operation that is focused on efficiency when exploiting an incumbent model.
Measured on	Innovation and adoption.	Scale, error reduction, productionizing, and growth.	Cost, profit, efficiency, and standardization.
Culture	Small and highly collaborative entrepreneurial teams able to adapt and innovate at speed.	Able to mature and improve capabilities to enterprise level; reduce waste and increase value and profitability.	Typically, outsourced, risk-averse with an operational focus on the reduction of complexity.
Simon Wardley's People and Cultures	Pioneers	Settlers	Town Planners

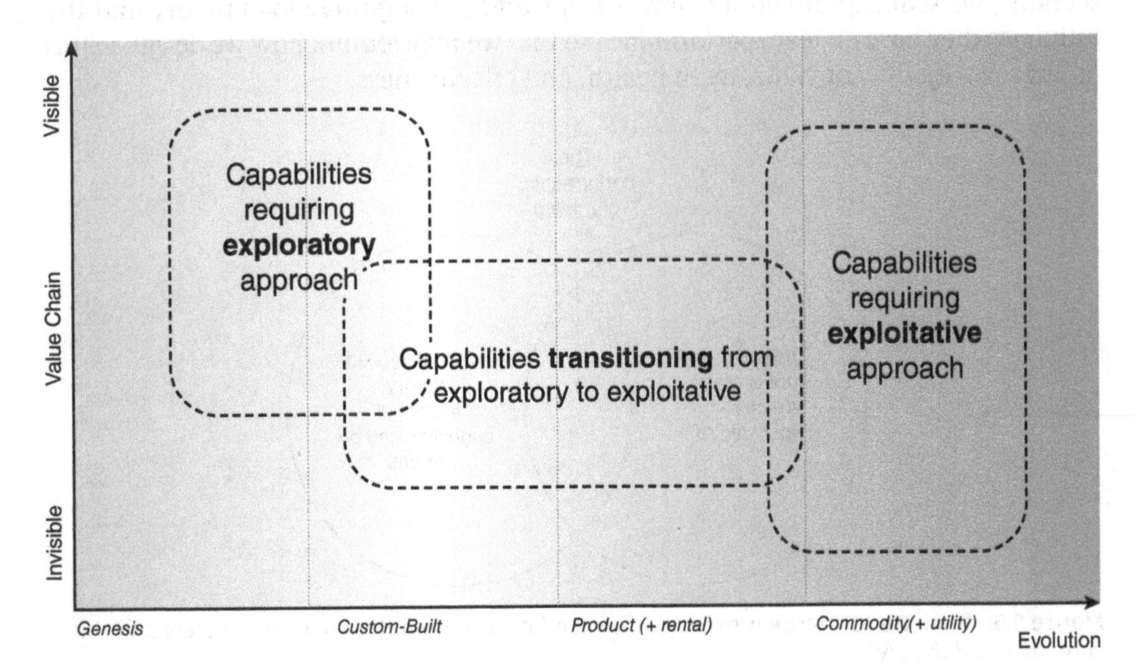

Figure 5.5: A Wardley Map and the three modes of operating to support an ambidextrous organization

Organizations today must adapt to change and innovate at speed to remain relevant, but they also must balance the need for stability as they scale. As businesses grow, the need for a hierarchy is increased, but one that will ensure communication lines and team setup is effective and supported rather than adding waste and bureaucracy. By blending the strengths of a hierarchy with a holacracy, we can create an organizational structure that enables clear guidance, control, and governance complemented by a set of networked teams, with their own context-specific processes and culture, that are able to adapt quickly in response to feedback and changes. This is how IT can contribute and support an ambidextrous organization.

Understanding the Influence of Conway's Law and the Cognitive Load Theory on Team Performance

In their book *Team Topologies: Organizing Business and Technology Teams for Fast Flow* (IT Revolution Press, 2019), Matthew Skelton and Manuel Pais highlight the importance that Conway's law and the cognitive load theory have on the effectiveness of teams. As shown in Figure 5.6, team performance is directly constrained by the architecture of software, itself influenced greatly by the organization of teams and the capacity of teams to manage domain and technical complexity. In the following sections, we will explore both Conway's law and the cognitive load theory and the influence they have on team performance so that we may rethink how we design teams to improve the flow of value, team health, and effectiveness.

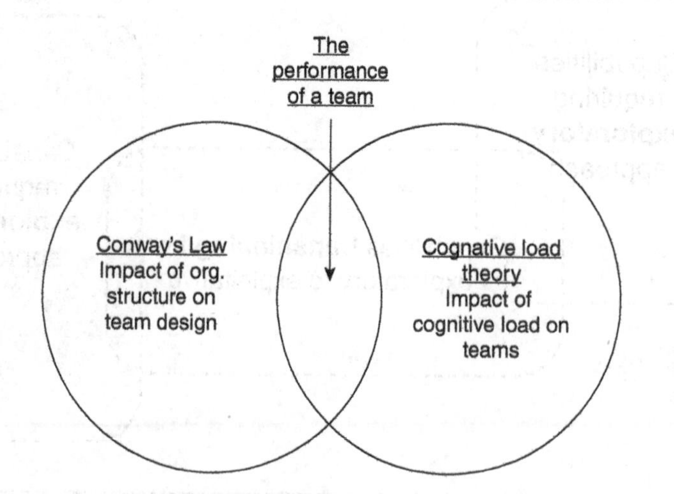

Figure 5.6: Team performance is greatly influenced by a combination of Conway's law and the cognitive load theory.

Conway's Law

In 1968, Melvin E. Conway wrote a paper on how reporting and communication lines have a large impact on the architecture of a system. In it he said, "Any organization that designs a system will produce a design whose structure is a copy of the organization's communication structure." In other words, the design of the software architecture is a result of the organization of teams. Consider the example in Figure 5.7; separate teams of database engineers, middleware specialists, and front-end designers will produce an architecture that mirrors how they are organized. This can result in many handoffs and a monolithic technical design making changes hard.

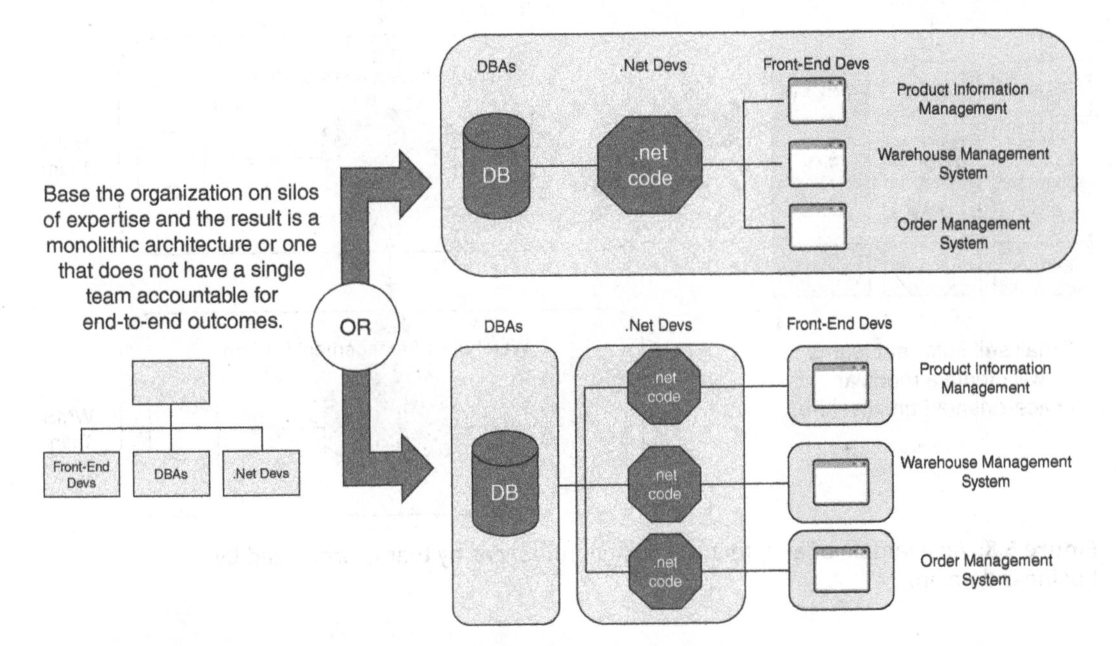

Figure 5.7: An example of architectural design influences by teams organized by function

With Conway's law in mind when designing the architecture of a system, we need to consider what Ruth Malian observed: "If the architecture of the system and the architecture of the organization are at odds, the architecture of the organization wins." Therefore, we should inverse or reverse Conway's law by designing the organizational structure to match the intended system architecture. Start with a blueprint of the software architecture and design the organization around it to produce the kind of architecture we intend. Again, consider Figure 5.8; instead of separating teams by function, we create an organizational structure with the sufficient skills in each team to achieve an outcome. In this case we have built teams for the architecture we want, and those teams are more likely to produce it with no handoffs due to the end-to-end responsibility.

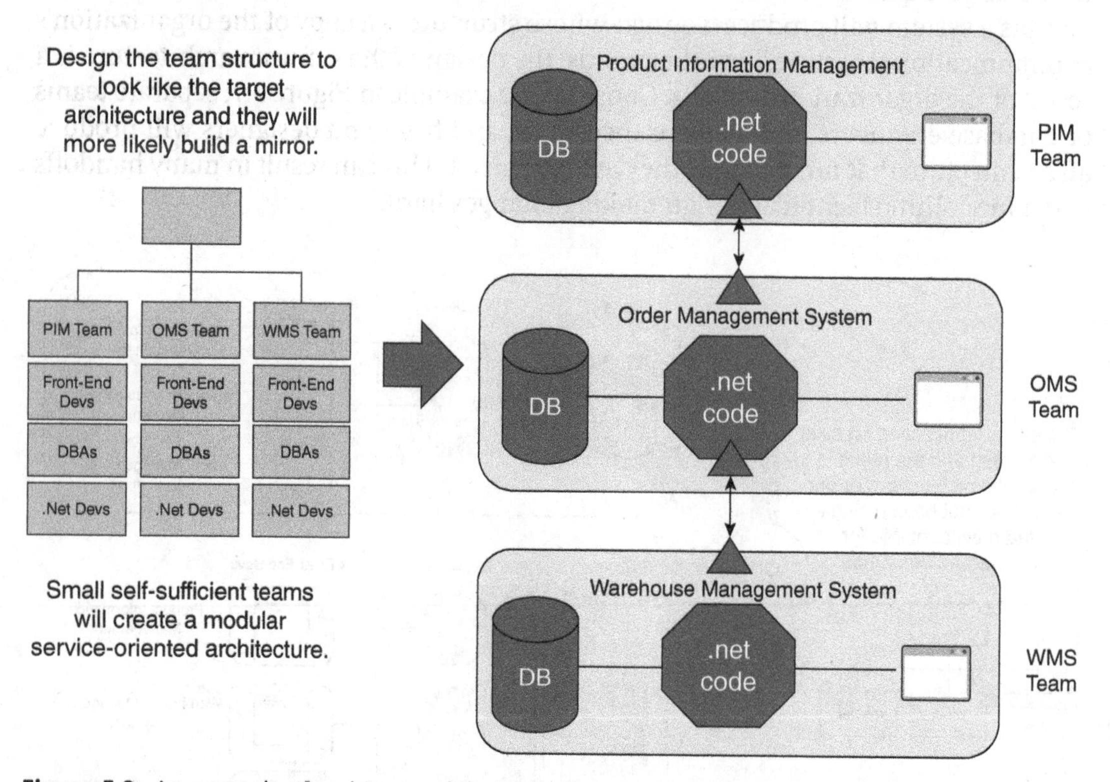

Figure 5.8: An example of architectural design influences by teams organized by business domain

Cognitive Load Theory

In 1988 John Sweller introduced the cognitive load theory, which refers to the total amount of mental effort being used in the working memory. As shown in Figure 5.9, there are three components to memory: sensory memory, working memory, and long-term memory. Where long-term memory is limitless, working memory is not. Cognitive load divides the load on working memory into three subtypes: intrinsic load, germane load, and extraneous load.

Intrinsic Load This refers to the inherent level of difficulty of the problem domain at hand. For example, the relative complexity between nuclear fusion versus an e-commerce shipping calculator.

Extraneous Load This refers to the way information around the intrinsic load is presented; this load does not add anything useful, but it is necessary. For example, when learning a new subject, you will have a high extraneous load if you have an ineffective teacher. In software development, extraneous load could also refer to the series of activities needed to deploy changes or set up a development environment.

Germane Load Refers to the work to integrate and retain new knowledge in long-term memory. For example, using process maps to capture complex flows or looking at patterns in new information to help retain knowledge.

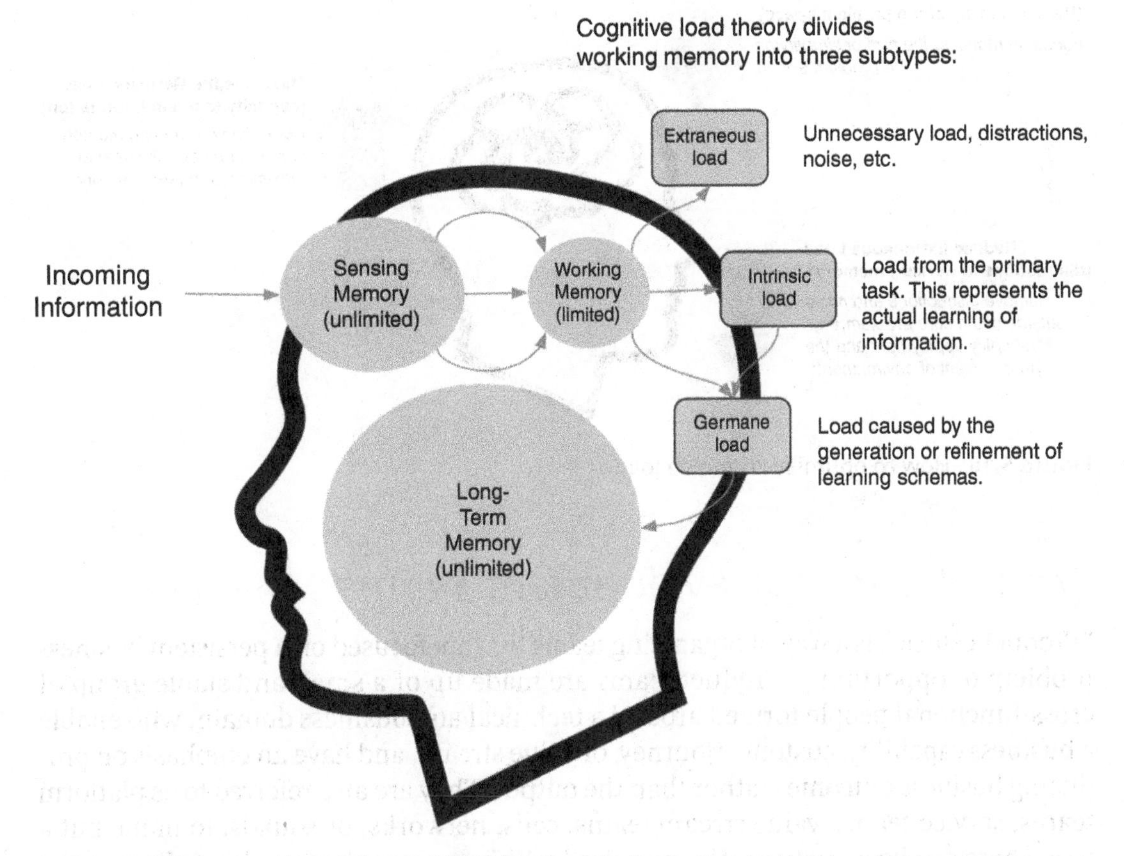

Figure 5.9: How memory works and the three types of cognitive load on working memory

As shown in Figure 5.10, if we overload people with too many distractions (extraneous load) or too complex a problem area (intrinsic load), then we end up with teams constantly context switching, unable to focus and master an area, and unable to commit any long-term learning. Therefore, to optimize team effectiveness, we need to save teams' working memory for what is core to the problem and reduce all other noise and distraction. Again, as show in Figure 5.10, if we can manage intrinsic load by giving teams a manageable amount of domain complexity—i.e., right-size boundaries of responsibility, reduce extraneous load by making it easy for teams to procure new environments, engage domain experts, and remove distractions—we will allow more capacity for Germanic load and the ability for teams to retain information in their long-term memories. This enables teams to master their domain and become effective problem solvers.

Simplify the Intrinsic Load (the complexity of the problem space)
Focus on reducing the size or simplifying the problem space.
Maximize the Germane Load (capacity to retain information)
Reduction on working memory can free up capacity here to maximize long-term memory.
Reduce Extraneous Load (distractions to working memory)
Remove distractions and noise outside of the core problem, e.g., simplify deployment and the procurement of environments.

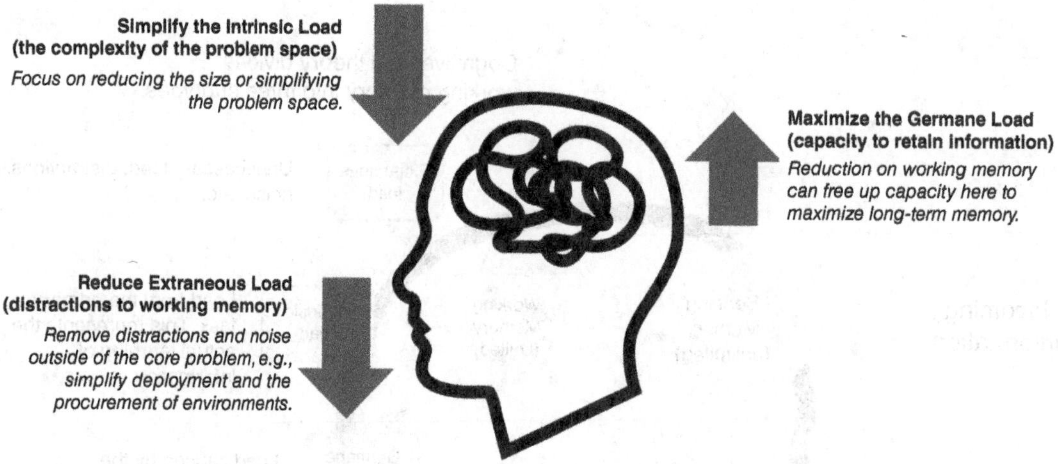

Figure 5.10: How to optimize cognitive load

Product-Centric Development Teams

"Product-centric" is a way of organizing teams that are focused on a persistent business problem or opportunity. Product teams are made up of a small and stable group of cross-functional people formed around a technical and business domain, who enable a business capability, customer journey, or value stream, and have an emphasis on producing business outcomes rather than the output. They are also referred to as platform teams, service teams, value stream teams, cells, networks, or squads, to name but a few. There can be a mixture of names used within an organization depending on the role a team plays. What is important is not to be too hung up on the name but instead obsess on the fundamental principles of product teams as highlighted in Figure 5.11:

- In contrast to project teams, product teams are long-lived.
- Product teams build/configure/integrate it and run it. They don't hand off to other teams.

- They are structured around technology that supports a business outcome that relates to customer experience, the business model, and business strategy.
- They have clearly defined boundaries of accountability, at both a technical and business level, enabling autonomy and end-to-end responsibility.
- They have a clear purpose, or outcome.
- They have a manageable amount of complexity, enabling them to master their problem area.
- Work is mostly value-driven, not plan-driven. Therefore, teams can adapt when they find a cheaper or faster way to achieve an outcome.
- Teams are perpetually funded, typically annually, rather than on a project basis.
- They are staffed with the necessary skills within the team to do the job.
- They are appropriately sized, based on a combination of social, business, and technical boundaries.

These are the hills to die on, not whether you call teams squads or cells.

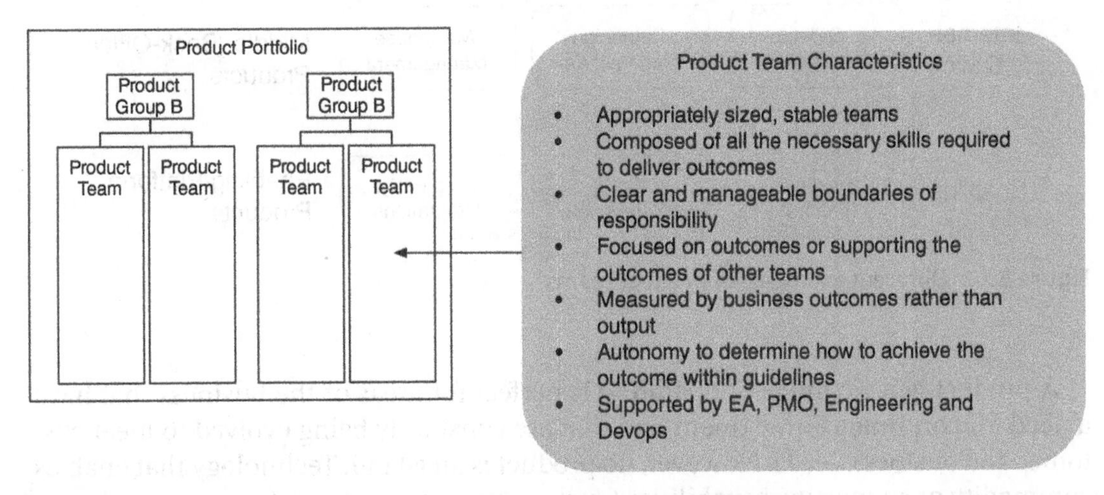

Figure 5.11: The characteristics of product-centric or product teams

Moving to product teams and a product-centric mode of operating represents a paradigm shift for organizations in terms of an operating model, both in its structure and execution. Paradoxically, team structures, their relationships, and communication patterns have always had a material impact on how effective an organization is at delivering value. However, designing the operating model around a product-centric mode of operating, in which the concept of a team becomes the fundamental building block of an organization, will lead to a better social and technical architecture.

The Definition of a Product

Although a product-centric mode of operating may have originated from pure play organizations where their value proposition was a digital product, this way of organising is applicable to all types of organizations that are investing in digital capabilities. As shown in Figure 5.12, the principles and values behind product teams can be applied to all capabilities throughout an enterprise, from the back office to the front. Even capabilities without a direct line of sight to real business customers can still be managed as products, catering to internal customers and supporting others with achieving business outcomes.

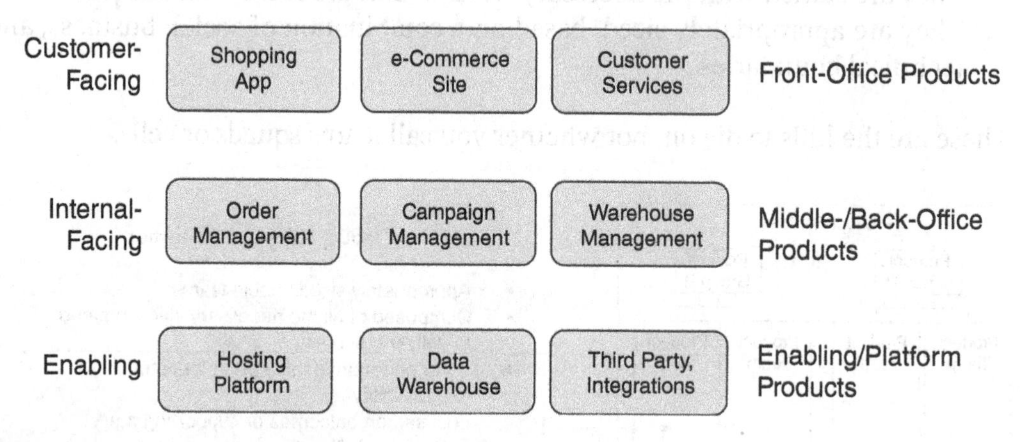

Figure 5.12: Different examples of product teams

A product-centric mode of operating is perfect for areas of the business that have a need for continuous investment and that are constantly being evolved to meet customer and business needs. However, no product is an island. Technology that enables commodity or supporting capabilities (such as human resources, finance, warehouse management, and workforce management), that doesn't change as frequently, and even if delivered with COTS software or managed by a third party, will still benefit from being managed as a product. Even if there isn't a need for a continuous investment in these areas, they are likely to be affected and need to change when dealing with large cross-cutting business initiatives and therefore also require a dedicated small team to manage and run. These products may have more of a focus on configuration, integration, and running rather than building and may likely require a different type of person than those products that are built and will rapidly evolve. The point is we can, and should, treat everything as a product, regardless how the cultures, structure, and processes may differ by teams.

Comparing to Project Teams

Traditional IT development teams are based around a project mode of organization, where scope is prescribed, time for delivery is all calculated up front, and teams are transient. The purpose of these project teams is to deliver an output. However, by following a product model and organizing a long-lived team around a business capability, experience, value stream, or customer journey, the team gains deeper relationships with their business peers and a deeper understanding of the problem domain along with its opportunities and constraints. This enables them to find better solutions and quickly, compared to temporary teams who are air dropped into work on a particular project with no affinity with the problem domain or the business colleagues that work within it. And because they are given outcomes to achieve, as opposed to output to deliver, they have the flexibility, and therefore the empowerment, to determine how best to use their time to achieve the directed outcomes. Table 5.2 highlights the main differences in product versus project teams. We will look at how governance is changed through the adoption of product teams in Chapter 7, "How We Govern."

Transient teams formed for specific projects is not an anti-pattern and is sometimes necessary for large one-off changes or, as you will read later, cross-team collaboration. However, this shouldn't be the default mode of organization if you want to be able to adapt quickly, deliver at pace, and have healthy teams that are intrinsically motivated.

Table 5.2: The difference between project and product teams

IT Project Teams Characteristics	IT Product Teams Characteristics
Predictive planning	Adaptive planning
Transient team	Persistent team
Output-focused	Outcome-and impact-focused
Funding for project	Funding for team capacity
Measured against plan	Measured against value
One-off delivery	Continual investment and improvement

Defined Boundaries of Responsibility

Product teams need to have clear boundaries of responsibility, at both a technical and business level. A lack of clear accountability for a technical boundary prevents a team from having the autonomy, and the ownership, to evolve and master their area. This can also lead to gray areas of technology that nobody feels responsible for, while unclear business boundaries can prevent alignment on the strategic initiatives that teams should focus on.

We want teams to work autonomously on their outcomes. This can only happen if we minimize the dependencies and avoid too many teams having joint responsibility for an outcome. Therefore, when defining team structure, we must ensure there is a cohesive relationship between team boundaries, both technical and business, and the desired strategic business outcomes. Dependencies will be impossible to avoid altogether. However, with clarity on ownership and relationships with other teams made explicit, we can manage cross-team collaboration and coordination easier.

Not only must teams have clear boundaries, but we must also ensure that the size and complexity within the perimeter that a team is responsible for is manageable. The more problem domains, especially if they are complex, that a team must deal with, the greater the impact on the team's cognitive load and therefore on its ability to execute effectively. The book *Team Topologies: Organizing Business and Technology Teams for Fast Flow* (IT Revolution Press, 2019), proposes an approach to manage a team's cognitive load by limiting the number of domains and their complexity. First, the authors, Matthew Skelton and Manuel Pais define and categorize problem domain complexity:

Low Complexity Domain Most of the work is straightforward because there is a correlation between cause and effect, so there is essentially a right answer. This is the equivalent to the Cynefin framework's obvious domain (introduced in Chapter 1, "Why We Need to Change the System"), which represents the "known knowns."

Medium Complexity Domain The team must analyze changes and then iterate a few times to understand the correct path forward because there may be more than one option on how to proceed—i.e., more than one right answer. This is the equivalent to the Cynefin framework's complicated domain, which represents the *known unknowns.*

High Complexity Domain The team needs a lot of experimentation and discovery because cause and effect can only be deduced in retrospect. This is the equivalent to the Cynefin framework's complex domain, which represents the *unknownable unknowns.*

Based on these categorizations, the authors propose the following:

- A team can manage two to three low complexity domains.
- A team can manage no more than one complicated or complex domain.

By limiting the amount of domain complexity (intrinsic load) we assign to a team, we will increase the capacity of a team to learn and solve problems (the germane cognitive load). We will also look, in the section "Appropriately Sized Teams", at how the complexity of domains has an impact on team size.

Clarity of Purpose

A product team is oriented to the customer or the consumer of its service; more specifically, it is focused on changing the behavior of the consumer to achieve a desired outcome. This is why product-centric is often referred to as outcome-oriented development. The outcomes that product teams are focused on will contribute to business impacts that will in turn contribute to business goals, all of which are achieved through an improvement in or the creation of a business capability. Having an IT strategy defined around outcomes requires us to have an organizational design structured around the same outcomes. As shown in Figure 5.13, improvements in the business capability of pricing will result in more competitive prices, which in turn should (based on our hypothesis or bet) contribute to an impact of greater conversion. In this example, the product is the organization's pricing capability and is made up of two technical applications, the competitor price scraper and the price management system.

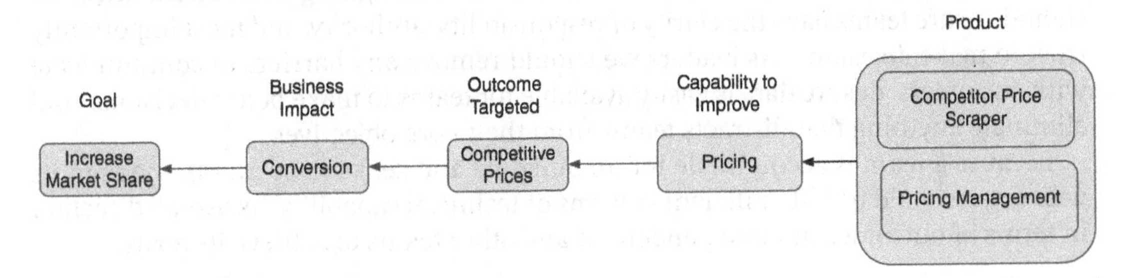

Figure 5.13: A product team based around a pricing capability

Products are defined by the outcomes they produce, the behaviors they can influence, and how they are consumed rather than the output produced by teams formed around them or the features available. In other words, you can say that the product teams themselves are defined by outcomes. Instead of project metrics (on-time, on-scope, and on-budget), teams are assessed on business outcome metrics, which is a true measure of team performance. Therefore, the metrics used for product team performance are the actual business outcomes desired as opposed to IT project output. We will explore team performance and measurement in Chapter 7. This helps to clarify the purpose of a team and how they contribute to business success. For example, a product could be measured by how competitive prices are—that is, how regularly we check and update our prices against a competitor.

Autonomous and Self-Sufficient

The structuring of product teams around business and technical boundaries that are related to strategic outcomes, as opposed to technical components only, allows teams to have full accountability for outcomes. The product teams themselves are free to self-organize as best they see fit to solve their set of problems. In addition, the hierarchy steers the teams with outcomes to be met—the what—and leaves the team with the task of determining the how in line with the target architecture. The hierarchy does not need to make all decisions. It merely sets direction, boundaries, and investment levels; the decision rights are pushed down to the product teams, who are invariably closer to the customer, to improve the ability to adapt and change direction based on feedback. Decision rights are given to the team through empowerment, which is enabled via trust; without trust there can be no real autonomy. If the product teams don't have the autonomy to make decisions for themselves and are required to request permission via a bureaucratic process and up through many levels of hierarchy, development will be blocked, opportunities will be lost, and constraints will persist, all resulting in slowing down the flow of value. Therefore, when designing team boundaries, we should ensure teams have the clarity of responsibility, authority, and most importantly trust to make decisions. As leaders we should remove any barriers to communicate with customers, ensure data is easily available for teams to make better decisions, and eliminate anything that distracts teams from their core objectives.

Because a team is accountable for an outcome and seeks to be autonomous in its delivery, it should be self-sufficient in terms of technical capability, its overall direction in terms of outcome, and independent of any other teams to achieve its goals.

Technical Capability From a technical standpoint, there is a need to ensure the team is staffed with all the necessary skills that will enable it to achieve its goals and work within their area of the business. What we are aiming for is the avoidance of teams having to hand off to another team to complete work—for example, hand off to the QA function for integration testing or hand off to IT Ops to deploy. In other words, cross-team dependency is important.

Outcomes and Direction Self-sufficiency is not only about technical capability within a team. A team should be clear on the outcomes it is responsible for and how they relate to overall business success as well as the relative importance against other goals. This will enable teams to organize their work, ensuring that they can prioritize their outcomes against helping other teams that are dependent on them—that is, on their strategic work as well as BAU backlogs for their capability area.

Cross-Team Independence Self-sufficiency is not about removing all collaboration and communication. Teams should collaborate when in discovery mode or when learning and sharing ideas. This communication should be encouraged, facilitated, and coached by leaders. However, overly chatty communication at the delivery mode will lead to highly coupled and dependent systems that are hard to change, and teams will lose their independence and ability to act autonomously. If this is the case and teams need to constantly communicate, then we must revisit how we have defined team boundaries to ensure team interfaces are correctly set to allow for autonomy. The more teams are independent, the more systems become decoupled and easier to manage, therefore cross-team independence is an important concept.

Appropriately Sized Teams

We want product teams to be appropriately sized, and size depends on the problem context that they will be operating in and how much the organization is willing to invest in an outcome of business capability. As the Wardley Map in Figure 5.14 shows, the evolution of a component will also have an influence on the size of a team.

- **Explorers (previously known as Pioneers):** On the far-left side of the map we have the uncharted space and products that are in their genesis. This is where there are many unknown unknowns and this is where we want to keep the teams the smallest, in the region of three to five people.
- **Villagers (previously known as Settlers):** In the middle of the map, we have custom-built and rental products. Here we will likely be leveraging off-the-shelf systems and configuring them to our needs as well as reducing waste in known processes. Teams can be larger, around 10 to 12 people, as while this area is complicated, we are able to plan on how to tackle problems.
- **Town planners:** On the far right we move toward commodity and industrialized products. These tend to operate in simple domains with obvious and clear ways of working. Here you can afford to increase the size of the team as there will be more well-known and well-trodden frameworks to lean on to manage work.

Regardless of the problem context a team is operating in, you want to keep them as small as possible. Small teams mean small perimeters of responsibility and lower cognitive load. Small teams will also lead to a more modular systems architecture, which is easier to manage. Larger teams result in a drop in communication quality due to the number of communication links between people. As shown in Figure 5.15, a team of 4 people will have 6 communication paths, 8 team members will have 28 paths, and 12 people will have 66 lines of communication, and therefore the larger the team, the higher cognitive load of team relationships.

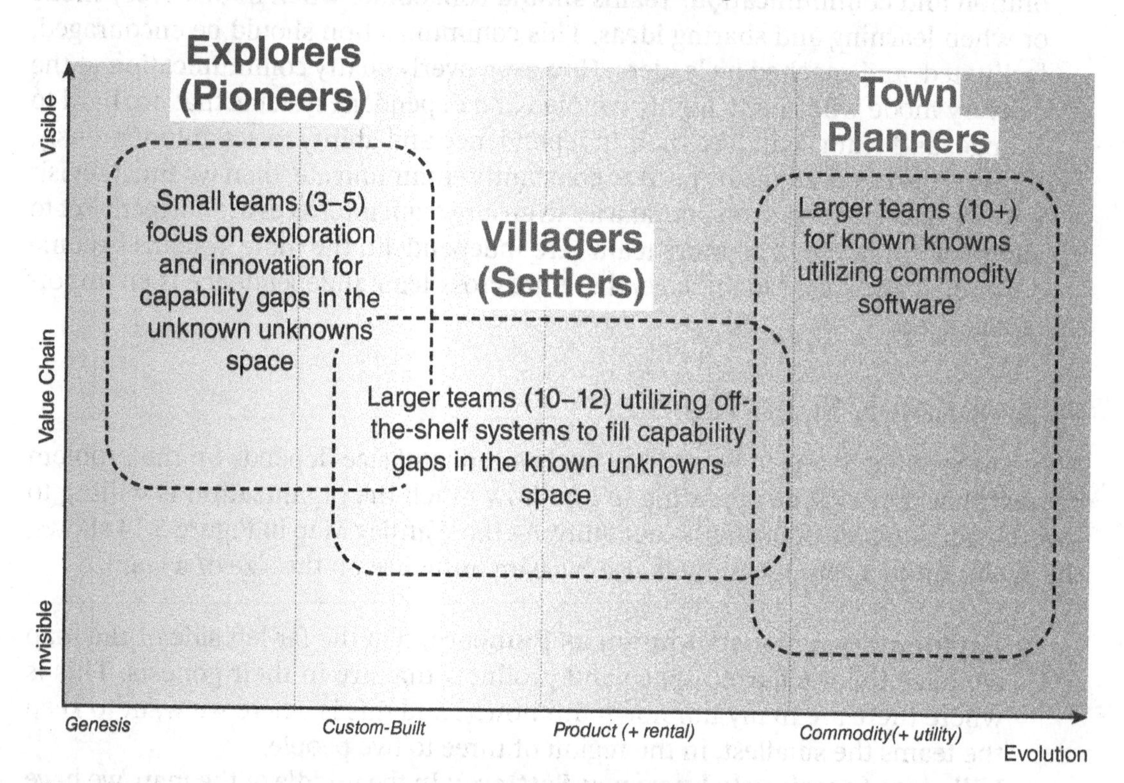

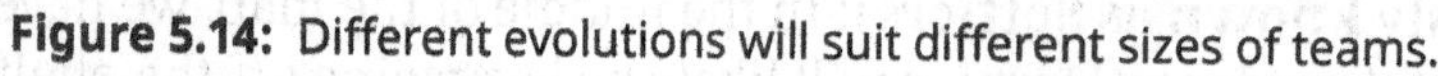

Figure 5.14: Different evolutions will suit different sizes of teams.

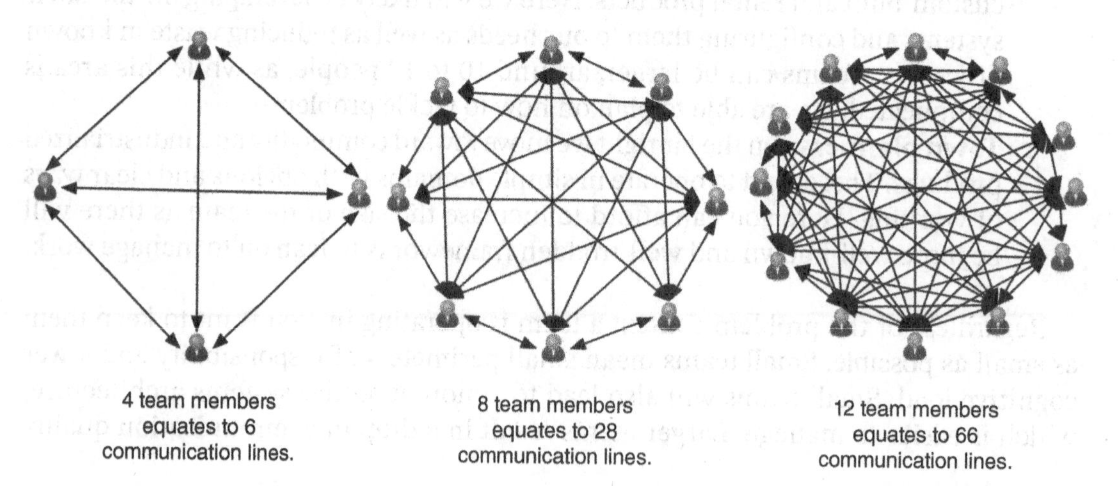

Figure 5.15: The comparison of the number of communication links between teams as they get larger

When sizing teams, we can also look at the work of Robin Dunbar. In his research, he found that 150 was the limit to the number of people one can have some form of an historical relationship with. Fifty was the number of people you can have active contact with and meaningful relationships. His study further showed that you can share a level of deep trust with around 15 people and have a close relationship with a group of 5. We can apply this research when defining team sizes, as shown in Figure 5.16. For a product team that requires high trust and close collaboration to achieve a shared set of goals, I recommend it to be from 5 to 9 people. For a collection of related product teams or a department to retain a meaningful relationship, the upper limit is 50 people, or 10+/- product teams. At a product portfolio level with a related collection of product groups or a company division, the limit on team size is 150.

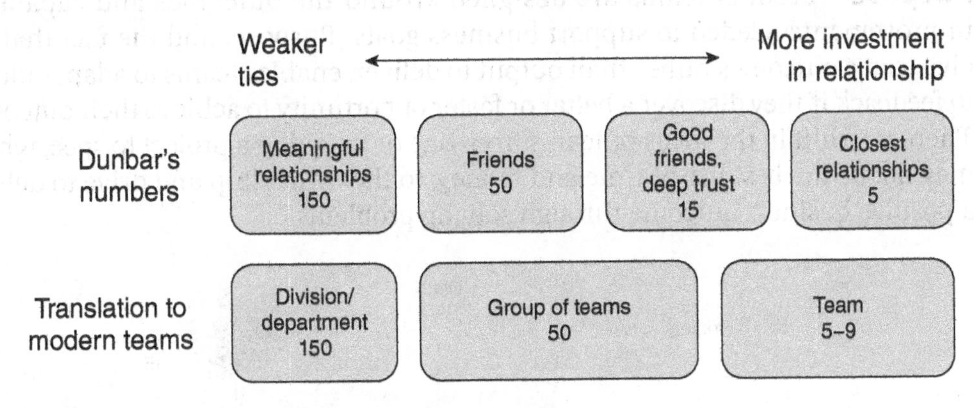

Figure 5.16: Dunbar's numbers translated to team sizes

When communicating within a team becomes difficult, you should reduce the need to communicate by creating collections of smaller teams. Figure 5.17 illustrates how you scale agility—by descaling the work and distilling products into small teams. Scaling agility is descaling. Since its early days, Amazon has based teams around the "two pizza rule," which mandates that no team should be so large that it cannot be fed with just two pizzas. For teams to remain small, you can subdivide big teams into smaller ones. Jeff Bezos, speaking at the Museum of Flight in Seattle in 2016, said a team of 10 or 12 people is "the perfect size to have natural human coordination without a lot of structure." He went on to say, "If you can arrange to do big things with a multitude of small teams—that takes a lot of effort to organize, but if you can figure that out, the communication on those small teams will be very natural and easy."

The Benefits of Product Teams

There are benefits for both the business and individuals when adopting a product-centric mode of operating. I have spoken about why autonomy, mastery, and purpose are key to an individual's intrinsic motivation. However, they also have great business outcome and delivery benefits.

Autonomy Self-sufficient teams can often deliver value faster due to the reduced need for handoffs to other teams. The reduction in dependencies can increase the flow of value.

Mastery Persistent teams with a consistent focus on a business area can iterate and evolve solutions to complex problems, increasing the chance of innovation. Focus and accountability for a single area can also lead to a higher level of code quality, meaning that code is easier and faster to change.

Purpose Product teams are designed around the outcomes and capability improvements needed to support business goals. Purpose, and the fact that we give teams outcomes rather than output to deliver, enables teams to adapt quickly to feedback if they discover a better or faster opportunity to achieve their outcome. There is a shift in the focus of teams from one of finishing a project to spec, which may not be the best use of time and money, to the ownership and drive to deliver a positive business outcome through solving problems.

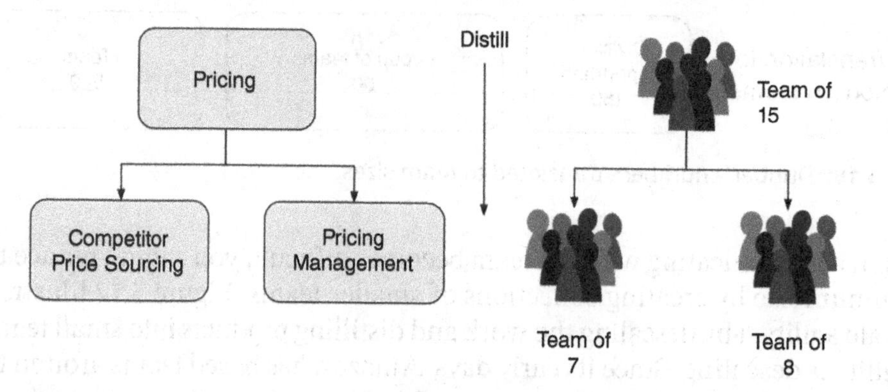

Figure 5.17: To keep teams small, distill them into smaller products.

Defining Product Team Boundaries

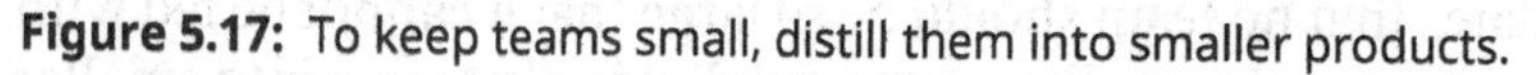

A move from a project mode of operating to a product mode starts with defining the products that exist in the organization and drawing appropriate team ownership boundaries. As discussed earlier in this chapter, when drawing boundaries, we should be aiming to limit dependencies and to optimize for an appropriate team size while ensuring we have a coherent product that is based around producing a meaningful business outcome. The product boundaries will be influenced not only by our business domains but also the technical and social domains at play, as shown in Figure 5.18. To define boundaries, we need to consider all the various seams at play in the software architecture and all that can influence it.

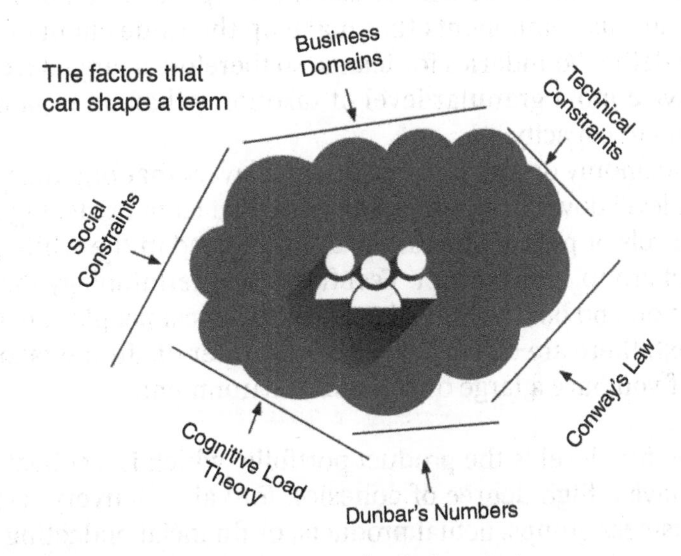

Figure 5.18: The factors that can influence team boundary design

The most common seam to use to define product boundaries is the business capability domains. Start by focusing on structuring product teams in relation to how they serve customers' journeys/experiences, value streams, or business capabilities, then look at how we need to adapt to support both technical (legacy systems, performance concerns) and social (team locations, communication lines) or other constraints (compliance, security, risk). For complex products, there may be a need to distill further and identify underlying platforms or complex subsystems that are a component of a product or support a product. In the following sections, we will look at how to structure a product portfolio using a product taxonomy; then we can dig deeper into how to define boundaries for teams using the team topology patterns.

Organizing the Product Portfolio Using a Product Taxonomy

The product taxonomy (the hierarchy for our holacracy), as described in the white paper *The Project to Product Transformation, IT revolution, 2019,* written by Ross Clanton et al., describes a technique to structure the design of a product portfolio to clearly show the collection of products, their hierarchy, and the relationships that exist within an organization. Just as you would group and nest the various product categories on an e-commerce retail catalog, we can do the same with the customer experiences, value streams, and business capabilities of a business. By using a product taxonomy, we can clarify all the products and their owners and boundaries within an organization, which is important for everyone in the business to understand for collaboration, information, and communication. While it is useful to distill the organization down to products to understand the various components that make up the value chain of an enterprise, the exercise is to define boundaries for teams, so therefore a large development team function will have a more granular level of taxonomy than an organization with a smaller development capacity.

The product taxonomy describes three distinct layers that organize the enterprise from the highest level down to where a single product team exists. Figure 5.19 shows the three main levels of product hierarchy as articulated in the white paper, namely portfolio, product group, and product. Feel free to use terminology that makes sense to your organization and both technical and nontechnical people can agree upon. As you will read next, there are also techniques to further distill the taxonomy beyond the three levels if you have a large or complex environment.

Level 1 The first level is the product portfolio, which is a collection of product groups that have a high degree of cohesion for value delivery. Typically, this is defined by business groups, actual products, or financial budgeting groups.

Level 2 The second level is the product group level, which organizes related products, that have a high degree of cohesion for value delivery. Typically, this will be grouped by business capabilities, value stream, or customer journeys.

Level 3 This is the level of the product team, formed around technology (software and or hardware) that enable the business capabilities, value stream, and customer journeys.

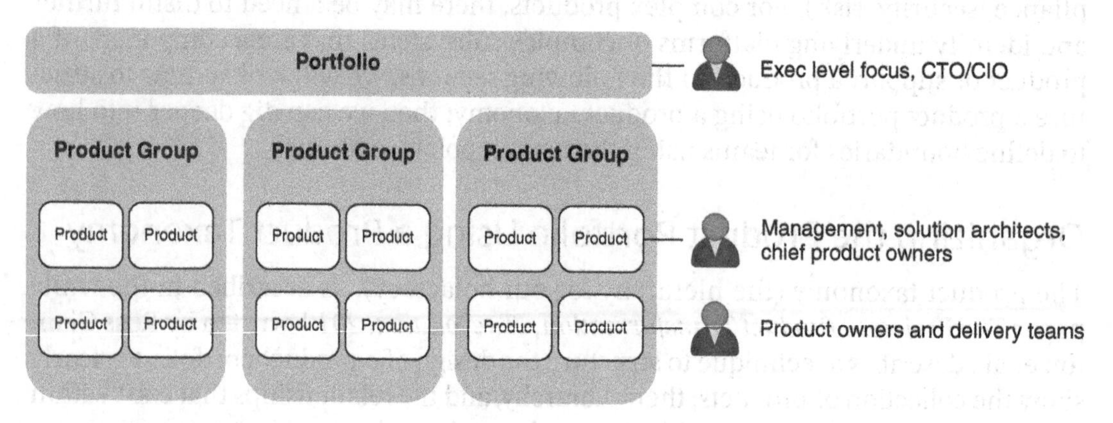

Figure 5.19: The three levels of the product taxonomy

Portfolio Level

Typically, but not necessarily always, the portfolio level is the collection of all the products or services (value propositions) offered by a company. As shown in Figure 5.20, while a small company may have a single portfolio (service or product), large companies may have many product portfolios supporting different business units. The business portfolios are a good starting point to anchor the product taxonomy portfolio level as this is usually well understood across the organization, and funding and investment is typically already set at this level.

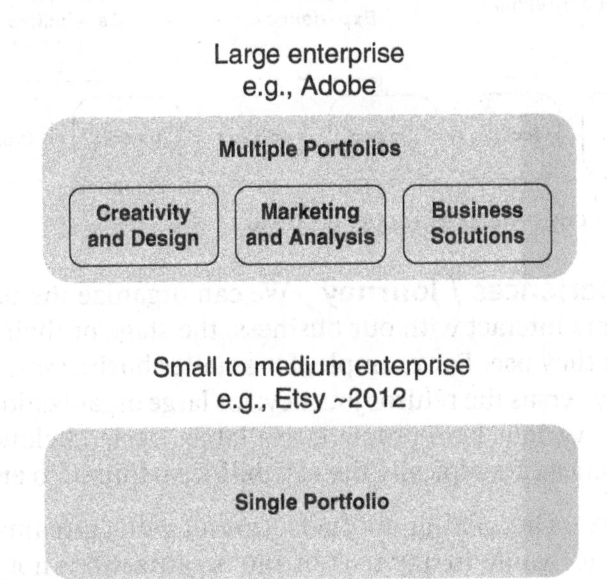

Figure 5.20: Product portfolios are large and small to medium enterprises.

Product Group Level

At the product group level is where multiple, interrelated products are managed. As shown in Figure 5.21, there are three approaches you can choose to group products, with each being driven by the business model: the customer experiences they offer, the value streams that produce value, or the business capabilities that support the value streams and customer experience. It is typical to use a mix of these three approaches; for example, you can organize some products by internal business capabilities, some by the external experience of a customer journey, and some by a value stream such as the supply chain flow.

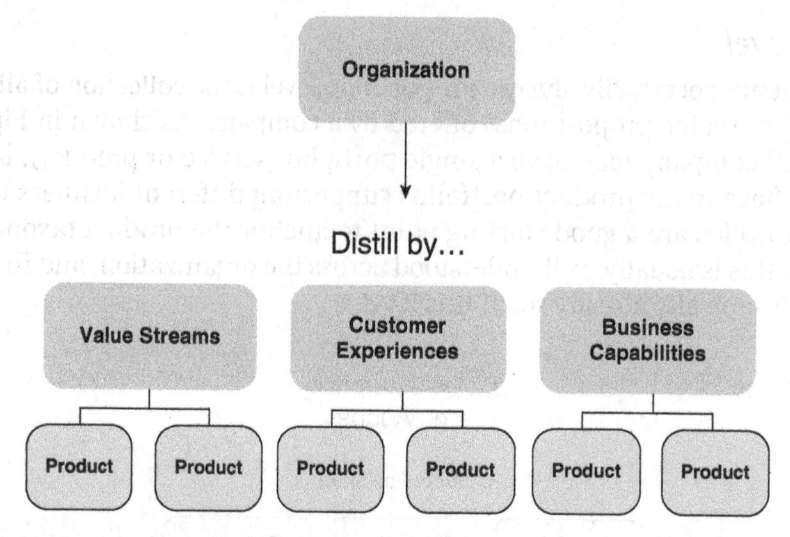

Figure 5.21: Product groups within a product portfolio

Customer Experiences / Journey We can organize the portfolio based on the way customers interact with our business, the stage of their journey with us, and the services they use. For example, for smaller businesses, this could be the purchase journey versus the returns journey; for large organizations such as banks, this could be the various loan products available. Nested below the grouping of customer experiences are typically the capabilities required to enable them.

Value Streams Organizing products around value streams is based on the journey to produce value in any part of the organization, not just for external customers. This could include parts of the supply chain and internal or external processes and services. Nested below the grouping of value streams are typically the business capabilities required to enable them.

Business Capabilities The alternative approach to group products is to look at the major business capabilities themselves that enable both the customer experiences and value streams and group products at that level. Capability models are hierarchical structures in nature and are a good starting point when defining groups of products as they will typically map to applications.

The product grouping emphasizes a cohesion between interrelated products. In large organizations, there can be multiple nested product groups, as shown in Figure 5.22. At each nested group layer there is a need for management and technical oversight to facilitate and coordinate cross-product team initiatives as well as distribute budgets in terms of product team capacity. Product grouping concepts should be defined around business concepts as opposed to specific technologies: for example, customer acquisition rather than the sales force app, marketing campaign management rather than the Adobe campaign, financial planning rather than sage. This is because the capability, the "what" a business does will remain static, but the "how" it is achieved will evolve over time along with the supporting technical solutions. By forming groups around business capabilities, value stream, or customer experiences, we can create a ubiquitous language that is understood throughout the business. It is advantageous to talk about investment, demand, and maturity at a business level rather than specific technology as this terminology is better understood and technology is only one part, along with people and process, that makes up a business capability.

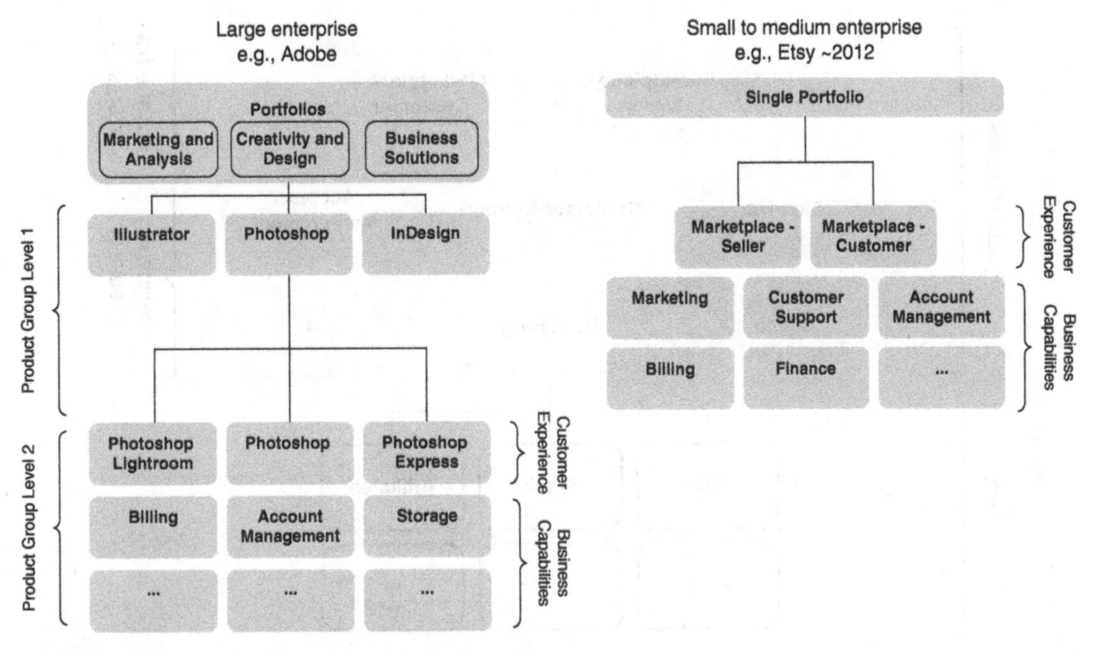

Figure 5.22: Examples of product groups at a large and small to medium enterprise

Product Level

The product level is where technology and people are formed into a persistent team to deliver an outcome or enable a capability. Figure 5.23 shows an example of the breakdown of products within a product grouping. Even though teams own the software or hardware at this level, we should still focus on the outcome rather than the technology itself. This is because technology evolves, but the product, aka capability, will likely remain as a consistent need unless there is a change to the business model and therefore a change in the business capability itself. In other words, understand and obsess about what problem a product is trying to solve in the first instance and then worry about how it is currently implemented.

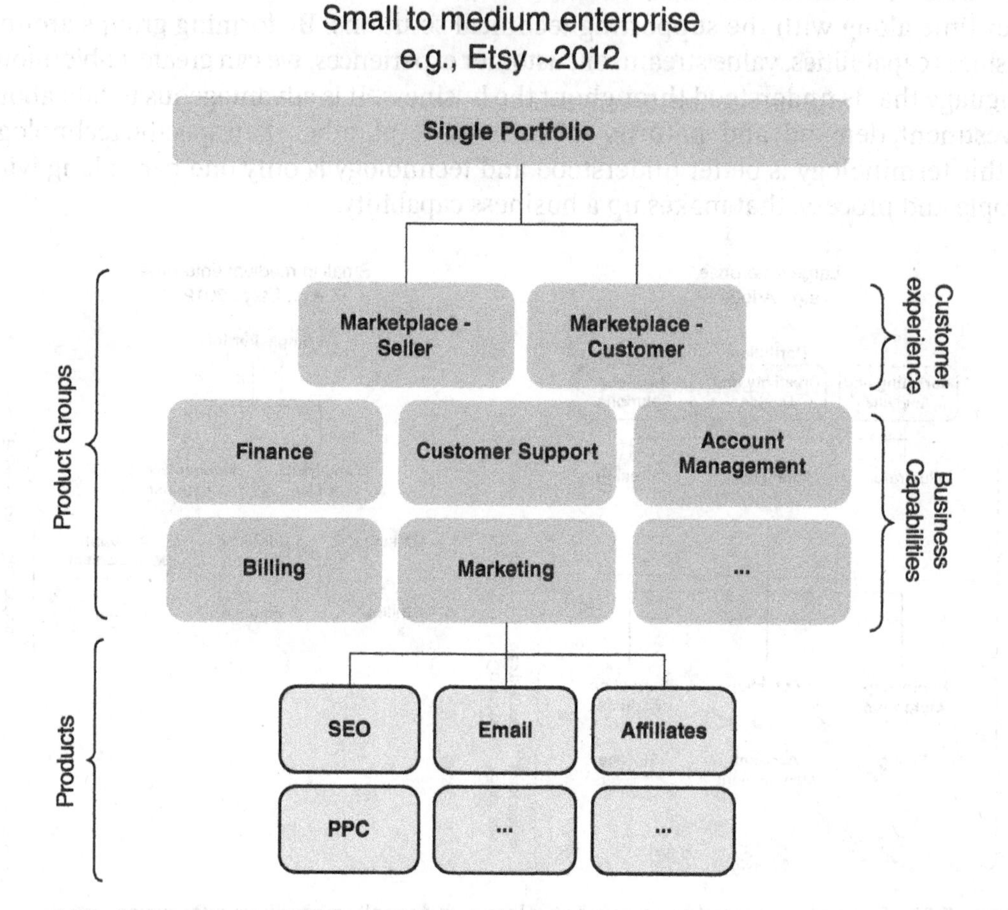

Figure 5.23: Product teams within a product taxonomy

A product itself can be distilled further if there is enough technical complexity, as shown in Figure 5.24. The next section discusses how we can employ the team topologies patterns and language to further distil a product into smaller component teams.

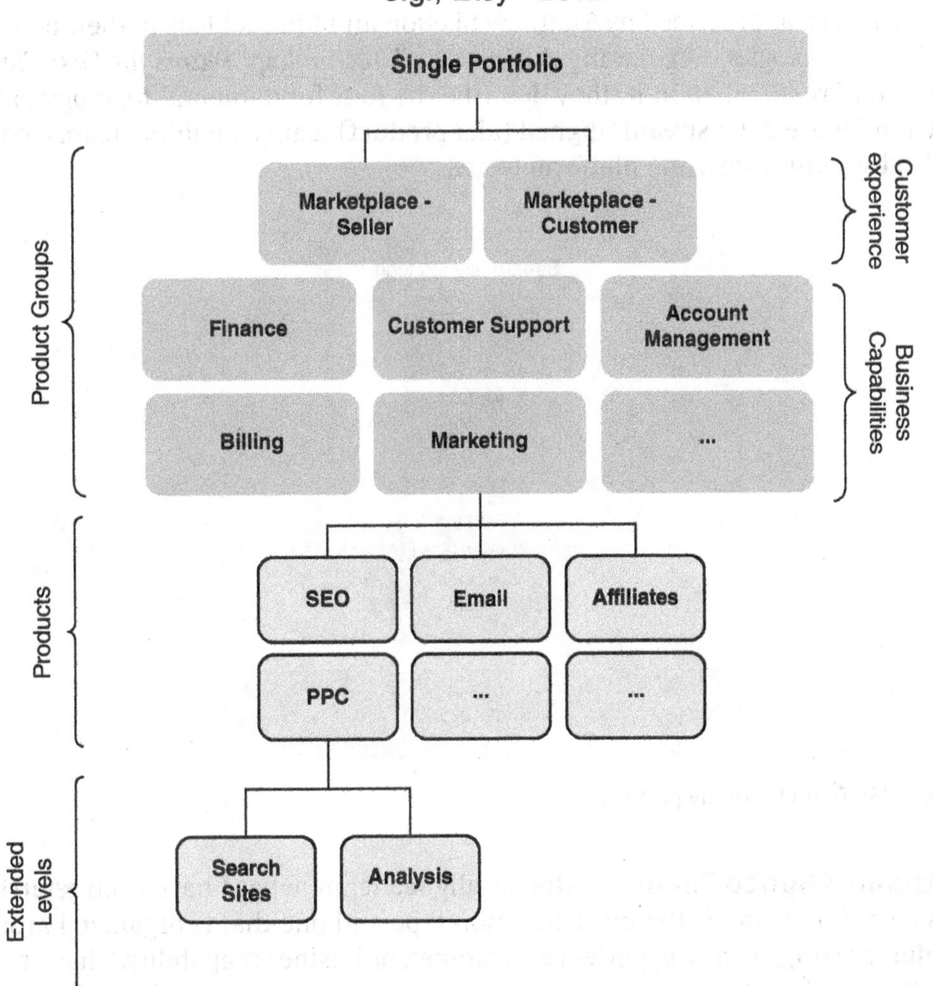

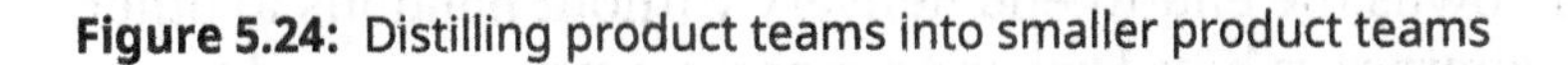

Figure 5.24: Distilling product teams into smaller product teams

Modeling Product Team Boundaries with Team Topology Patterns

Not all teams will be formed around a distinct product with a clear outcome. Some specialist teams may focus on a subset of a product to alleviate some of the cognitive load of product teams. Some may provide foundational capabilities in the form of a platform to multiple other teams, and some will help a product team to achieve its goals. We can further model teams and their relationships using the team topologies design patterns as described by Matthew Skelton and Manuel Pais in their book entitled *Team Topologies: Organizing Business and Technology Teams for Fast Flow* (IT Revolution Press, 2019). In it, they describe the four fundamental topology patterns shown in Figure 5.25: stream-aligned (aka product) teams, enabling teams, complicated subsystem teams, and platform teams.

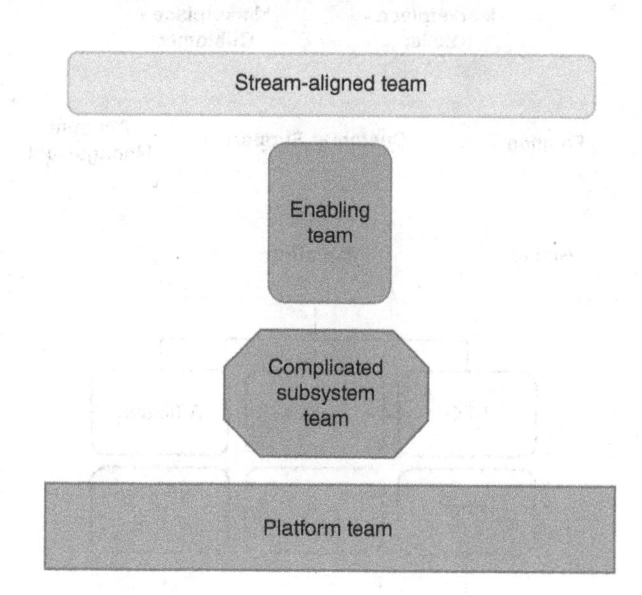

Figure 5.25: Team topology patterns

Stream-Aligned Team A stream-aligned team, what I have been referring to as a product team, is the most common type and one that is organized around a value stream, journey, experience, customer, or business capability. They are close to their end customers, autonomous, and empowered to use direct feedback to deliver value. The team owns the end-to-end delivery without handoffs to other teams. All other team patterns are based around supporting stream-aligned teams to deliver value and reduce cognitive load.

Enabling Team The role of an enabling team is to assist a stream-aligned team to overcome challenges and fill in any gaps in capability. An enabling team typically comprises specialists that can support a stream-aligned team with deep expertise in a particular technical area, thus avoiding the need for a stream-aligned team to have experts in an ever-growing portfolio of technologies. An enabling team allows the stream-aligned teams to focus on their primary and core problem areas, thus allowing them to manage cognitive load while also offering enough support for the stream-aligned team to close gaps and become self-sufficient to prevent long term reliance on an enabling team.

Complicated Subsystem Team Where there is a persistent need for significant specialty skills or deep expertise in a component, a separate team can be formed to look after a subsystem of a product. The goal of the complicated subsystem team is to reduce the cognitive load on a stream-aligned team by looking after an area of high complexity or one that requires specific skills and knowledge.

Platform Team A platform team is organized around the development and support of foundational capabilities that provide common services to many stream-aligned teams to accelerate development. The stream-aligned team is still responsible for the end-to-end delivery and ownership of their service, but it can leverage internal services provided by the platform team. As well as minimizing the cognitive loads of the stream-aligned team's non-core activities, another benefit is the consistent services to end users (internal and external) that platform teams can provide, such as a single method for identity management across products and journeys.

The Need to Constantly Evolve Teams

Organizational design is not a one-off exercise; it is something we must constantly review as it is the single biggest influence on system design and software architecture. Melvin Conway states "Because the design that occurs first is almost never the best possible, the prevailing system concept may need to change. Therefore, flexibility of organization is important to effective design." What this means is that we must be conscious of the fact that our environment will grow and change, and we need to ensure that our organizational design is explicitly set up to match our target system architecture. Likewise, we must regularly check in with teams on their cognitive load levels. Are teams uncomfortable with the size of their area of responsibility? Are they constantly context switching and are unable to retain information on areas that they look after? If so, then it may hint at the need to review the structure.

An Example of a Team Topology

Figure 5.26 shows an example of how we might organize teams within a simple retail commerce business. The teams in the front office are all designed around steps in the customer journey.

- The demand team is focused on upper funnel content and SEO to drive organic demand.

- The discover team's purpose is to help customers find the right product. They are focused on customer engagement.
- The order team manages the checkout process with an eye on conversion of customers who enter the checkout funnel.
- The bike builder team is a complicated subsystem team that provides a simple interface for the order team to offer customers the ability to build a bespoke product.
- The service team manages all the customer post-booking needs, ensuring that customers can self-serve.

The customer journey teams are supported by two enabling teams:

- The SRE team supports with site performance and reliability.
- As the customer journey teams are all build teams, they are supported by a cloud team. This team supports by managing the cloud environments and setup.

The back office is composed of mainly COTS systems, so teams are focused more on run and configure:

- The product information management (PIM) and content team, merchandising team, order management and fulfilment team mange the enterprise systems.
- The forecasting team is a separate subsystem team that looks after the many API interfaces with hotel aggregators, flight providers, transfers suppliers, and excursion partners that all require specialist knowledge.
- The marketing team is composed of developers that utilize an enterprise platform to generated demand for the site.
- The finance team is a third-party service provider that supports the finance set of systems.
- The data and BI team acts as a platform to provide data to other teams and self-service reporting to the rest of the business.

Evolving to Business and IT Fusion Teams

As technology and IT have moved from being a back-office department to a critical part of most organizations, we have seen the need to become more aligned to delivering outcomes rather than output to contribute to positive business impacts. The move from transient project teams to persistent product teams is an evolution in how the IT department is organized. The next step of that evolution is to become completely embedded in business teams rather than simply collaborating with them on outcomes. Realigning teams around outcomes must not be limited to IT; we must start to blend nontechnical people into core teams that are tasked with removing constraints and discovering new opportunities. This is not limited to business colleagues merely providing feedback on the IT team's efforts; they must actively contribute to solutions and explore options on tactics and initiatives that will produce the desired outcomes. Highly collaborative teams composed of a diverse set of people with different backgrounds and experience and from different parts of the business are advantageous to discovering innovative ways to solve problems and unlock opportunities. This helps with the "us versus them" or the business versus IT level of thinking of the past. Therefore, the next transition of the IT organizational design is to become embedded in and part of a business capability team.

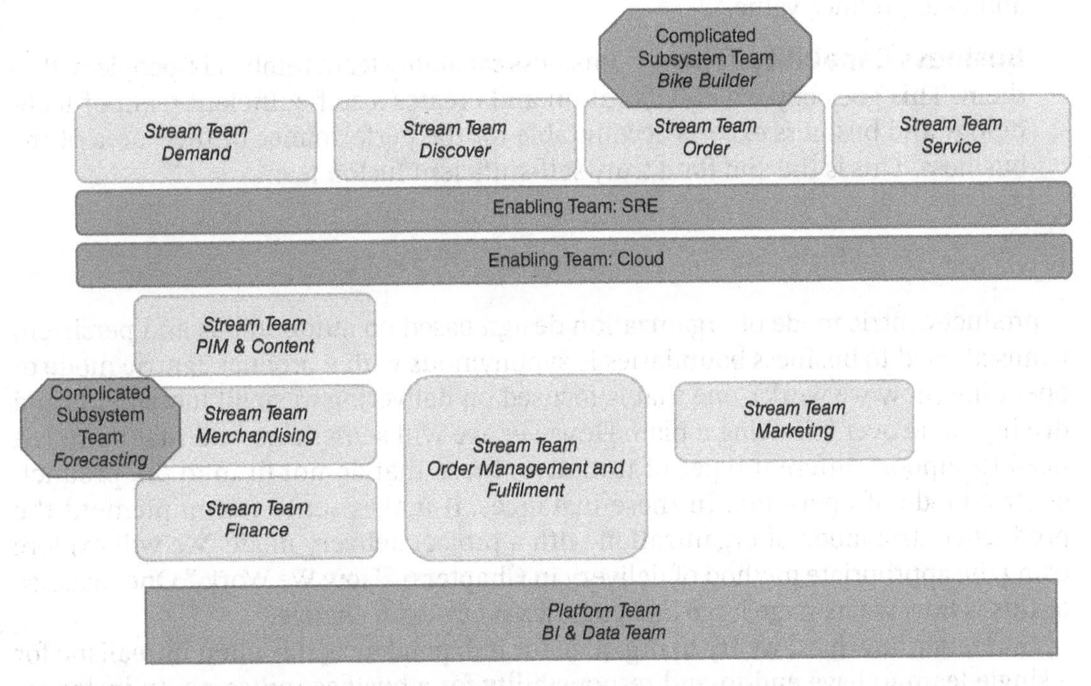

Figure 5.26: An example of a product team setup at a small retail e-Commerce business

Business capability teams, the fusion of IT and business experts, are the next evolutionary step for teams:

Project Teams IT project teams have moved from functional to self-sufficient teams, although they are transient in nature, forming for a particular task before disbanding when it is complete. The project teams tend to focus on delivering to an agreed list of requirements or upfront scope. This was an effective way to manage obvious problems but fell short when dealing with unknown unknowns. As scope is given to teams in the form of requirements, the only way to measure performance was conformance to scope, time, and budget.

Product Teams Product teams, which we have focused on in this chapter, are the next evolutionary step. Teams are persistent and focus on a single area of business and technical domain. They are aligned to business goals and collaborate with business peers on achieving desired outcomes through the improvement and/ or creation of business capabilities. They are measured by and responsible for the value they produce, not on what output they deliver; therefore, they are naturally more customer-focused than project teams. Teams now have more "soft skills," or business-facing skills, in the form of product owners that increase the team's ability to produce value.

Business Capability Teams Business capability teams embed IT people within them. This goes beyond collaboration and creates a self-sufficient team of technology and business experts accountable for the performance of their area of the business. This is the aim for a truly self-sufficient fusion team.

Managing Cross Team Dependency

A product-centric mode of organization design based on autonomous and persistent teams aligned to business boundaries is synonymous with a product-centric mode of operating (how we work), one that is focused on delivering in small increments and driving value over following a plan. However, we will sometimes find that there is a need to support different types of initiative, those that do not fit in to the product-centric mode of operating. In these instances, it makes sense to complement the product-centric mode of organization with a project delivery mode. We will explore using the appropriate method of delivery in Chapter 6 "How We Work." One instance of this is how we manage large and complex cross-team change.

No matter how hard we try to organize for independence, it is often unrealistic for a single team to have end-to-end responsibility for a business outcome. In instances where multiple technical areas need to be modified, requiring many different product teams to collaborate to achieve outcome, as shown in Figure 5.27, we can use the role of an outcome owner. This is a leader that sits outside the team to assist in solving problems, acting as a bridge across organizational boundaries and managing cross-team dependencies. Their role is to facilitate and aid collaboration rather than the operational delivery side, which should be the responsibility of each of the product teams.

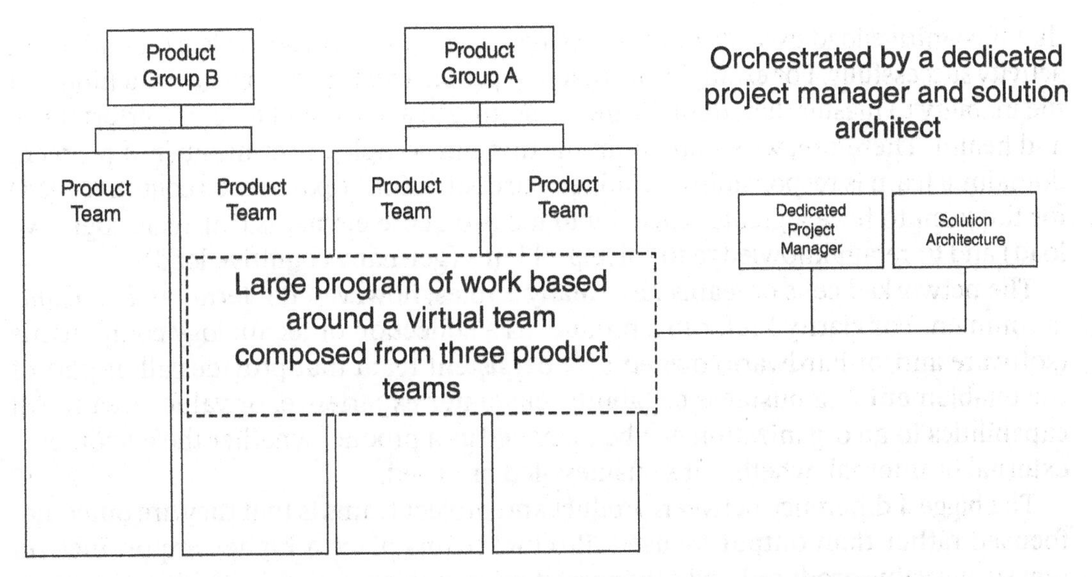

Figure 5.27: Projects are used to coordinate across product teams for complex change.

Summary

The two extremes for structuring the design of an organization are the holacratic structure based on a network of teams and a strict hierarchy based on teams siloed by function or geographical or product division. Neither one of these is ideal; what we need is a balance between the two. The strengths holacratic network teams coupled with the control hierarchies bring can be complementary. Network teams can act autonomously with the power to adapt and change the business at speed based on customer feedback, while hierarchies can manage communication, steer, and set boundaries and constraints to operate within. By using aspects of both a hierarchy and holacratic design, we can produce a structure that is greater than the sum of its parts.

When designing network teams, it is important to understand the material impact that both Conway's law and cognitive load theory can have on team performance. Conway's law states that the system, in our case the software architecture, is a result of the organization of teams, or as Ruth Mulan puts it, "If the architecture of the system and the architecture of the organization are at odds, the architecture of the organization wins." To avoid this and ensure we are delivering to the target architecture needed to support the IT strategy, we can inverse Conway's law by starting with the organization of the teams to match the target software architecture. Cognitive load theory suggests that if cognitive load exceeds our processing capacity, we will struggle to complete an activity successfully. For example, overloading people will lead to context switching and the inability to master their domain area, resulting in a detriment to team performance and health. Therefore, we should limit the domain complexity or number of problem domains a team is responsible for and reduce distractions (extraneous cognitive load) for the team to have a greater capacity to focus on core elements (intrinsic cognitive load) and to retain knowledge to solve problems (germane cognitive load).

The networked cells of teams have many names; however, the term *product teams* is common. For clarity I refer to a product as a collection of technology components (software and/or hardware) owned by a persistent team that provides all or part of the enablement to a business capability, customer experience, or value stream. All capabilities in an organization can be managed as a product whether the customer is external or internal, whether it's business-led or IT-led.

The biggest difference between product and project teams is that they are outcome-focused rather than output-focused. Product teams place a higher importance on measuring value produced and solving customer problems over the tracking of scope, budget, and time metrics. They are persistent and accountable for a business problem domain, owning and running all the related technology. They are self-sufficient, containing all the necessary skills and capabilities required to achieve their outcomes. They are autonomous, empowered, and trusted to use their time as they see fit to solve problems they have been tasked with. To enable them to focus on a core problem and gain deep expertise, we can limit cognitive load by keeping teams small and domain complexity manageable. Amazon refers to this as the "two-pizza teams" — from 8 to 12 people. This is in the range of Dunbar's number of people with whom you can share a level of deep trust and have a close relationship. As we grow, we scale by descaling and sub-dividing problem spaces down into small teams.

When defining boundaries, there are several factors that can influence the seams of product teams. Technology, compliance, and social constraints will impact how you split teams; however, the best place to start is with business domain boundaries. How you structure teams should be designed in direct response to the strategic needs of the business. Copying the structure of other organizations without understanding the contextual reasons and specific situational awareness that informed them is foolhardy and will likely result in suboptimal performance. We can organize the product portfolio based on several starting points, such as customer experiences, value streams, or business capabilities. It is not uncommon to see product hierarchy based on a mix of internal capabilities and external customer experiences. A product can be further distilled beyond the business domain using the patterns covered in *Team Topologies: Organizing Business and Technology Teams for Fast Flow* (IT Revolution Press, 2019).

The way you organize the IT department and the way it works is a very visible change to the way you operate. However, it is the principles and practices that are equally as important to the shape of teams. Also remember that while this may be the biggest visible change, it is only one component of the total IT operating model. Both how we work and how we govern must complement the changes to the organization structure.

6 How We Work

Without data, you're just another person with an opinion.

—*W. Edwards Deming*

If I had an hour to solve a problem, I'd spend 55 minutes thinking about the problem and five minutes thinking about solutions.

—*Albert Einstein*

The greatest danger in times of turbulence is not the turbulence—it is to act with yesterday's logic.

—*Peter Drucker*

The How We Work component of the operating model defines how work is carried out to achieve the goals of the business. In Chapter 2, "Philosophies for a New System," we introduced the philosophies of lean, agile, and design thinking. In this chapter, as highlighted in Figure 6.1, we will look at the frameworks and methodologies that we can leverage to put those philosophies into practice to contribute to improving or creating business capabilities. We will also look at the more supporting IT management frameworks that cover the full spectrum of capabilities internal to IT.

The core principle behind how we work is to move beyond simply asking for requirements towards discovering how we can generate outcomes that make business impacts that in turn contribute to a strategic objective. In other words, instead of asking teams to "build this feature," we instead ask them to "deliver this outcome." This fundamentally shifts IT from the role of an order taker to a contributor of business impacts.

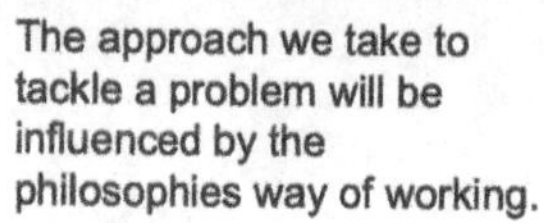

Goal
Philosophy
Approach
Methodology
Different philosophies are useful for different contexts, e.g., Lean to eliminate waste and Design Thinking to discover new value.
The approach we take to tackle a problem will be influenced by the philosophies way of working.
The appropriate methodology gives us the frameworks, tools, principles, and practices to turn the philosophy into action.

Figure 6.1: Methodologies put philosophies into practice.

To define the outcomes that will potentially achieve a goal, there is a need to conduct an appropriate level of discovery to understand the problem space, the opportunities and/or constraints, at a deeper level. This prevents us from addressing the symptoms of a problem rather than the root causes. We need to understand our customers and users, empathize with their pain points and identify the obstacles that are preventing a goal from being achieved. Once we have done this, we can turn these pain points into outcomes for teams to deliver.

At the solution level, there is a need for continued discovery as teams are not handed requirements; they are handed desired outcomes. Discovery is focused on generating ideas on how to achieve an outcome. How teams approach solving a problem will largely depend on the type of problem teams are facing. Some solutions will be obvious, and we can be certain of their success. However, many of the ideas will address complex problem spaces and based on assumptions and hypotheses. Therefore, there is a need to apply a fit-for-purpose approach to suit the problem context teams find themselves within. Agile is not a panacea. Being adaptable also applies to the methods you use when approaching a problem rather than using the same approach for all problems.

IT Management Frameworks

An IT framework is a series of documented processes, based on best practice, to define policies and procedures around the implementation and ongoing management and control of such areas as governance, security, service delivery, and compliance. These industry standard frameworks act as a blueprint providing structure and guidance to ensure all areas of the operating model are being controlled. While the frameworks can be somewhat heavyweight, they can be extremely useful for guidance in specific contexts. There are areas of the business in which we need to take an innovative approach and experiment with, and there are areas where we can leverage best practice, especially if the capabilities are generic to all organizations, such as security and compliance. However, as with many things, it's best to tailor any framework to your context and size of business rather than trying to follow them verbatim. Figure 6.2 shows the capabilities of IT using the IT4IT reference architecture that we will look at in detail in Chapter 12, "Tactical Planning: Deploying Strategy." Overlaid are the groups of frameworks, with Table 6.1 detailing a description of what each framework covers and examples of popular frameworks. It is worth noting that many frameworks cross over each other, but I have endeavored to point you in the direction of example frameworks that are useful to the specific area, even though parts of them may also be applied to other areas of the operating model. As with any best practice, it is useful to understand a framework fully and then make an informed decision on what parts will be helpful to your context, customizing to suit your needs.

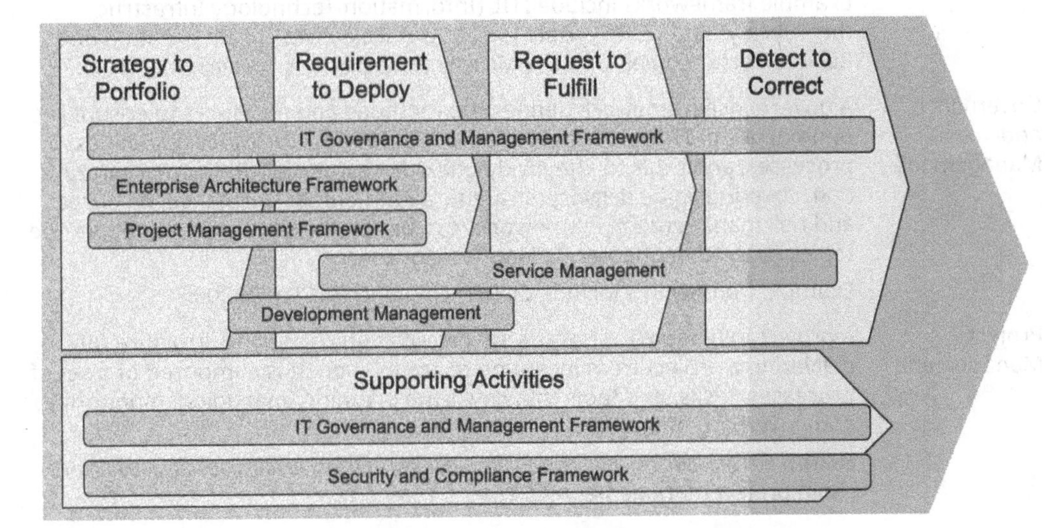

Figure 6.2: Frameworks that are applicable to the various capabilities of IT

Table 6.1: IT frameworks

Area	Framework
Security	An IT security framework is a series of controls used to define policies and procedures around the protection of enterprise systems. The framework acts as a blueprint for managing risk and reducing vulnerabilities. Example frameworks include ISO 27000 series, the Center for Internet Security (CIS) framework, and NIST which stands for the National Institute of Standards and Technology.

Continues

Table 6.1: IT frameworks (*Continued*)

Area	Framework
Compliance	A compliance framework or standard is a structured set of guidelines and best practices that detail the controls that an organization needs to put in place to meet regulatory requirements. Examples of compliance standards include the Payment Card Industry Data Security Standard (PCI DSS) and the General Data Protection Regulation (GDPR). PCI DSS is an information security standard for organizations that handle debit and credit cards. GDPR is a regulatory compliance framework for the processing of personal information of EU citizens.
Service Management	IT service management is a set of policies, processes, and procedures for managing the delivery, improvement, operation, and support of customer-oriented IT services. Example frameworks include ITIL (Information Technology Infrastructure Library) and COBIT, short for Control Objectives for Information and Related Technologies, and the Microsoft Operations Framework (MOF).
Governance and Management	A governance framework defines the methods and measures to ensure the optimal use of IT investments in an organization. It provides guidelines, processes, and tools to effectively utilize IT resources within an organization, covering value delivery, strategic alignment, performance, resource, and risk management. Frameworks can be leveraged to explore deeper the topics covered in Chapter 7, "How we Govern." Example frameworks include COBIT, ITIL, and ISO/IEC 38500.
Project Management	A project management framework provides guidance and structure for delivering a project from inception to completion. It is composed of a set of processes, tasks, and tools that cover the planning, managing, monitoring, and governing of projects. Example frameworks include Prince 2, Disciplined Agile Toolkit, and critical chain project management (CCPM).
Enterprise Architecture	An enterprise architecture (EA) framework is the collection of processes, templates, and tools that are used to describe an organization's architecture. The EA is composed of the business, information, application, and infrastructure architectures. The EA frameworks are used to model the desired, or target, state to achieve the business and strategic goals. Enterprise architecture frameworks are useful when devising IT's strategic contribution. Example frameworks include The Open Group Architecture Framework (TOGAF) and Zachmen. We will explore EA in more detail in Part 3, "Strategy to Execution."

How to Solve Problems from Discovery to Delivery

As illustrated in Figure 6.3, there are three phases we must go through to address a business need:

1. **Establish the goal and apply the appropriate problem-solving frameworks to solve it.** Clearly frame the goal (the opportunity or constraint to be removed), ensuring everybody has a shared vision and understanding of what we are trying to achieve. What is the strategic intent? Do we want to remove waste? Or explore a new channel?
2. **Discover more about the problem and define where to focus effort.** We begin with deep discovery into a problem to define the specific outcomes we need to deliver to achieve it.
3. **Select an appropriate approach to solution delivery.** We then develop solutions to try to realize those outcomes. As some of the outcomes may be assumptions or unproven hypotheses, we will need to validate with tests before doing the work to deliver a full solution. In addition, we may have several options to provide a solution to achieve an outcome. At each step there is a need to gather feedback from customers to ensure we are on the right path. This process can also be used to determine the operational actions when determining how best to solve a tactical initiative.

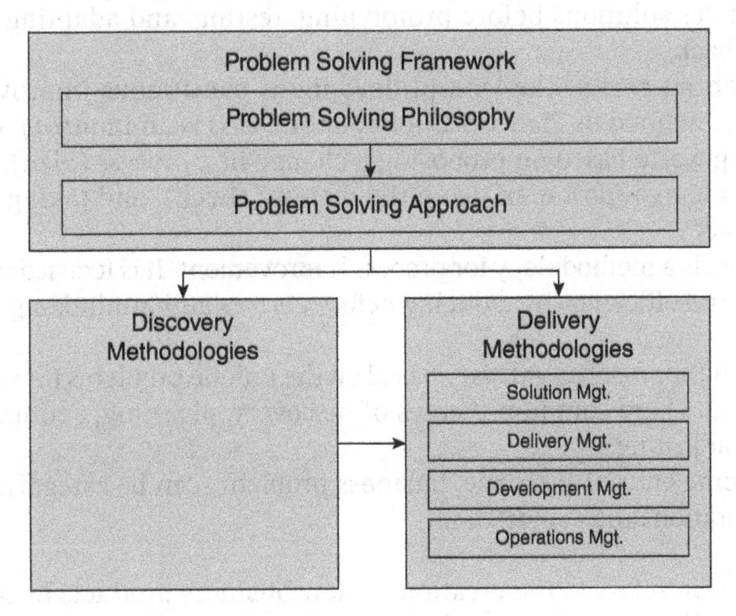

Figure 6.3: The three main components of problem solving

Throughout the rest of the chapter, I will cover the various frameworks and methodology options that are available for each of the three phases that culminate in addressing customer or business need. Every problem is unique to you and your organization's context, and as different contexts will require different strategies, there can be no perfect pattern or combination of problem-solving framework, discovery tool, and solution management approach. What I will present in this chapter can be viewed as the purist options. There will be a requirement to mix and match, to experiment to discover what works, and to utilize aspects of several different approaches to create a set of tools and approaches that are appropriate to your needs and the situation that you are facing.

Problem-Solving Methodologies

Problem-solving frameworks provide a structured and systematic approach to tackling a business problem. In the following sections, we will look at three problem-solving frameworks:

- The Double Diamond framework, an implementation of the design thinking philosophy, provides guidance for dealing with wicked problems. It empathizes with users and their problems, defines the outcomes that need to be addressed, and ideates solutions before prototyping, testing, and adapting them based on feedback.
- An essential part of the lean philosophy is continuous improvement. The scientific method of Plan-Do-Check-Act (PDCA) is an improvement cycle for reducing waste based on proposing a change in a process (plan), implementing the change (do), measuring the results (check), and taking appropriate action (act).
- Six Sigma is a methodology for process improvement. It is focused on increasing quality and efficiency by reducing defects, errors and minimizing variation.

While specific approaches will vary based on the unique problems they are designed to solve, they all share common themes of discovery, planning, acting, measuring, learning, and adjusting.

At the extreme ends of the scale, business problems can be categorized into two buckets, exploration, and exploitation.

- Exploration refers to the creation of new business products or services. It is based on discovery of innovative value propositions through experimentation and refinement.
- Exploitation refers to the optimization of existing products or services through the continuous improvement cycle of waste removal.

As shown in Figure 6.4, to guide our thinking on what problem-solving approach would suit each problem context, we can map these three approaches to the various evolution states based on a Wardley Map that was introduced in Chapter 2, and we will look closer at this in Chapter 12. This is of course a purist view; there is no right or wrong approach. The problem-solving frameworks you choose should be based on your context. Any approach that delivers value is correct. You are best to determine what is most suitable for your situation.

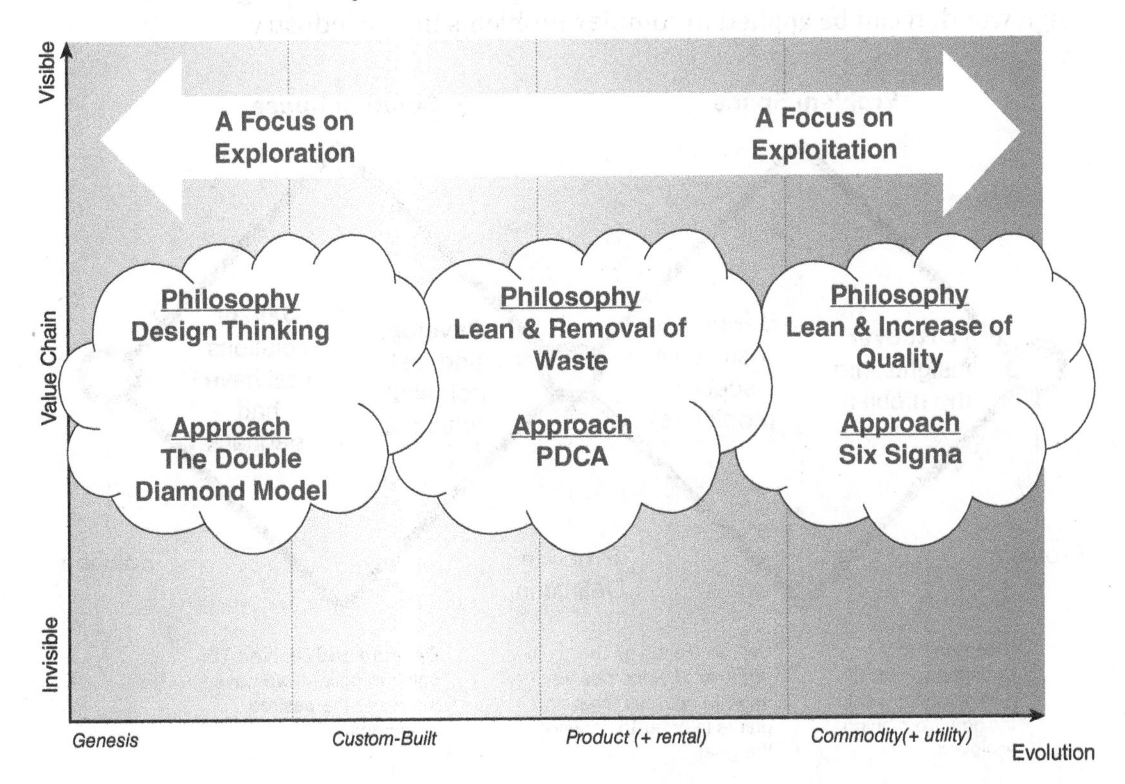

Figure 6.4: Problem-solving methodologies mapped to the evolution stages on a Wardley Map

Focused on Exploring: Design Thinking and the Double Diamond Model

The Double Diamond model is a design thinking process that uses critical thinking and reflective practice to deliver creative solutions and ideas. The model is based on the framed innovation approach by Kees Dorst, professor of design innovation at the University of Technology. As shown in Figure 6.5, the design process is divided into a problem space and a solution space. In the problem space, the first step is to discover insight into a problem so that we have a clear definition of the problem itself.

The next step is to define which areas to focus on to solve the problem. In the solution space, the third and fourth steps are to develop a solution and deliver it to solve the problem. For complex problems, we may develop and test prototypes before delivering a full solution; in clearer problems we may spend time engaging with third parties to compare COTs products. Although the Double Diamond model originated from the design world, it can be applied to complex problems in any industry.

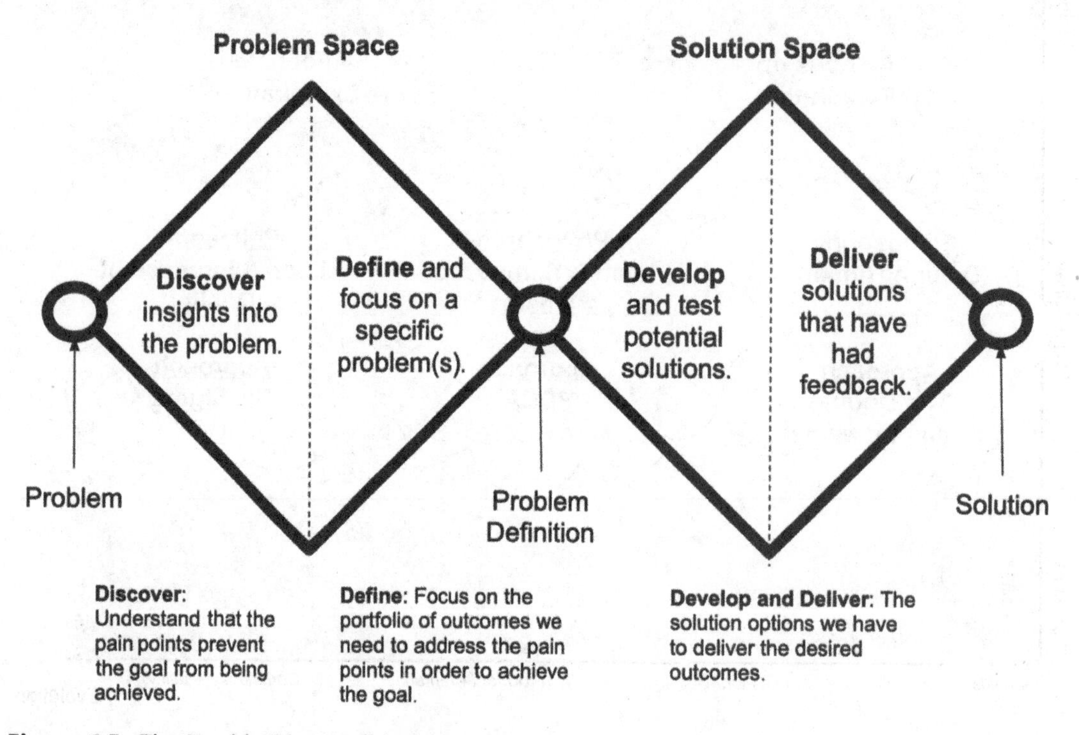

Figure 6.5: The Double Diamond design process

It's important to note that the diamond model is not a linear process; in fact, discovery is a continuous process, in both the problem and solution areas. This is an important point to stress because we are not looking for a product or service to create; we are instead looking for the shortest path to achieve a goal. In other words, by obsessing on the problem, we can avoid being wedded to a single solution and therefore be open to more impactful ways to achieve a goal. In this regard, we can say that discovery is never done. As shown in Figure 6.6, the more research, empathy, and understanding we gain in the problem space, the greater the chance of breakthroughs in revealing the most likely outcomes that will achieve a goal. As we explore solution spaces on how to achieve an outcome, we gather real feedback from customers, again leading to new insights as well as the confirmation or the refinement of outcome hypotheses that can be fed back into problem discovery. Throughout the process of problem and solution discovery, we must engage with customers early and often to get feedback that will enable us to refine ideas and reduce the effort and time to produce business impacts.

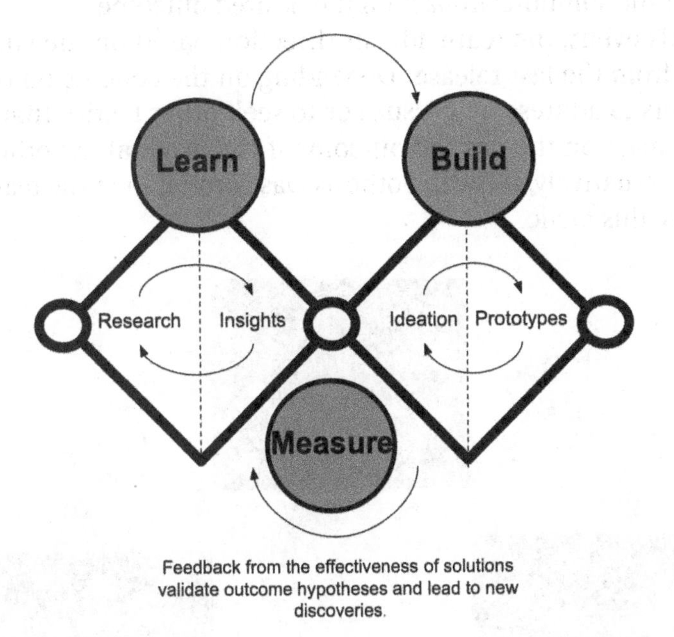

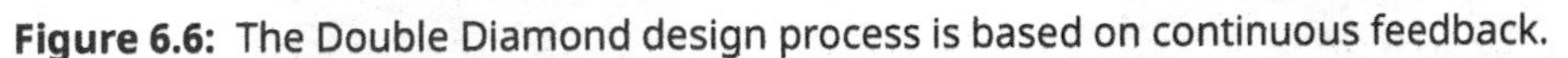

Figure 6.6: The Double Diamond design process is based on continuous feedback.

Focused on the Removal of Waste: Lean and PDCA

In contrast to exploring new opportunities for products and services with design thinking and the Double Diamond framework, many problems will be focused on process efficiency. Lean focuses on the improvement of existing value streams, business models, and processes by removing "waste" to make them more efficient and more valuable for customers. Here we are typically looking at exploiting what we have rather than exploring new opportunities.

To support a focus on improvement and the reduction of waste, we can leverage the PDCA (Plan, Do, Check, Act) improvement cycle as shown in Figure 6.7. PDCA, also known as PDSA (Plan, Do, Study, Act), was introduced by W. Edwards Deming and is based on the scientific method of problem solving; it comprises the following steps:

- **Plan** The team comes up with a hypothesis on what is the most valuable and achievable tactic to reach a desired outcome based on the learnings of the previous iteration.

- **Do** The team executes the plan.
- **Check (or Study)** After the solution has been released to the customer; feedback is gathered by the team. Data and customer experience is used to evaluate the success of the solutions impact on the desired outcome.
- **Act** In this event, the team adjusts direction based on the study of feedback and data from the last release. Depending on the conclusion of the study, the team pivots to address new issues or to seek other tactics that will result in a greater impact on the desired outcome if the original hypothesis was proven wrong. Alternatively, if the hypothesis was correct, then the team can continue to improve this tactic.

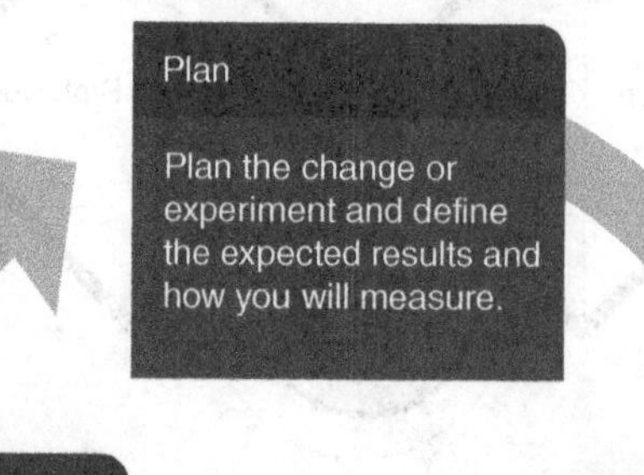

Plan
Plan the change or experiment and define the expected results and how you will measure.
Do
Carry out the change or experiment.
Check
Check the results of the change. What have you learned? What worked? What failed?
Act
Based on the results, decide how to act. If successful, rule out the change. If you fail, run the cycle again.

Figure 6.7: W. Edwards Deming's PDCA cycle

Focused on Quality and Consistency: Six Sigma and DMAIC

Lean and the PDCA cycle together make up a great lightweight and continuous improvement methodology to optimize existing products and services. However, if you're looking to reduce variability and risk and improve quality and strive for perfection, then an approach with Six Sigma might be better suited for your needs. It is best used for analyzing processes and reducing error rates and defects. We can adopt a Six Sigma approach in IT to the obvious problem domains that have very prescriptive and clear plans or frameworks to follow. For example, security frameworks such as NIST, CIS security frameworks, PCI, and other compliance frameworks suit a Six Sigma approach to ensure we are not deviating from the control lists that these frameworks offer. The Six Sigma process shown in Figure 6.8 comprises the following steps:

- **Define** The first step is to define the goal of the process or service. In other words, what is the level of performance you are trying to achieve?
- **Measure** After the goal is defined, capture data on the current situation by mapping the process to determine areas of waste and opportunities to improve.
- **Analyze** Once you have captured data, analyze it to get to the root of the problem. Understand real constraints and areas with the highest leverage for improving the product or service.
- **Improve** Next, determine, test, and implement the solutions to address the waste.
- **Control** Once the improvement is made, ensure that it does not degrade. Continue the cycle until you have reached and maintain the optimum level of performance.

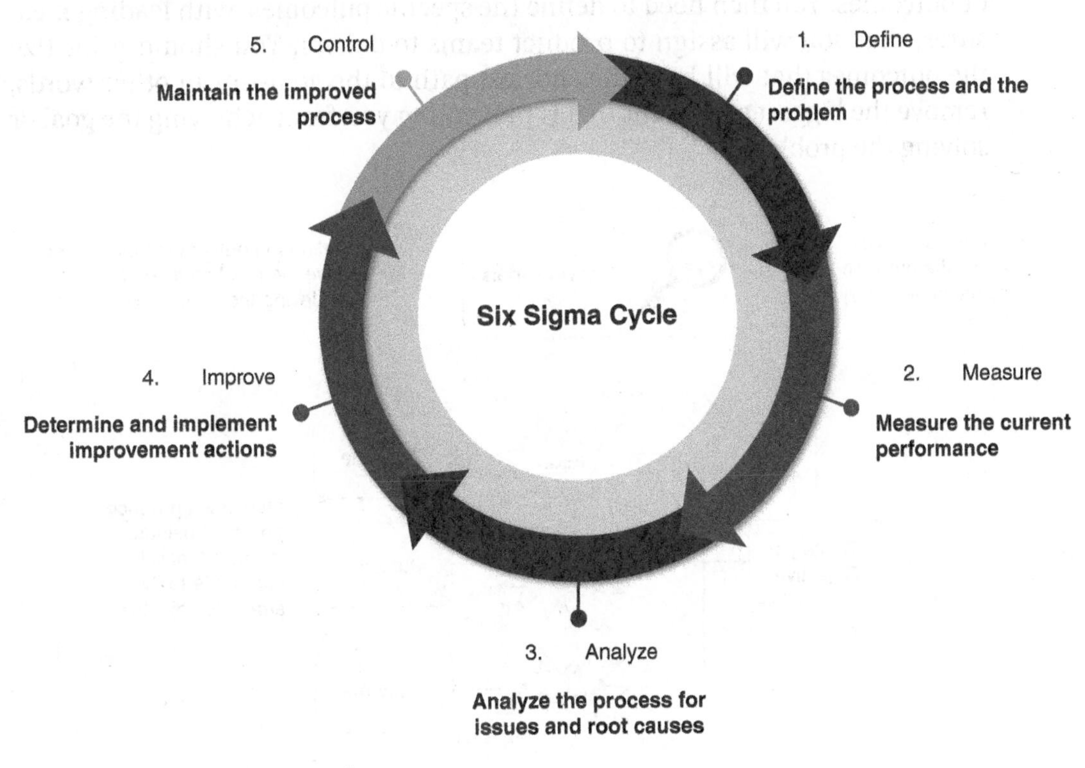

Figure 6.8: The Six Sigma cycle

Discovery Tools for Understanding the Problem Space

The first stage of solving a problem or achieving a goal is to understand it at a deeper level and identify the highest leverage points, the issues, or opportunities that, when addressed, will have the greatest impact on achieving the goal. As introduced in Chapter 2, the theory of constraints encourages you to identify the most important limiting factors and target those to improve the entire system rather than addressing symptoms of problems. As shown in Figure 6.9, there is a need to perform deep problem discovery before we define the areas we need to address that will likely achieve the goal.

- **Deep Discovery about the Problem:** Deep discovery is about understanding the customers and/or business needs, what they're trying to achieve, and the constraints and pain preventing them from completing their jobs. It's about gathering data and analysis and then uncovering opportunities to create business outcomes that will likely remove customer and business barriers and result in the business impacts we need.
- **Define the Outcomes to Focus On:** After discovery, you will have a portfolio of outcomes. You then need to define the specific outcomes, with leading measures, that you will assign to product teams to deliver. You should prioritize the outcomes that will have the shortest path to the goal—or in other words, remove the biggest constraint that is preventing you from achieving the goal or solving the problem.

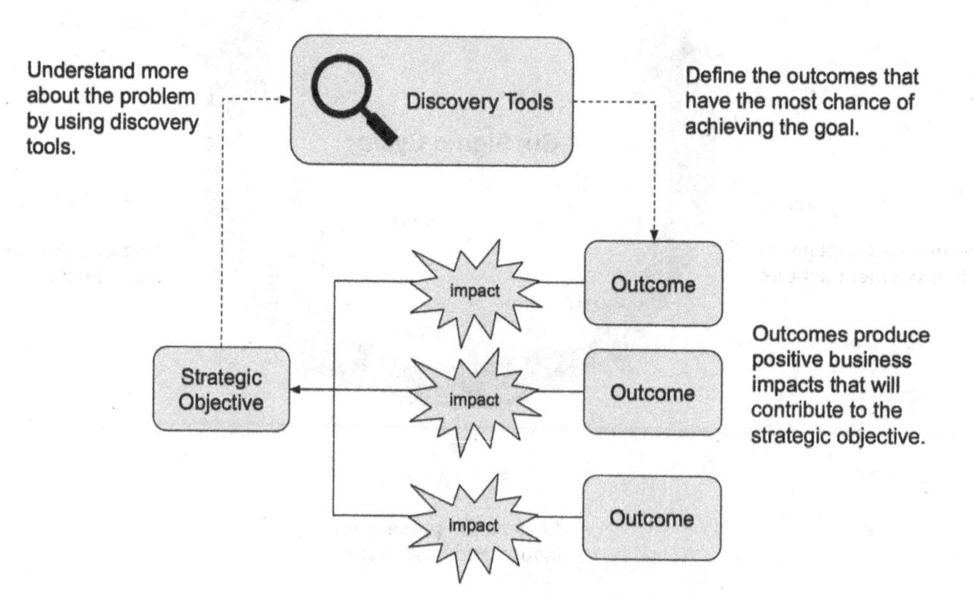

Figure 6.9: Problem discovery identifies the outcomes required to achieve a goal.

I have grouped the discovery tools and approaches based on focus:

- **Internally:** A focus on understanding the inefficiencies, waste, and constraints impacting the effectiveness of your processes.
- **Externally:** A focus on understanding the pain points, challenges, and opportunities of the usability of your product or service.

There is no right or wrong way to discover. You will need to apply the most appropriate tool based on your unique context. In many instances, problems will require internal as well as external investigation tools to identify the internal process optimization required to address customer experience pain points.

Discovery Tools to Address Customer Experience

When dealing with complex problems such as how to improve a customer experience, how to increase satisfaction or loyalty, or how to change customer behavior to improve conversion or increase average order value, we must first begin by empathizing with customers. The aim of discovery here is to learn about your customers' problems: how they occur, why they occur, the impact to customers and the knock-on impact to the business. By understanding the barriers and obstacles that prevent customers from doing the jobs they need to do, you are in a much better position to offer impactful solutions. We will look at three popular methods of discovering customer pain points and frustrations:

- **Jobs to be done:** Understanding the problems that customers are trying to solve or the jobs they are trying to complete.
- **User journey mapping:** Understanding your customers' emotions and how they feel when using your service. Here we can map problems or opportunities to improve their experience. We will later look at how to link business processes to customer pain points so that we can uncover the root causes of customer frustration.
- **User research:** Both quantitative and qualitative, to reveal who your customers are and what they need, desire, and value from your value proposition.

Jobs to Be Done

Capturing the jobs that customers are trying to complete as stories is another powerful way of empathizing with the needs of your target consumers. The jobs to be done (JTBD) framework was invented by Tony Ulwick and popularized by Professor Clayton Christensen. Professor Christiansen describes JTBD as "a tool for evaluating the circumstances that arise in customers' lives." It focuses attention on customers and the thought processes that would lead those customers to "hire" your product or service to complete a job. Ultimately, customers make decisions about what products and services to use because they have a problem that they need to solve. As Theodore Levitt famously used to tell his students, "People don't want to buy a quarter-inch drill. They want a quarter-inch hole!" By shifting focus, we can gain a deeper understanding of the underlying needs and problems that are trying to be solved and therefore offer an optimized product, service, and experience to meet their needs. This helps to keep the customer at the center of what you are doing rather than focusing on your product and trying to convince customers to use it.

Perhaps the most famous example used to illustrate the approach of JTBD is a case study on how it was used to help McDonald's increase revenue and customer satisfaction for milkshakes. McDonald's wanted to improve performance in its milkshake line but despite huge amounts of data from user research and customer profiling, it was unable to produce a positive impact in terms of increased revenue or profit. Professor Christensen and his team approached the problem by looking to understand the job, or need, behind customers that led them to buy a milkshake. When conducting research, they observed that half of milkshakes were sold before 8:30 in the morning and that it was the only item the customer bought before driving off on their commute.

His team asked customers why they were buying milkshakes in the morning, what job they were trying to get done, and what alternatives they had tried. The responses were surprising. Customers mentioned trying other snacks for their breakfast commute but complained that they were hard to eat while driving, the experience was over too quickly, they created a mess, or they didn't fill the customer up until lunchtime. However, it was the milkshake that was easy to consume while driving, took a while to consume, and kept customers full. The need the customer was articulating was to have something to eat during a long commute that would keep them occupied and full throughout the morning; the milkshake was hired to fulfil this job.

While not always strictly necessary, a helpful trick is to capture a JTBD using the following structure:

"When _____" "I want to _____" "So I can _____"

- "When _____" focuses on the situation.
- "I want to _____" focuses on the motivation.
- "So I can _____" focuses on the outcome.

For example, the McDonald's milkshake job story could be captured as follows:

***When** I'm commuting in my car to work, **I want to** have something easy to eat/drink **so that I can** feel full enough to get through the morning until lunchtime.*

As you can see, the job story format encodes the context behind what a customer wants as well as the customer's ultimate desired objective. We can continue to come back to this job statement to test if our product or service satisfies the need.

Higher-level JTBD formats, including your organization's value proposition, are long-lived as they represent a fundamental customer need rather than how it is currently being fulfilled—what the customer wants versus how they want it. For example, people have always had a desire to listen to music to enhance or alter their state of mind. As technology has evolved, how customers have gotten this job done has moved from record players, to radio, to the Walkman, on to MP3s, iPods, and streaming platforms. Therefore, as technology evolves and new possibilities are enabled, we need to review a customer's JTBD to ascertain if we can optimize the experience of the job they have hired us to do. This is opposed to seeing how we can leverage technology solely for the benefit of the business or because it adheres to a trend. Think about how many organizations wanted an app because others had one instead of thinking about what job the app would do for a customer. The JTBD framework keeps us honest and focused on resolving real customer problems. If this is achieved, then we stand a much greater chance of achieving business goals.

User Journey Mapping

A customer journey map represents a customer's experience with your product or service. It is a highly visual tool used to build knowledge, consensus, and most importantly, empathy for customers' pain points, needs, and wants. Journey maps can highlight hot spots on a customer journey based on the insights obtained through customer research using the techniques highlighted earlier. These problems can then be turned into opportunities for teams to optimize a customer experience to lead to business benefits.

As shown in Figure 6.10, a customer journey map typically consists of the following elements:

- **A customer persona and scenario.** A customer persona is a representation of a market segment, and a scenario is an example of a job to be done. It is important to have a user and a scenario in mind when creating a journey map because it makes it easier to empathize and model both behaviors and feelings. A journey map can be created for each major customer segment and selection of jobs to be done.
- **Customer's journey steps.** Once we have a persona and scenario, we can then identify the steps customers take in their journey to complete their job. The stages represent a customer's need throughout their journey. For example, the general phases for B2C customers are typically awareness, consideration, purchase, retention, and advocacy.

- **Customer touch points.** A touch point is any moment a customer interacts with your brand, whether that be seeing a magazine or TV advert, responding to an email, visiting on your website, or downloading and interacting with an app. Touch points from all channels should be captured on a map because they form the overall picture of your customer's experience with your product or service.
- **Customer pain points, emotions, and feelings.** Perhaps the most important section of the map is to capture the customer's emotions and feelings. This is where we capture the pain points, wants, and needs of a customer. This section can be supported by the data you gather when conducting user research.
- **Opportunities for improvement.** Last, we can capture the ideas and insights to optimize the experience of the customer and overcome the obstacles in their path. Here we should also look for measurements that help us clearly see how we can turn a pain point into a positive customer and benefit.

User Research

When dealing with customers, both internal and external, we can gather data, analysis, and insights through a combination of formal and informal methods. The aim of this form of market research is to summarize the voice of the customer. This will help us understand more about the opportunities and constraints that can lead to breakthroughs in how we provide products and services to customers to ultimately achieve a business goal.

Following are some of the approaches that can be used to gather data from customers:

- **Customer interviews.** The number of teams that I have met that have never spoken to a customer of their product or service is staggering. Talking to customers who use your product is one of the easiest ways to gauge their feelings about your value proposition. When customers leave a negative review, or indeed a positive one, we can follow it up and ask them about their unique context and the experience they had to identify opportunities for improvement that will result in us moving closer to a goal.
- **Focus groups** are guided discussions between a team and a diverse collection of potential customers or users. You can partner with third parties to talk to segments of customers to understand and gather analysis on what products and services they are interested in.
- **Site intercept surveys** are short feedback questionnaires that can be presented to customers while they are interacting with a website or app. They can be displayed at key points in the customer journey to capture customer emotions, frustration, and feelings toward a product or service.

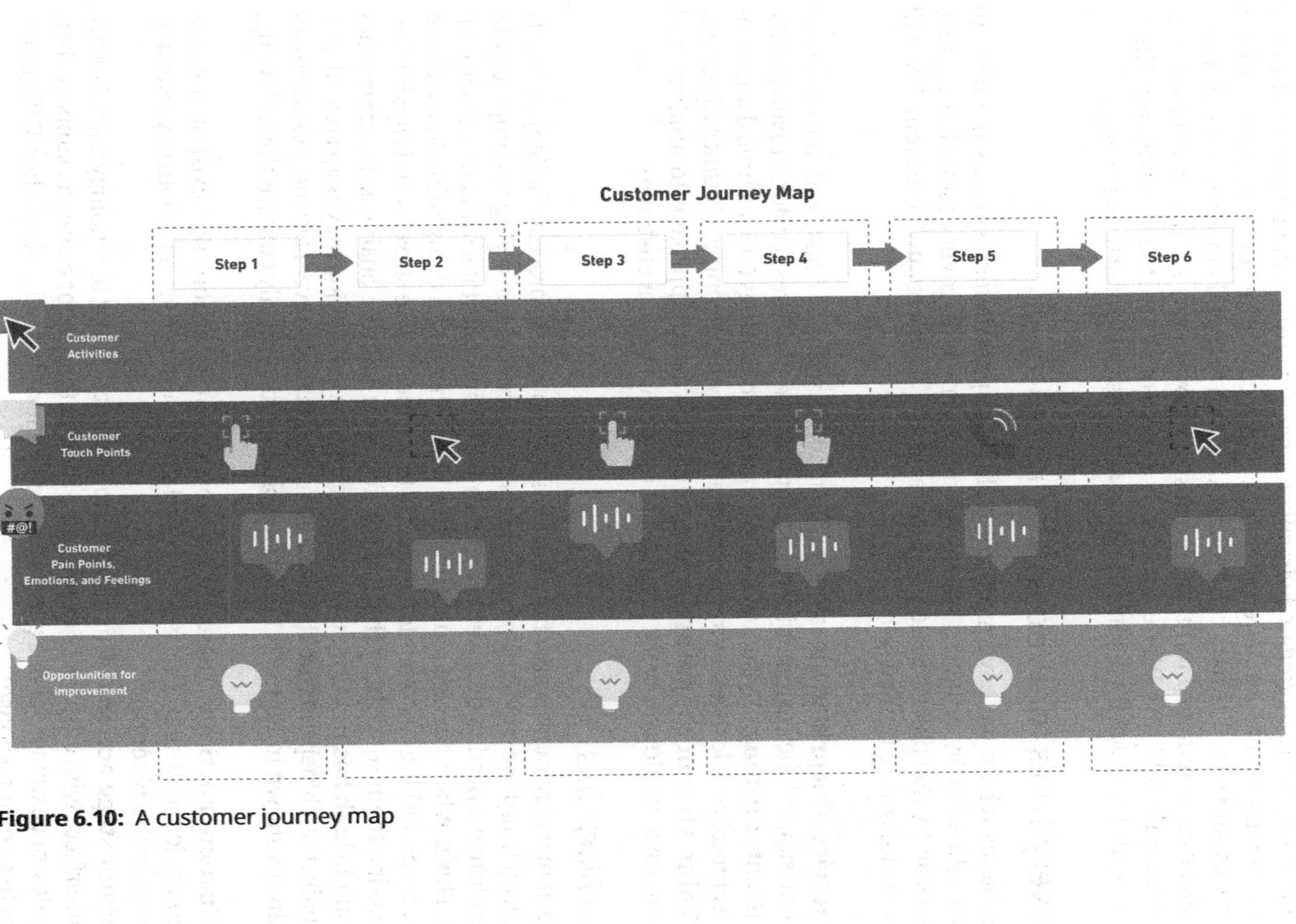

Figure 6.10: A customer journey map

- **Heat maps** are aggregated visualizations of website interactions used to discover customer usage patterns. Maps capture cursor movements, page scrolling, clicks, and taps. The data can then be analyzed to reveal any barriers or obstacles that can be removed to make business improvements such as conversion rate or revenue add-ons. While not strictly a voice of the customer technique, this is very useful to see if customer actions are the same as customer words.

Discovery Tools for Process Optimization

For problems where we have more control of the parameters, such as the need to increase efficiency by reducing waste, we can leverage discovery tools that capture processes and value chains to determine waste and problems in the system. We will look at the following frameworks:

- **Service blueprints:** Visualize the pain points with the organizational processes that support the customer journey to optimize and address customer pain points.
- **Event storming:** A visual workshop practice that brings both technical and non-technical people together to explore challenges within complex business domains.
- **Value stream mapping:** A lean manufacturing approach to analyze and improve the flow of work required to deliver value to a customer.

Service Blueprints

Poor customer experiences are often due to internal business process challenges. While user research and the frustrations customers experience when trying to complete jobs may be clear, we need to discover the root cause of these obstacles and customer pain points. This is where service blueprints can help. They are a companion to customer journey maps, linking customer touch points to business value streams and capabilities, thus making them something that we can focus on to drive business outcomes that will result in both business impacts and customer benefit. If journey maps reveal customer pain, then blueprints are the treasure maps that help reveal business weaknesses and pain. As shown in Figure 6.11, service blueprints typically contain the following:

- **Customer actions.** These are the same customer actions that exist on the customer journey map. This is your correlation point between the journey map and the service blueprint.
- **Front stage actions.** This is the employee actions or technology that directly interfaces with customers to complete their actions—for example, support chat, sales, or customer service agents on the phone or indeed a phone line IVR (interactive voice response).

- **Back stage actions.** These are the activities that systems or employees do to complete a customer action. For example, this is customer service agents using internal systems to give a customer information or fulfill their request.
- **Support processes.** These are the business processes that enable and support the front stage and backstage actions. For example, if a customer contacts us for a banking loan, these support processes could be the internal automated checks to see if the customer qualifies for a loan.
- **Business measures.** It is useful to add metrics to provide additional context and evidence to the service blueprint, especially if it supports something that is a pain to customers. An example may be the time spent on various processes, or the number of errors or rework often required in a process. Having benchmarked business metrics will help you to measure any improvements or to see if problems have become worse.
- **Business pain points.** Just as we captured customer pain points on a user journey map, we should capture business pain points on the service blueprint. In addition to the business pain, we can capture employee frustration. Highlighting where we have issues, backed with the measurement evidence, can help us to identify where to focus our efforts on improvement and type that will align to and result in customer benefit as well.

Domain Discovery with Event Storming

Event storming is a workshop activity created by Alberto Brandolini that is designed to quickly build an understanding of a problem domain in a fun and engaging way for cross-functional teams. Groups of domain experts, the ones with the answers, and development team members, the ones with the questions, work together to build a shared understanding of the problem domain to discover the most effective areas for improvements. It is particularly useful in assessing the health of an existing business process or value stream. The process of event storming occurs in an open environment that has plenty of space for visual modeling, be that lots of whiteboards or an endless roll of brown paper. The process, as shown in Figure 6.12, is not linear and is more organic in its flow, but the process is as follows:

- Start with **Domain Events**. The problem domain is explored by starting with a domain event, that is, events that occur within the problem domain that the business cares about. A post-it note representing the domain event is added to the drawing surface, and then attention is given to the trigger of that event.
- Then add concepts when needed to model the process:
  - An event could be caused by a user action that is captured and added to the surface as a **command**.

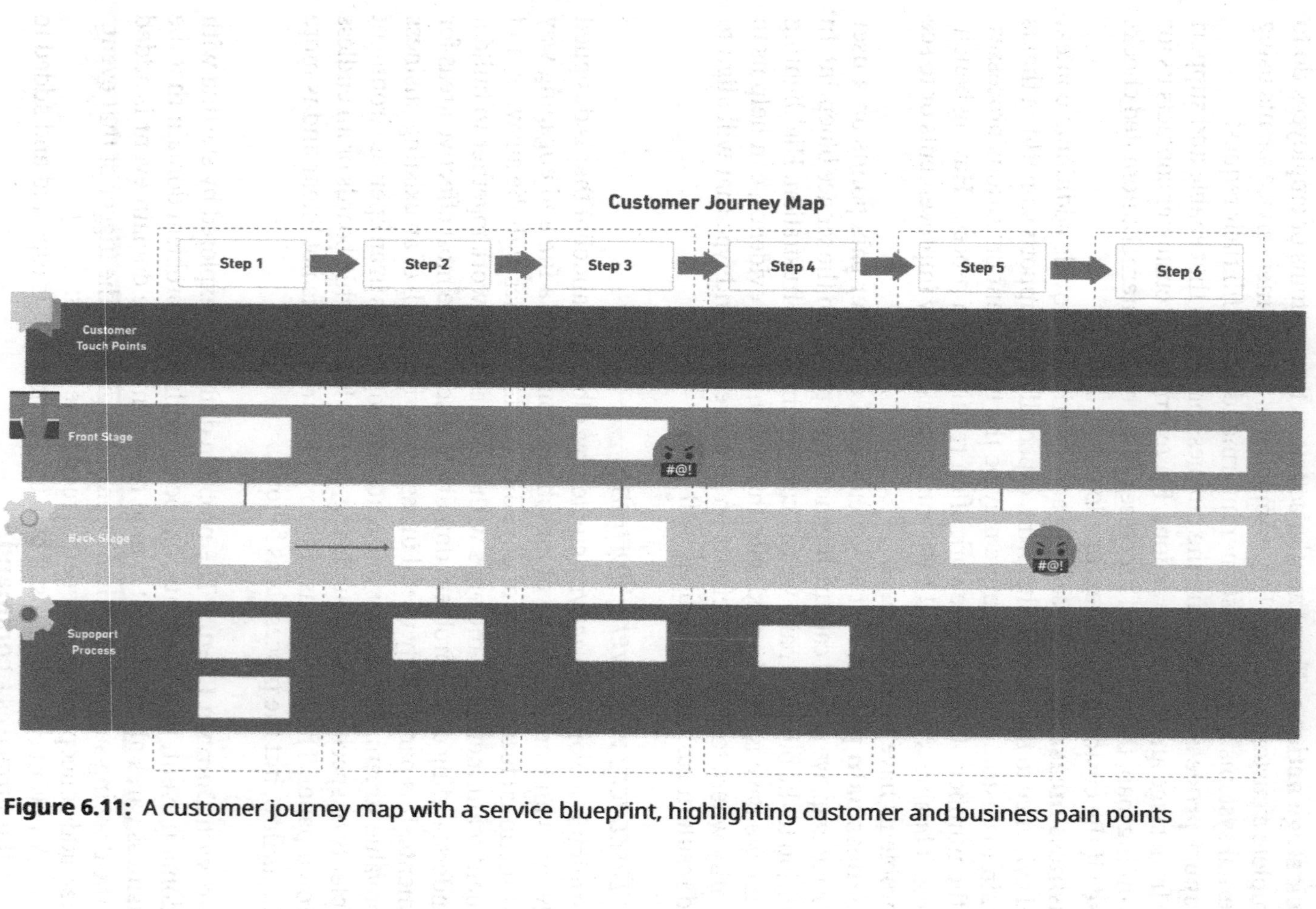

Figure 6.11: A customer journey map with a service blueprint, highlighting customer and business pain points

- Commands can be issued by an **actor** (customer/user) triggering the action.
- For an actor to decide and issue a command they need information. This is captured as a **query model**.
- Sometimes commands are generated by the system in the form of a **policy** (an automated or manual process), which itself was a reaction to an event. For example, a policy may state that "when X event happens, issue command Y."
- An external **system** could be the originator of the event; they are also used to describe the source of an action.

- Last, identify areas of waste, confusion, or opportunities for improvement with **hot spots**. Highlight areas where there are inconsistencies between different perspectives and competing goals, bottlenecks, or impediments to the flow of work.

This activity continues until there are no more questions. The team is left with a highly visual representation of the value stream or process, highlighting the areas where there is an opportunity to improve. We can then estimate addressing each hot spot to build a collective understanding of areas to focus on and a consensus around the next actions. The relatively simple yet sophisticated and highly collaborative practice of event storming helps to facilitate the discovery process with a cross-discipline team with different backgrounds.

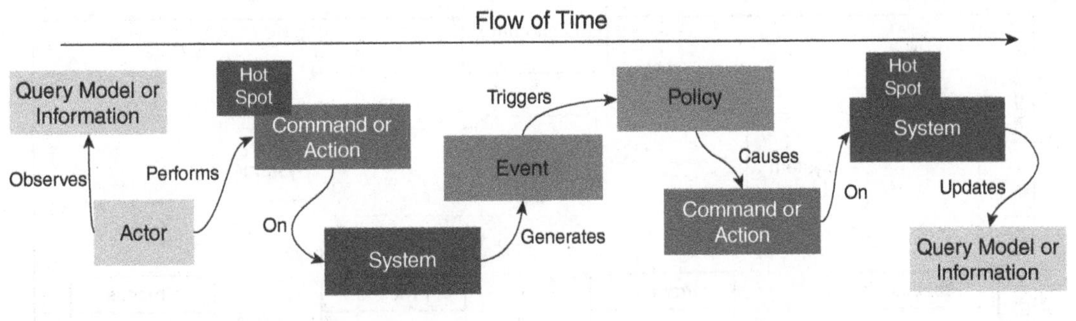

Figure 6.12: An event storming map

Identifying Waste with Value Stream Mapping

Value streams are the sequence of operational activities that transform a customer request into the delivery of customer value, be it a service or a product. Core value streams are those that are focused on delivering your primary value propositions to customers; however, an organization will also have several supporting or enabling value streams that have internal customers. Value stream mapping, then, is a technique to map and visualize these flows of value across a business, the primary purpose of which is to identify opportunities for improvement. This is achieved by visualizing the flow of information, materials, and work across functional silos and determining waste captured as delays, rework through poor quality, and non-value-adding activities. While born from the manufacturing sector, value stream mapping can be applied to knowledge work in the form of service flows: the steps involved from customer request to the delivery of value and IT's delivery pipeline. The point of value stream mapping is to visualize work as a system, the flow of value across a business, to determine the direction of focus for maximum impact. This means it operates at the macro or strategic level rather than examining the processes of individual departments.

The archetype map of a value stream heavily influenced by the manufacturing domain where it originated is shown in Figure 6.13. However, as covered in depth in *Value Stream Mapping: How to Visualize Work and Align Leadership for Organizational Transformation* by Karen Martin and Mike Osterling (McGraw Hill, 2013), value stream mapping is applicable to information flows as well as physical flow. Mapping knowledge and service value streams requires a change in approach from the traditional manufacturing context, but the principles are the same. In the manufacturing context, we measure the time and quality of a physical product across a flow; in knowledge and service work we focus on the quality and flow of information.

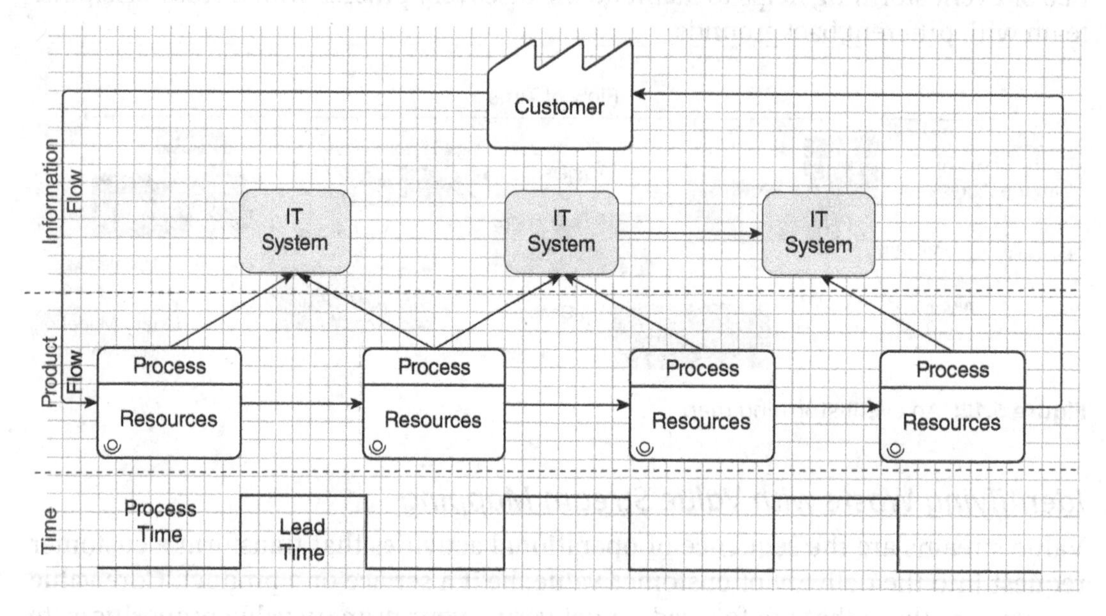

Figure 6.13: An example of mapping a value stream

The steps for mapping a value stream are as follows:

1. Determine the scope and boundary by deciding the problem you are solving for. Where to start will largely be driven by the outcome you are looking to achieve and your strategic direction to reach your business and strategic goals. To demonstrate the process, I will use the example of a request-to-return value stream. This is the series of activities that take place to return and refund a customer for a physical product. Our goal is to improve the efficiency of this process and reduce the time it takes for a customer to be refunded.
2. The second step is to ensure the team improving the value streams consist of those that have the authority to make meaningful changes.
3. The next step is to capture the AS-IS flow. You cannot do this in isolation; you must go and see the actual process at the place of work, otherwise you will fall into the trap of modeling how you think the process operates or how you would like it to operate. The first pass is to model the flow, the sequence of activities as shown in Figure 6.14. For each step in the value stream, we capture the high-level activity and the team that performs it. In our example, we can see that a customer requests a refund by contacting the CX team; the team then asks the customer for more information. Once the CX team has the info, they hand off to the operations team to book the return with a delivery partner. Once that's complete, they pass back to the CX team to inform the customer. The OPs team will regularly check manifests generated daily by the delivery partner to see if the item has been collected. Once it has, then the OPs team can let the CX team know so that they can refund the customer. How you distill the value stream into blocks of activities or steps is perhaps more art than a science but aim to map anything between 5 and 15 activity steps. You don't want too much detail, otherwise it will turn into a complex process mapping exercise. This is why the mapping exercise is best performed on a whiteboard with sticky notes rather than in electronic form as it allows the right level of abstraction to be arrived at as you refine the value stream.
4. The second pass, as shown in Figure 6.15, is to capture the data on process and information flow, wait times, lead times, and quality of information flowing through the value stream. We start by identifying the applications that each process step interfaces with. All applications and the flow of information should be mapped, including any home-grown spreadsheets that are used to store data or drive decisions as well as any automated flows.

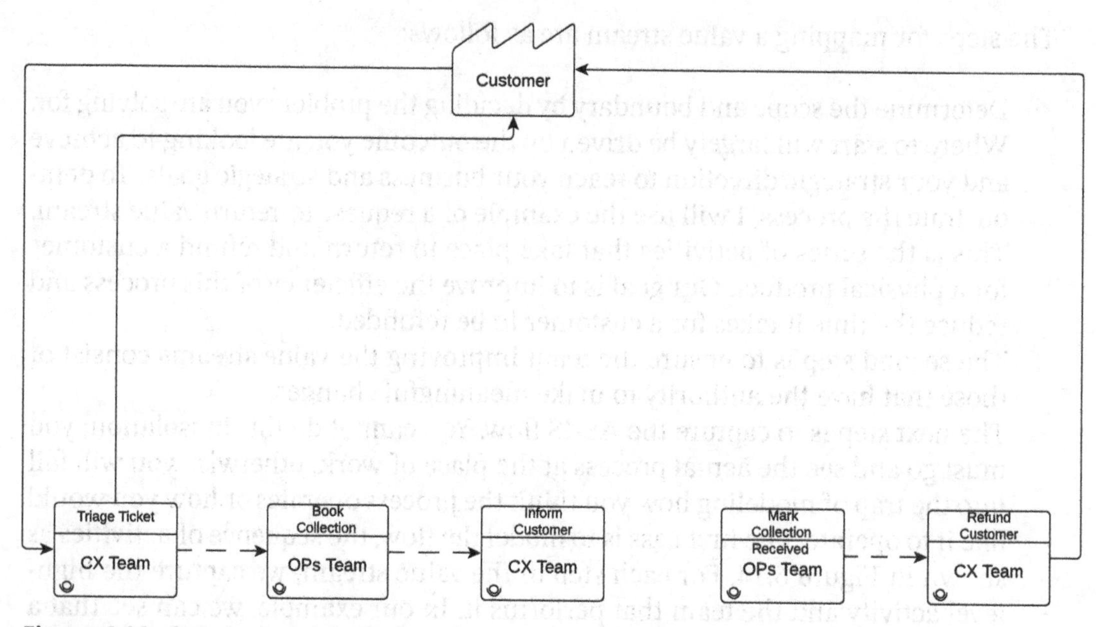

Figure 6.14: Step 1, model the flow or the sequence of activities.

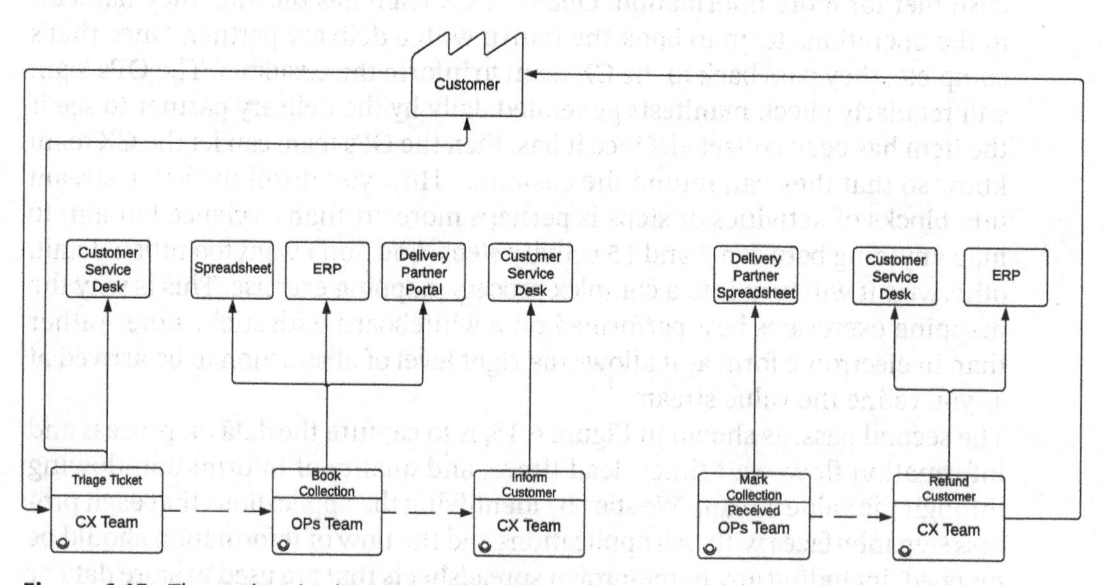

Figure 6.15: The updated value map with related applications

As shown in Figure 6.16, we should also measure lead time (LT), process time (PT) and % of work that is complete and accurate (%C/A). Lead time is the amount of time work waits before being handled, process time is the amount of time it takes to complete an activity, and the % of complete and accurate is the % of errors we have that require rework. The %C/A can be measured by asking downstream teams how much information they receive that is not usable or missing and therefore needs correction.

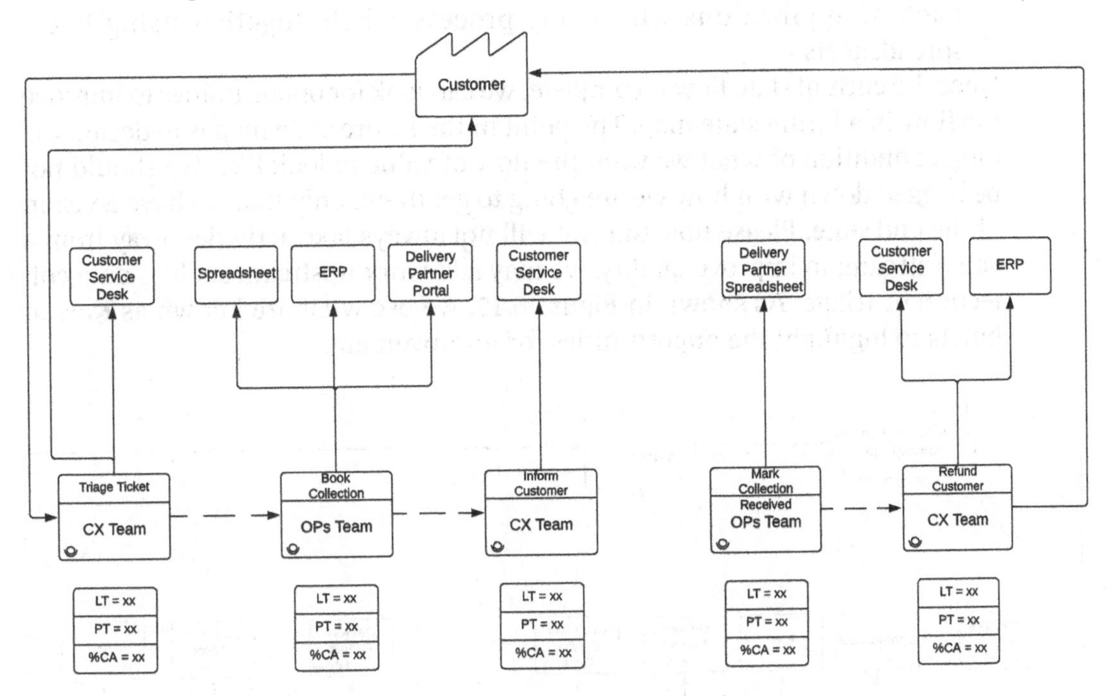

Figure 6.16: The updated value map with data on processing time

It is only when we take the time to map the value stream from end to end across departmental silos that we truly begin to grasp the amount of waste for the first time. The waste typically found in the process of value stream mapping includes the following items:

- Multiple handoffs between teams and departments.
- A high level of waste and rework due to errors in or a lack of information. No feedback upstream of quality problems in order that they may be addressed. A lack of information systems ensuring data quality at point of input or systems contributing to poor quality in data.
- No documented or standardized process to follow leading to variation in the quality of work.
- Lack of accountability and ownership on some processes, or rather the gaps between the processes.

- Activities that offer no value and are superfluous to the value stream.
- Multiple or overly complicated information systems in use requiring multiple data input per process. Systems not talking to each other or are not aligned to the actual process activities.
- Lack of applications with many processes held together using Excel spreadsheets.

5. Once the current state flow is complete, we can look for opportunities to improve the flow in a future state map. The point of the future state map is to design the target condition of what we want the flow of value to look like. We should not be bogged down with how we are going to get there, only that we have a vision of the end state. Please note that we will not always take activities away from a value stream; to improve quality, we may add work upstream such as data collection or triage. As shown in Figure 6.17, we use what are known as Kaizen bursts to highlight the opportunities for improvement.

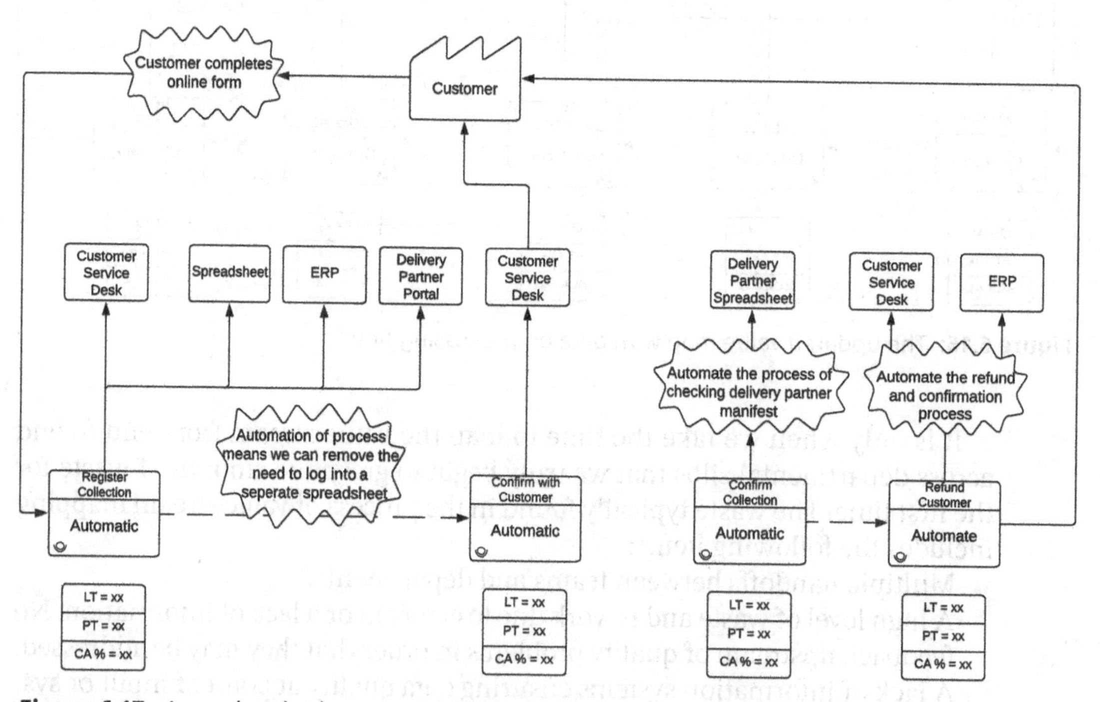

Figure 6.17: An updated value map, highlighted with opportunities for improvement

6. As we have captured the data behind waste in the flow, be it lead time or process time of % complete and accurate, we can prioritize the biggest waste in terms of its overall impact on flow. This is how we apply the Theory of Constraints, as introduced in Chapter 2, to focus direction on the biggest impediment to the flow.

Once the current state of a value stream is mapped, you will often find that people are surprised by the delays between process steps, the waste caused by poor quality information upstream that causes rework downstream, and the bottlenecks and constraints that reduce flow. Specifically, from an IT perspective it is revealing to see how many information systems, be they spreadsheets, bespoke apps, SaaS-based applications or on-premises ERPs, are involved in each process step. Indeed, it is also interesting to see and understand gaps in systems support and where information of automation could radically improve steps in the flow. Understanding the disconnect between IT systems and the capability needs of each activity helps IT leaders make informed technical decisions on how to consolidate and simplify the architecture to ultimately improve the flow of value.

It is worth noting that value stream mapping sits firmly in the exploit camp where there are existing customer value streams. It is an ideal tool for complicated problems but not ideal for complex ones. As with all the tools presented in this chapter and indeed this book, treat them as a "means to the end," not the end itself. Value stream mapping is not about producing pretty diagrams; it is instead about facilitating a group of people that have knowledge of a domain and the authority to make changes to improve the flow of value within. It is the process of visualizing, communicating, and discovering opportunities for improvement rather than the maps themselves that is important. Value stream mapping encourages a systems thinking approach and the focus on improvement in the system as a whole over local optimization. Value stream mapping is a powerful tool to bring consensus and baseline our understanding of key flows of value across the business. It creates a shared understanding, increasing alignment and empathy, of where we are and where we want to get to.

Root Cause Analysis with the Five Whys and Cause-Effect Diagrams

How do we effectively analyze and solve problems? As discussed in Chapter 3, "How to Change the System," systems thinking encourages us to look beyond the symptoms of a problem to discover the underlying causes. To do this we should tackle problem solving from the point of view that all problems are systemic, in that it is the consequences of the environment we design that causes problems and not people. People can only work within the confines of the system we define. People are fallible and will make mistakes. If mistakes lead to problems, then it is the fault of the system we design, because the system assumes that mistakes will not be made, rather than the fault of a person.

As leaders, it is our job to remove problems for our teams, and we are in the unique position to change the system of work to prevent the cause of problems and their impacts disrupting the flow of value in the organizations. Therefore, it is important that we identify and fix the root causes rather than just the symptoms of problems. If we do not, then we should expect to have to fix the symptom of a problem again and again as it appears. Our first action, however, is to determine if a problem is worth solving in the first place. If a problem does not impact our ability to achieve our, or our team's, goals, then it is merely an annoyance and spending time and effort on it will be of little value.

A simple but surprisingly effective technique to root out causes of problems is the Five Whys. The *Five Whys* is an iterative technique used to interrogate the relationships between cause and effect to reveal the root cause of a problem. The technique is deceptively simple; the method is based on asking why repeatedly until the real cause of a problem is discovered. As shown in the example in Figure 6.18, each answer to why a problem has occurred is the basis of the next question. We can iteratively ask why and drive deep into a problem until we find the root cause rather than a symptom of a problem. In this way the Five Whys is a process of systems thinking; we are looking for root causes and not symptoms. The process is called the Five Whys due to an anecdote that it took five iterations to reach the root cause, but the number of iterations needed to discover the root cause will vary based on the complexity of a problem.

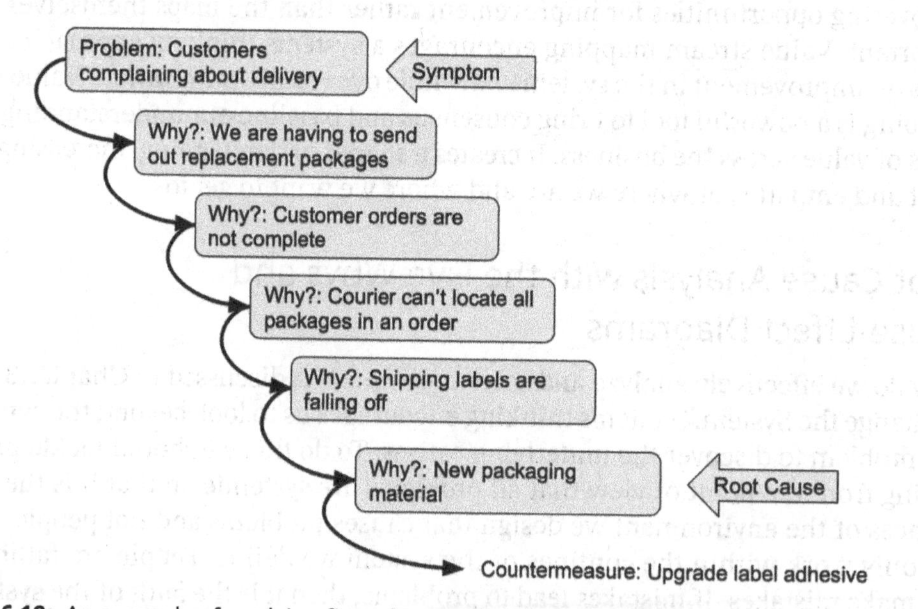

Figure 6.18: An example of applying five whys to root cause analysis

However, complex problems don't tend to have a single root cause. To map all the root causes that contribute to a problem, we can combine the Five Whys with a cause-effect diagram, a tool used to visually organize all the causes for a specific problem. A popular type is the fishbone, or Ishikawa, diagram as shown in Figure 6.19. This diagram builds up on the Five Whys by uncovering the multiple root causes resulting in the negative effect.

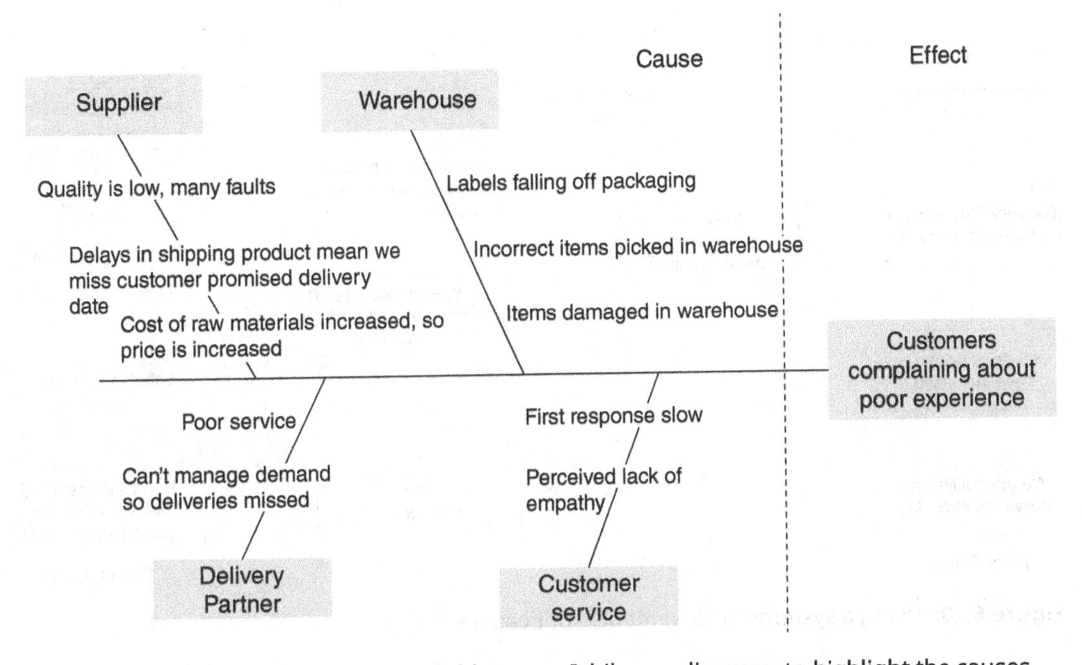

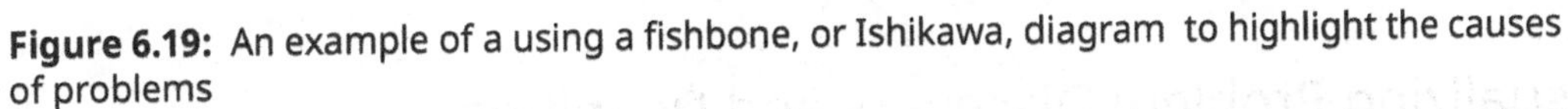

Figure 6.19: An example of a using a fishbone, or Ishikawa, diagram to highlight the causes of problems

The challenge with the fishbone diagram is that it does not show the relationships between problems, or what systems thinking calls reinforcing or feedback loops. A different method of modeling cause-effect diagrams with a systems thinking focus is a current reality tree diagram, which is part of Goldratt's ToC thinking processes. Shown in Figure 6.20, in this diagram we can see the systemic loops that are reinforcing the root cause of problems more clearly.

The method you choose to employ will largely depend on the complexity of your problem. I typically start with the Five Whys and evolve into systems thinking diagrams if the problem is complex. These root cause methods themselves are very lightweight, providing enough structure to help facilitate conversations rather than arming you with a list of killer questions that are guaranteed to reveal the underlying root cause. However, what these tools do well is they facilitate the problem-solving approach itself, giving you enough guardrails and a framework to anchor your investigation to the reasons behind problems. As a leader, you will constantly be faced with complex problems. Having an effective mechanism to distill cause from effect and reveal the underlying issues is vitally important to removing blockers and constraints.

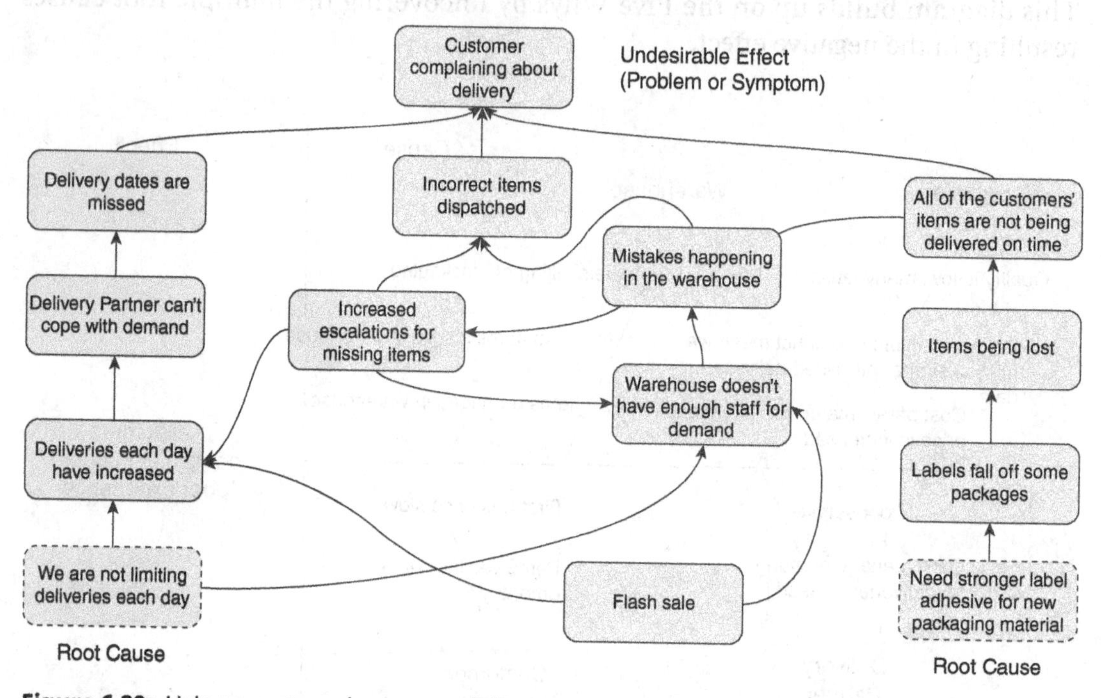

Figure 6.20: Using a systems loop to model root causes

Visualizing Problem Discovery and Definition with Impact Mapping

How do we plan for the features and business outcomes that will contribute to and have the biggest impact on business goals? Impact Mapping, introduced by Gojko Adzic in his 2012 book *Impact Mapping* (Provoking Thoughts) and as shown in Figure 6.21, is a lightweight, collaborative planning technique for teams to discover the outcomes or changes in behavior that will contribute to business goals. It is a highly visual method based on the practice of mind mapping, allowing teams to clearly show scope and underlying assumptions on how deliverables result in user behavior change, which in turn contributes to business impact.

By starting at the output level with upfront requirements and plans, we can lose connection and line of sight to the goal we are trying to achieve. However, if we employ impact mapping, we start at the goal level, then look at the outcomes (i.e., whose behavior we can change) that would contribute to the goal before lastly looking at what we need to deliver to achieve the outcomes. This process ensures that output is understood in the context of how it contributes to a goal, which helps with understanding the relative importance of one feature from another. This simple method makes it very easy for everyone to contribute to planning and understand the reasoning behind deliverables.

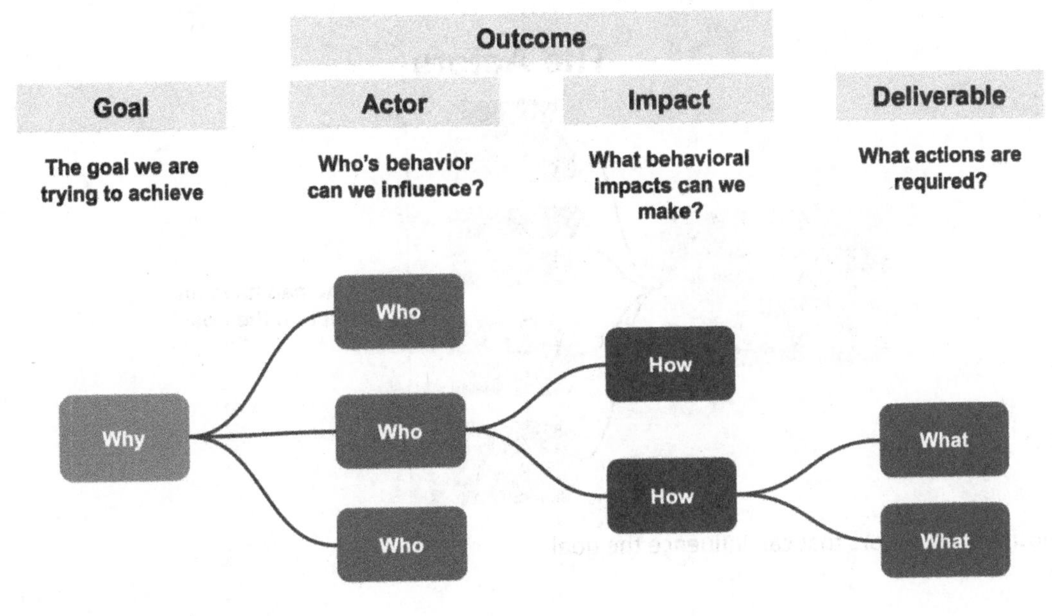

What behavioral impacts can we make?
What actions are required?

Figure 6.21: An impact map

These are the steps to build an impact map:

1. As shown in Figure 6.22, we start with the goal, the problem we are trying to solve, and the reason we are doing this exercise. We state the objective and a way to measure success. By explicitly stating the goal, we ensure focus on the objective. It may seem obvious, but knowing why we are doing something, the ultimate purpose, enables us to make better decisions. Too often IT teams are only handed outputs and have no real grasp for the desired outcomes.

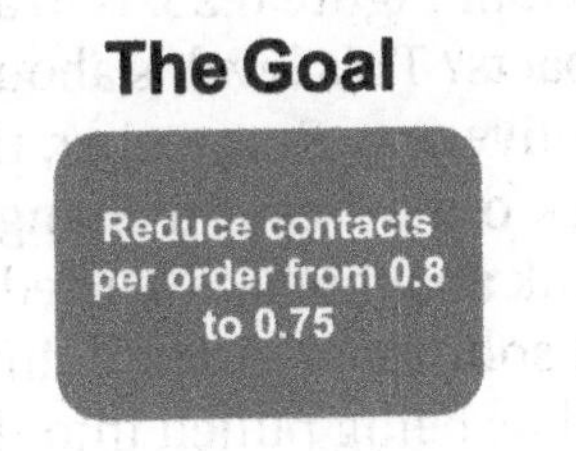

The Goal
Reduce contacts per order from 0.8 to 0.75

Figure 6.22: Clarifying the goal

2. As shown in Figure 6.23, the next step is to determine who's behavior we can change to achieve the goal. We need to map out all the people or personas that can contribute to or hamper the goal. We can think about the consumers or users of our product, those who will be impacted by it or those who will manage or sell it. For example, we may map "Online Customers," "Repeat Customers," "Sales Agents," "Package Creators."

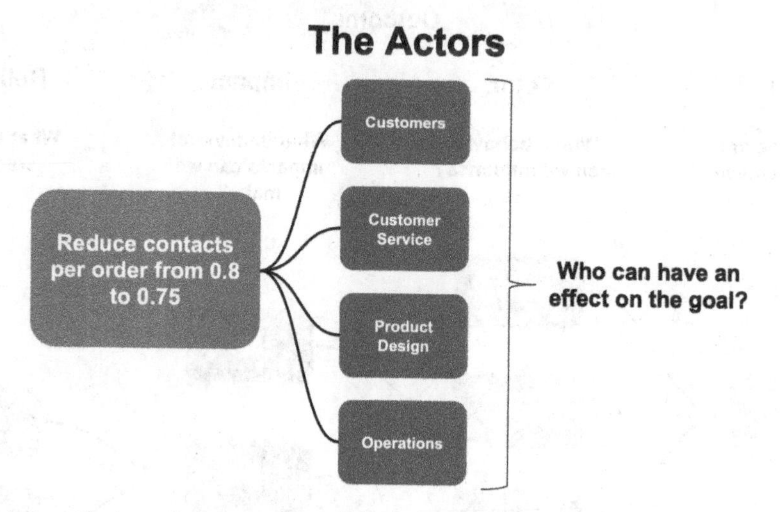

Figure 6.23: Actors that can influence the goal

3. We can then map the behaviors we want from our actors by asking how each actor can contribute to the goal. The combination of an actor and a change in their behavior is referred to as an outcome. Often, I will combine an actor and behavioral change into a single outcome, but it does help to start by thinking about actors first and then understand what jobs our actors want to get done. As shown in Figure 6.24, it is also useful to put in a measure to understand what a future state may look like; for example, to contribute to the goal, a user should be twice as productive. By asking for desired outcomes, actors plus behavior change, rather than a requirements list, we make it much easier to collaboratively plan with nontechnical people.
4. The fourth step, as shown in Figure 6.25, is to ask what we can do to support or enable the required impacts? This level is about scope, where we want to focus effort in building and delivering. Remember, this doesn't always have to result in a code. We should ask ourselves if a change in a process could enable the impact? At this level, think about what we need to do rather than how we would do it. Don't think about solutions; instead, think about high-level capabilities required. This will avoid us being pulled into the details.

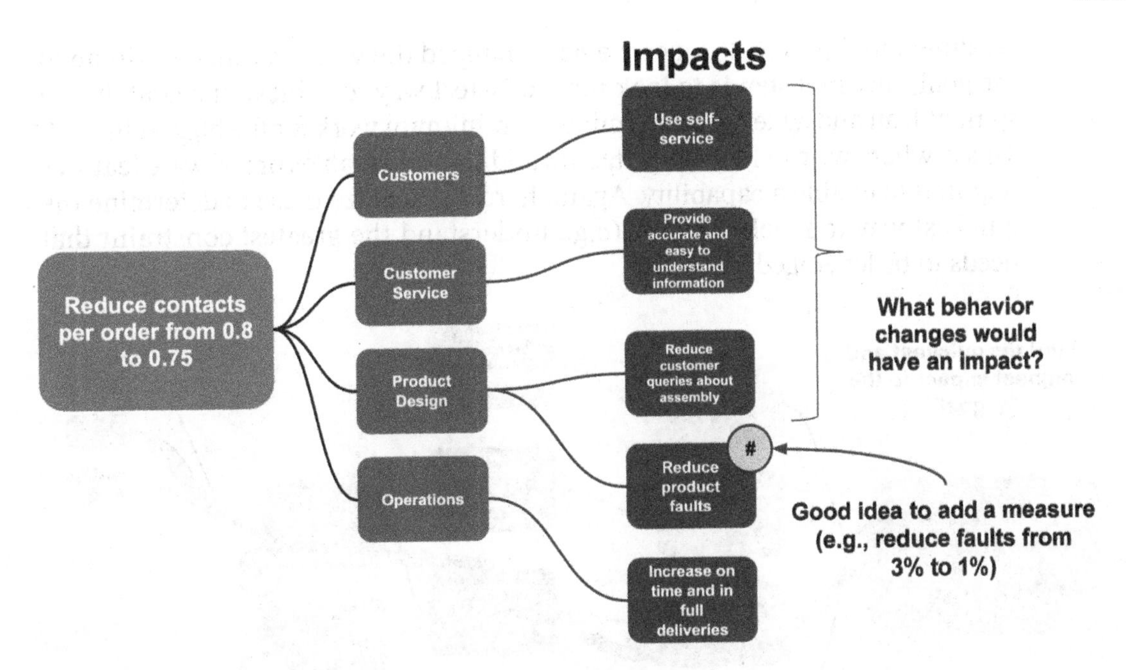

Figure 6.24: The impacts actors can make to influence a goal

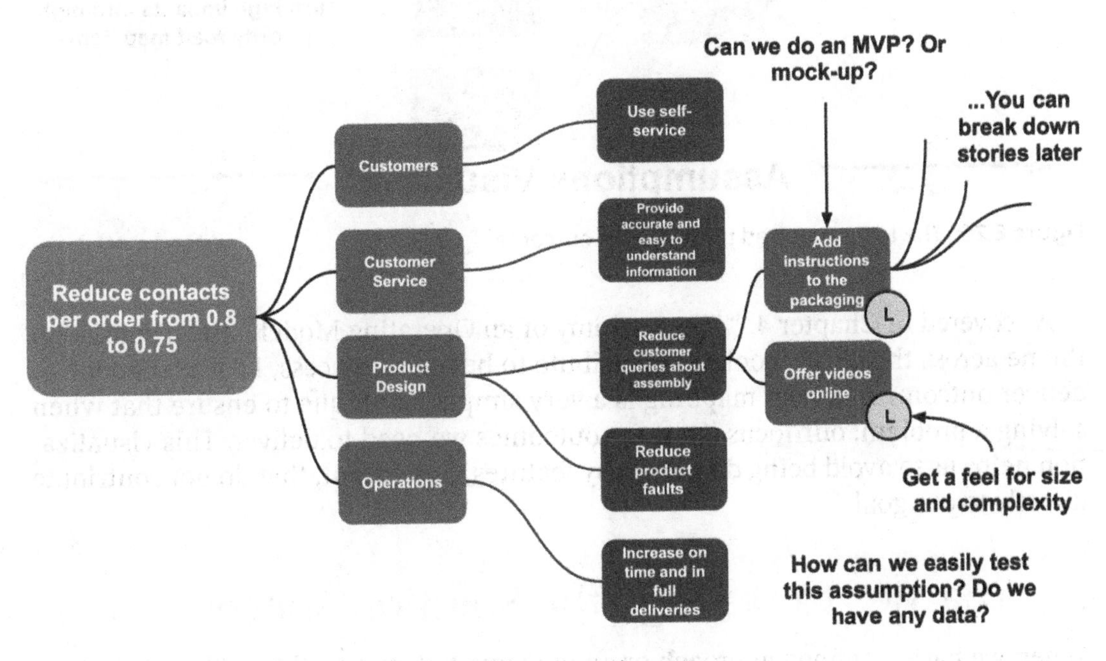

Figure 6.25: The outcomes required to achieve actor impacts

5. As shown in Figure 6.26, once we have mapped the ways we can contribute to the goal, the final step is to look for the fastest way to achieve the goal, in the spirit of lean and agile, and determine the minimum work for the biggest impact. This is where we can look at the tactics and the deliverables or software features required to enable a capability. Again, here we can size these to determine the quickest way to reach the goal (e.g., understand the greatest constraint that needs to be leveraged).

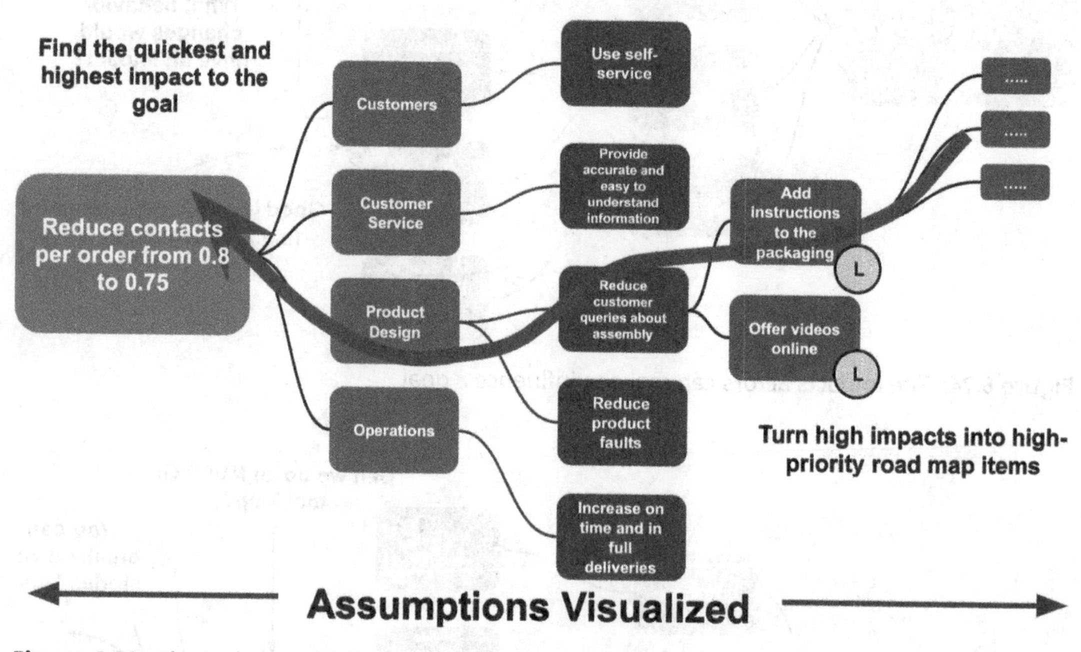

Figure 6.26: The tasks required to deliver an outcome

As covered in Chapter 4, "The Anatomy of an Operating Model," and a consistent theme across the entire book, to contribute to business success, IT leaders need to deliver outcomes. Impact mapping is a very simple technique to ensure that when solving a problem, our focus is on the outcomes we need to deliver. This visualization helps us to avoid being distracted by features, our output, that do not contribute directly to our goal.

Approaches to Manage the Solution Space

When we have a defined approach or an outcome to focus on that will contribute to removing a problem or achieving a goal, we can turn to the solution space. As shown in Figure 6.27, there are three main components that need to be considered when working in the solution space:

Solution Concept Before we begin to think about delivering a solution, there is a need to ensure that it has been through an appropriate level of governance. For example, we need to ensure that it aligns with the enterprise architecture target vision (e.g., build versus buy, which will be explored in detail in Chapter 12). The other forms of governance, such as portfolio planning, alignment to business strategy, prioritization, and funding, will be covered in detail in Chapter 7, "How We Govern."

Solution Delivery Life Cycle Once the solution has been approved, we need to select the most appropriate development project methodology to manage the delivery based on the problem domain of the solution.

Solution Delivery Delivering the solution consists of three components:

Delivery Management: How we will manage creating the solution.

Development Methodologies: How we will manage the software.

Operations Methodologies: How operations and delivery teams can work together to reduce hand-offs.

Just as in the case of the problem space and the problem-solving methodologies and discovery tools, there are many options for each component of the solution space. Each has its own strengths and weaknesses based on the various technical and social contexts that will exist in your organization.

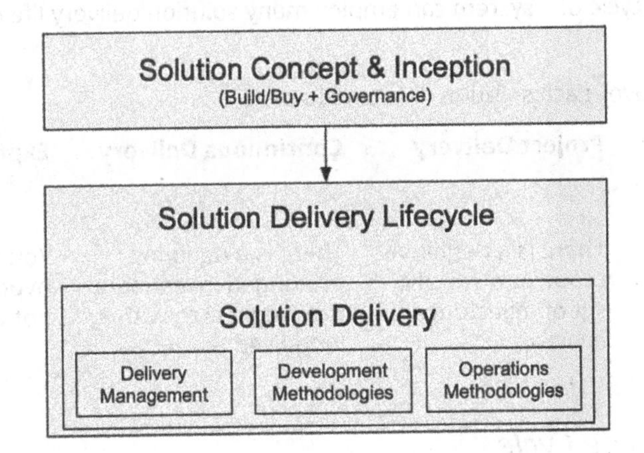

Figure 6.27: The anatomy of the solution space

Solution Delivery Life Cycles

The way we approach delivering a solution to solve a problem or reach a goal is varied. Disciplined Agile Delivery (DAD), the software development portion of the Disciplined Agile Toolkit, calls the different approaches solution delivery life cycles. Different contexts will require different solution delivery cycles, or in other words, different ways of working. The solution life cycles we will take a detailed look at are project delivery (creating something to fill a scope), continuous delivery (removing all the waste and improving performance), and exploratory (exploring the feasibility of something). As shown in Figure 6.28, a solution delivery life cycle exists within a system life cycle. A system life cycle spans the initial concept for a technical solution to a product or service through construction to its eventual retirement. A system will evolve many times before it retires and will likely use many different solution delivery life cycles. In the example highlighted in Figure 6.28, a system started as a hypothesis that needed to be tested for feasibility (exploratory life cycle). Once this was proven, the system was built with a fixed scope for an initial launch (project delivery life cycle). Then, over time further enhancements are made (continuous delivery life cycle). Table 6.2 gives an indication of when each solution delivery life cycle approach could be considered.

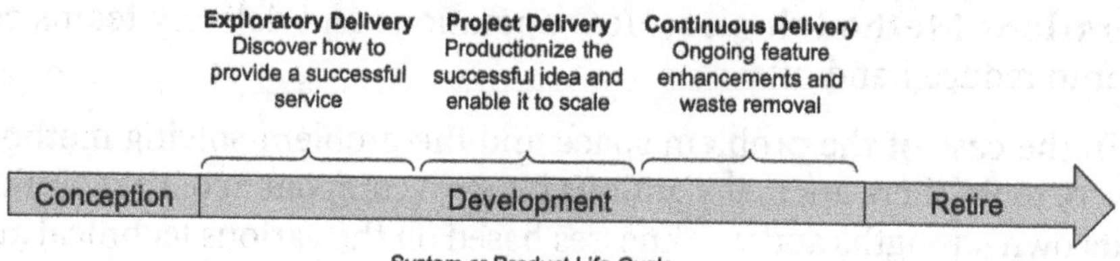

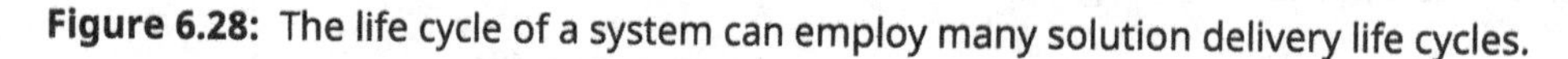

Figure 6.28: The life cycle of a system can employ many solution delivery life cycles.

Table 6.2: When to favor each solution delivery life cycle

Solution Delivery Life Cycle	Project Delivery	Continuous Delivery	Exploration
Consider when. . .	There is a definitive scope or a specific set of milestones to achieve.	There is a need for prolonged investment in a single area with a long-lived team.	You need to validate a hypothesis in an area not well known.

Project Delivery Life Cycle

The project delivery life cycle, as shown in Figure 6.29, is the sequence of phases that a project goes through from initiation to planning and execution to closure. The project delivery life cycle is useful when you have clear milestones, scope, or a defined path. For example, integrating a warehouse management system or a payment gateway is a good example when it is appropriate to adopt this life cycle. Once the system is in place, the project ceases to exist.

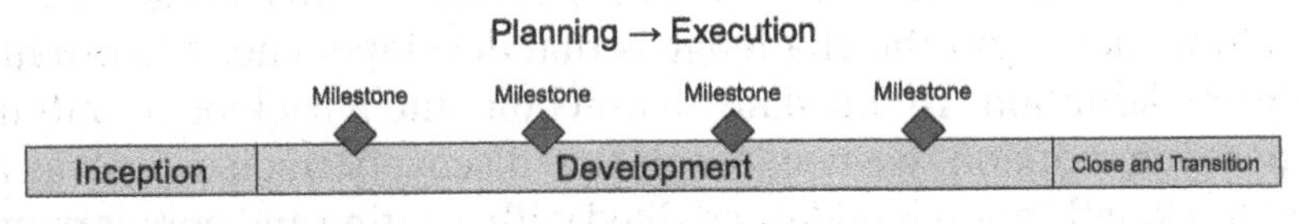

Figure 6.29: The project life cycles

Prince 2 is an implementation of the project delivery life cycle that provides a structured approach for project delivery. As shown in Figure 6.30, a project can be delivered in an iterative manner as well as the traditional liner approach. SAFe is another project framework structure, although focused on much larger projects where there is a need to coordinate many teams.

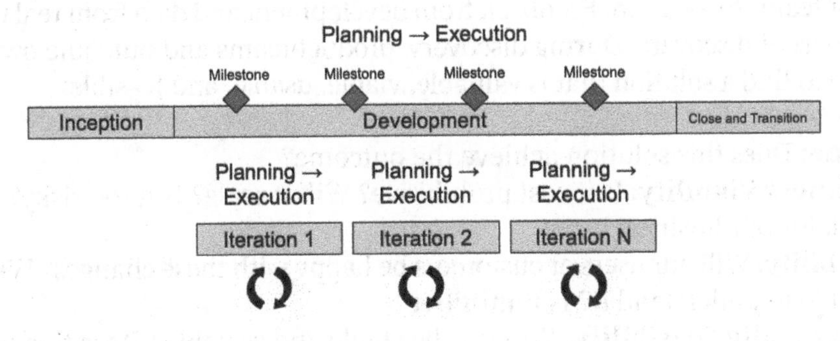

Figure 6.30: Prince 2 and iterative delivery

Continuous Delivery Life Cycle

Where the project delivery life cycle has a definitive end, the continuous delivery life cycle, as shown in Figure 6.31, represents a continuous stream of development that ends only when investment is withdrawn or a goal is achieved. It is suitable for teams focused on a problem or product for a long period of time. For example, a team could be tasked with improving conversion of customer checkout or sign-ups.

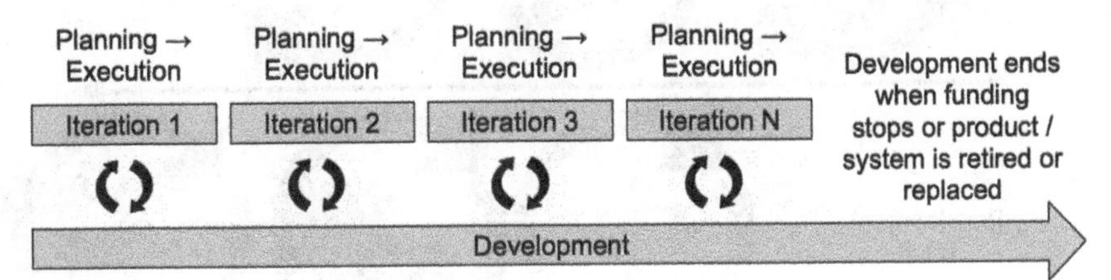

Figure 6.31: The continuous delivery life cycle

Dual track development is an implementation of the continuous delivery life cycle. The goal of the product teams is to deliver an outcome, but as shown in Figure 6.32, to achieve this the team will identify sub-outcomes. Product teams will generate a backlog of outcomes to work on, but the challenge is that developers need to produce output based on feature definition. To translate an outcome into a backlog of output features that developers can work on, we need a solution discovery track as well as a solution development track. Jeff Patton is widely credited with creating and popularizing the concept of dual track development. As shown in Figure 6.32, product teams must engage in two separate parallel streams of work, one focused on outcome solution discovery and the second on development. The discovery track is about determining the most appropriate approach to achieve an outcome by quickly identifying potential solutions and validating them. The development track is about delivering those solutions for customers. The discovery and delivery tracks run in parallel, usually with discovery feeding into the next development iteration so that there is a well-thought-through backlog of features for teams to focus on. Feedback from development and data from real users will also form part of discovery. During discovery, product teams and outcome owners will collaborate to find a solution that is valuable, viable, usable, and feasible:

- **Value:** Does this solution achieve the outcome?
- **Business Viability:** Is it cost prohibitive? Will it scale? Is it too risky? Will this work for our business?
- **Usability:** Will our users or customers be happy with these changes? Will people be able to understand it? Is it intuitive?
- **Technically Feasibility:** Will this be stable and scalable? Does this introduce risk? Do we have the right skills and technology? How long will this take?

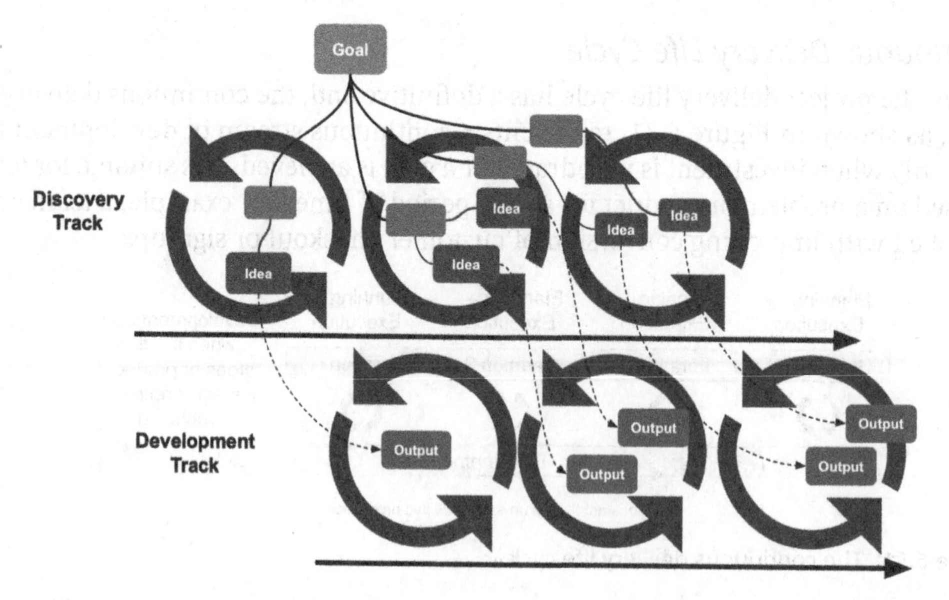

Figure 6.32: Dual track development.

Exploratory Life Cycle

The aim of the exploratory life cycle is to minimize upfront investments in solutions to solving wicked problems where there is a high level of uncertainty. An exploratory life cycle favors making small experiments to obtain fast feedback and learning to validate and adapt solutions to ensure that teams have a viable product or service before making any further investments. It is useful when addressing complex problem domains where there are unknown unknowns, such as changing customer behavior or launching a new product or service, and where there is an allowance for experimentation, flexibility, and evolution.

An example of the exploratory life cycle is the Lean startup. *The Lean Startup* Crown Publishing Group (2011) is a book by Eric Ries that introduces a new methodology for developing product ideas that has a strong emphasis on validated learning achieved through fast build-measure-learn cycles as shown in Figure 6.33. Lean startup solves complex problems and tests solutions using a scientific method. It is an experimentation-based approach, and one that seeks to test a hypothesis and gain real customer feedback before committing to build a feature, service or product. The *lean* in the name comes from the focus on eliminating wasteful activities such as trying to design the perfect product and creating detailed planning in favor of focusing on generating a viable product, service, or solution. The result is a reduction of unpredictability, which is a characteristic of a complex problem domain.

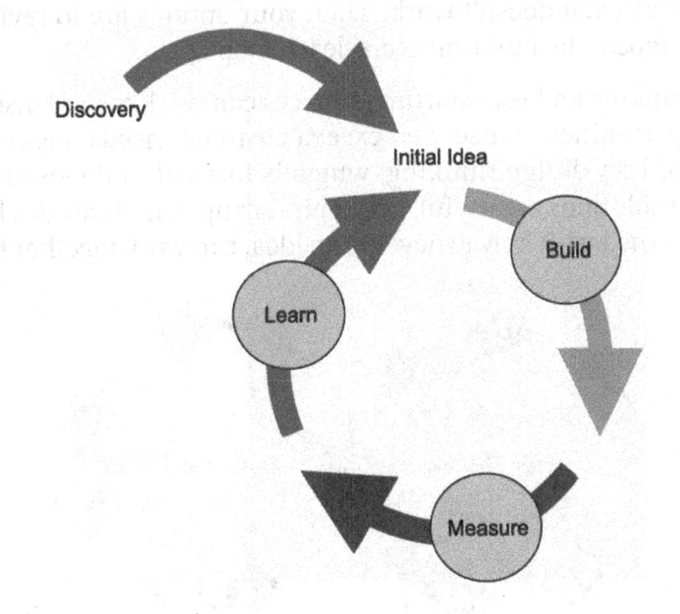

Figure 6.33: The lean start-up cycle

The lean start up cycle is based on the following 3 steps:

1. Build

   The goal of the build phase is to test your hypothesis or idea by delivering the smallest possible product that meets the user's need, which is known as a minimum viable product (MVP). The MVP allows a minimum marketable product (MMP) to be launched to gain real-world feedback of the validity of the idea. The point of an MVP is to determine the viability of a business idea. It needs to function well enough to be useful to early adoption customers.
2. Measure

   By launching an MVP and MMP, you can begin to gain real-world customer feedback and data on the effectiveness of the product with minimum investment and time. Measuring the customer impact is integral to the process and ensures that you don't invest time in features that are not valuable to end users.
3. Learn

   After analyzing the data from the MVP and comparing the evidence to your original hypothesis, you will have one of two ways forward. If your hypothesis was proved right and you have validated your original assumptions, you can continue the next iteration of build-measure-learn to ensure you are still validating your ideas after each incremental build. If your original idea has been proven to be unsuccessful, you can go back to the drawing board with valuable experience of what doesn't work. Then your options are to revisit your hypothesis and repeat the build-measure-learn loop again.

Both design thinking and lean startup practices can work toward discovering value and opening opportunities to meet and exceed customer needs. Figure 6.34 shows a simplistic view of how design thinking, which is focused in the problem space and discovering what solutions are useful, and lean startup, which resides in the solution space and is concerned with how to develop an idea, can work together to create value.

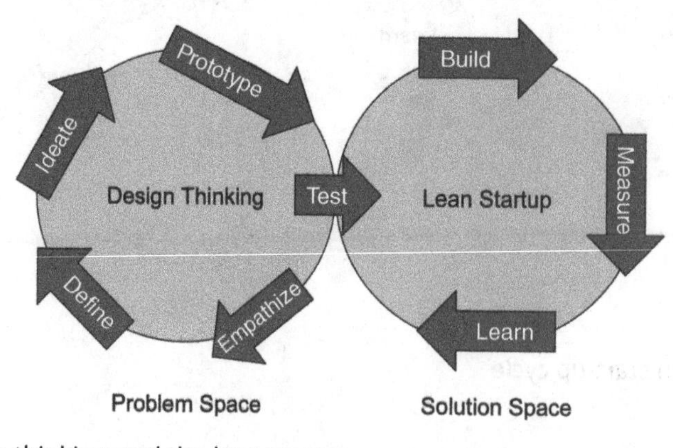

Figure 6.34: Design thinking and the lean startup

The hardest part of this loop is the build part. To quickly ascertain if a new product or feature will be successful without wasting time and resources, we need to think in a fast, inexpensive, restrained, and elegant (FIRE, formerly FIST) manner. A build for an experiment calls for the development of a minimal or basic product for testing at first. A minimum viable product (MVP) is a product, feature, or service that is introduced to customers in its early stages of development to ensure that it is a viable offering. The aim is to prevent ourselves from falling into the build trap and spending all of our time and resources developing a fully featured product before we have even proved that it has a high chance of success. The following are some methods that can simulate a product, feature, or service without fully building it to validate assumptions before committing to developing a product fully:

- **Concierge testing** is a qualitative interview-based method for testing the value of a product or feature with potential customers in their native environment. By asking customers about a proposed service, or even providing it manually, we can gauge the amount of interest a customer would have in using it. For example, before Uber Eats or Just Eats was a concept, you could conduct interviews with customers in local takeaways on the value of ordering food for delivery on an app from the comfort of their own home. You could even provide a basic service with limited restaurants and no technology that would simulate the real process to gain valuable feedback. Concierge testing gives an early indication on the value customers see in your product or service without the need to spend time and resources building a fully fledged service.
- A **usability study** focuses on a customer's ability to achieve a goal or complete a task while using a prototype of your product. The prototype could be a series of screen mocks or paper sketches or even a mocked-up interactive website or app. It won't work, but it will give a sense of whether a customer will find the provided functionality useful or valuable. Again, this is qualitative feedback that teams can use to discover opportunities to improve the existing experience.
- **Wizard of Oz tests** give a customer the impression of a fully functional product or service by faking certain interactions or processes behind the scenes to gauge demand or test the value of your product or feature. For example, before investing heavily in an automated booking system for a complex product online, we could instead build a product that appears to book everything online but has a human manually completing the booking. If the booking system is a success, only then do we invest in automation. While the test is manual and so will not easily scale, it can give quanatitive data and analytics around customers who believe they are using a real product.
- **Fake door testing** is about quickly and cheaply validating a feature by asking customers if they would use it. For example, if we wanted to offer a gift-wrapping service for our products, we could add a button or check box that allows customers

to select it. However, the service would not actually exist. If selected, we would advise that this feature is perhaps "coming soon." By measuring how many of the potential customers selected the option, we can make more informed decisions on the next action to take. Again, this is a low-effort and quantitative method of gaining feedback.

- **A/B testing** refers to an experimentation process that shows one of two or more versions of a web page or element of a page to different segments of visitors to determine which version has the greatest business impact. Visitors in group A, the "control" group, are shown the original page element. Group B visitors, the "variation" group, are shown the new page or element. This is particularly useful for website conversion optimization or to ensure that the introduction of a new feature has a higher impact than the incumbent before fully rolling it out.

Delivery Management

When it comes to managing the delivery of a solution, we have many options. I will cover three of the most popular delivery management approaches:

- **Waterfall / Big Design Upfront (BDUF)** is a linear approach to project delivery management with a focus on detailed planning. This traditional delivery management works well where we are certain and clear on the outcome that we are after, and there is a high degree of certainty on how we will achieve it. It is also essential when dealing with many teams or where there is a critical need to get something right the first time.
- **Scrum** is focused on short iterative development cycles known as sprints; this allows teams to perform just enough planning to enable them to quickly adapt based on new information. This is useful for complex projects where new information requires plans to be adapted.
- **Kanban** is an extremely lightweight and flexible approach to delivery management that is focused on a continuous flow and limiting work in progress. It is useful when it is hard to plan due to frequent changes.

Each of these approaches is useful in a particular context, Figure 6.35 gives an indication where one should be considered over another. As with all the frameworks and methodologies in this chapter, there is no wrong or right approach, only one that produces results.

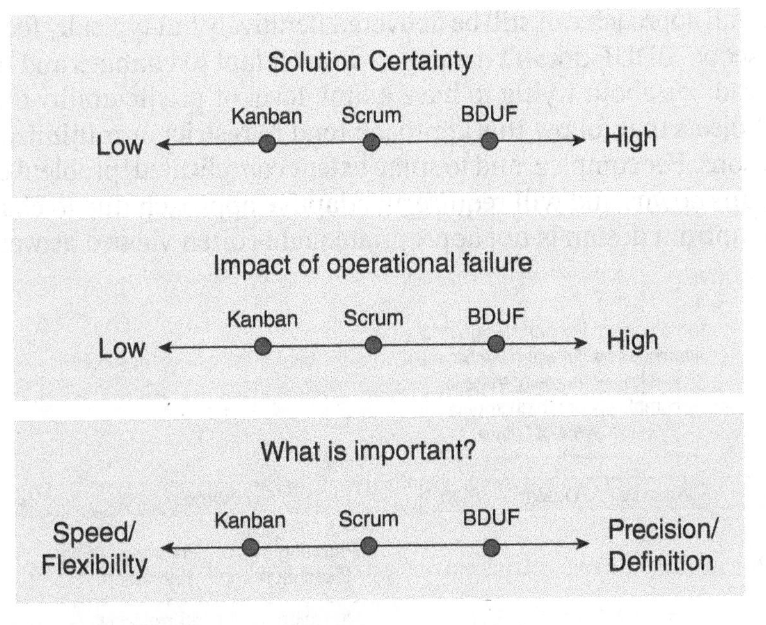

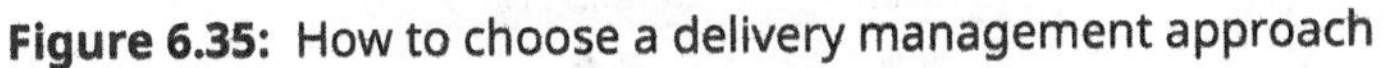
Figure 6.35: How to choose a delivery management approach

Waterfall / BDUF

The objectives of a big design upfront (BDUF), or waterfall, approach is driven by the need for predictability and dependability, hence why there is a need to spend a significant amount of effort on the initial requirements and design phases. As a result, delivery is executed in a linear manner, as shown in Figure 6.36.

A BDUF approach is an appropriate choice in the following scenarios:

- Requirements are largely determined in advance and remain relatively stable, i.e., where there are few unknown unknowns.
- Where there are dependencies on workstreams outside of the technical domain.
- Where there is a large-scale project requiring the coordination of many teams. However, this often doesn't scale well to larger groups of teams that can benefit from designs, plans, and documentation to aid communication and coordination. Hence the need to keep teams in a complex problem domain smaller and distill teams as they grow.
- There is a significant level of criticality to get it right the first time or hit a date, perhaps for software for domains such as aerospace, military, medical or nuclear.

The waterfall approach can still be delivered iteratively but typically focused on milestones and scope. BDUF doesn't mean you won't adapt to changes and blindly follow a plan. Instead, it's about trying to have a high level of predictability on the delivery schedule. Projects that follow this approach tend to restrict or minimize any changes for good reasons. For complex, and to some extent complicated, problem domains, that have many unknowns and will require an adaptive approach due to changing needs, this level of upfront design is not appropriate and is often viewed as wasted effort.

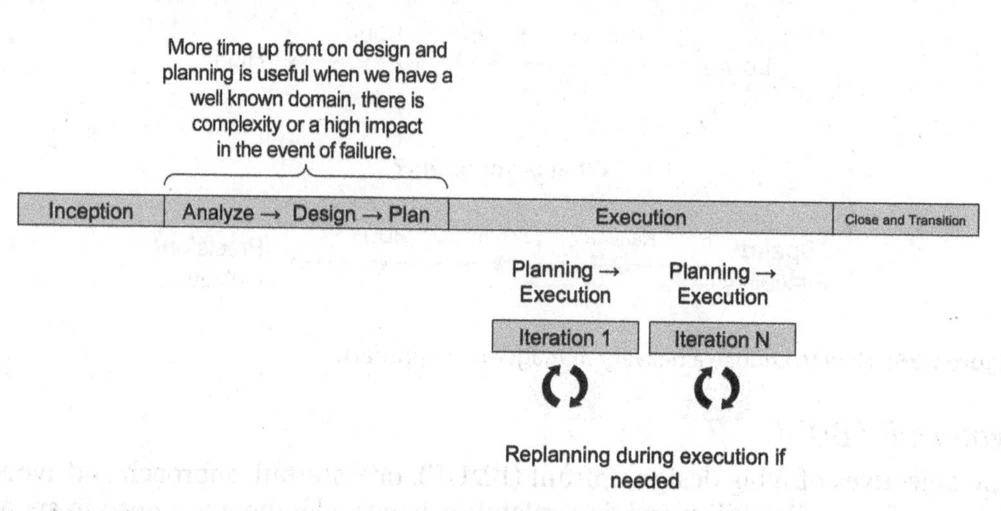

Figure 6.36: The waterfall, or BDUF, approach to delivery management

Scrum

Scrum is a lightweight agile project management framework ideal for complex and complicated problem domains. It is designed for teams of 10 or fewer members, based on the same reasons as covered on team size in Chapter 5, "How We Are Organized." The following is a high-level overview of the process of Scrum, as shown in Figure 6.37:

- **Product Backlog:** The product backlog is a list of features that will contribute to release. This list is maintained based on research and feedback from the development team and customers. The process of backlog grooming is to prioritize the list of items so teams always have a well-thought-through list of important features to pull from. It shouldn't ever be a wish list of items.
- **Sprint Planning:** The work that the team will commit to delivering is determined during sprint planning and forms what is known as the ***sprint backlog***. The team agrees on a goal for the sprint and comes to a consensus on how much work they can commit to based on the complexity of the features and the team's capacity.

- **Sprint:** A sprint is the period of time where the scrum team works to complete the sprint backlog. This can be anywhere between one and four weeks in length depending on how complex the work is and how fast teams want feedback. During this period, the team works through the committed sprint backlog. However, new items can come in if they are deemed more critical, but they will replace other work. The idea is to keep this to a minimum to avoid context switching.
- **Daily Stand-Up:** This is a daily, typically morning, quick meeting. The goal of the daily stand-up is to ensure alignment for everyone on the team and to provide an opportunity for any problems or concerns people have with the sprint goal.
- **Sprint Review:** At the end of the sprint, the team showcase their work to customers for feedback and ideally release something into production. The insight based on feedback will be used to prioritize the product backlog and potentially focus development and discovery in new areas.
- **Sprint Retrospective:** Where the sprint review is a meeting for teams and customers, the retrospective is internally focused, and the topic is continuous improvement. The team discusses what worked and what didn't work in the last iteration, be that the work or the current ways of working. The meeting is about continuous improvement so is less about blaming and more about improving ways of working in the future.

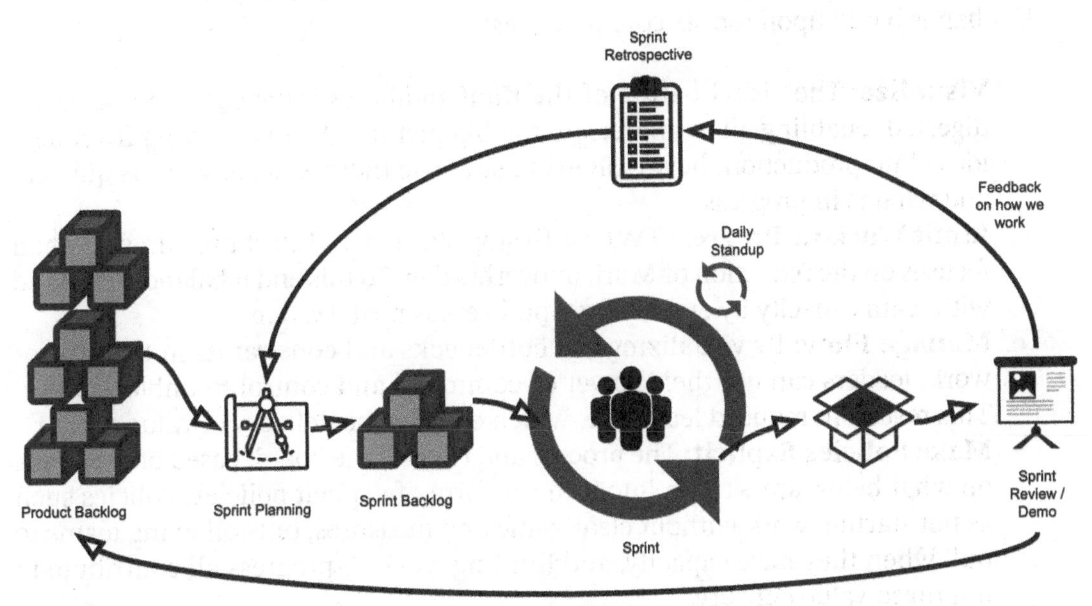

Figure 6.37: The scrum approach to delivery management

The key factor to learning is the constant delivery of working software. Finishing small features and putting them into production is the only way you will validate or disprove assumptions and hypotheses. Therefore, to adapt quickly from learning, teams must focus on frequent and iterative delivery; that is, to deliver working software every two, three, or four weeks. By delivering something complete and of value in short cycles, we can easily adapt and change course. Design and planning is an ongoing process, not a stage. To get feedback to enable redesign and replanning, you must finish things. The best way to finish things is to keep them small. The best way to keep things small is to work in small iterations that produce something of value. This enables teams to adapt faster.

Kanban

Kanban is a simple but effective framework that was developed in 1940s Japan by Taiichi Ohno. It is based around a manufacturing efficiency practice called Just-in-Time (JIT) designed to control the flow of inventory. The goal is to limit the buildup of excess inventory and thus avoid waste at any point in production. The Kanban methodology has been adapted to project management, replacing inventory with tasks, and matching the flow of work to the capacity of a team. The Kanban method revolves around the Kanban board, as shown in Figure 6.38, which visualizes the flow using cards to track work, clearly highlighting blockers and issues that prevent flow.

Kanban is based upon the six core practices:

- **Visualize:** The visual nature of the Kanban allows information to be easily digested, enabling all to appreciate the big picture of work flowing from new ideas into production, how it aligns to strategic themes, what work is queued, and what is in progress.
- **Limit Work-in-Progress (WIP):** To stop starting and start finishing, Kanban focuses on the reduction of Muri, or overburden. To this end it balances demand with team capacity by employing a pull versus push system.
- **Manage Flow:** By visualizing the bottlenecks and constraints in the flow of work, leaders can use their power of command and control to unblock work. This results in reduced lead time, which enables the delivery of value sooner.
- **Make Policies Explicit:** The process and rules of the board, based on feedback on what helps, are written into it in the form of explicit policies. Policies such as not starting work without clear value and measures, only allowing teams to pull when they have capacity, and limiting work-in-progress all contribute to improved value delivery.
- **Implement Feedback Loops:** A regular cadence of feedback provides visibility into how well investments are progressing toward desired outcomes. This review enables organizations to respond to internal and external feedback and make decisions on whether to preserve inflight work or pivot to new opportunities.

- **Improve Collaboratively:** The visual nature of a Kanban board makes it an effective communications device, which can facilitate and focus discussions around attacking the work, not the person.

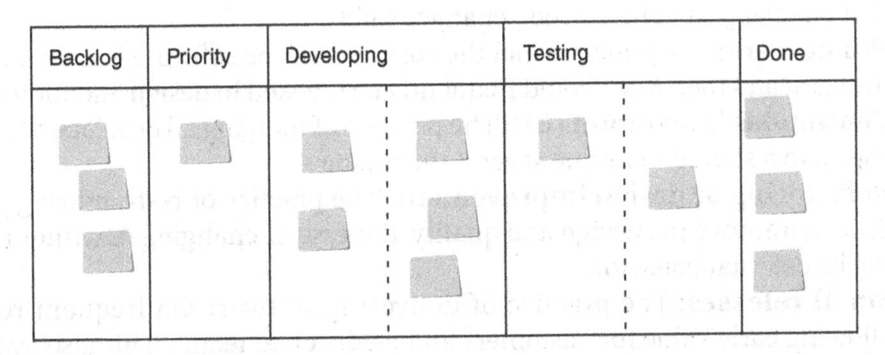

Figure 6.38: A Kanban board

Development Methodologies

The purpose of a software development methodology is to improve the quality and agility of software. While there is an overlap with some delivery management approaches, the software development methodologies focus on software construction as well as management practices. In the following sections, we will look at Extreme programming (XP) and lean principles in software development.

Extreme Programming (XP)

Extreme programming (XP) is a software development methodology intended to improve software quality and adaptiveness to emerging customer requirements. It shares several values with lean around human-centered design. It strongly recommends short development cycles and frequent releases to adapt to feedback and improve working practices. This is why it is so often paired with scrum. The core of XP is the interconnected set of software development practices:

- **Pair programming:** A technique in which two programmers work together at one workstation to simultaneously focus on both the tactical and strategic nature of the code.
- **Planning game:** As XP is about just-in-time planning, there is a need for continuous planning to guide the product into delivery. Release planning is focused on determining what features are included in which near-term releases. Iteration planning focuses on the tasks of the developers.

- **Test-driven development (TDD):** TDD is a development technique based on the principle of first writing a test that fails before writing any new features or code. The benefits of TDD force you to work in small iterations, tackling one requirement at a time on an as-needed basis and leaving you with a test suite that enables you to make code changes safely.
- **Whole team:** The principle that the customer of the software is an extension of the team and therefore should frequently be invoked in design and for feedback.
- **Continuous integration (CI):** The practice of merging all developers' working copies to a shared mainline several times a day.
- **Refactoring or design improvement:** The practice of restructuring existing code to improve the design and quality, the ease of changing it, without changing its external behavior.
- **Small releases:** The practice of delivering software via frequent releases, allowing early value for customers and feedback to teams. This also avoids big bang releases.
- **Coding standards:** The set of rules that the entire development team agrees to adhere to throughout the project, ensuring code quality and self-describing code.
- **Collective code ownership:** The principle that states everyone is responsible for all the code and that everybody is allowed to change any part of the code.
- **Simple design:** The practice of refactoring code to keep it simple to understand and change without making it simplistic.
- **System metaphor:** The practice of ensuring that the naming of each class or method is self-descriptive, making it easy for the entire team to understand the code base.
- **Sustainable pace:** The practice of teams working at a steady pace and not burning the candle at both ends, which can lead to burnout, mistakes, and quality problems.

Lean Principles in Software Development

Mary and Tom Poppendieck, in their book *Lean Software Development: An Agile Toolkit* (Addison-Wesley Professional, 2003), identified seven core principles from the lean manufacturing philosophy and translated them for the practice of software development. These principles are eliminate waste, build quality in, create knowledge, defer commitment, deliver fast, respect people, and optimize the whole.

Principle 1: Eliminate Waste A key tenet of lean is to eliminate waste to maximize customer value. Mary and Tom Poppendieck took the seven wastes defined in lean manufacturing and applied them to software development:

- Avoid writing unnecessary code or feature bloat as this increases lead time and delays feedback loops.

- Limit work in progress, stop starting, and start finishing work to reduce complexity and context switching.
- Reduce delays in the software development process—cycle time and wait time—as this increases lead time and delays feedback.
- Clarify needs and vision with customers instead of asking for requirements. This will reduce rework, constantly changing requirements, and increase motivation.
- Reduce the need for overly bureaucratic and complex governance models that introduce delays.
- Improve communication channels for fast feedback and decision-making to avoid delays and frustrations.
- Avoid partially done work. Only work that is complete and in production provides value for the customer and real experience for the team to learn from.
- Reduce defects and quality issues that require rework and lead to delays and poor customer satisfaction.
- Avoid task and context switching, which result in poor work quality, delays, and low motivation.

Principle 2: Build Quality In Avoid the need for reworking by making it everyone's responsibility to build quality in from the start. These are examples of some of the ways teams can build in quality:

- Pair programming and peer reviews to improve quality and understanding.
- Test-driven and behavior-driven development to ensure focus on the outcome versus output and to enable safe refactoring (the improvement of readability and flexibility without altering behavior) of code.
- Automated regression and smoke testing.
- Incremental development.
- Regular customer feedback and show and tells.
- Automated continuous integration and deployment.
- Reduce context switching, delays for decisions, and knowledge gaps.
- Ensure teams are aligned on the goals of work in process.

Principle 3: Create Knowledge Software development is a design process where continuous learning is fundamental to the process. Incorporating the following techniques will help to retain valuable knowledge and amplify learning.

- Knowledge sharing sessions
- Documentation
- Code / Peer reviews
- Training
- Pair programming

Principle 4: Defer Commitment Defer commitment and decision-making to the last responsible moment to keep options open while you learn more about a problem space. This avoids teams falling into the trap of creating wasteful detailed upfront plans and rigidly following them before you have a full understanding of the problem and how you will be able to solve it.

Principle 5: Deliver Fast Deliver at pace by limiting work in progress and focusing on smaller batch sizes. Limit features to the minimum that can deliver value for a customer to get useful feedback as quickly as possible. Avoid overengineering solutions before you know that a solution is valid; often "good" is good enough. Amplify blockers and impediments to leadership quickly and urgently to avoid delays. Delivering faster isn't about working faster; it is about working smarter. It's all about eliminating waste and maximizing the work not done.

Principle 6: Respect People At the center of software design are people. It doesn't matter what framework you use, organizational design you employ, or software methodology you follow, it all boils down to how motivated and passionate the people are that are doing the job. Respect for people is a common theme running through every aspect of lean. Respect for new ideas, for challenges to old ideas, for new members of the team and people outside the team is critical. A positive mental attitude and the ability to get on and collaborate is fundamentally essential to succeed.

Principle 7: Optimize the Whole Like lean production, we need to focus on the end-to-end value stream, which is the sequence of events that move an idea from code into production that meets the customer's real need. The goal is to deliver value to customers at pace, to achieve what we need to optimize the value stream and eliminate waste. One of the most effective ways of doing this is by keeping the end-to-end value stream as the responsibility of a single team that is empowered and accountable for its delivery. Organizing a team around an outcome and giving them the trust and autonomy to deliver is extremely powerful and intrinsically motivating.

Operations Methodologies: Devops

While development is more akin to a design process than a production process, there are still opportunities to reduce waste and optimize the development value stream. Devops is a philosophy that focuses on the reduction of lead time between software development (dev) and IT operations (ops) processes by eliminating any non-value add work from the IT value chain. Devops is a combination of cultural change and practices focused on moving from siloed working and over the fence throwing of requirements, which has historically taken place between development and operations departments, to a systems view of delivery improved by fast feedback and close collaboration.

Devops is based on the Three Ways, as laid out in both The phoenix project a novel about it devops and helping your business win (IT Revolution Press 2015) and the The DevOps Handbook: How to Create World-Class Agility, Reliability, & Security in Technology Organizations (IT Revolution 2021). These principles underline all the practices of the DevOps way of working.

1. **The principles of flow:** A focus on the lead time of the delivery of work through the IT value chain to the end customer.
2. **The principles of feedback:** A focus on fast and amplified feedback loops to ensure constant improvement to flow.
3. **The principles of continual experimentation and learning:** A focus on a high trust working culture and ongoing organizational improvement to improve the flow of work.

The first way is to take a systems approach to the end-to-end delivery of work between IT and operations. Optimization should focus on the flow of work through the IT value stream as opposed to just optimization at a local level—in other words, it doesn't matter how good a process the development team has if work is blocked from getting into production by poor operational processes. Unlike in manufacturing, where you can easily see the buildup of inventory, it can be less easy to see the accumulation of knowledge work; therefore, it is essential to visualize all work in the system in order to understand constraint and measure throughpout. As well as visualization, another way to improve flow is to reduce the work in progress (WIP) and stop IT teams from having to multitask in other ways to start to finish work before picking any new work up. Technical practices can help to automate the pipeline where possible to reduce lead time. The fundamental goal of the first way, as shown in Figure 6.39, is to identify wasted time and eliminate it to improve flow of work throughout the system.

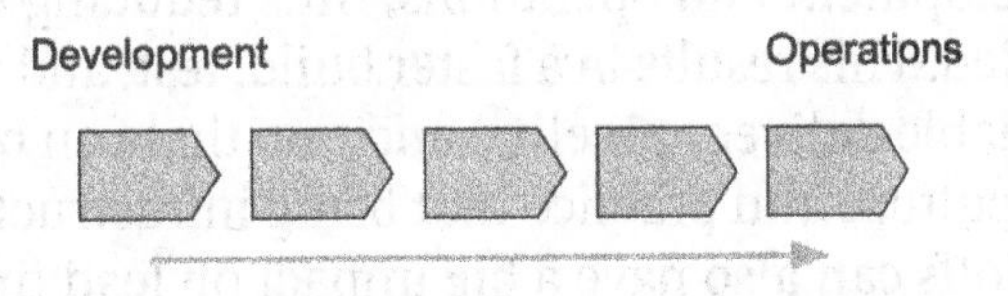

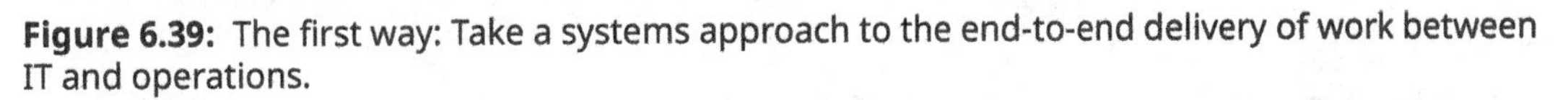

Figure 6.39: The first way: Take a systems approach to the end-to-end delivery of work between IT and operations.

The second way, as shown in Figure 6.40, is about ensuring that there is constant feedback to optimize the value stream. Amplifying information and providing fast feedback on what is not working as well as what is working is vital to improving the flow of delivery to the customer. Information on blockers, constraints, or wasted time in the flow as well as feedback from the customer on the delivered product all helps to improve value delivery.

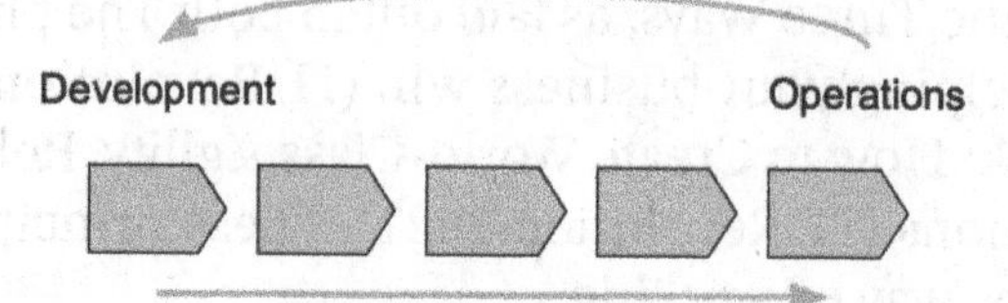

Figure 6.40: The second way: Ensure that there is constant feedback to optimize the value stream.

The third way, as shown in Figure 6.41, is about creating an environment and culture that is always pushing to improve. We need to allow experimentation, including taking risks with the safety to fail, and instill a culture of that mastery that requires continuous and deliberate improvement.

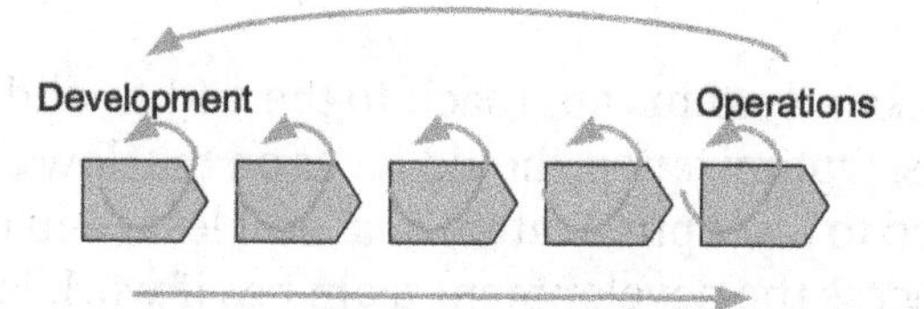

Figure 6.41: The third way: create an environment and culture that is always pushing to improve.

Two examples of DevOps tactics are continuous integration (the practice of automating the integration of code changes) and continuous deployments (the practice of automating deployments), as illustrated in Figure 6.42. By applying continuous deployments and continuous integration, we can move quality to the left; that is, ensure we don't pass on a defect to the next stage and automate repetitive manual processes and handoffs between development and operations, thus reducing the amount of rework and manual work needed. This results in a faster build, test, and release process as well as ensuring a more reliable delivery pipeline and one that can be rolled back easily in the event of failure. A culture and practice that bring infrastructure and development together to avoid handoffs can also have a big impact on lead time and quality issues.

Summary

The Ways of Working component of the operating model is about how and when we adopt the frameworks and methodologies built upon the philosophies of Chapter 2 to guide how we operate. We initially looked at the various frameworks that covered all the capabilities of an IT department from security to project management. However, a large majority of your time will be focused on contributing to both strategic and operational essential business needs.

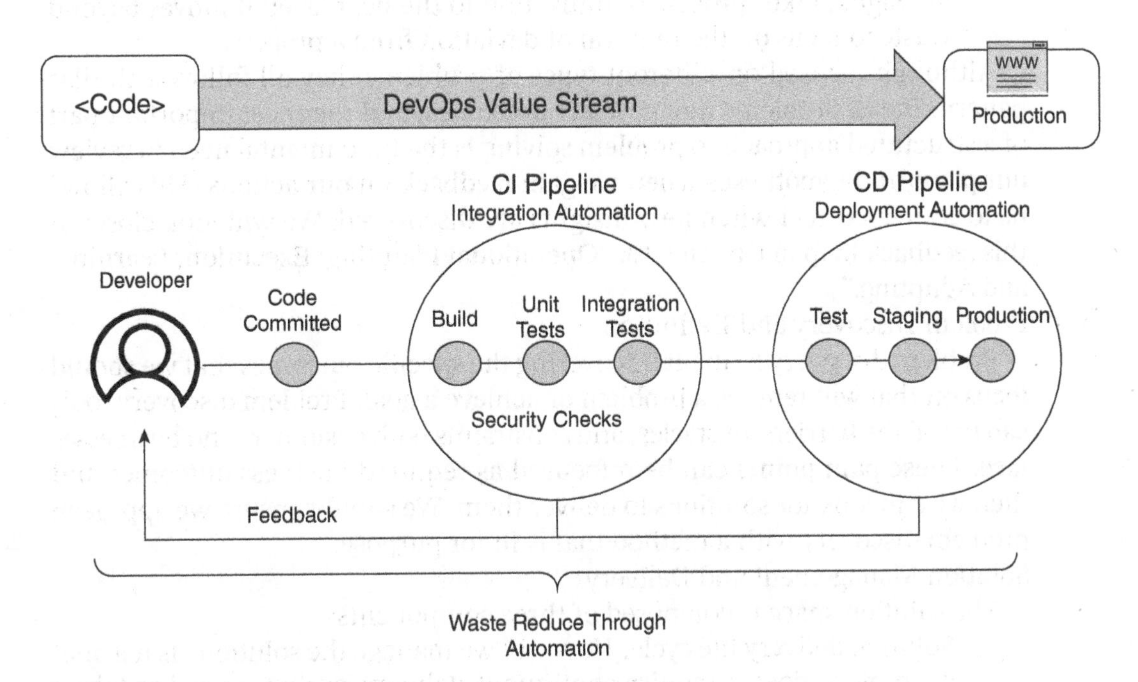

Figure 6.42: CI and CD pipelines

A modern IT department, one that understands the art of the possible, needs to become an active contributor to business success. To accomplish this there is a need for a shift in our ways of working to deliver outcomes that solve real customer problems and that will directly contribute to realizing goals rather than producing output and delivering to an agreed scope or plan. To achieve this, we need to understand how we solve problems.

Problem solving is split into three parts: the problem approach, distilling the problem space, and managing the solution space. Each of these parts requires a context-specific approach.

- Problem-Solving Methodologies.

  To help structure our thinking and guide our problem-solving efforts, we need to employ an appropriate problem-solving methodology. We looked at three in this chapter:

  - Design thinking and the Double Diamond framework is a useful approach when you are exploring a new product or service.

- Lean and PDCA are lightweight frameworks that focuses on the removal of waste and inefficiencies in existing products and services.
- Six Sigma takes process optimization to the next level; it moves beyond waste to focus on the removal of deviation from a process.

Although focused on different types of problems they all follow a similar pattern. Break problems down, select outcomes, and the most important part of a structured approach to problem solving is the fundamental need to review our plans and hypotheses when we gain feedback on our actions. This allows us to course-correct when new insights are discovered. We will look closer at this feedback loop in Chapter 13, "Operational Planning: Execution, Learning, and Adapting."

- Problem Discovery and Definition

  Problem discovery is about uncovering the specific outcomes that we should focus on that will remove a problem or achieve a goal. Problem discovery tools can reveal the barriers, obstacles, and constraints both customers and businesses face. These pain points can be refocused as required business outcomes and then we can look for solutions to deliver them. We should ensure we approach problem discovery with a method that is fit for purpose.
- Solution Management and Delivery

  The solution space is composed of three components:
  - Solution delivery life cycle: How will we manage the solution? Is it a one-off project, does it require continuous delivery, or do we need to take a more exploratory approach?
  - Delivery management: How will we manage the delivery? Do we require big upfront design? Should we tackle the delivery in an iterative manner so we can manage change, or is there so much change that we should just limit the work in progress?
  - Development and operations methodologies: How we manage the technical change, at both a development and operational level.

  We must apply a suitable "fit for purpose" approach for each of the solution components, depending on the unique context of the solution space. You know the saying, "if all you have in your toolbox is a hammer, then everything looks like a nail." In short, it is not appropriate to apply an agile methodology to all problem contexts.

In a nutshell, there is no one way of working. To be agile you need to adapt your way of working and use appropriate methods based on your problem context. Businesses will win because of the design thinking and lean startup ways of working that help to explore new business models and solve real customer problems. Organizations can use agile methods to develop products further based on feedback. However, businesses can lose if their supporting business processes are costly and wasteful. Especially as customer expectations on what good customer service looks like are set very high thanks to the Amazons et al of this world. As organizations grow, they may be faced with large, complicated problems such as warehouse moves, or manufacturing plant setups that require upfront thinking and a plan-driven approach based on the number of teams involved and the criticality to business performance. Just as important businesses will need to adhere to best practice for security and compliance, and not deviate (Six Sigma) from tried and tested processes and practices.

7

How We Govern

Productivity is meaningless unless you know what your goal is.

—*Eli Goldratt*

Tell me how you will measure me, and then I will tell you how I will behave. If you measure me in an illogical way, don't complain about illogical behavior.

—*Eli Goldratt*

Control the organisation with arbitrary measures and, in fact, you actually diminish control. Or control the organisation with measures related to purpose used where the work is done, and you will achieve genuine control and what's more, your people will innovate.

—*John Seddon*

Fundamentally, governance is about "doing the right things" and "doing the things right." The traditional governance methods are predominantly focused on control and compliance rather than value creation, and on the fundamental premise that work is predictable. While suitable in simple problem contexts, this is not an effective way to manage work in a rapidly evolving environment increasingly defined by complex problems where work is emergent in nature. The result of this rigid approach is that an organization does not have visibility on how effective its investments are, only how well it is at adhering to a forecasted plan, budget, and scope. Fixed long-term planning denies us the opportunity to change direction at speed if our context changes or our assumptions turn out to be incorrect. This inability to quickly adapt is the failure of a traditional portfolio management approach.

If we want true organizational agility to adapt to changes in the business environment and the ability to manage work in complex problem domains effectively, we need to move from annual to continuous planning and budgeting, plan work around long-lived teams, focus on outcomes and impacts rather than being tied to particular solutions, and measure performance based on business value rather than output only. To optimize our investments, either by persevering with what we are doing or pivoting and moving funds to other opportunities, we need to frequently review the value that is being delivered based on real customer feedback as well as understand the impact of any changes in our business environment. This will give us the chance to challenge the assumptions our strategy was based on and thus allow us to correct our course where needed. This requires a mental shift from focusing on compliance and control, or more accurately the illusion of control, to an approach that is comfortable with uncertainty and based around practices aimed at supporting teams to adapt and experiment in their efforts to deliver value.

In this chapter we will look at how to adapt our governance based on how we are approaching a problem. You will understand how governance applies to the three main levels of an organization—strategic, tactical, and operational. We will look at how we align work, ensuring that higher-level goals cascade down to lower-level action. How we capture and manage demand. How we prioritize and fund work. How we measure the value of work, understanding it primarily in terms of value to customers, rather than business benefit only. How we empower people to make decisions at each level of the organization. And finally, a better way to review the performance of work, moving from measuring the progress of a plan to a frequent cadence of monitoring value delivered and lessons learned to determine if we should pivot or persevere with the work.

What Is Governance?

IT governance is a set of processes that ensures IT investments are effective and efficient at delivering enterprise value. At its most fundamental level, IT governance can be thought of as two principles:

Do the right things: How we prioritize IT investments.

Do things right: The oversight processes to ensure that investments are delivering.

Table 7.1 details the processes that sit under the two principles, which will be explored in detail throughout this chapter.

Table 7.1: The processes that support doing the right things and doing the things right

Do the Right Things	Do the Things Right
■ **Ensuring strategic alignment:** How we ensure work is aligned to the business strategy. ■ **Managing demand:** How we capture and manage demand. ■ **Prioritization:** How we prioritize demand. ■ **Investment:** How we fund work. ■ **Measurements:** How we measure work.	■ **Decision rights:** Who is empowered to make decisions. ■ **How we review performance:** When and how we review the work. ■ **How we manage risk:** How we ensure technical, security, compliance, and business continuity.

As you read in Chapter 6, "How We Work," different types of problems require different approaches. To complement each different approach to work, the methods of governance, namely investing, prioritizing, reviewing, and measuring the work, also need to adapt. After all it does not matter how agile your development teams are if the system they work within, and which governs them, is working against them. So, the question is how do you design a governance framework that can control both problems in a simple context where there is predictability and problems within a complex or wicked context with unknown unknowns? How should we budget for initiatives for which we cannot determine the scope or end? How can we ensure that investments are allocated to the most important work that will have the biggest impact on business success and will deliver the strategic goals rather than those with the most polished business care? How can we adopt a more flexible approach to planning that can adapt to feedback? How do we ensure the way we invest works with and not against the other components of the operating model that are changing to support a more agile method of operation? To achieve this, we need a lean approach to governance, one that is balanced, adaptive, and considerate to the way of working as detailed in Table 7.2.

Table 7.2: A balanced approach to governance

Governance Area	A Balanced Approach to Governance
Alignment	**Ensure there is an explicit and visible link between work and strategic objectives.** If work deviates from our strategic direction, then ensure the work is justifiable with a short-term payback.

Continues

Table 7.2: A balanced approach to governance (*continued*)

Governance Area	A Balanced Approach to Governance
Demand management	**Reduce the information required for new ideas.** Make it easy for anyone to submit an idea with the minimal amount of information. Build a lean business case for ideas that we want to advance and invest in. **Visualize the management of demand.** Ensure we don't start more than we can finish by limiting work in progress and allowing teams to pull work in rather than pushing work onto people.
Priority	**Prioritize business outcomes using good enough data on relative priority or the cost of delay.** Prioritize with good enough data to move fast. Sacrifice precision for speed and work with data that is good enough and complement with human judgement. **Increase the frequency of the prioritization feedback loops.** Have a regular cadence to reprioritize plans. Ensure incorrect prioritization decisions do not live long, based on the cadence of the volatility of your organization.
Investment	**Fund mostly for product team capacity.** For complex or long-lived problem spaces, fund team capacity to work on initiatives. For solutions that have a defined scope or known milestones, fund as a project. **Use the most appropriate method of funding initiatives based on problem context.** For problems in a complex context, take a venture capitalist, value-led approach to budget allocation. For simpler problem contexts, favor a more traditional project-led funding model. Use traditional business cases to assess the financial viability for large IT investments in commercial off-the-shelf software.
Measurements	**Articulate value first in customer terms.** Work should have a clear measurable value that should be articulated in terms of customer value before business benefit. By balancing genuine value for customers with business benefits and return, we can avoid losing sight of our customer proposition and why we exist at all. **Cascade purpose and measures into the work.** By being clear on value, we define purpose. We can then specify the measures that show how we are delivering against the value statement and use the measures to define the most appropriate method of action. We cascade the purpose and measures, the direction, into the work, not the method. **Use measures that show progress.** When determining work to deliver business outcomes, we should use leading indicators to get fast feedback to confirm our assumptions so we can invest more or to show us that we were on the wrong path and afford us the ability to quickly test an alternative path.

Governance Area	A Balanced Approach to Governance
Decision rights	**The leadership team sets direction and intent.** The exec determines the strategic objectives and impacts and prioritizes these against "business as usual" (BAU) needs. At this level, investment is focused on a team's capacity. The three things assigned to teams from leaders are outcomes, boundaries, and investment. **Senior leads and product group owners coordinate focus.** At a tactical level, there is a focus on coordinating the delivery of outcomes across one or many product teams. At this level, investment is focused on team capacity and giving funding initiatives. **Trust team to manage time.** At an operational level, product delivery teams determine how best to create or improve capabilities to achieve outcomes while balancing the BAU needs of their area of responsibility. Teams are closest to the work and are best placed to make just-in-time calls on the relative priority of BAU vs. strategic endeavors. Give the autonomy to manage unplanned features and do what is right but balance with alignment by cascading the desired strategic outcomes.
Performance	**Have a frequent cadence for planning and reviewing.** To adapt plans in response to feedback or changes in the business context, we must employ a frequent cadence of performance review. **Adapt performance review methods based on context.** Understand what context you are operating in and then apply the appropriate level of governance. For obvious problem context, favor plan-driven work and review progress against the schedule and required features. For complex problem domains, favor a value-driven method of work and review performance by looking at value delivered, and lessons learned from feedback, over following a plan.
Risk and compliance	**Differentiate between good failure vs. bad failure.** Jeff Bezos says, "I always point out that there are two different kinds of failure. There's experimental failure—that's the kind of failure you should be happy with. And there's operational failure." **In complex problem domains, failure is expected.** When exploring new or complex problems, failure is expected and helps build understanding. However, operational failure is bad and should be avoided. **Operational risk should be limited.** To avoid operational failure as well as risk, security, and compliance, we should utilize best practice and follow one of the frameworks introduced in Chapter 6.

Alignment: Linking Work to Strategic Intent

We can use a goal tree to visualize the alignment of IT investments to strategic objectives. A goal tree, as shown in Figure 7.1, is a logical expression of all the critical success factors required to achieve a goal. Both goals and the critical success factors are expressed in terms of outcomes, whereas the necessary conditions are stated as actions required to accomplish the critical success factors or parent necessary conditions, as shown in Figure 7.1, necessary conditions can be nested. For large complex problem contexts, there may be many levels of necessary conditions to be met.

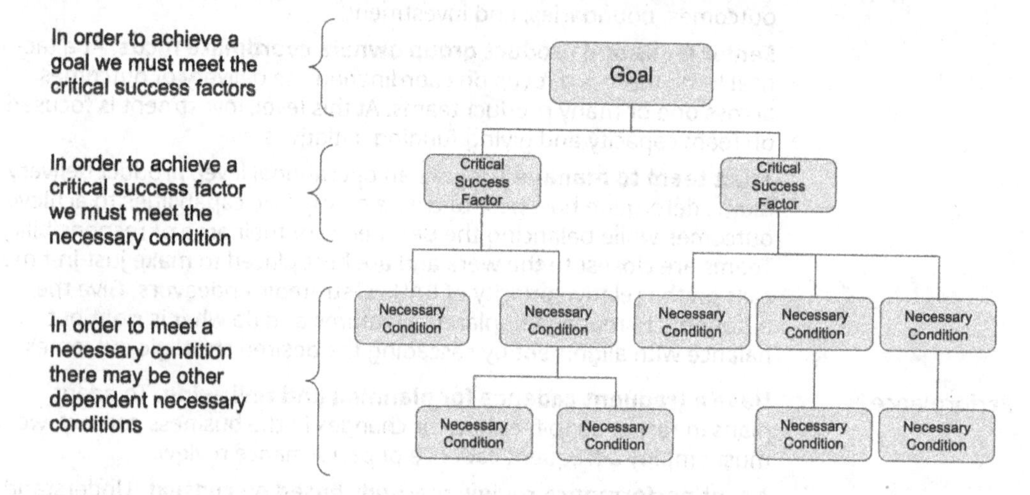

Figure 7.1: A goal tree

The benefit of using a goal tree is the cascade of goals down to the ideas and work that will realize them. By utilizing a goal tree, we can clearly visualize the strategic alignment of teams, ensuring all effort is focused in the right direction. As the goal tree is designed around a cascading relationship built upon necessity it is very easy to see any effort that does not contribute directly to critical success factors that will deliver the overall goal. Because of this, the goal tree helps keep work honest and focused on the most important and critical initiatives and prevents effort going astray on pet projects or local departmental optimization at the expense of the overall goals. Due to the clarity of strategic direction and focus of the goal tree along with the transparency of cascaded necessary conditions, it becomes a very useful tool to help with decision-making. When looking at investment options, it is much easier to agree on priority because it is clear on which initiatives contribute to goals and which do not.

An alternative version of a goal tree based on the same principles is the Lean Value Tree (LVT), shown in Figure 7.2. The LVT uses a different terminology to reflect the uncertainty of dealing with complex problems. To achieve goals, we may not be certain of the critical success factors as they may be hypotheses, so instead, in the LVT we use the terms *bets* to make it explicit that this is a hypothesis and may not work (i.e., we may lose the bet). Instead of a necessary condition, the LVT uses the term *initiative*.

Initiatives define what should be done, the concrete executable actions, to test the value hypothesis of the bet.

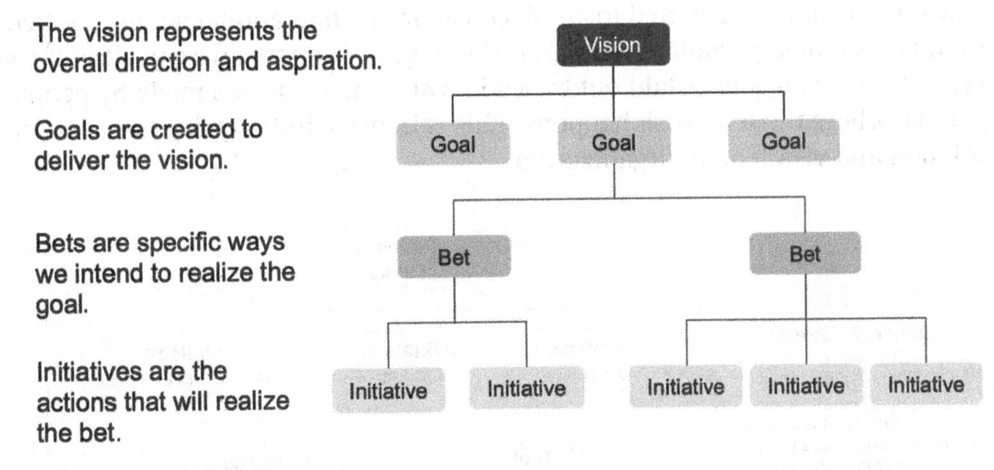

Figure 7.2: A Lean Value Tree

There are many ways of articulating the various levels of work in the organization. You should always use terminology that works for your context. The important point is not what you name the concepts but instead what they stand for. Again, like many decisions, it is whatever makes sense for your context. The language I am most familiar with, and use consistently throughout this book, are *strategic objectives*, *tactical initiatives*, and *operational actions*, as shown in Figure 7.3:

- **Strategic objectives** are what we will focus on to deliver on the business aspirations and goals and thus turn the vision into a reality. These are the choices of how we will win, and where we will play, written in a manner that makes clear the objective and how we will quantify when we have achieved it.
- **Tactical initiatives** represent the ideas and hypotheses of how we could achieve a strategic objective. These are the programs of work or long-term investments on product teams to solve a problem.
- **Operational actions** are the execution level features, projects, or outcomes we delegate to teams to deliver.

The utilization of a goal tree or LVT gives a clear and simple view of strategic intent cascading down to operational execution. The alignment of objectives, tactics, and actions are made transparent, ensuring everyone understands how what they are doing contributes to business success. It is a lightweight planning tool that can be easily updated as and when plans change and adapt to changes in context, more experience, and data. It's easy to understand and, due to its visual nature, articulates our assumptions. The value of the goal tree comes from its ability to keep focus on what's important to fulfil the vision and the purpose of a system, ensuring that any initiative that is not directly tied to an objective or any operational action not tied to an initiative is robustly challenged. This is the tangible way that we can close the gap between strategic objectives laid out by leaders and the decisions made by people at the gemba, where the real work happens. This is how we link the day-to-day work to the mission and vision of an organization.

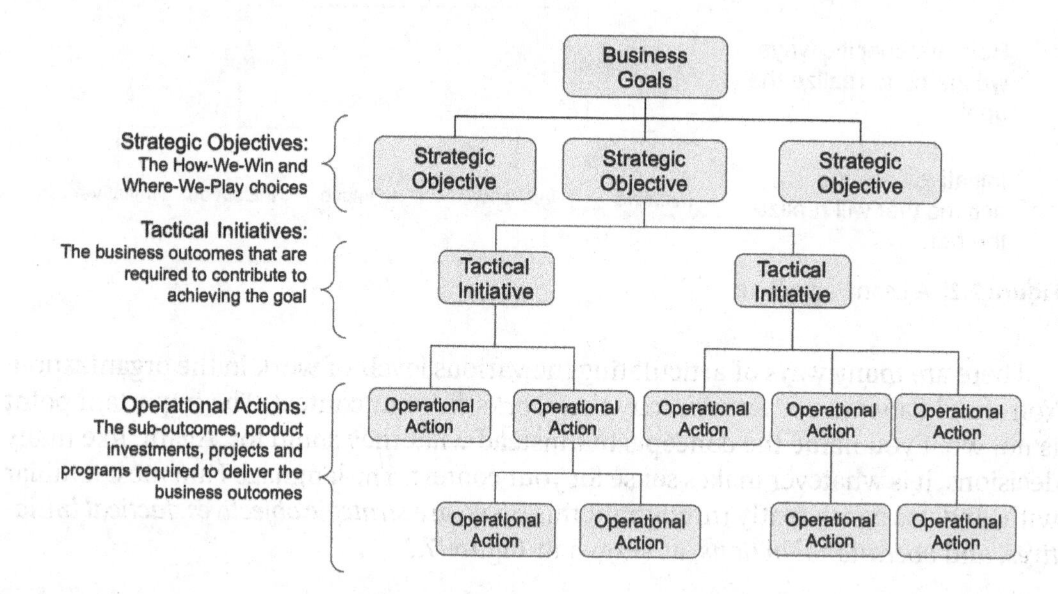

Figure 7.3: A revised "take" on the goal tree

Doing the right things and doing things right applies to each level and each node of the goal tree. As we look at governance throughout this chapter, understand that it is applicable in the same manner to all levels of the organization, from strategic through tactical and down to the operational level.

Managing Demand: Visualizing Work

To control work at any level you need an effective way to capture it and organize it. When it comes to capturing ideas, we need to have no more detail than is appropriate, but we must make clear the intent. When it comes to managing that demand, unlike in a manufacturing context where the buildup of inventory is easy to see in knowledge work, we can't see the work and therefore we need a visual way to organize and clarify it.

Capturing Demand

Whether we are talking about strategic objectives, tactical initiatives, or operational actions, capturing demand at each level works in the same way. We need to capture the intended value and measurement to simplify governance. As highlighted in Figure 7.4, at a minimum we should capture the following:

- The expected enterprise value/business outcome/ impact we are trying to make.
- The context behind why we are proposing this. What does it relate to further up the tree? What is the value of doing the work?
- How this work will mitigate a constraint or capitalize on an opportunity (i.e., the approach, plan, or solution outline).
- How the work will be measured, ideally leading measurements as well as lagging. Do we have a baseline measure? What measures will we use to show progress against the outcome?
- The high-level effort and cost.

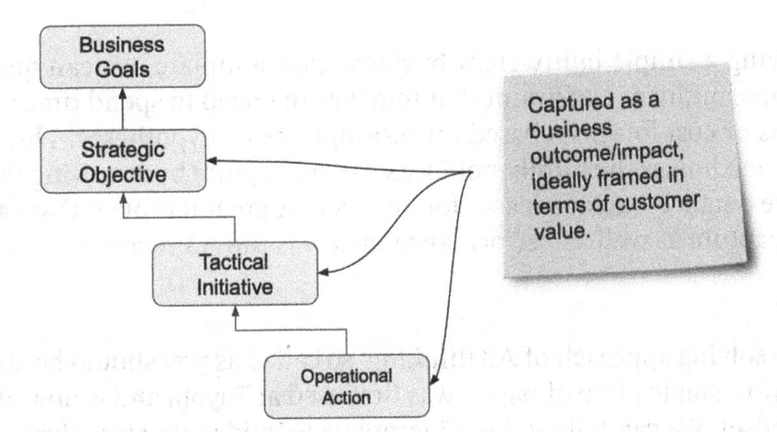

Figure 7.4: Capturing demand at any level

We don't want an overly complex or bureaucratic process to become a blocker for idea generation, but on the flip side, we also can't afford to have no process at all. What we need is a streamlined, clear, and simple process for capturing ideas with the appropriate level and precision of information requested at the right time. As shown in

Figure 7.5, at the initial stage we just want to capture core information about an idea; we don't need detailed financials, just how we can overcome a constraint or capitalize on an opportunity—even a single line is acceptable. This is important as we want to generate new ideas from anyone in the business without the need for masses of analysis and research. For ideas that have merit, we can perform more of an analysis to determine if we should invest effort in pursuing the opportunity. Depending on the risk, effort, and whether we need to make a big upfront investment, we can increase the level of scrutiny and information needed to determine if we will ultimately invest in making the idea a reality.

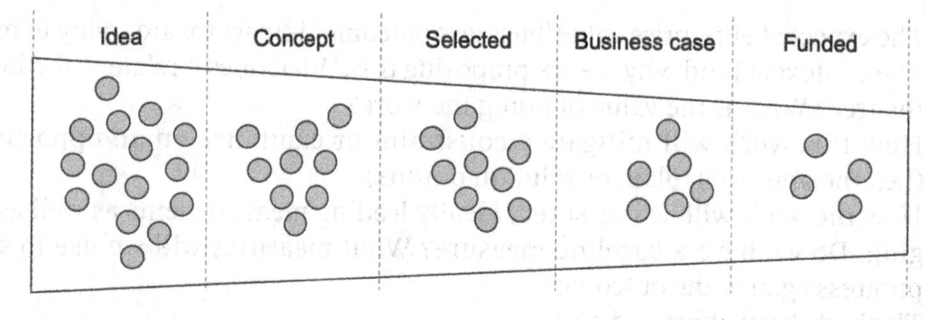

Figure 7.5: The ideas pipeline

By employing a simple lightweight business case template, we can quickly get a feel for an opportunity or constraint that removes the need to spend time on creating detailed plans or cost forecasts based on assumptions or hypotheses. This lean portfolio process maximizes the number of ideas we can capture by removing the waste of having to give detailed business cases for any idea. A great template that can be used for demand capture as well as performance review is the A3 report.

A3 Reports

The problem-solving approach of A3 thinking, so called as you should be able to distil a problem onto a single piece of paper, was first used at Toyota and is now at the heart of a lean mindset. We can follow the A3 template to guide ourselves through a series of questions to make sure we fully understand a problem to avoid jumping immediately to the keyboard and delivering an inadequate solution. There are two parts to the A3 template, as shown in Figure 7.6, the sections on the left column are focused on the problem, defining the goal, how we will measure it, and the reasons behind the problem. The sections on the right are dedicated to the solution, the set of hypotheses that will contribute to achieving the goal, the impact they are having, and next actions based on feedback. As is also highlighted in Figure 7.6, the A3 template maps to the

PDCA loop as introduced in Chapter 6. The A3 can also be used throughout a project because it is best used as a living and breathing document to summarize progress and next actions.

The A3 report structure consists of eight sections:

- **Theme:** Concise statement of the focus area of this report.
- **Background:** Context behind why this is an important topic, perhaps how it aligns to business success, including any relevant historical data and information.
- **Current Condition:** A detailed description of the current situation with data clearly supporting the problem statement. Typically you would show flows, charts, and data to clearly visualize trends such as increasing errors, dilution of profits, decreasing conversion, or subscribers not renewing. This is the AS-IS situation.
- **Goal Statement/Target Condition:** This is the TO-BE state. This is the goal, importantly with measures to specifically show the target state.
- **Analysis:** This describes the root cause of the underlying current condition. This is the result of discovery and analysis to determine the factors affecting the current state.
- **Countermeasures:** A summary of the outcomes, with owners and measures, that we will look at targeting that will contribute to achieving the goal. In essence, this is the set of outcomes we are targeting to move from the current state to the desired future state.
- **Check Results:** This is impact following the plan of countermeasures. This will be regularly updated to show progress on the countermeasures and the impact they are having on the goal measures.
- **Follow-Up:** These are the actions based on the impact that the countermeasures are having. This summary details if we will change our plans or continue investing with the current plan. In addition, we will add any other notes and lessons learned.

DIBB and Amazon's Press Release

Other examples of a lightweight business case are the elevator pitch and Amazon's Press Release, the latter of which is based on a mock press release targeted at the end customer that announces the finished product and details the problem and solution. A third example presented at a talk called "Spotify Rhythm" that Henrik Kniberg gave in 2016 at Agile Sverige, which was based on Spotify's approach (at the time) to strategic alignment. In the talk, and as shown in Figure 7.7, he spoke about Spotify's using the DIBB framework for business cases, DIBB standing for Data-Insight-Belief-Bet. In the simple example he gave to demonstrate the process, data taken from the website and mobile apps was analyzed, which led to an insight on how people listened to music. The beliefs about these insights were that people were moving to mobile devices and this was the future. From this insight, bets can be formed; in this instance the bet to invest in mobile development capability. Henrik described it as an argument framework that made the chain of reasoning from data to decision explicit. Henrik went on to explain that the DIBB boards were used at team levels as well as a strategic level.

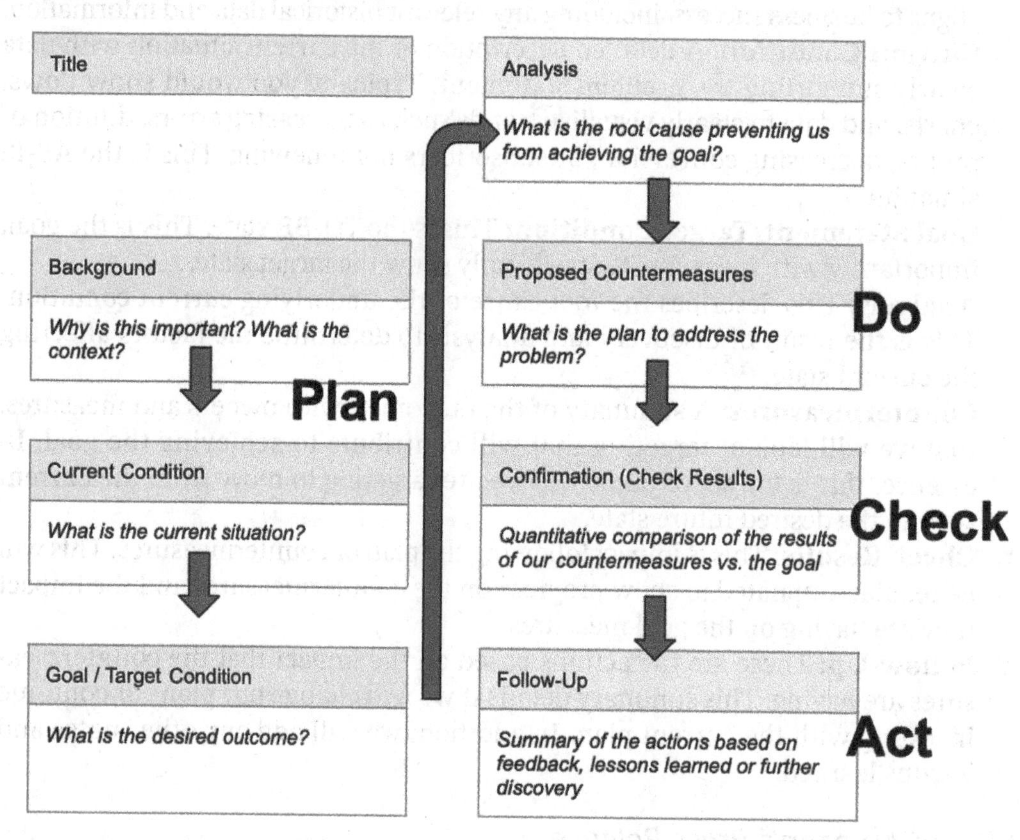

Figure 7.6: The A3 template follows the PDCA loop.

Visualizing Work Using Kanban

In order to support a lightweight process of managing the flow of work from idea to production, we can employ the kanban method. While we will focus on the kanban's process in terms of managing an organization's project portfolio, the concepts can be applied to all levels of the organization.

DIBB - An Argument Framework			
Data Feedback and measurements from the internal and external context	**Insights** Interpretation and analysis of the feedback and measurements	**Belief** What the insights lead us to believe is true	**Bet** How we will test our belief
• Mobile user devices increase • Lack of capability in mobile development	• Mobile devices will continue to be used to access our product and service • We won't be able to service the way users want to access our service because of lack of expertise	• Need to focus on a mobile first strategy	• Hire mobile development capability • Develop mobile platform • Retrain existing development resource into mobile devs

Figure 7.7: DIBB argument framework

The portfolio kanban provides a highly visible overview and a way to manage in-flight and queued work. It can aid the delivery of value by balancing demand with capacity, focusing on finishing rather than starting new work, narrowing the focus of work to what is important, and highlighting constraints in the flow of work. It encourages frequent feedback, offering early opportunities to change investment decisions, thus avoiding sunk costs.

The portfolio kanban framework is based upon the six core practices of kanban:

Visualize: The visual nature of the kanban allows information to be easily digested, enabling all to appreciate the big picture of work flowing from new ideas into production, how it aligns to strategic themes, what work is queued, and what is in progress.

Limit Work in Progress (WIP): To stop starting and start finishing kanban focuses on the reduction of Muri, or overburden. To this end, it balances demand with team capacity by employing a pull versus push system.

Manage Flow: By visualizing the bottlenecks and constraints in the flow of work leaders can use their power of command and control to unblock work. This results in reduced lead time, which enables the delivery of value sooner.

Make Policies Explicit: The process and rules of the board, based on feedback on what helps, are written into it in the form of explicit policies. Policies such as not starting work without clear value and measures, only allowing teams to pull when they have capacity, and limiting work in progress all contribute to improved value delivery.

Implement Feedback Loops: A regular cadence of feedback provides visibility into how well investments are progressing toward desired outcomes. This review enables organizations to respond to internal and external feedback and make decisions on whether to persevere with in-flight work or pivot to new opportunities.

Improve Collaboratively: The visual nature of a kanban board makes it an effective communications device, which can facilitate and focus discussions around attacking the work, not the person.

Visualize

If the goal of portfolio management is to oversee the IT investments of an organization in order to meet its strategic and business goals, then by definition it has to be able to take a lot of information and communicate it effectively. Kanban's focus on visualization and transparency, coupled with its flexibility in terms of a board template, makes it both intuitive and easy to communicate investments at each stage of the portfolio funnel. A project portfolio funnel can be thought of as the organization's "strategy to portfolio" value stream; ideas or concepts are gathered, defined, prioritized, invested in, and delivered to produce value for customers, which in turn generates revenue, or cash, for the business. By mapping and visualizing the activities and the queues between activities, people can appreciate the flow of work in the value stream and, more important, the constraints and bottlenecks. The reason visualization and transparency are important is that they make the need for action obvious—action to complete tasks to move work through the funnel, action to remove constraints where work is stuck in the system, and action on the system itself if there is a need for a change in portfolio process. The visibility of queues and work in progress also helps people appreciate why work cannot start immediately.

Figure 7.8 is an example of a portfolio kanban board. The flow of work is from left, idea, to right, production (i.e., done). This is the visualization of the strategy to portfolio value stream. Each card represents a business outcome, the template of which will be discussed later in this chapter. The first column on the left contains five rows. Rows 1, 2, and 3 are the three strategic goals, row 4 represents mandatory projects, and row 5 shows business improvements. The ideas lanes are for business outcomes in the early stages of development. You will read later the explicit policies of how and when cards move across the board. Once an idea is fleshed out, it is put in the priority lane, along with other fleshed-out business outcomes ideas. Once agreed, these ideas move into triage. Triage is where we determine which teams will be involved in the work to deliver the outcomes. The triage step is important as it translates the business outcome into the team trying to achieve it. However, try as we might, work does not always align to a single team, but the board can clearly communicate this. The business outcome is then added to the team(s) backlog of work and then only pulled into "in flight" when the team has capacity to limit work in progress, again which we will examine later in this section.

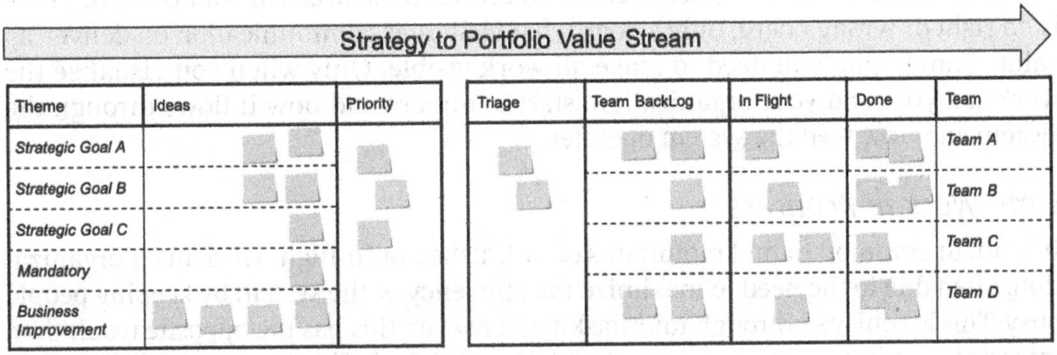

Figure 7.8: Using a kanban board for project portfolio management

The benefits of using a kanban board to visualize the flow of work for all to see cannot be understated. Here are just a few advantages:

- In the preceding board example, I explicitly define an intake stream for the three strategic objectives versus mandatory and business improvement outcomes. This gives a visual cue to see how much work is being focused on the strategic intent versus reliable operations. An alternative to this is to use different colored cards to track where focus is being placed throughout the board.
- It is harder to ignore and allow pet projects or unimportant work to be prioritized due to the clarity of how work aligns to strategic themes. This enables us to stop them before too much effort is wasted.
- As the board clearly shows the queues of work and the work in flight, it can help to manage expectations on new work. Just because someone requests work does not mean it will be picked up immediately; work needs to wait for team capacity. In addition, it can help to avoid wasting time in creating business cases for new work if we understand the queue of more valuable work in the team backlog and the capacity of a team.
- Blockers and impediments to the flow of work are easily highlighted. We will look closer at the flow of work later in this chapter.
- The priority lane enables work to be evaluated in the context of other ideas rather than on its own merit.

Figure 7.8 shows one way of constructing a portfolio kanban and it represents only a view in time. When first constructing a board, you should start with what you are doing now and model your current workflow. This means you can implement kanban quickly and without a change to your ways of working. As part of feedback and collaborative improvement, the board can then evolve to the needs of your context. There is no right or wrong board, only a board that facilitates communication on delivering value. Simply put, you need to make all work visible. Only when you visualize the work can you, and your organization, start to understand how it flows through the system and how well the system operates.

Limit Work in Progress

It is not uncommon to find mountains of unfinished or in-flight work in an organization, often due to the need to maximize the efficiency of the system by keeping people busy. This is achieved through multitasking; however, this has the opposite result than intended, as often more work is started than is finished. This is a case of focusing on cost efficiency over customer responsiveness in that we would rather keep teams busy, even though multitasking will add delays into the delivery of customer value. Lead time increases when we multitask due to the need to constantly context switch between different problem areas. To deal with the buildup of work and the slowdown in delivery, we can increase capacity by adding more people, but according to Brooks's law, this makes the work even later, especially if we must hire people from outside the organization with no prior domain knowledge. Simply put, multitasking does not work. What we need to do is focus on what's important until it is complete. We can achieve this by limiting the work in progress to prevent overburdening and slowing down the flow of value.

To reduce WIP, we need to balance workload with the capacity of the system. As delivery capacity is fixed, it is predictable; therefore, the work should be pulled by teams as capacity permits rather than work being pushed into teams. This way we will only start work when we have the capacity. As shown in Figure 7.9, teams pull from the portfolio backlog when they have capacity rather than demand being pushed onto them. This ensures we don't overload teams and maintain WIP capacity limits. We bring the work to the people, long-lived teams, not the people to the work.

As shown in Figure 7.10, we can impose a WIP limit at any stage of the value stream. The WIP limits themselves should be based on feedback on how the system is performing. For example, you can impose a WIP limit on the "Ideas" to not waste time talking about new ideas when there is no capacity to take on any new work.

Manage Flow

To reduce the time from having an idea to seeing it delivered into production, we should focus on improving the flow of work. As shown in Figure 7.11, lead time is the total time between having an idea and it being delivered. The time to develop the solution, the cycle time, is only part of the process. Typically, the non–dev time, or wait time, can be greater than the cycle time. Therefore, we should measure the time it takes for a proposed business outcome to move across the entire value stream until it is delivered. By tracking lead time, we can visualize how much time projects spend waiting or blocked and determine ways to improve the process.

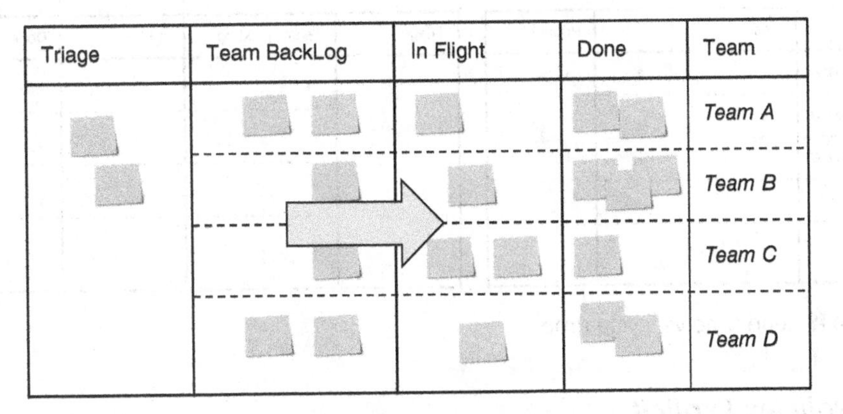

Figure 7.9: Teams pull work in when they have capacity.

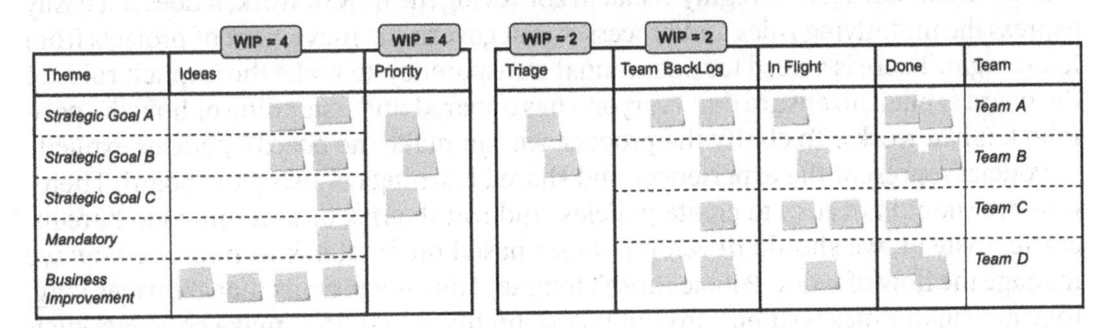

Figure 7.10: Setting work limits

One method to ensure the steady flow of value is for teams to have a backlog of work they can pull from. This means ensuring all the exit criteria to move work into a team's backlog is complete, such as having clear measures of success so that they can pick up new work when they have capacity. Perhaps the most fundamental way to manage flow is to focus on the board from right to left. We should swarm on problems with the in-flight work to remove constraints and bottlenecks to reduce wait times. Spend time on the items on the right, and finish what is in progress before focusing on the next item, otherwise you will fall into the trap of multitasking.

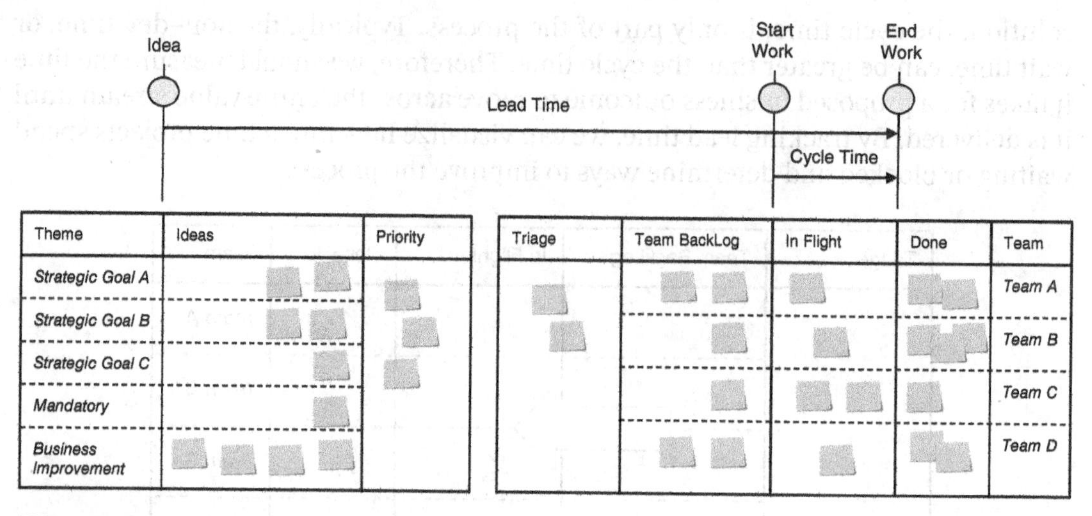

Figure 7.11: Lead time vs. cycle time

Make Policies Explicit

While the kanban board is highly visual in conveying the flow of work, it does not easily express the underlying rules and processes that govern the movement of projects from left to right. There is a need for operational transparency, to make the implicit rules of the board explicit, to ensure that everyone has a shared understanding of how the portfolio kanban works. To clarify the process, we can make the board's policies explicit.

Policies represent the experiences and shared learnings of using the board. Therefore, we should not rush to create policies, and run the risk of becoming too bureaucratic. Instead, we should develop policies based on feedback to improve how we manage the flow of work. Policies aren't long separate documents that are pages long; they are small notes written into the board, highly visual, that make concepts such as exit policies clear so that people know what is required to move a project from one stage to the next. As shown in Figure 7.12, we can add exit policies to the kanban board for each column or where required. Again, we don't start out by defining policies; we create them based on feedback when they are needed to improve the flow of work. As an example, for a business outcome to move from ideas to the priority lane, it we will need to have a lightweight business case, which was discussed earlier in this chapter.

Implement Feedback Loops

Without feedback on the work in the system we can't make decisions on whether we should pivot to a new opportunity or persevere and continue to invest in what is in flight. The process of reviewing in flight, prioritizing, and funding work on the board should happen on a regular cadence, typically between once a month to once a quarter. However, feedback will occur at different intervals depending on if we are reviewing objectives, initiatives, or operational actions; however, all of these items feed into, and influence, the portfolio view, either from a top-down or bottom-up perspective. Table 7.3 shows the typical frequency of reviews at each level of the organization.

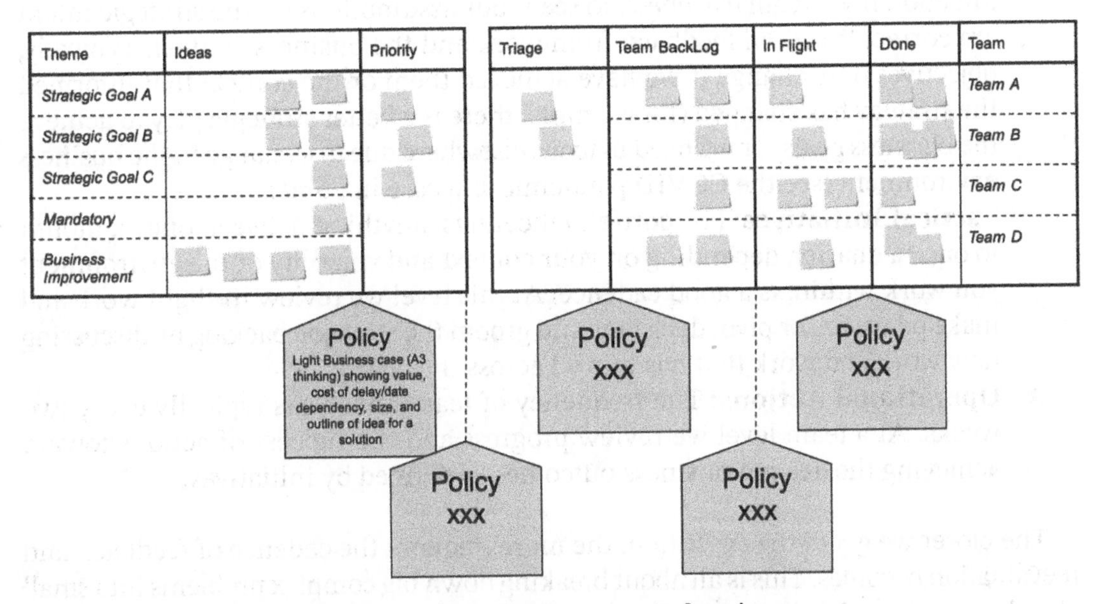

Figure 7.12: Define exit policies to govern the movement of work.

Table 7.3: Feedback loops applied at different levels of the organization

	Description	Plan	Review Frequency
Business Goals	**Financial health** Measures of how the business is valued	Up to 3 years, broken down into annual goals	Quarterly - Annually
Strategic Objectives	**The choices we make to achieve goals** Our how-to-win and where-to-play choices	Up to 3 years, broken down into annual goals	Quarterly - Annually
Tactical Initiatives	**The behavior we want to see to achieve the strategic goals** Strategic outcomes, operational and mandatory needs	Up to 18 months	Monthly - Quarterly
Operational Projects, Programs, and Product Investments	**Ideas and hypothesis on how to achieve the desired outcome** What we hope will change behavior	6-12 months	Fortnightly - Monthly
Teamwork	**Operational projects** What we need to do to prove hypothesis, e.g., feature X, feature Y	Plan for development iteration, Weekly - Fortnightly	Weekly - Fortnightly

- **Strategic Objectives:** We should not expect the strategic or business goals to change frequently, but we can greatly reduce the risk of heading in the wrong direction if we regularly check to see if our assumptions on the strategic intent are correct based on feedback from work and the business context. Typically, objectives only change if we have achieved them or the context that informed the strategy has changed (for example, there is a better strategic play to achieve the business goals) or we need to focus elsewhere due to a change in the business environment (see the COVID pandemic as a case in point).
- **Tactical Initiatives:** For portfolio meetings, anything between once a month to once a quarter, depending on your context and volatility of the environment you work within, is a good cadence. At this level we review in-flight work and make persevere or pivot decisions and groom the strategic backlog by discussing new work and work that has moved across the board.
- **Operational Actions:** The frequency of team reviews is typically every two weeks. At a team level we review progress and the success of actions toward achieving the desired business outcomes as defined by initiatives.

The closer we get to the execution, the more frequent the cadence of feedback and reevaluation becomes. This is all about breaking down big complex problems into small chunks to reduce the time to feedback so that we can frequently review our priority decisions. As the process of prioritizing will be more frequent, we need it to be fast, even at the cost of precision. However, this is acceptable because the decision will be short-lived, and we will have the opportunity to change direction within a short period if our assumptions prove incorrect. This fast and regular feedback helps with control in that we have opportunities to course-correct if effort is not being focused in the desired areas. We can stop projects that are not delivering value early, avoid any unnecessary sunk costs, and invest in the things that are contributing to our strategic objectives and business goals. Therefore, if we do choose the wrong hypothesis or our assumptions are proven to be incorrect, then we want our decisions on priority to be short-lived.

As well as feedback on the work, we can also use retrospectives during portfolio review and planning sessions to ensure that any new policies or template amendments can be incorporated as the board and process evolves to meet our needs.

Improve Collaboratively

The portfolio kanban board should be highly visible to all in the organization, ideally represented as a physical board in an office accessible to all if possible. However, due to the remote nature of modern work, an online board is also appropriate. At a team level, people will be interested to understand the backlog of ideas and the outcomes aligned to their own backlog. It's worth pointing out, though, that many of the ideas would have originated from members of the team themselves. Seeing ideas from other people and what other teams are working on can help collaboration and build understanding of what may be required from them to help other teams achieve their outcomes.

Product, value stream, or customer experience leaders will want to understand the outcomes related to their area of the business, understanding how they can contribute to the strategic goals as well as having visibility of where the relative priority of local improvements to their areas stacks up against more strategic aligned work. For the leadership team and the board, they will want to see how much time and effort is being invested into strategic endeavors and what new opportunities they have in order to make recommendations on further funding and to ensure that money is being spent in the right areas.

You should think of the board as a communication and a facilitation device, one that is a catalyst for discussions, clarification, questioning, and hopefully new ideas. It should give clarity to all about the strategic focus of the organization as well as the operational and tactical initiatives that are being invested in to achieve the strategic and business goals.

Prioritization: Focusing on the Things That Matter

An organization's strategy details all the areas in which it should focus its efforts to be successful. However, organizations cannot do everything at once. There are limits on funds and capacity to manage change. One of the hardest things about planning at any level is what to focus on first, but perhaps more important, on what to leave out.

Prioritizing Strategic Objectives

Prioritizing strategic objectives at the company level is an art more than a science. The priority is a reflection on how the leadership team wishes to invest in the strategic choices to achieve the organization's vision and meet business goals balanced with the needs of keeping the incumbent operation running. Some objectives will have very short-term payback, while others may be longer term. For example, an objective to expand geographically may not have as strong a short-term ROI case compared to an objective to reduce the cost of manufacturing. However, global expansion offers access to a bigger addressable market and will also be a big factor in how potential investors value the company. This is why we can't simply look at ROI alone; we need to look at the enterprise value of work. Similarly investing in a new product could be riskier than increasing investment in an incumbent product; but without investing in the future, we could be exposed to competitors that do. After all, if we are going to be disrupted, who better to do it than ourselves.

Prioritizing Tactical Initiatives

In an ideal world, we would focus all effort on initiatives directly linked to the strategy. However, we do not live in an ideal world. Many of the commodity and supporting capabilities that exist within an organization's value chain will require investment. Software needs to scale and may need to change to meet new legal and compliance requirements. These goals may not directly impact the business strategy but may have dire consequences if they are not funded and addressed. As well as understanding and prioritizing what is needed to fulfil our strategic outlook, we must also be aware of what is stopping us from maintaining normal day-to-day performance. What problems do we have? What nonstrategic capability needs investment. Nonstrategic initiatives can be derived from the needs of capabilities that are not directly linked to enabling the business strategy. However, any focus on initiatives that are not anchored to the strategy will not contribute to achieve its overall business goals and fulfil its vision. Therefore, we must ensure any work that is undertaken that is not in relation to enabling strategy is carefully considered even if it is an improvement. Or in the words of Peter Drucker, "There is nothing so useless as doing efficiently that which should not be done at all." This all means we must be ruthless in where we direct our effort to be effective.

An organization has needs outside of its strategic endeavors. The strategic initiatives describe what needs to change to fulfil the strategic position of where-to-play and how-to-win. However, we also need to run a business and keep the lights on, which can often incur large effort and costs. There will be times when we need to focus on operational needs above strategic endeavors. After all, there is little point in focusing on strategic objectives when the fundamentals of a business do not scale or are found to be inadequate and causing customer frustration. When the need is appropriate, large nonstrategic initiatives need to be centrally directed at the enterprise level and will likely interrupt strategic work. For example, with the introduction of GDPR (the General Data Protection Regulation), there were new requirements to manage customer personally identifiable data that would likely have crossed across multiple product teams. This type of work requires explicit priority and direction to coordinate at an organizational level. Along with mandatory needs such as those based on compliance and law, there are often operational needs that may not be directly impactful to strategic goals but nonetheless are in need of attention. We will look at smaller BAU items of work later in the chapter and how the team *can* manage them against strategic work. However, from the point of view of prioritizing larger work an organization undertakes, as shown in Figure 7.13, we can categorize tactical initiatives into four buckets:

- **Strategic Initiatives:** This is the deployment of strategy, broken down into annual, quarterly, and even monthly outcomes and measures. They don't tend to change a lot quarter to quarter unless there is big change in the wider context, a new opportunity presents itself from internal feedback, or the goals are achieved. This should be a small and focused list. As shown in Figure 7.13, a strategic initiative directly relates to the strategy.

- **Mandatory Initiatives:** These are mandatory projects that must be invested in. These projects are typically for compliance reasons such as GDPR and PCI or for legal reasons such as adherence to the financial regulations of a new geographic market—see Brexit and COVID for examples. In addition, there are emergency projects, which if not undertaken will have a sufficient impact on the organization's ability to trade, will have negative impacts on profit or increase risk.
- **Operational Essential Initiatives:** These are tactical projects that will improve the capability of the organization to produce value. These include process efficiency and simplification, cost reduction, quality improvements, and risk reduction. These can be considered BAU and perhaps the cost of a growing business. However, this is separate to the product team level BAU, which is focused on small enhancements, technical debt, and bugs. These differ from strategic initiatives as they may bring a business benefit, one that is required to maintain operations but not aligned to and in support of the business strategy.
- **Architectural Initiatives:** Purely technical optimization, risk, or cost savings initiatives.

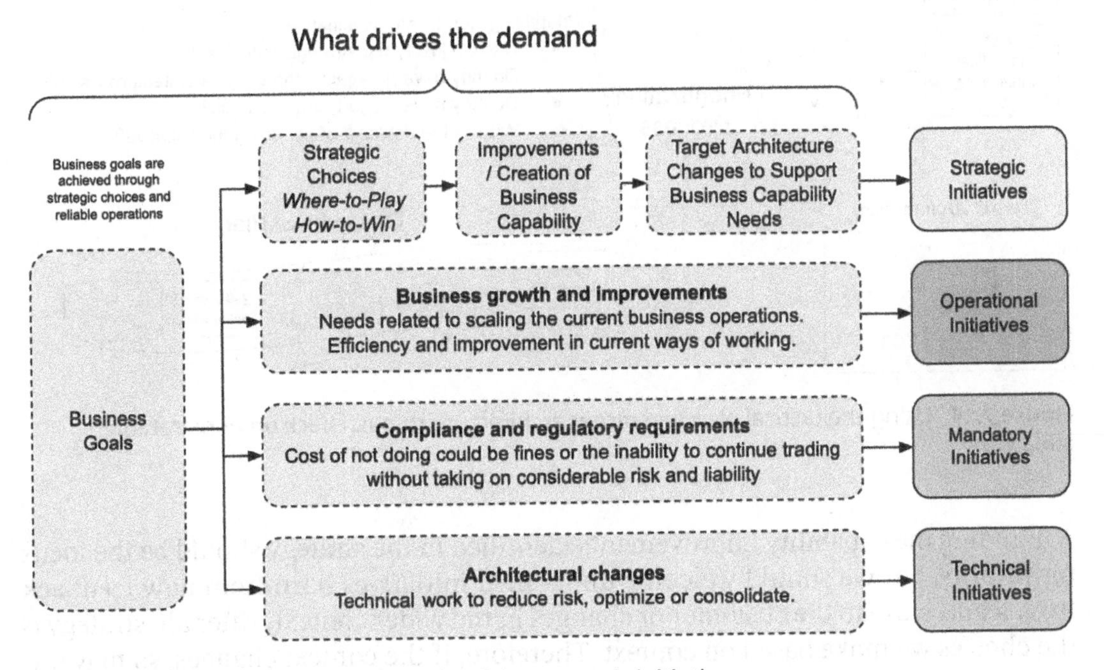

Figure 7.13: The demand on IT outside of the strategic initiatives

The Importance of Having a Clear Strategy and Tactical Plan

The IT contribution to business strategy identifies the investment in technical capability that is required to achieve a company's vision and goals. This is why being clear on strategy and in turn the areas we need to focus on is important as it makes the process of prioritization so much easier. Organizations don't lack for good ideas, but without the clarity of choice that strategy brings, we have no hope of focusing on the right things, or better, the right outcomes. Without such a framework, improvements will be made of course, but most likely they will be local optimization at the cost of system-wide improvements. With an IT strategy and a tactical road map built to enable critical business capabilities, we will be better placed to make informed decisions on what we choose to invest in and what to push back on, as illustrated in Figure 7.14.

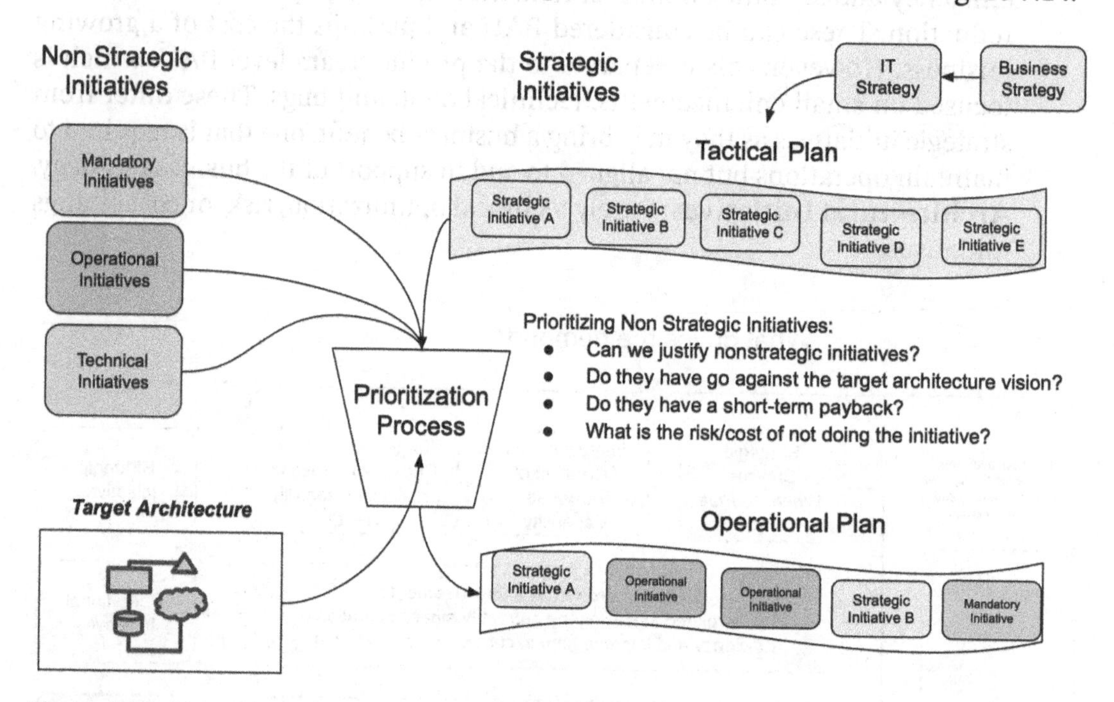

Figure 7.14: Using the tactical plan and target architecture to push back on nonstrategic initiatives

Funding the capability improvements identified in the strategy should be the focus for priority, but we should welcome unplanned initiatives born from new feedback from teams close to the customer or changes in the wider context. After all, strategy is the choices we make based on context. Therefore, if the context changes, so may our strategy and therefore our choice on what and what is not important. For example, before the COVID pandemic, organizations may have had very clear strategies.

However, when the pandemic hit, they may have had to drastically pivot to deal with the constraints or opportunities that nationwide lockdowns and travel restrictions presented on their day-to-day operations. As well as initiatives that will further our strategic objectives, we will need to work on nonstrategic initiatives. Initiatives that are not in line with the IT tactical plan or contribute directly to the strategy should have a strong business case to either address a risk or a compliance need or to generate immediate business benefit (revenue, profit, or cost saving) to justify their investment. In addition, nonstrategic initiatives may delay IT landscape optimization and consolidation plans, thus incurring technical debt. Therefore, as part of the priority we must also understand the alignment of the proposed initiative to technical principles and target architecture suggestions to understand the risk of deviating from the tactical plan.

Prioritizing Operational Action

When it comes to priority at an operational level, for teams managing work in simple contexts with a project-led method of working, they simply need to follow a plan or follow the best advice of a systems integrator who has done similar projects many times before. For teams in complex problem domains, they need to determine the best hypotheses or bet that will enable the desired outcome. In Joshua Seiden's book *Outcomes Over Output*, he refers to "the Magic Questions" that can be applied to determine where best to focus effort to achieve a desired business outcome in a complex problem domain. These questions are as follows:

1. What are the user and customer behaviors that drive business results?
   (This is the outcome that we're trying to create.)
2. How can we get people to do more of these behaviors?
   (These are the features, policy changes, promotions, etc. that we'll do to create the right outcomes.)
3. How do we know that we're right?
   (This uncovers the dynamics of the system, as well as the tests and metrics we'll use to measure our progress.)

As shown in Figure 7.15, teams can leverage the impact mapping tool, as covered in detail in Chapter 6, to visualize these questions and to determine the fastest way to prove a hypothesis with the tactics forming the team's backlog of work. By visualizing the outcome, teams can focus on the problem rather than a particular solution as well as quickly determine priority of action. This helps to encourage the behavior of constantly changing operational actions if they do not produce the expected results or if they discover new hypotheses that require less effort to validate.

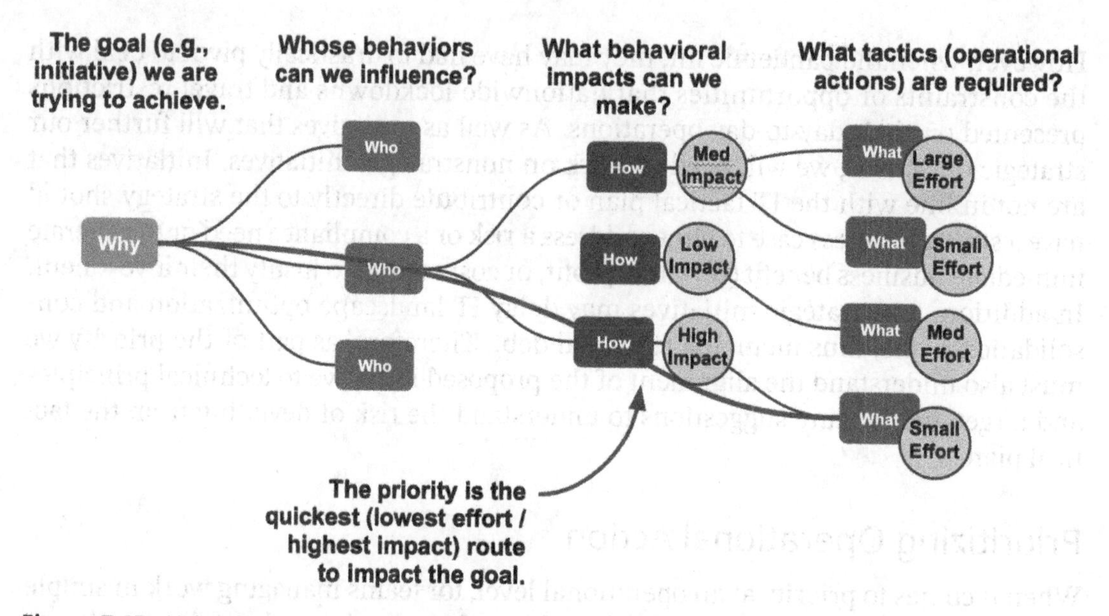

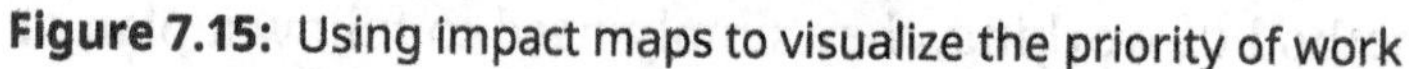
Figure 7.15: Using impact maps to visualize the priority of work

Helping Others

In addition to a focus on the teams' own outcomes, often a team will be required to support other product teams. As much as we can try to ensure teams are completely autonomous, objectives will cross team boundaries and some teams will require the support of others. Therefore, when it comes to team-level prioritization, it is important to also understand the organization's focus. With a clear stacked rank of organizational objectives, teams can make priority decisions on who to support and where best to spend their time. This helps to strengthen the autonomy teams have over prioritization challenges balanced with the alignment of a clear list of strategic objectives.

Prioritizing BAU vs. Strategic Work

One of the principles of basing teams around products means that they need to build and run the product. This means the continuous enhancements of the capability area in addition to any strategic objectives directed by the leadership team. Smaller BAU jobs that are local to the product team are too small to warrant adding the strategic backlog for leadership prioritization. Smaller business as usual tasks fall into the following three categories:

- Maintenance, aka bug fix
- Small feature enhancements
- Technical debt

As shown in Figure 7.16, to avoid the different priority challenges between the strategic needs and the, often necessary, BAU activity, teams can employ demand shaping. Demand shaping can help the balance between unplanned work (BAU) and the planned work (directed from the leadership team). Teams can reserve capacity during iterations to focus on planned work, and the rest of the capacity can be used for unplanned work. Again, this unplanned work can in turn be split into reserved capacity for technical debt versus bugs and small features if the team feels that this is beneficial. This ensures teams have capacity to focus on centrally directed outcomes but the flexibility to address local concerns as well. If the team is not involved in strategic work for an iteration or two, the strategic capacity can be used to direct efforts toward local optimization against the KPI measures of their business area. On the other hand, if there are no critical bugs, important small features, or technical debt that needs to be paid back, teams can choose to expand the amount of strategic work for an iteration. Again, this gives teams the autonomy to flex to the needs of the business and prioritize to focus on the right thing.

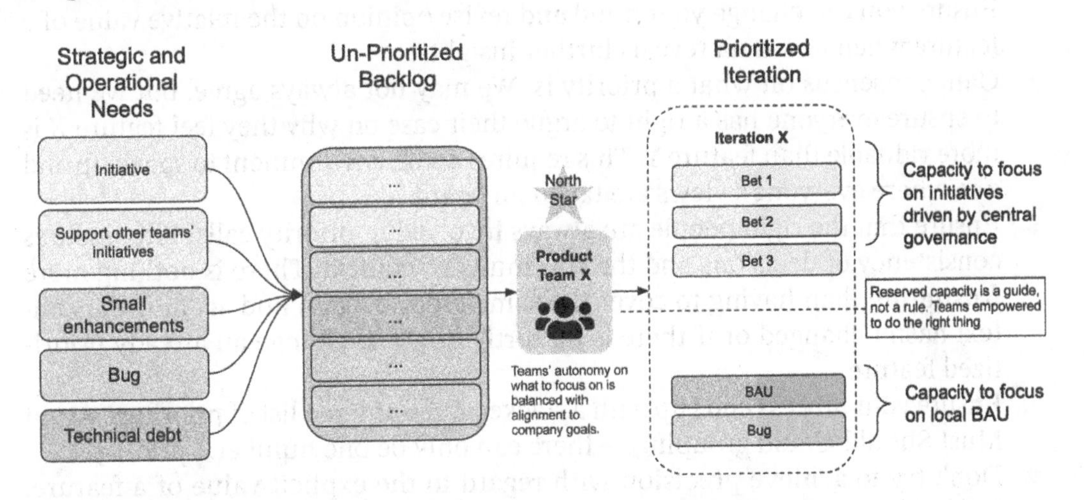

Figure 7.16: Teams employ demand shaping to manage smaller BAU work.

Methods of Prioritization

When it comes to the tactical (initiative portfolio) and operational level, there are several methods that can be used to help with prioritization. We will look at two here, namely *Highest Value For Lowest Complexity* and *Weighted Shortest Job First, or CD3.* In truth neither is perfect. Regardless of what method you use, there are some useful heuristics to follow when prioritizing what teams should focus on:

- Where possible, rely on data over opinion, ideally real data from real customers. But don't rely on data solely; balance it with team judgement.
- Be ruthless when prioritizing. Focus on value over politics and local optimization. This means having the courage to say no to features that don't contribute to your strategic goal and are not an operational necessity.
- Have the courage to prioritize quality when it is required. This means paying down technical debt a little at a time and often. Make time for this even though it is sometimes hard when compared to a feature that has a clear line of sight to creating customer value. Sometimes it is necessary to sacrifice quality to meet a fixed date that is out of your control or to capitalize on future-mover advantage. However, when trading quality for speed, ensure that this is an explicit decision and one that you will address later.
- Use the wisdom of the team. Persistent teams gain deep knowledge on both the problem and solution spaces. They understand their customers and can contribute useful insight and experience in situations where there is a lack of data.
- Ensure you can change your mind and revise opinion on the relative value of a feature when new data reveals further insights.
- Gain consensus on what a priority is. We may not always agree, but we need to ensure everyone has a right to argue their case on why they feel feature X is more valuable than feature Y. This requires a safe environment to speak up and one where everyone's views are taken on board.
- Ensure that the right people are always involved in priority calls. This ensures consistency of decisions and the continuity of context. There is nothing more frustrating than having to cover the same ground again and again if the context hasn't changed or if there is no further data to change an already prioritized feature.
- Ensure your process ends up with a stacked rank-ordered list of priorities. Avoid Must-Should-Could groupings—there can only be one number 1 priority.
- Don't try to achieve precision with regard to the explicit value of a feature. Instead, compare on relative value for a quicker method to resolve priority.
- When prioritizing effort, take on board the full cost of implementation, including the impact to other departments and any risks. For example, if you were to scrape a competitor's site for price checking, the effort may be minimal but there is a risk you will be accused of denial of service; there is also the overhead of keeping the scraper up-to-date as you are reliant on the UI template, which could change at any time.

There is no perfect prioritization algorithm for basing decisions at the portfolio or team level. Balancing the BAU operational needs with centrally directed objectives and supporting other teams with their objectives is difficult. However, by having one input stream, the team backlog, we can balance needs by using reserved capacity and the clarity of a stack-ranked list of company objectives to determine where efforts should be focused. However, with multiple demands on a team, a decision on priority can always be escalated to a leader if they feel that they are unable to make a call. As you will read in Chapter 9, "How We Lead," leaders can use command and control to unblock disputes on priority and give more context that teams may not be aware of.

Highest Value for Lowest Complexity

A simple way to determine the best next action for investment is by looking for the highest-value bet with the lowest effort to implement or test. As shown in Figure 7.17, we determine the business value of a bet or tactic and then divide it by the estimated effort. This leaves us with a ranked list of relatively prioritized bets.

For estimated effort we can look at measures such as these:

- Cost
- Time
- Complexity
- Risk

For business value we should use the leading measures that contribute to the desired outcomes. These measures are specific to the problem you are trying to solve, as in the following examples:

Desired outcome: Improve customer services

- Time reduced for answering customer questions.
- % of questions customers can self-service.
- Reduce issues to prevent new customer services enquiries.

Desired outcome: Increase online transactions

- Increase conversion.
- Increase volume into the online funnel.

We can then ask each member of the team to score both sets of measures in the effort and value tables using T-shirt sizes (small, medium, or large), Fibonacci numbers, or a simple scale of 1 to 10. This is very similar to planning poker, a development team activity that uses the experience and knowledge of the team to reach consensus to assign a complexity value to a development story or task to determine the capacity of a development iteration. As with planning poker, highest value for lowest effort is a lightweight and fast method of executing priority decisions that involves the entire team and one that promotes discussion if there is misalignment between individual value or effort scores.

Bet/Tactic	Time Reduced	Customer Self Service	Issue Reduction	Customer Satisfaction	Value Score
Customer FAQ portal	5	1	0	2	8
Customer service knowledge base	4	0	0	2	6
Prevent incorrect orders	5	0	5	5	15
....					

Bet/Tactic	Value Score	Effort Score	Value over Effort
Customer FAQ portal	8	6	1.3
Customer service knowledge base	6	5	1.2
Prevent incorrect orders	15	9	1.6
...			

Bet/Tactic	Cost	Risk	Complexity	Effort Score
Customer FAQ portal	3	1	2	6
Customer service knowledge base	3	1	1	5
Prevent incorrect orders	5	1	3	9
....				

Figure 7.17: Visualizing highest value for lowest effort

We can visualize this in a different way, as shown in Figure 7.18. The initiatives are mapped on to the matrix based on the following criteria:

- **Impact:** On the x-axis we have impact from low to high. It is important to be explicit on how you measure impact. Do you want to compare the impact something has this year or over three years? Also think of compliance or security initiatives in terms of cost or risk avoidance.
- **Effort:** On the y-axis we have effort from high to low. This gives an indication of the time and complexity required to deliver the initiative.
- **Cost:** The size of the circle representing the initiative indicates the relative investment required. The bigger the circle, the more funding is required. Again, this should show the total cost of ownership for the same period as the impact.

Based on the position of an initiative on the matrix, we can summarize its priority.

- **High Impact-Low Effort:** At the top-right above the matrix, we have the initiatives that are considered the "low-hanging fruit." We should prioritize these first due to the high return on investment opportunity.

- **High Impact-High Effort:** In the bottom-right corner, we have the initiatives that should be focused on next. While there is more effort required to deliver these initiatives, they will still have a high impact.
- **Low Impact-Low Effort:** In the top left-hand corner, we have initiatives that are low effort but have minimal impact. You should consider if a collection of these initiatives is less effort overall but similar impact as a high effort, high impact initiative.
- **Low Impact-High Effort:** In the bottom left we have initiatives that are high effort but low impact. These are the last things you should prioritize due to the poor return on investment.

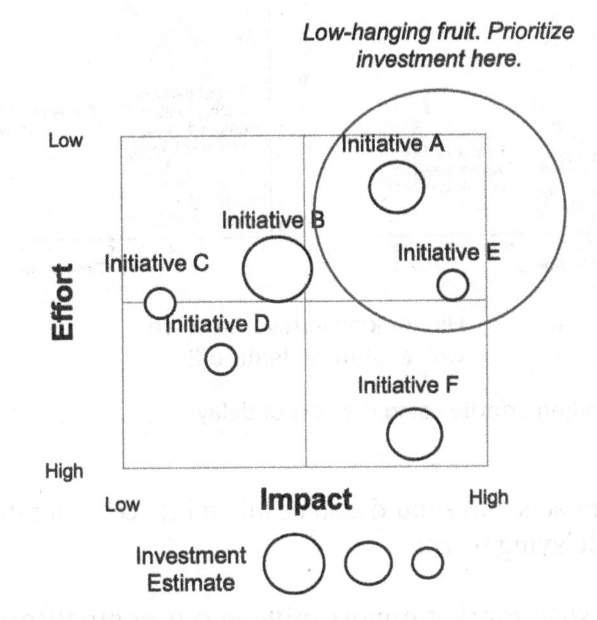

Figure 7.18: Plotting initiatives to determine priority

Weighted Shortest Job First or CD3

Another method to order prioritization is to order by the cost of delay. The cost of delay defines how much value (for example, costs saved or increased revenue) you will sacrifice from putting off a development. It differs from the Highest Value for Lowest Effort method by focusing effort on feature duration only rather than total effort, which includes risk, cost, and complexity as covered in the previous section. In the book *Principles of Product Development Flow, Celeritas Pub; 1st edition (29 May 2009)*, Don Reinertsen employs a model that visualizes the cost of delay. For instance, as shown in Figure 7.19, if you delay the start of feature 1, which will return £800 worth of revenue a week with a 10-week lead time to build, over feature 2, which will return £600 of revenue a week with a 5-week lead time, then you will end up with a cost of delay that's worth £4,000. However if you choose to start feature 1 first, then the cost of delay in terms of lost revenue is £8,000. Therefore, all being equal with risk and complexity, you are better off to focus on feature 2 to maximize revenue.

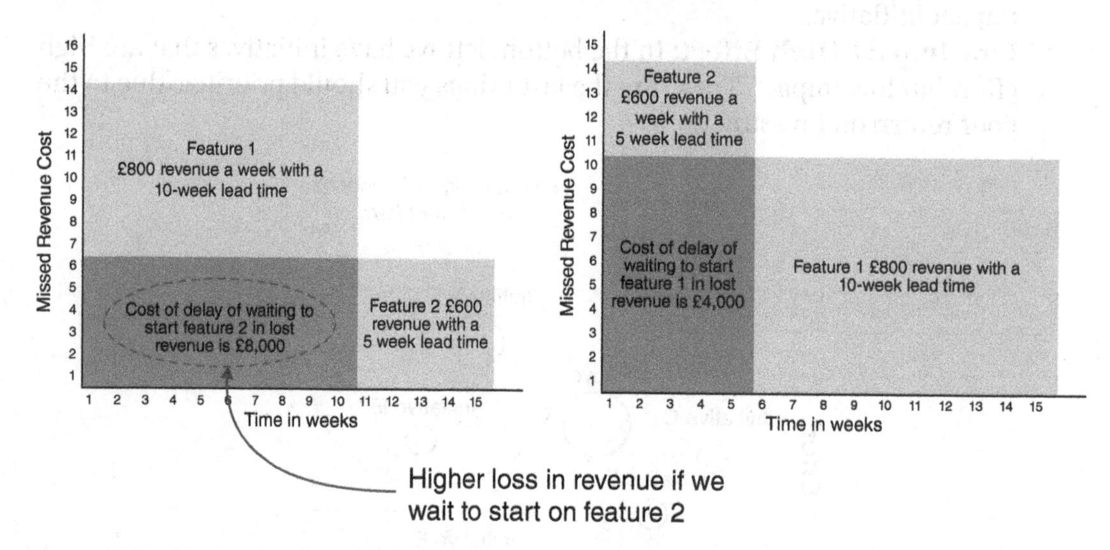

Figure 7.19: Understanding priority using the cost of delay

There are other impacts that should also be taken into consideration when we consider the impact of delaying work:

- We have a first-to-market opportunity as our competitors don't offer this service/feature.
- The feature is in support of another's team and is a dependency.
- Legal or compliance date; if delivery is delayed, then fines will be incurred.
- Fixed date related to trading; delivery beyond this date will result in no value as the feature is only short-lived.

Technical Debt and Risks

When it comes to architectural initiatives that are purely focused on IT-specific issues that can't be incorporated in a business sponsor IT initiative, we need to prioritize in a different manner. As with the business-sponsored IT initiatives, we can evaluate the critical architectural initiatives using a matrix as can be seen in Figure 7.20. The initiatives are mapped to the matrix, like a risk register, based on the following criteria:

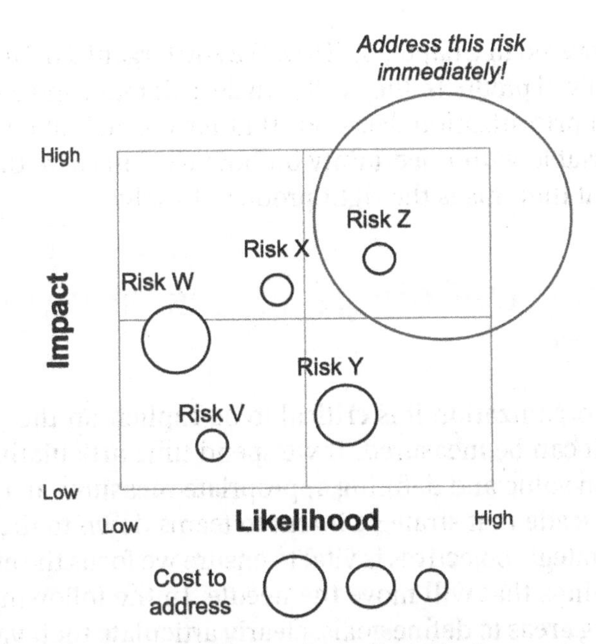

Figure 7.20: Plotting IT capability initiatives to determine priority

- **Impact:** On the y-axis we have impact from low to high. When architecture evolves over time without due care and attention, problems can emerge that have an impact on our ability to execute. We can model this in terms of an impact. Examples of impact are slower throughput if there is complexity, poor customer experience if there are quality or performance issues, higher costs if there is duplication, risk of incident if security concerns are not addressed, and fines if compliance needs are not fulfilled.
- **Likelihood:** On the x-axis we have how likely the risk will have an impact, from unlikely to realized. In the case of risks that are now issues, the position should be reflected as realized.
- **Cost:** The size of the circle representing the initiative indicates the relative investment required to address the risk or issue.

Using Judgement as Well as Data

We must be comfortable with the fact that we will never have enough certainty in anything other than simple problem contexts. When working in complex problem domains that are nondeterministic, we need to operate on good enough data, sacrificing precision for speed of feedback on a decision. Other than simple problems, we may not have enough data to be 100 percent confident of an idea before we decide to do it. Therefore, we need to balance a process that makes use of both the available data and human judgement (namely, instinct and insight). When it comes to human judgement, as we covered in Chapter 5, "How We Are Organized," if we form persistent teams around long-lived products, they will gain deep domain and customer knowledge that can be used in prioritization decisions that have a lack of data. Fundamentally we must be comfortable with uncertainty on how we will solve the problem, but we need to be confident that this is the right problem to solve.

Measurement: Defining and Cascading Value and Measures

At all levels of the organization it is critical to be explicit on the purpose, the value of work, and how it can be measured. If we spend time articulating and structuring goals, being clear on value and defining appropriate measures at a strategic level, we make it easier to cascade that strategic intent to teams down to the operational level. Aligning work to strategic objectives is vital to ensure we focus the efforts of our scarce resources on the things that will move the needle. In the following sections we will look at the following areas to define goals, clearly articulate their value, define appropriate measures, and cascade those measures into work:

- How we can describe the value of a goal using the three archetypes of customer value, business benefit, and risk reduction/cost avoidance
- How to apply the appropriate measures of work to ensure we are achieving desired outcomes rather than simply tracking output or activity
- How we can structure and articulate goals to drive the correct behavior and align effort to moving the measure
- How we can align strategic objectives with team-level actions by cascading the purpose and measure into the work so that all the work undertaken is for the improvement of the targeted outcome

Types of Value

Ideally goals should focus primarily on delivering customer value. After all, value is why customers seek products or services from our business, and revenue and profit are the benefits you receive in return for providing them. By focusing on value to customers (i.e., fulfilling a customer need), we stand a greater chance of bringing benefits for the business. However, not all goals can be directly focused on customer value, and indeed not all will be aligned to the strategic intent. Some goals will be focused on ensuring a reliable business operation. While we should strive to measure all work in terms of value to a customer, the reality is that we also need to invest in business operations and therefore value that is beneficial only to the business. We may need to improve operations as part of scaling the business, implement security controls, and adhere to compliance measures to avoid costs and reduce risk; but these important and necessary actions do not directly add to customer value. As shown in Figure 7.21, we can understand value in three buckets: customer value, business benefit, and risk reduction and cost avoidance.

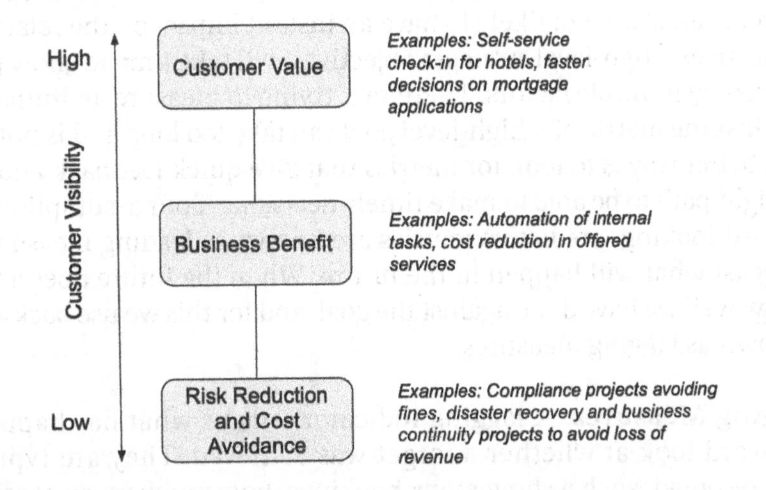

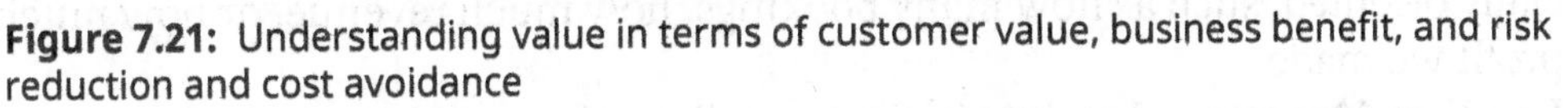
Figure 7.21: Understanding value in terms of customer value, business benefit, and risk reduction and cost avoidance

It is worth pointing out that there is a need to balance customer value and business benefit. We should not be sacrificing customer value for short-term profit in order to maximize our shareholder value as this will have negative long-term impacts. However, at the same time, we need to ensure that there is a sustainable connection between the value we provide to customers and the business benefit we receive. This balance is sometimes hard to measure. For example, offering a better customer service experience won't lead to an immediate business benefit for top line growth, but a satisfied customer will likely return and act as a promoter. This is precisely why the net promoter score and customer satisfaction measures, while imperfect, are important as forecasting indicators as there is a connection between customer value and business benefit. In the next section, you will learn about leading indicators, which enable us to get early feedback on the connection between delivering customer value and forecasting a business benefit, allowing us to adapt and pivot our approach to balancing the delivery of customer value while ensuring we receive business benefits. Being explicit on whether a goal has value for a customer or is just valuable for a business can help ensure organizations to not lose sight of their core value proposition and the need to continually solve problems and offer value for their customers.

Types of Measure

Shortly we will look at how to effectively connect goals to work, cascading and aligning on value from the business and strategic objectives down to the operational actions. However, we need to ensure that the measurements we give to low-level actions are aligned and clearly contribute to high-level business goals. The operational actions that teams deliver against are not likely to have an instant impact on the related high-level goals or objectives. High-level strategic objectives will take time to show progress and are influenced by multiple factors. Therefore, trying to measure an initiative's impact directly against the metric of a high-level goal can take too long and is not particularly accurate. A better way is to look for metrics that give quick feedback and signal if we are on the right path to be able to make timely decisions if our assumptions are wrong. These forward-looking, predictive metrics are known as leading measures; they will help to forecast what will happen in the future. When the future does arrive, we can measure how well we have done against the goal, and for this we use backward-looking metrics known as lagging measures.

- **Lagging Measures:** A lagging indicator tells us what has happened. It is a backward look at whether a target was achieved. They are typically financially oriented, such as how many bookings, how much revenue, or how much profit we made.
- **Leading Measures:** Leading indicators tell us what may happen in the future. They act as a prediction, looking forward to future outcomes. Good leading indicators will have a strong causal relationship to lagging indicators—for example, how many calories you consume versus burn a day can help to predict how much you will weigh at the end of the month.

As shown in Figure 7.22, we can use leading measures to help tell us what will happen rather than solely relying on the lagging business and strategic objective measures that tell us what has happened. As you can see, as we break the goals down to initiative outcomes and then to the operational action, we move from lagging to leading indicators as the feedback on work is faster. This cascade can continue down if work still has meaning, is independent, and has value.

The challenge is to be able to correctly identify the right leading measures that have a significant influence on whether we will achieve the strategic objectives. Leading measures are good at giving fast feedback and making decisions on where to focus effort; however, they are difficult to get right. It is easy to fall into the trap of selecting metrics that are easy to measure but have little impact on the overall goal. To combat this, ensure that there is a logical-sense check that demonstrates the relationship between the leading indicator and how its accomplishment contributes to the strategic goal.

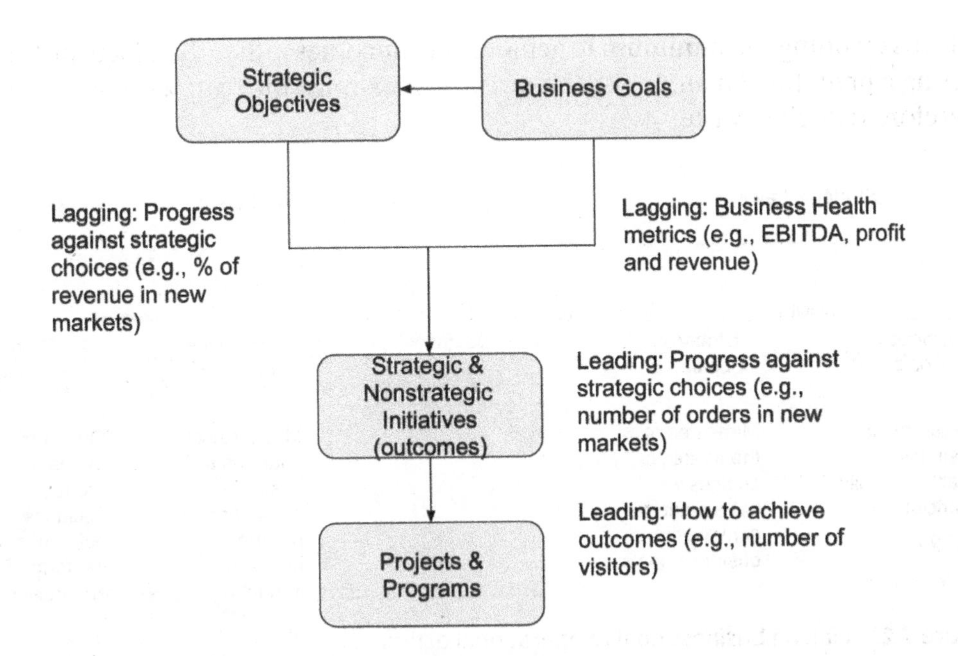

Figure 7.22: Where to use leading vs. lagging metrics

Cascading Value into the Work

To break the large business goals and strategic objectives down into meaningful, aligned, and independent chunks of work that teams can deliver, we need to cascade the direction of purpose and the measures of value into the work. That is not to prescribe what to do but instead cascade the outcomes we wish to see along with how they are measured. Cascading purpose improves our ability to guide teams, support innovation, enable autonomy, and ensure alignment. We can create the alignment between strategy and execution by moving to a portfolio of required initiatives framed in terms of business outcomes over a portfolio of projects for anything other than a simple context or mandatory operational need. As you can see in Figure 7.23, outcomes link business goals and strategic objectives to the team output.

Thinking in terms of outcomes presents a mental shift for leaders; instead of asking teams to deliver a specific solution, they instead articulate the problem to be solved in terms of the value they desire and define it with meaningful leading measures of success. By stating contribution to strategic objectives in terms of desired outcomes rather than solutions, we can remove any constraints on how they will be achieved, thus allowing for innovation, creativity, and importantly, ownership. As leaders we should provide direction in terms of clear value opportunities to teams rather than prescribing specific solutions for anything other than simple problem contexts. By focusing on outcomes over output, we get out of the build trap and make it possible to focus on doing the minimum to achieve the outcomes rather than focusing on completing a predefined plan. In other words, we maximize the stuff we don't need to do, therefore reducing waste.

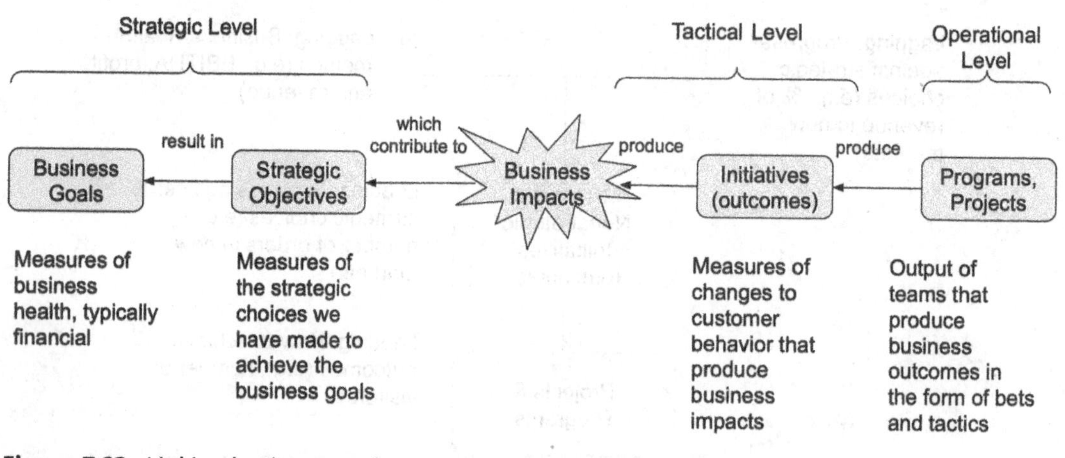

Figure 7.23: Linking business goal to operational action

This mental shift in how to manage the system of work is documented in John Seddon's book *Freedom from Command and Control: Rethinking Management for Lean Service, Productivity Press; (12 Aug. 2005),* in which he introduces his systems thinking approach to redefine the purpose, measures, and methods of work. John Seddon states that "there is a systematic relationship between purpose (what we are here to do), measures (how we know how we are doing), and method (how we do it)." This relationship is shown in Figure 7.24. We can compare the traditional methods and challenges of putting either measure or methods first against the benefit of putting purpose first.

- **Measure-Purpose-Method:** We target our employees to hit a metric. Purpose by default becomes the need to hit the measure no matter what. The method is then constrained to hit a predetermined target regardless if it has any reflection on actual value. This way of thinking invites people to game the system.
- **Method-Purpose-Measure:** We want to install a new ERP, or we want to do project X because we should be doing it or because everyone else is doing it. Again, the purpose is relegated to simply completing the project. Measures are reduced to completion metrics or shoehorned to align with strategic intent.
- **Purpose-Measure-Method:** Start with purpose, in our case the value we want to generate, ideally defined in terms of customer value. Then define how we can

measure that we are delivering that value. Last, determine the best method to achieve it, using the measure as feedback to adjust the method.

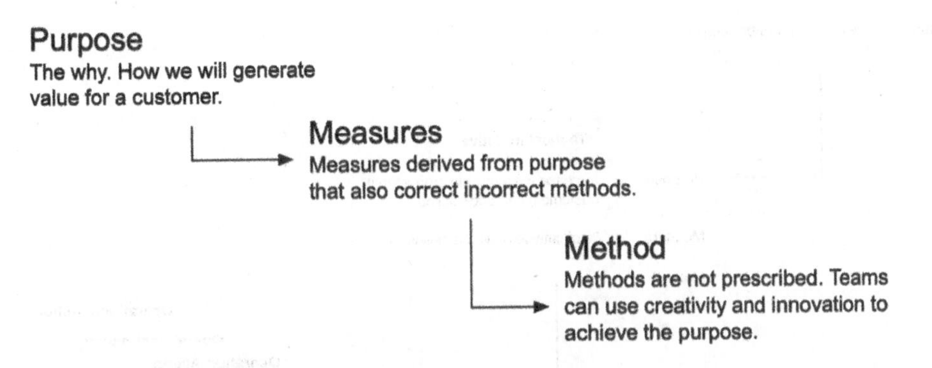

Purpose
The why. How we will generate value for a customer.
Measures
Measures derived from purpose that also correct incorrect methods.
Method
Methods are not prescribed. Teams can use creativity and innovation to achieve the purpose.

Figure 7.24: Methods are designed to achieve measurements that are derived directly from purpose.

In traditional thinking we lose sight of the most important point—to deliver customer value to generate business benefit. By putting purpose first, while defining the appropriate measures, it makes it much simpler to define what operational action is required. Leaders control the system of work, which means they control performance. Therefore, as leaders we must ensure that measures are derived from purpose and that the appropriate method is left to the team to determine to improve performance and the delivery of value.

As shown in Figure 7.25, we can use the pattern of purpose-measure-method to cascade the value and measurements to each level of the organization. This cascade approach uses the measure of the level above to determine work at the level below, ensuring all work is aligned and connected to the overall goal at the top of the tree. This drastically simplifies our ability to measure and track the progress of work as there is a clear connection from strategic goal to tactical execution.

We can employ tools such as impact maps or goal trees to give us a visual and consistent method to connect work to goals, ensuring both purpose and measures cascade down into operational work, which in turn contributes back up to the business and strategic goals. By cascading direction, we can break work into small manageable chunks while ensuring we do not lose its connection with the purpose and at the same time not tying ourselves to a particular method of execution. This process provides the discipline to focus on the opportunities or problems without the danger of falling into the trap of becoming wedded to a particular solution or thinking of a solution before we understand the problem. To understand the flow and cascade of measures from business goals to operational output, we will look at how values are defined at each level of the organization.

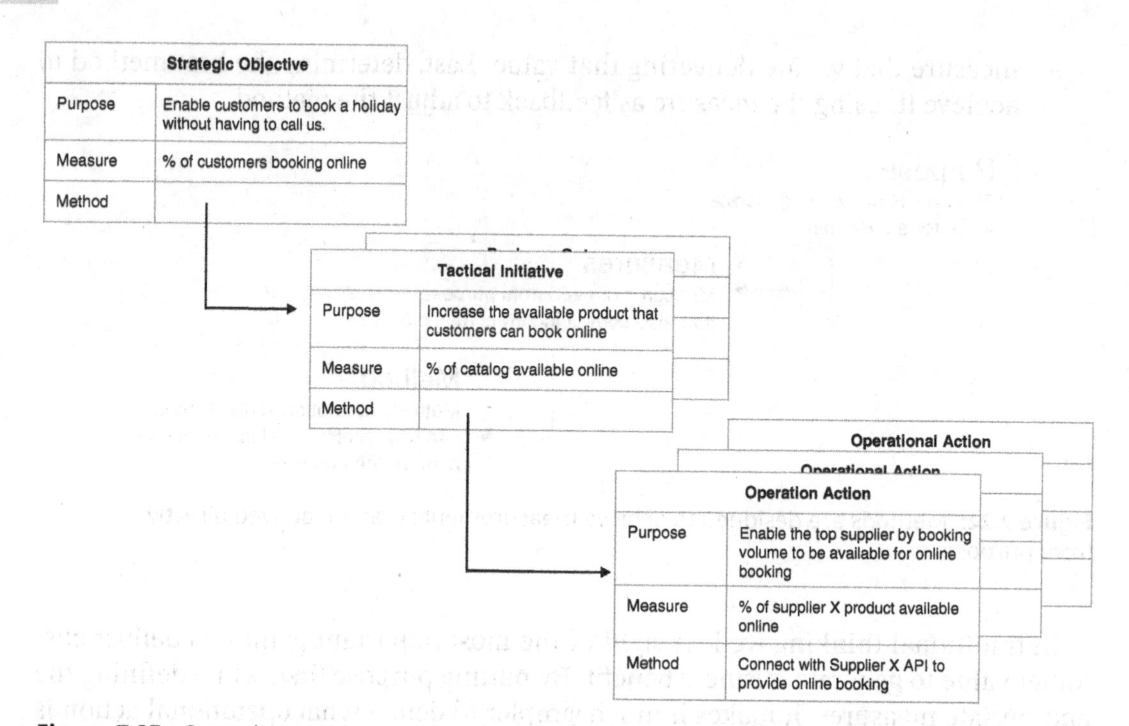

Figure 7.25: Cascading purpose and measurements down the organization's levels

Defining Value

How we define value at each level of the portfolio, whether strategic objectives, initiatives, or operational action, speaks volumes on what we value and what we mean by value. Of course, "for profit" businesses are in the business of making money, but by framing goals in a different way we can help show delivery teams the purpose behind the strategic intent, allowing teams to attach an emotional feeling to the business's chosen direction for success. Intrinsic motivation requires a purpose greater than increasing shareholder value. People want to help people, be they external customers or internal business colleagues. As shown in Figure 7.26, asking to "Drive 25% of transactions online" is less inspirational and meaningful than to say, "Enable customers to book a holiday without having to call us, at a time that suits them." Therefore, articulate the goals in terms of customer value to tie them back to the strategic intent and thus clearly communicate the importance. This also helps to keep focus on solving the real underlying customer problem, keeping the customer at front and center of our thoughts as we develop solutions.

As well as ideally being driven by customer value, a goal must also be SMART. SMART goals are Specific, Measurable, Ambitious, Realistic, Time-bound. Consider Figure 7.27, which adds a smart goal to the strategic intent of moving customers online, clarifying on the increase in transactions and by when. The strategic goals should be ambitious and provide guidance for operational day-to-day activities but avoid short-term thinking that can result in teams losing focus of the big picture of the strategic choices behind the goal. It is worth remembering that the organization's strategic objectives are hypotheses, and as with any forecasted metric, we should expect to adapt or modify the target depending on internal feedback and any changes in the business context. As we will cover in Chapter 13, "Operational Planning: Execution, Learning, and Adapting," strategy is not static and strategic objectives should be reviewed at a regular cadence, often in time with the portfolio review rhythm.

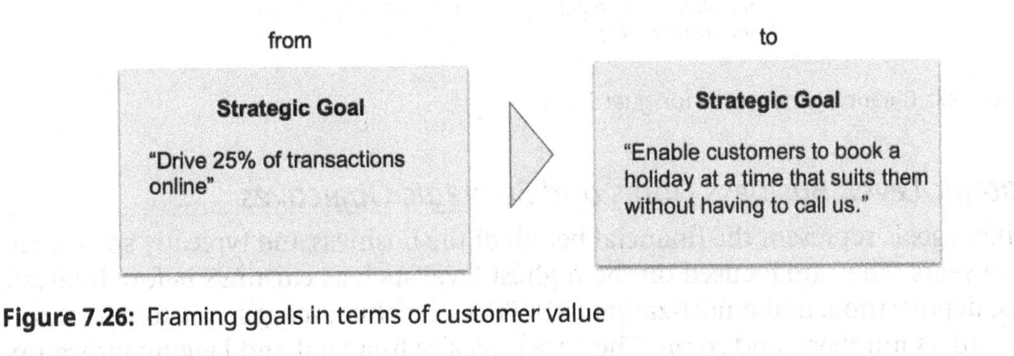

Figure 7.26: Framing goals in terms of customer value

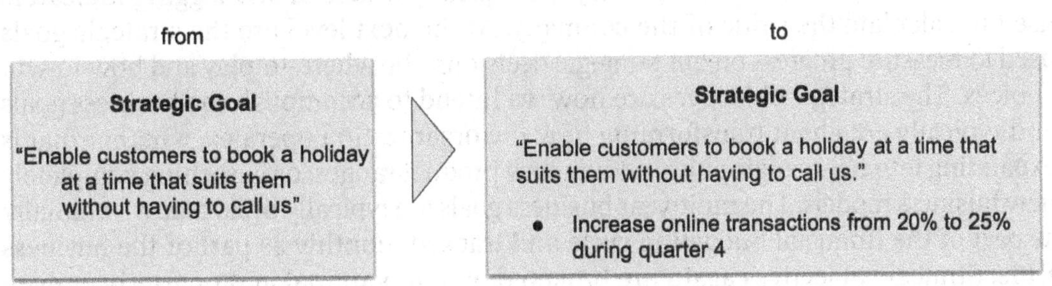

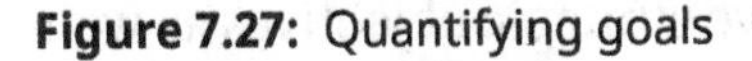

Figure 7.27: Quantifying goals

It is sensible to have more than a single measure of a goal to ensure we are not achieving something at the cost of something else (e.g., gaming the system). For example, a single measure of speed of delivery may be achieved at the expense of quality. To avoid this, we could balance the goal by ensuring that there is a secondary measure to limit quality problems, thus ensuring all work is focused on speed of delivery while not negatively affecting quality. Figure 7.28 shows examples of balancing measures of goals, also known as guardrails.

Strategic Goal

"Enable customers to book a holiday at a time that suits them without having to call us."

- Increase online transactions from 20% to 25% during quarter 4
- Maintain average order value of £5,000
- Maintain average margin of £950
- Maintain 0.1 contacts per order to customer service for online orders

Figure 7.28: Balancing measures for guardrails

Strategic Level: Business Goals and Strategic Objectives

Business goals represent the financial health of the business and typically span three to five years. They are focused on the highest level such as earnings before interest, taxes, depreciation, and amortization (EBITDA), revenue, margin, customers numbers, orders numbers, and so on. They are typically financial and lagging indicators used to calculate the value of the company. At the next level are the strategic goals used to measure progress on our strategic decisions, the where-to-play and how-to-win choices. The strategic objectives are how we intend to accomplish the business goals and typically are about transforming how the organization operates, whether that is expanding into new geographies, adding new product ranges, or launching completely new business models. The multiyear business goals are typically broken down annually as part of the financial budgeting cycle and tracked monthly as part of the business KPIs. Strategic objectives again are broken down into annual and quarterly or even monthly targets. Work needs to address the transformative strategic objectives as well as the underlying business goals directly. If the strategy is to launch into new markets, it does not mean there won't be anything to invest in the existing market; after all, we need at least to maintain a reliable operation.

It is very hard for teams to have an impact with strategic objectives as they operate at too high a level with many factors that could influence them. A more useful goal for teams to deliver against are business outcomes, articulated so that they clearly contribute to the strategic goal but use measures that the team can influence.

Tactical Level: Business Outcomes

The purpose of the business is to achieve its strategic objectives, the customer value measures, which will in turn result in business benefits being achieved—in other words, our business goals. However, as shown in Figure 7.29, strategic objectives are too large to guide teams as many factors outside of our control can influence them. They represent the impact we are seeking and don't offer teams direction of focus—that is, how will we achieve the goal? On the other hand, output does not help either. Output is too prescriptive, especially for complex problem domains where there are many unknown unknowns. Giving teams output (or in other words, projects) with predefined scope and plans will not help in complex problem domains. What we need to give teams are outcomes to tackle. A business outcome is an improvement to a business capability that delivers a benefit to the organization in line with the business strategy. Josh Seiden, in his book *Outcomes over Output: Why Customer Behavior Is the Key Metric for Business Success* (independently published, 2019), defines an outcome as "a change in human behavior that drives business results." He goes on to explain that "outcomes are the changes in the customer, user, employee behavior that lead to good things for your company, your organization, or whomever is the focus of your work." Therefore, where we are less certain on the outputs that will resolve a problem, as in the case of complex problem domains, we should focus on articulating desired outcomes. However, in the case of more obvious problem domains, where we know exactly what will deliver what we need, then we can focus on output and be explicit on what we need to reach a result.

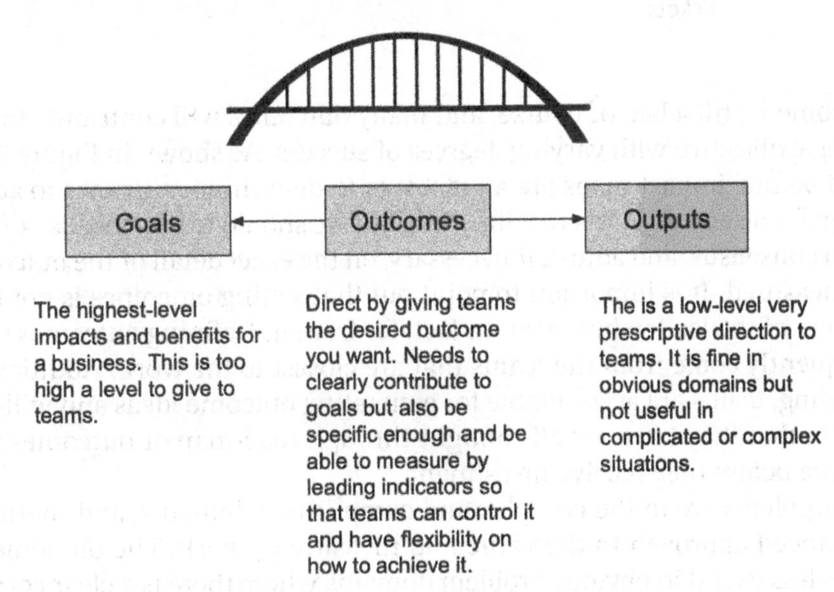

Figure 7.29: Tactical initiatives, aka business outcomes, bridge strategic objectives to operational action.

Outcomes contribute to strategic objectives and business goals, but more importantly they are things we can measure and influence. This is because we can measure customer and user behavior. Table 7.4 shows how the outcomes are based on changes to behavior and how they align to strategic goals.

Table 7.4: Examples of leading tactical initiatives contributing to strategic objectives

Tactical Initiative (Leading Business Outcome)	Business Impact (Lagging Contribution to the Strategic Objective)	Strategic Objectives (Lagging)	Business Goals (Lagging)
Reduce the % of flights fulfilled by humans	The avoidance of adding operations staff required to manually book flights	Avoid adding X heads by self-service and automation equates to £xxx in margin improvement per order	EBITDA Goals of £xxxm
Reduce the % invoices manually processed	The avoidance of adding operations staff required to manually book and check invoices		
Reduce the % of tickets processed manually	The avoidance of adding operations staff required to manually process tickets		

An outcome is still a bet, of course, and many outcomes will contribute to achieving a strategic objective with varying degrees of success. As shown in Figure 7.30 and mentioned before, impact maps are a great way to determine outcomes to achieve a business goal. The outcome sets the direction, but we should expect feedback from the teams, gain consensus, and adjust, if necessary, on the exact detail of the outcome and how it is measured. It is important to point out that setting outcomes is not the sole responsibility of the leadership team. In fact, far from it. Defining outcomes to invest in will frequently come from the teams that are closest to the work. Additionally, to secure funding, teams are accountable for generating outcome ideas and will need to demonstrate that they have a well-thought-through road map of outcomes in their business area before they receive investment.

Not all problems are in the complex and complicated domains, and therefore we need a balanced approach to the value and measure of work. The outcome-based approach is less useful in obvious problem domains where there is a clear correlation between cause and effect and therefore a high degree of confidence that a particular solution will work. In these instances, it makes sense stating the output required and using a plan-led approach. In complex problems domains, full of unknown unknowns, where the choice of action is unclear, an outcome approach is favorable to focus effort.

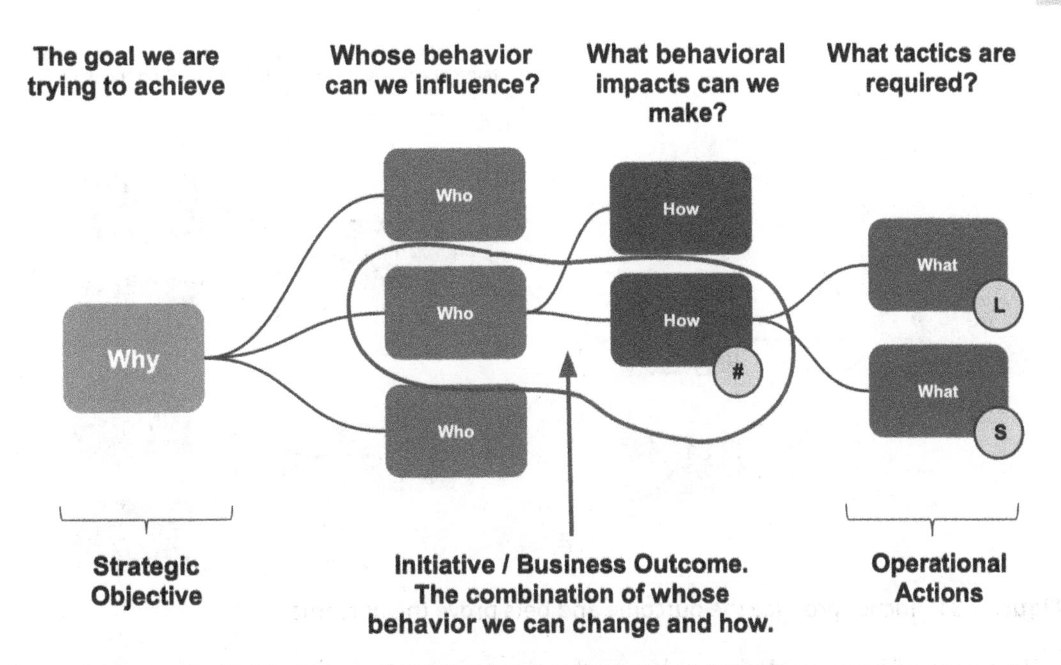
The goal we are trying to achieve
Whose behavior can we influence?
What behavioral impacts can we make?
What tactics are required?
Who
Who
Who
How
How
#
What
L
What
S
Why
Strategic Objective
Initiative / Business Outcome. The combination of whose behavior we can change and how.
Operational Actions

Figure 7.30: Tactical initiatives can be thought of as business outcomes.

Operational Level: Sub-Outcomes, Programs, and Projects

As shown in Figure 7.31, at the team level work is designed to achieve the desired outcomes. There may well be more than one option on how to achieve an outcome; we call these bets or sub-outcomes and the work to prove the bet the tactic or operational action. Just as a goal can have many outcomes associated with it, so too an outcome can have many bets—as they say there is more than one way to skin a cat. The reason for using the term *bet* is that, as with outcomes, these are hypotheses on what we think will provide the desired outcome. Each bet can be thought of as an experiment or a smaller hypothesis to the larger hypothesis of the outcome. Here you can use techniques such as A/B testing and Wizard of Oz, as covered in Chapter 6, to quickly prove out an idea without committing lots of effort.

Bets and tactics represent the changes to our business capabilities. Therefore, they involve changes to people and processes as well as technology. In fact, we may come up with ideas whose tactics do not require any technical change at all. By focusing on the outcome, we can maximize the work we don't do. If we can achieve an outcome through a process change rather than a large technical endeavor, then this is preferable. The point is to focus on the desired outcome and not technical solutions—in other words, obsess over problems and keep all options open for solutions.

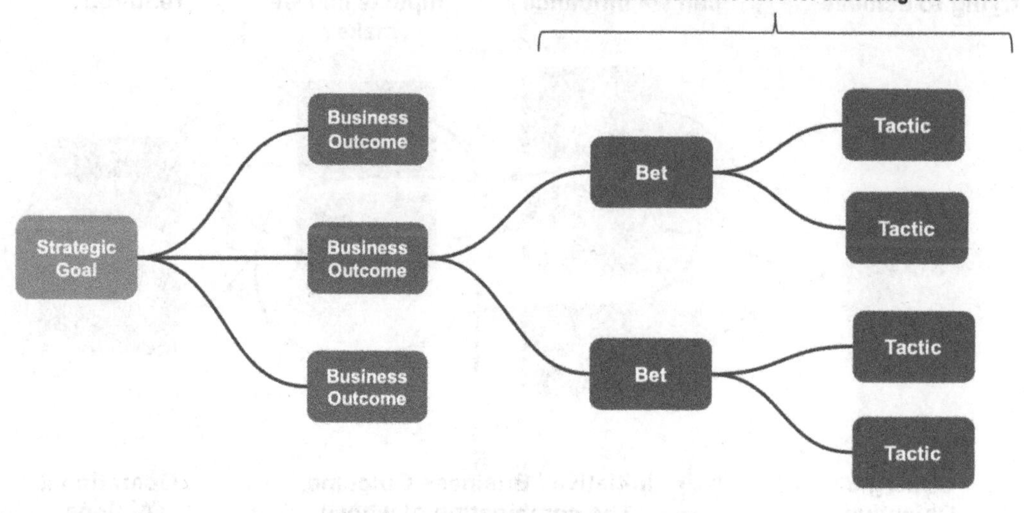

Figure 7.31: Tactics produce the outcome and bets prove the outcome.

In terms of how we define value and measurements:

- **Bets:** A good bet should clearly show how the hypothesis contributes to achieving the business outcome. It should also have measures that work as leading indicators.
- **Tactic:** At this level we measure the tactic as output or activity. By working on a tactic, we produce output that proves a bet, which itself, hopefully, will contribute to a desired outcome.

Table 7.5 builds upon the example in the previous section, showing how the bets and tactics contribute to and align with business outcomes.

Table 7.5: How bets and tactics contribute to and align with business outcomes

Operational Action (Leading Sub-Outcome or Bet)	Tactical Initiative (Leading Business Outcome)	Business Impact (Lagging Contribution to the Strategic Objective)	Strategic Objective (Lagging)	Business Goals (Lagging)
Allow customer to self-service flights	Reduce the % of flights fulfilled by humans	The avoidance of adding operations staff required to manually book flights	Avoiding the addition of X heads by self-service and automation equates to £xxx in margin improvement per order	EBITDA Goals of £xxxm
Automate flight booking				

Operational Action (Leading Sub-Outcome or Bet)	Tactical Initiative (Leading Business Outcome)	Business Impact (Lagging Contribution to the Strategic Objective)	Strategic Objective (Lagging)	Business Goals (Lagging)
Scan PDFs	Reduce the % invoices manually processed	The avoidance of adding operations staff required to manually book and check invoices		
Use API where possible				
Scrape support websites				
Customers print own tickets	Reduce the % of tickets processed manually	The avoidance of adding operations staff required to manually process tickets		

Investment: Funding for Outcomes

The primary goal of governance is to determine how an organization will allocate its limited resources to the work that will have the greatest chance of delivering its business goals and aspirations. This allocation is defined by a combination of the organization's investment strategy (whether that be focusing on objectives that explore the development of new product or service ideas or objectives that exploit and enhance existing ones) and the relative benefit of each proposed initiative. The priority and sequence of initiatives are determined by business benefit, dependency, and the investment strategy. As you have read in Chapter 4, "The Anatomy of an Operating Model," there is a need for an adaptable operating model and therefore by definition the investment model must also be equally flexible. The funding model must be able to support a quick response to changes that result from internal feedback and events in the external business environment.

The investment process is made up of four stages, as illustrated in Figure 7.32:

- **Setting Investment Targets:** The executive team sets investment targets for objectives and goals based on the strategic direction of the company. Investment can be targeted by business unit portfolios, product portfolios, or strategies. Investment targets act as guardrails on how the organization intends to allocate its investment budget. An organization could choose to target short-term payback investments, known as run and grow, that exploit the current business model versus long-term transformable investments that are more speculative but are vital to ensuring the business avoids being disrupted. Other methods to define investment targets could involve geographies, specific products, or customer segments.

- **Allocating Funding to Initiatives:** Most of the investment at this level is the funding in team capacity, how many people and teams within each given product group, plus any budget for software. For cross teamwork, an initiative can also be funded as a project and a one-off investment. The executive team funds product groups based on the criticality of their product area (business capability, value stream, or customer journey) in relation to achieving the outcomes detailed in the initiatives. In addition, the exec team will also fund based on the capacity needed to maintain existing capabilities to continue to run the business.
- **Allocate Funding to Team Capacity to Deliver the Initiatives:** This splits investment into two phases. First, the organization determines how much it wants to invest in each area of the business in terms of team capacity to action change—the capacity investment level. Second, there is an investment in business outcomes we want the different areas to focus on based on the business strategy.
- **Allocating Capacity to Projects and Programs:** After each product group has been funded, the product group and product owners will allocate product teams' time based on the features they believe are needed to deliver the initiatives in the strategic plan. We will explore this in detail in Chapter 13.

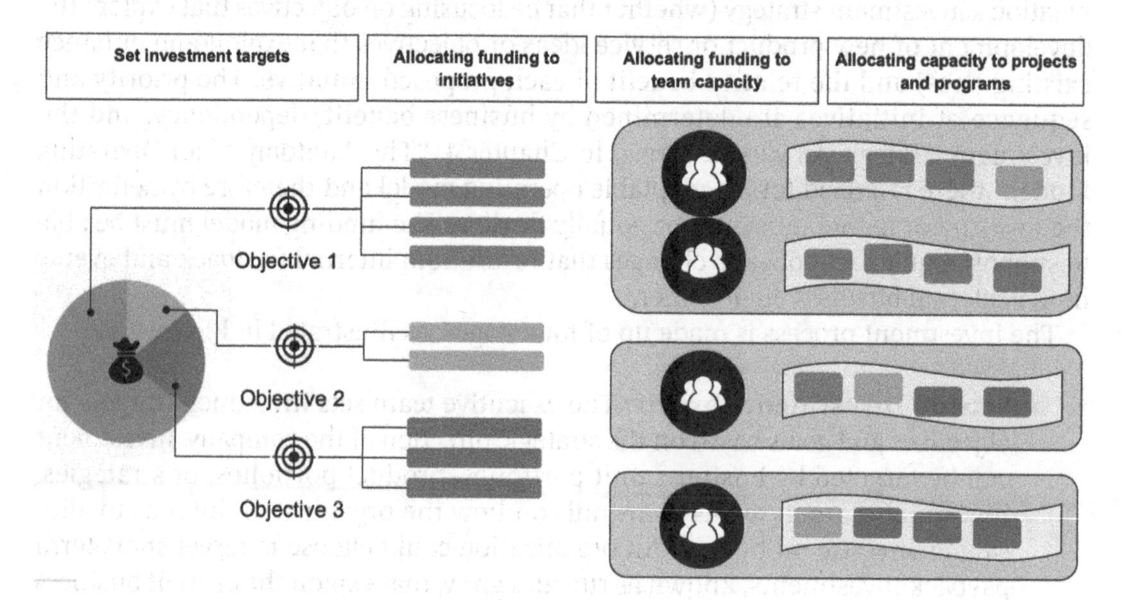

Figure 7.32: The four levels of investment strategy

Setting Investment Targets

Before teams can be allocated funds, there is a need for the exec team to allocate investment targets. As illustrated in Figure 7.33, the exec team determines what percentage of the investment budget should be spent on each strategic objective over the coming year. The exec team's investment strategy is based on how best to achieve the business vision and not a pure return on investment calculation. For example, the exec team may choose to invest 10 percent in discovering new products and be prepared to have little to no return on investment in this financial year. They may decide to invest in geographical expansion, which could have a lower ROI than if they invested in the core market. However, their thinking could be based on how more appealing a business looks if it can operate in a larger market. Alternatively, they could choose to focus investments on the unit economics of their core market and exploit existing products to consolidate cash if there are unfavorable market conditions for exploration. This is why it is key to understand the business context as we will cover in Chapter 10, "Understanding Your Business." You need to understand the policies behind decisions and especially the future intentions of the board.

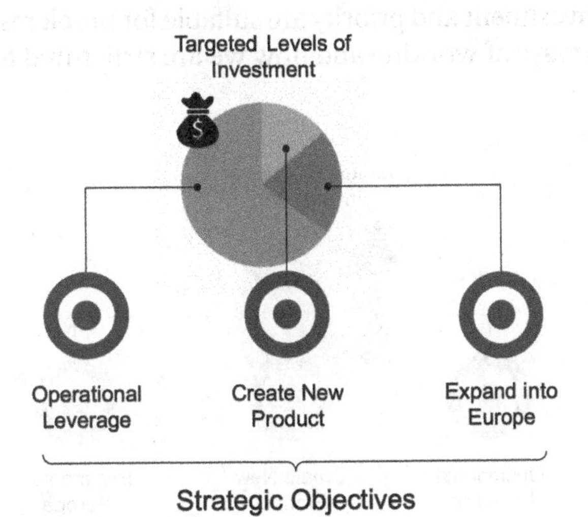

Figure 7.33: Targeted levels of investment for strategic objectives

Allocating Funding to Initiatives

At the tactical level, we fund initiatives as shown in Figure 7.34. The outcomes are those that we believe will contribute to the strategic objectives. The conventional IT budgeting process is based around the concepts of transient teams, fixed plans, fixed costs, and fixed scope. For anything other than simple problems with obvious solutions, this simply does not work. The way we fund work should complement how we work, the context of the work, and how we are organized. Traditional methods are based around funding a project model and prioritizing on the strength of a business case with a forecasted ROI. It is predicated on the assumption that work is deterministic and therefore scope, budget, and time can be fixed. This makes sense for very simple projects in well-understood obvious problem domains or for exploiting well-known business areas. However, awarding investment based on the strength of a business case and the details of the plans is not applicable in all problem contexts. We simply do not have that level of precision upfront for every initiative.

For problems in complex contexts, there are simply too many unknown unknowns to predict the exact activities, costs, and scope that are required to solve them. Prioritization planning methods that are held annually are slow to react to changes in the business environment. They demand detailed ROI projections and detailed data on opportunities, which is not effective for complex problem contexts and business environments of high volatility. Work in complex contexts by its very nature is emergent; therefore, we will not get to a level of precision until we start the work. While traditional methods of investment and priority are suitable for problems in obvious contexts, they are barriers to ways of working and how we are structured to tackle problems in complex contexts.

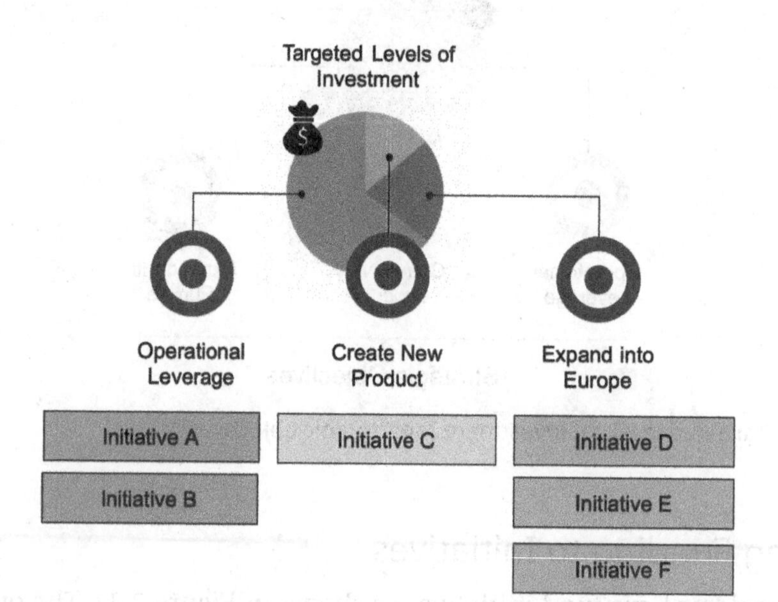

Figure 7.34: Initiatives are funded to contribute to achieving the strategic objective.

What we need is a way of adapting funding that is sympathetic to work in complex domains in addition to traditional methods of funding that are suitable for obvious problem contexts. We need to use the appropriate method of funding based on context. We need to fund teams to work on delivering value without being tied to a prescriptive plan or upfront scope so that they are not restricted in their ability to change tactics to meet goals as their understanding of a problem area increases. We also need a fast method of priority that employs good enough data, human judgement, and relative prioritization based on value. However, by sacrificing precision for speed, we also need to fund and prioritize on a more frequent cadence, allowing us to review, change, or amend priority decisions in reaction to internal feedback as well as the external context, thus reducing the impact of making the wrong choices.

Using the Right Investment Method for Initiatives

As illustrated in Figure 7.35 and highlighted on the Wardley Map in Figure 7.36, the way we fund initiatives depends on the problem context and how we will approach solving it. Teams working on novel and unique capabilities versus teams working on improving waste and immaturity in commodity capabilities need to be funded in different manners. For commodity or simpler contexts where requirements are fixed and don't require an agile way of working, we can employ a project- or plan-focused method of investment For teams that are working in a new, novel area with complexity with high uncertainty, it makes sense to opt for more of a venture capital approach from the start, where funds are released over time in small drops in line with value generated. This investment method is commonly known as product funding. The difference between the traditional method of funding and this new approach can be seen in Table 7.6.

Table 7.6: Project vs. product funding

	Project Funding	Product Funding
What are we funding?	A project with the scope defined upfront.	A team focused on a set of business outcomes.
What does the funding cover?	Lifetime of the project.	Lifetime of the team.
How do we sign off on investment?	Business case detailing costs, scope, and benefits.	The team showcases a road map of themes and leading metrics that will be used to achieve the desired business outcome.
What type of project suits this method of funding?	Plan-led projects that reside in obvious problem domains.	Value-led projects that reside in complicated or complex problem domains.
When do we stop funding?	When the scope has been delivered to plan.	When we have achieved a satisfactory level of value and want to switch to another outcome.

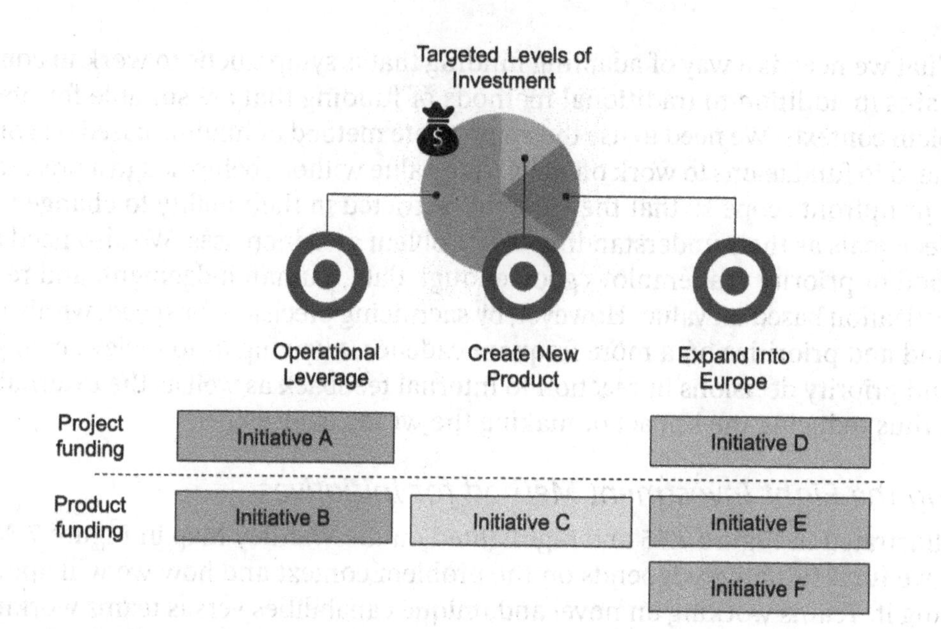

Figure 7.35: Different initiatives will require different methods of funding.

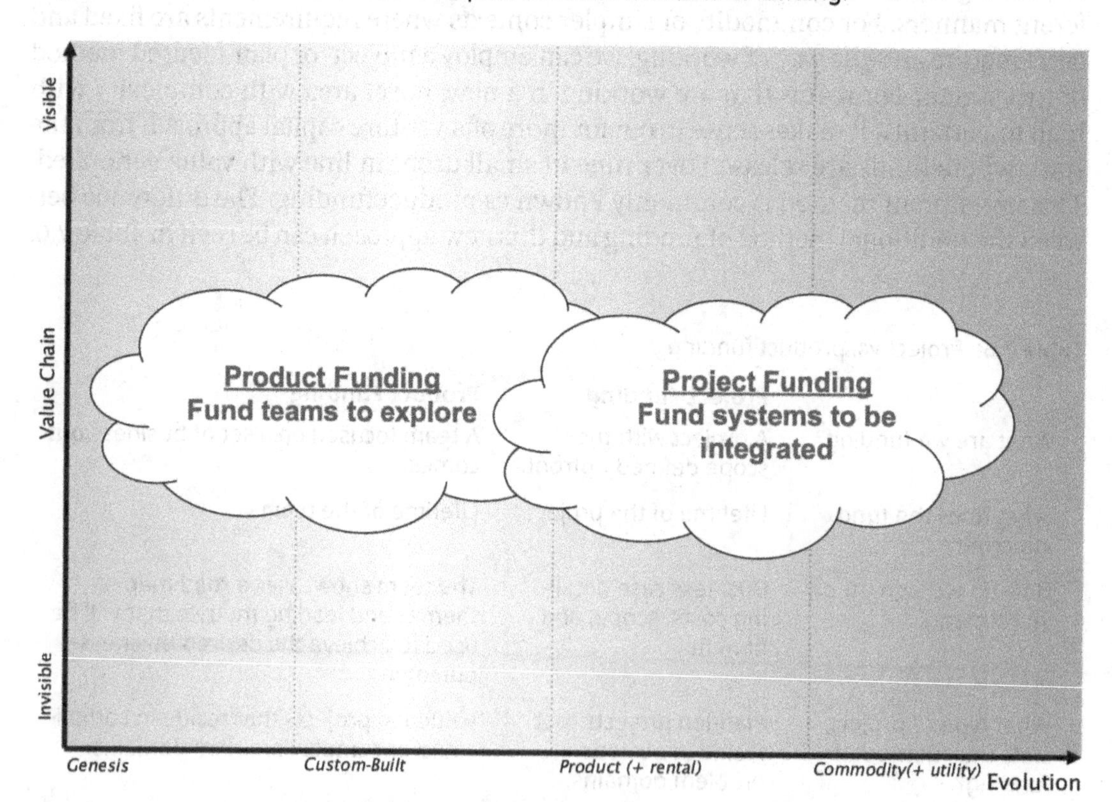

Figure 7.36: Wardley Map showing how the evolution of an area can affect how we fund

Product Funding: Using a Venture Capital Approach to Investment in Complex Problem Domains

We can think of product investment in complex contexts in the same manner as a venture capitalist would approach a new product investment. We start by giving teams a business outcome to focus on and provide some seed money. Further funds are released to the team as value is delivered, value being a measurable progression toward the business outcome. As shown in Figure 7.37, teams look at multiple bets to have the best chance at achieving an outcome. Once they have proved that a hypothesis produces value, they can continue to invest in it until they reach the point of diminishing returns. At this point, they can pivot to a new bet. This funding model prevents sunk costs in that we only fund for the bets that work as opposed to the project model, where we fund for full scope and deliver it in its entirety whether it provides value or not. We can avoid investing too much in the wrong idea by having regular feedback via short delivery cycles and basing progress on leading outcome-based metrics that have a direct link to the desired outcome.

This venture capital–style approach is in stark contrast to the traditional project-led model of execution, which only measures value at the end of a project, if at all. This is all made possible due to the short development cycles, which in turn allow for releasing funds in shorter budget cycles. So, while we allocate budgets annually at a capacity level, we can manage them monthly, quarterly, or in whatever cadence we choose at a team outcome level. This funding process contributes to the mind shift of a focus on delivering incremental value rather than delivering output and features. Because of the regular cadence of plan-fund-do-feedback, teams are ruthlessly focused on ensuring that investment brings value in the shortest time frame. The result is a more flexible method of allocating investment, one that can adapt to changes in the context and can reinvest in bets that result in capability improvements that contribute to business outcomes and stop funding ideas that are not contributing to strategic goals.

It is far easier to fund at a coarser level where there is predictability (i.e., capacity for change) as opposed to funding discrete projects at a granular level where there is uncertainty in how effective they will be. We do not lose any governance in this funding model because we are able to review regularly and report on progress toward each business outcome, making any correcting actions such as increased investments, removal of impediments, and direction setting as necessary. We remove the need to spend time and effort checking that we are adhering to a budget or a plan. Instead, we align funding capacity, teams, and leadership around the organization's most important business problems and make governance decisions and accountability around the KPIs that measure the outcomes we truly care about, value delivered, rather than any other vanity project metrics such as features shipped.

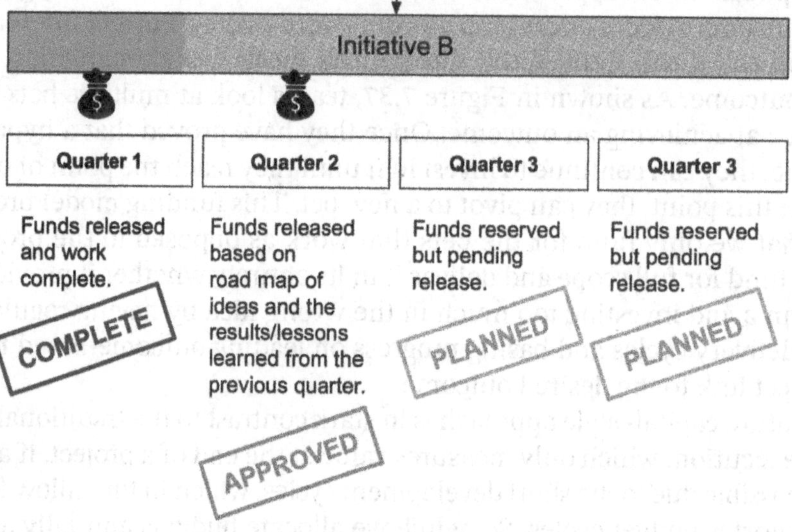

Figure 7.37: In product funding, investment is released based on value delivered.

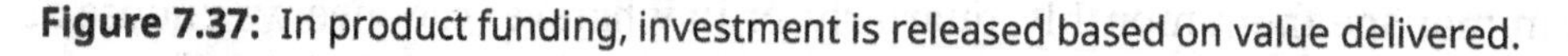

Project Funding: Using a Project-Led Approach in Simple Contexts

While most of the funding will be for the capacity of long-lived product groups and teams focused on delivering outcomes, some will be assigned for discrete projects that don't require a consistent and permanent team to invest in their evolution. For commodity capabilities and where we are looking to leverage COTS software products usually in conjunction with a systems integrator, we can use a more traditional plan-driven project approach. This enables us to have more certainty about total costs, and we can perform more due diligence to ensure that if we are signing up to a three-year deal with a software supplier with integration and ongoing costs, we are backing the right horse. As mentioned, these capabilities may need to be project led for the initial implementation of a COTS but then revert to a venture capital approach when the platform is integrated.

Investing in Team Capacity

As illustrated in Figure 7.38, an approach to investment that can reduce the risk of trying to predict unknown unknowns, offers greater flexibility to move investment quickly when the context changes, and ensures that investment is allocated to the most important areas is to fund the capacity to change at a product level. Rather than trying to spread the budget across a disparate set of projects, we can explicitly invest in the products that are critical to delivering tactical initiatives, aka business outcomes. This represents a move to a top-down investment strategy by prioritizing investment, in the form of teams, in the most impactful areas and then setting outcome targets for teams to determine how best to achieve them. By making this change we can control investment allocation more easily and move the funding decisions to the critical capabilities and to people closest to the work rather than to a central committee judging business cases.

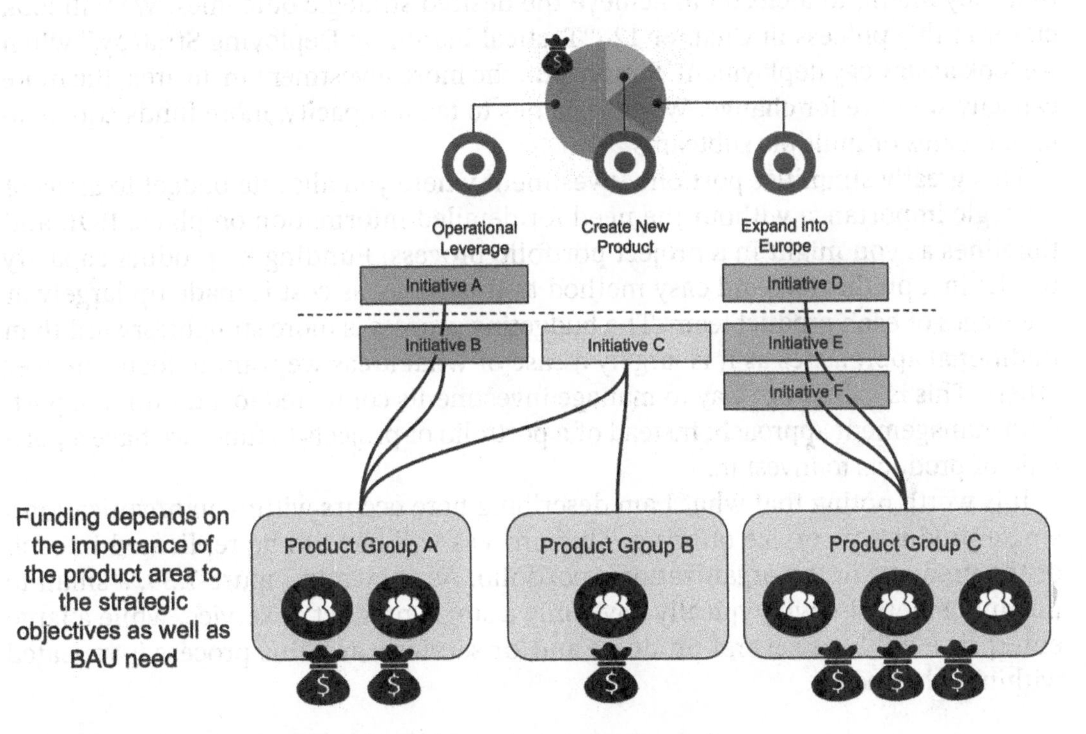

Figure 7.38: Product teams are funded based on strategic need.

Annual Investment in Product Team Capacity Based on Strategic Need

Capacity investment is drastically different from the traditional annual operating planning budget process. However, there's still a need to allocate funds and size product teams and their budgets, and this still makes sense on an annual basis, in line with overall financial budgeting activities. During annual investment cycles, funding is allocated by the leadership team. We can determine at a high level the capabilities required to support the strategic goals based on the detail covered in the IT strategy (see Chapter 11, "IT Strategic Contribution"). By funding annually and basing funding around fixed strategic goals and business capabilities, we have the best chance at keeping teams stable.

Funding is based on the relative importance of the business capabilities that each product group supports—in other words, how much contribution we think the teams will make to the strategic goals. Leaders will determine relative investment levels based on the strength of road maps of outcomes that will contribute to business goals. Rather than a bureaucratic and wasteful governance process requiring detailed plans to receive investment, teams define a road map of high-level themes and problems that they intend to focus on to achieve the desired strategic outcomes. We will look closer at this process in Chapter 12, "Tactical Planning: Deploying Strategy," when we look at strategy deployment. Simply put, the more investment in an area, the more capacity we have for change. When it comes to team capacity, more funds equate to larger teams or multiple subteams.

This greatly simplifies portfolio investment where you allocate budget to areas of strategic importance without the need for detailed information on plans, ROI, and timelines as you might in a project portfolio process. Funding by product capacity results in a predictable and easy method to track cost, as cost is made up largely of the wages of each product team. The budgeting process is more straightforward than traditional approaches as it is largely a case of what areas we want to focus on over others. This is a different way to manage investments compared to a traditional portfolio management approach; instead of a portfolio of projects to fund, we have a portfolio of products to invest in.

It is worth noting that what I am describing here occurs within an organization's single product or service offering. This process will need to be replicated in each of the products in the organization's portfolio. As shown in Figure 7.39, a small to medium enterprise will typically have only a single product or service, while a large enterprise will have several products and/or services, and this process is repeated within each product.

Invest in the Capacity to Manage BAU as Well

When determining the funding allocation to product groups, the exec team needs to consider not only the portfolio of initiatives (strategic, operational essentials, architectural) that sit within a product group's area but also to include funding for the capability to manage BAU portfolios, where decision rights and prioritization are democratized to the teams themselves. Remember in the organizational structure as covered in Chapter 5, there is no dedicated small change, support, or BAU team. In this new method of investment, there is no separation between build and run activities. Product teams own all the applications and technology that relate to their business area. They design solutions to problems; they build it and they run it with no handoffs. This means that one team deals with strategic business outcomes as well as BAU items, where BAU covers small enhancements, technical debit, and bugs.

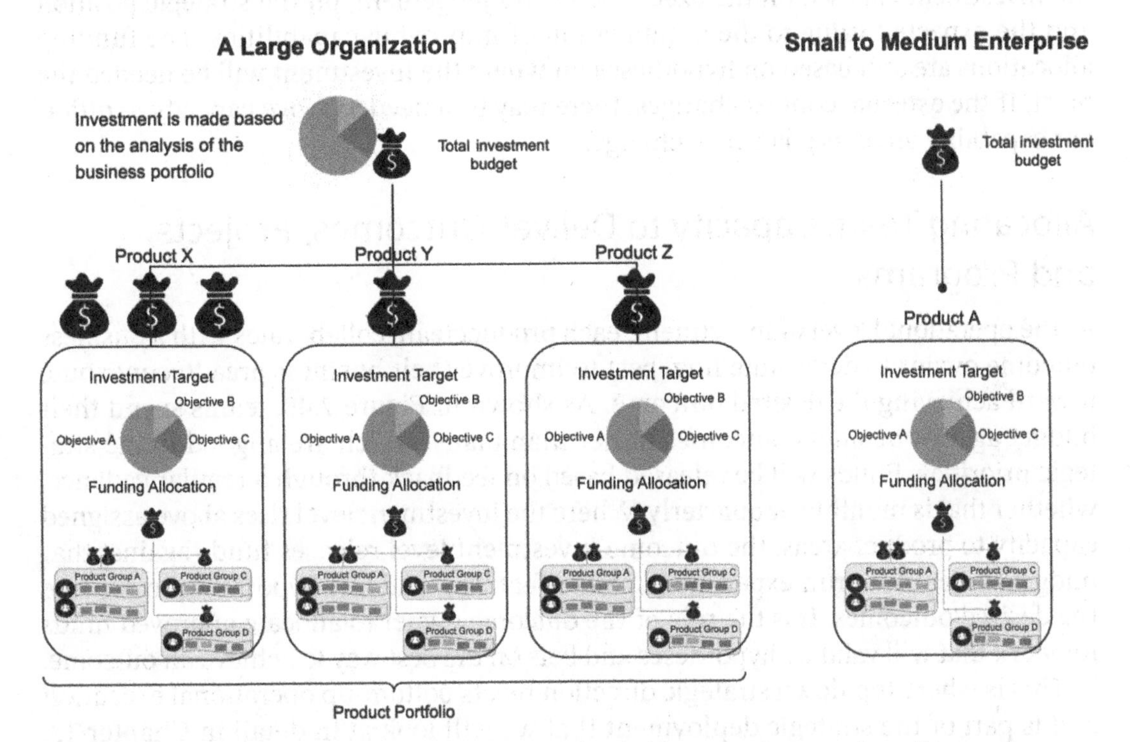

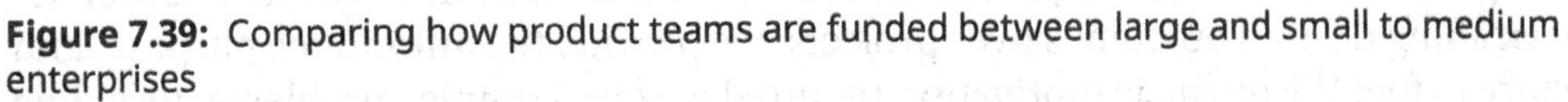

Figure 7.39: Comparing how product teams are funded between large and small to medium enterprises

The knock-on effect for investment is that we need to fund capacity not only for strategic outcomes but for lights on operational needs as well. All capabilities need to be covered and owned by a team even if that team is kept to a minimum. Capability areas that are covered by COTS solutions should not be exempt either. While this may seem that we are funding and investing in areas of no strategic importance, we only need to invest the minimal capacity. It means we can retain team knowledge and architectural integrity to be responsive to changing needs. If we compare this to the alternative project mode of operating, we have transient teams that have no intrinsic motivation to invest in the code base that they are working on before they hand it over to another project or maintenance team. The project model results in a loss of deep domain knowledge and a code base that is ever more difficult to change every time a project ends and a team is broken up.

Reviewing Funding Allocation

Funding is typically set at the start of the financial year, but capacity allocation can be reviewed on a regular basis and be adjusted over time if things change. At this level the investment is based on the executive team's judgement, on the strategic position and the expected value to the organization of improving capabilities. The funding allocations are still based on hypotheses on where the investment will be needed the most. If the external context changes, there may be a need to move capacity to different capability areas as goals may change.

Allocating Team Capacity to Deliver Outcomes, Projects, and Programs

At the operational level of investment, each product team collaborates with a business outcome owner to determine how best to improve their business area to contribute toward achieving the desired outcome. As shown in Figure 7.40, teams spend their budget against business outcomes rather than plans, which are aligned to the strategic priorities. Funds will be released based on feedback through a regular cadence, whether that is monthly or quarterly. Where the investment level rises above assigned capacity to product areas, the outcome investment level releases funds against that budget for teams to run experiments, build features, and test hypotheses to achieve the desired outcomes. It is the role of the outcome owner to allocate approved funds for work that will validate hypotheses and bets on the best way to achieve an outcome.

This is where top-down strategic direction meets bottom-up operational execution and is part of the strategic deployment that we will look at in detail in Chapter 13. By cascading desired outcomes over projects, we provide the autonomy, purpose, and mastery to fuel the intrinsic motivation required to solve complex problems. By giving strategic direction in the form of desired outcomes but not prescribing how to tackle them, we achieve accountability and alignment to balance team autonomy. From a funding position, outcome owners and product teams are completely accountable for how they use their time and how they prioritize work in the pursuit of outcomes.

Quicker to Adapt

As we are funding outcomes rather than projects with fixed scope, teams can pivot at will, evolving their road map of work (hypothesis if you will) constantly as they gather more data on the business outcome problem or indeed if the context changes and they need to target a new outcome. As product teams discover more about their problem domain, they will come up with new ideas. If they find a promising solution that is delivering results, they may continue to invest in that over experimenting with other ideas. Because we fund product groups and not projects, product group leaders have the autonomy to allocate time and investment where they feel it will have the biggest chance of delivering the desired business outcomes. If we give teams the accountability for outcomes, then we must give them the autonomy to use their budgets as they see fit without having to keep going back to get these things approved at a higher level. This enables those people closest to the problem to make quick decisions and drive real business value.

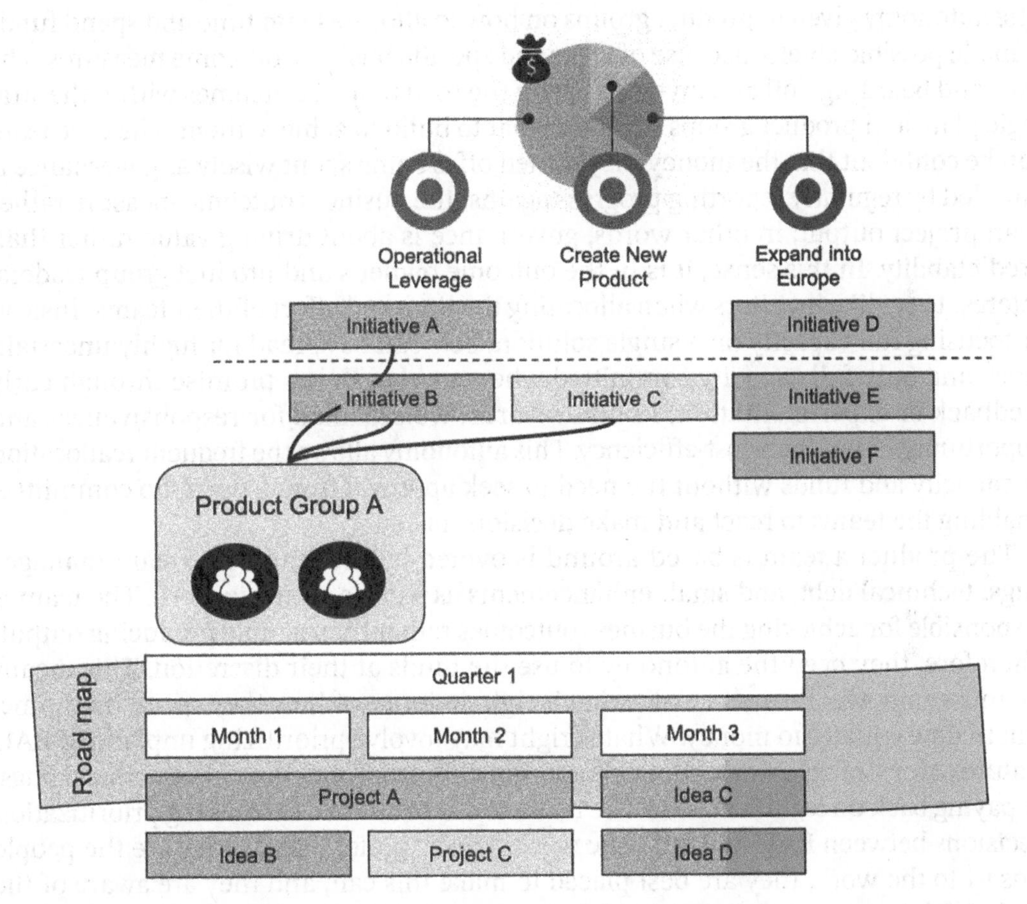

Figure 7.40: Product teams "spend" their capacity on delivering the features, projects, and programs required to achieve an initiative.

Funding in this manner is simpler because, instead of asking, "What project should we prioritize based on assumptions and predictions of the future?" we can ask ourselves, "How much money do we want to invest in solving this problem or achieving this outcome?" "What can we achieve by this date?" "How much should we invest in this hypothesis to achieve the outcomes before we look at alternative ways?" This makes far more sense when dealing with complex problems because we can control the controllables (i.e., how much we want to spend on a problem) and task a team with prioritizing the next best action to reach the outcome and regularly review to determine if we want to continue to invest or move funding somewhere else.

Trusting Teams to Manage Funds

The autonomy given to product groups on how to allocate team time and spend funds is made possible chiefly because of clear and specific business outcome measures. The exec and board sign off and invest based on the road map of outcomes within the strategic plan, and product groups work on what to build to achieve them. The exec team can be confident that the money they signed off is being spent wisely as governance is handled by regularly reporting progress against the business outcome measure rather than project output; in other words, governance is about driving value rather than predictability. In this sense, it is in the outcome owner's and product group leaders' interest to act like investors when allocating the time and effort of their teams. Instead of focusing full capacity on a single solution, bets can be spread for highly uncertain work and only full capacity committed when an idea shows promise through early feedback or experimentation; in other words, we organize for responsiveness and opportunity rather than cost-efficiency. This autonomy allows the frequent reallocation of capacity and funds without the need to seek approval from a portfolio committee, enabling the teams to react and make decisions faster.

The product a team is based around is owned by the team. The team manages bugs, technical debt, and small enhancements as well as strategic work. The team is responsible for achieving the business outcomes rather than simply producing output. Therefore, they need the autonomy to use the funds at their discretion. This means we must empower the team to do what is right in terms of how they spend their time, where time equates to money. What is right may involve prioritizing unplanned BAU features over strategic work, such as trade-sensitive small enhancements, critical bugs, or paying back on technical debt. The teams are accountable for making prioritization decisions between BAU and strategic work. This is preferable, as they are the people closest to the work, they are best placed to make this call, and they are aware of the trade-off between operational and strategic needs. This is why we need to balance team autonomy with the alignment to strategy, alongside regular review, and steer sessions to avoid a total focus on local optimization. Work and progress is transparent and the cadence of feedback regular, which avoids the need for micromanagement or detailed reporting.

Trusting teams, giving them both autonomy and accountability, to use their time and funds in the most effective way is part of the mental model change required to instill an intrinsic motivated workforce. We don't prescribe projects for teams; instead, we delegate outcomes and support teams to solve it their way while also balancing the BAU needs of their business area. We don't lose any control over funding in this model. How success is measured is still defined by leadership from defining the strategic outcomes. With autonomy comes accountability. To build trust and continue receiving funds, teams are accountable for value. They regularly show progress toward outcomes using business-focused metrics over activity metrics to continue to secure funding.

Decision Rights: Empowering People

Up to now I have largely focused on product teams, specifically the teams that deliver the technology, but the principles that we have discussed in Chapter 5 apply to teams at all levels in the IT organization. The principles of focusing on outcomes and being self-sufficient and autonomous are applicable to IT leaders down to IT delivery teams. What will change is what they are accountable for, but what is the same is how they make decisions.

Figure 7.41 shows the cascading delegation, ownership, and responsibility from the strategic down to the operational, with outcome-based teams at various levels of the nested product groups all aligned with the level above. It is important to note that the teams are roles, not people; for example, the same individual may be responsible for a product team and be the technical lead on an outcome. Depending on the size of your organization, you may have people fulfilling multiple roles or a person per role. If it is the former, then apply the principles around cognitive load that have been discussed in Chapter 5 to have a manageable amount of complexity to deal with. For each team there is a different set of accountabilities related to the level that they are working. However, what is common is the need for trust and autonomy for teams to make decisions.

At each level of the organization there is a focus on producing outcomes. The product teams based around these outcomes will be made up of various people in the business, but each will require a technical representative. Just as a product team will collaborate with business experts to improve or create a business capability, various levels of the IT organization will collaborate to achieve business success. The CTO/CIO will collaborate with the rest of the exec team on which goals to focus on and the impacts required. IT leadership, architects, and product group owners will collaborate with their peers on the best outcomes to achieve the required impacts and coordinate any cross teamwork. Each team, from vision to delivery, is a multidisciplinary team, made up of technical and business expertise. In *Sooner Safer Happier: Antipatterns and Patterns for Business Agility, IT Revolution Press (11 Dec. 2022)*, Johnathan Smart et al. refer to this as a "triumvirate of roles," three key roles within each team. As shown in Figure 7.42, these roles are a value outcome lead, whose responsibility is to focus on what needs to be achieved with an outward view of the customers; a team outcome lead, whose responsibility is inward facing to help the system of work (this can be PMO support); and an architecture outcome lead, whose responsibilities are to focus on the high-level technical implementation.

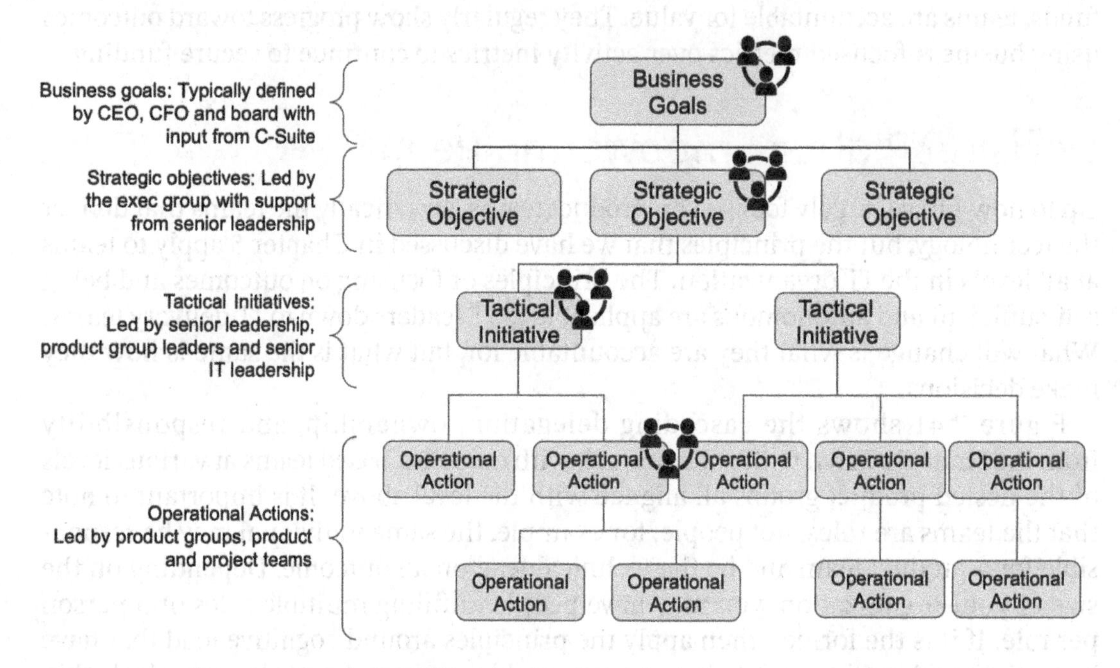
Business goals: Typically defined by CEO, CFO and board with input from C-Suite
Strategic objectives: Led by the exec group with support from senior leadership
Tactical Initiatives: Led by senior leadership, product group leaders and senior IT leadership
Operational Actions: Led by product groups, product and project teams
Business Goals
Strategic Objective
Strategic Objective
Strategic Objective
Tactical Initiative
Tactical Initiative
Operational Action
Operational Action
Operational Action
Operational Action
Operational Action
Operational Action
Operational Action
Operational Action
Operational Action

Figure 7.41: The areas of governance

Strategic Level: Setting Intent

The leadership team is responsible for the overall strategy; for prioritizing the strategic objectives against operational essential, mandatory, or technical needs; and for funding team capacity. As shown in Figure 7.43, for each strategic objective there is an owner and team assembled to determine how best to tackle the problem and what impacts will likely result in achieving the goal. The team consists of a senior leader along with members of all the business capabilities, including IT, that are part of the value stream that will collaborate on delivering the objective. The job of this team is to create a portfolio of outcomes, aka tactical initiatives, that will contribute to achieving the strategic objective, along with determining the relative level of investment it wants to make in each capability area. The impacts and measures cascaded down will guide other teams on how to contribute to business success. Strategic objective team leaders work closely with the product group leaders and product owners closest to the area of influence and form a consensus on the best way of achieving strategic objectives and what outcomes to pursue.

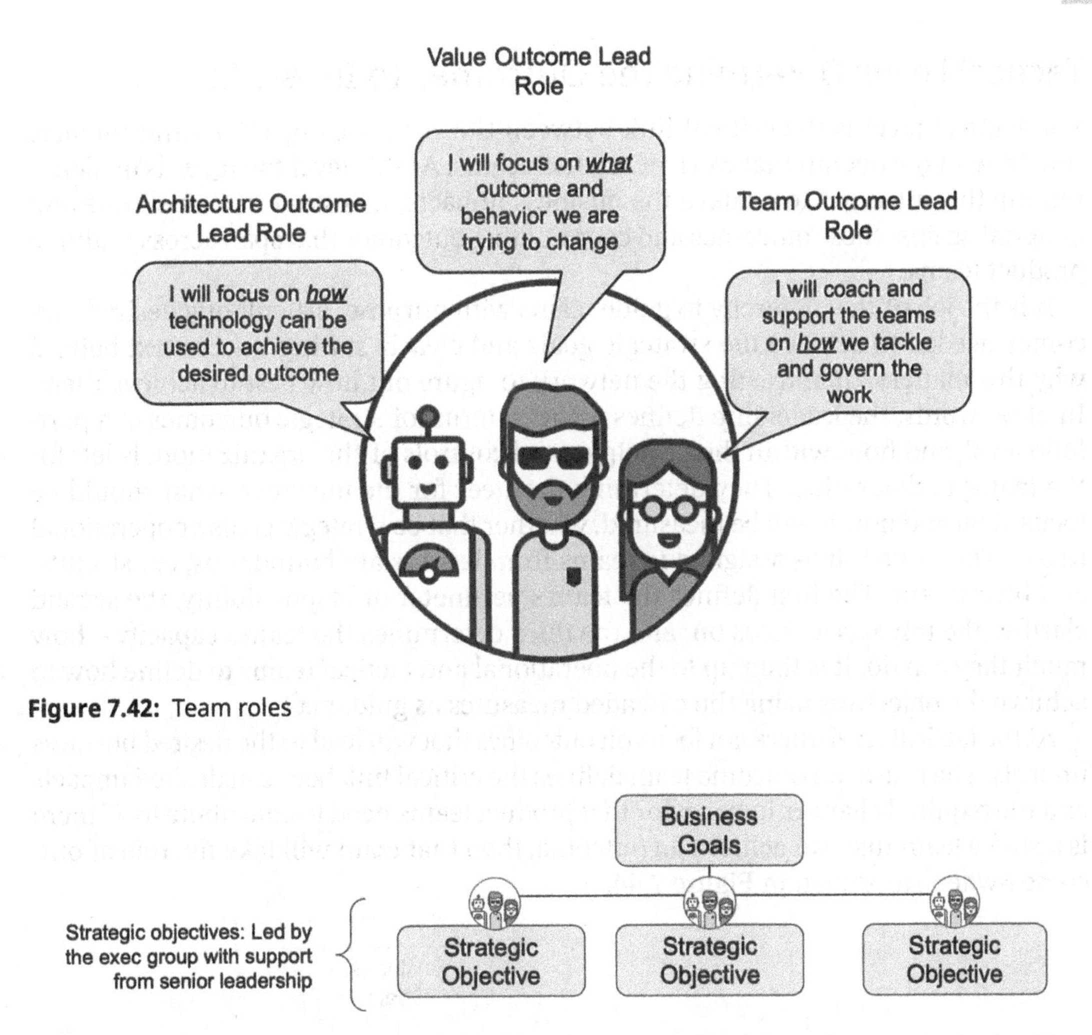

Figure 7.42: Team roles

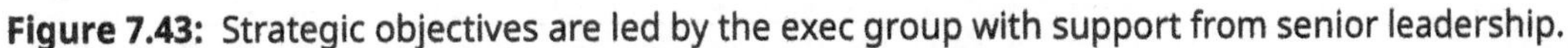

Figure 7.43: Strategic objectives are led by the exec group with support from senior leadership.

Objectives that are expected to have a large impact on a business and require a large investment and large organizational effort (such as creating a new business model or revenue stream or launching in a new country) will require a dedicated leader to succeed. Objectives related to improving existing capabilities and business models do not need a dedicated person to focus full time; they will need only the role of an objective owner. Amazon refers to dedicated goal owners as Single-Threaded Owners. This is a leader that will be completely focused on the goal with no distractions, as they will have no other competing priorities and thus the cognitive load capacity and energy to drive a large initiative. Single-Threaded Owners have the authority to act as a bridge across many product groups and product teams.

Tactical Level: Determine the Outcomes to Invest In

The tactical level is the critical link between the strategic objectives and impacts sought with the operational work of product teams. At this level the focus is on determining the outcomes to achieve the business impacts, investing in teams' time and material against those outcomes and coordinating outcomes that span across multiple product teams.

It is the job of the hierarchy to guide teams with purpose, articulating desired outcomes needed to achieve the strategic goals and clearly stating the context behind why this matters, then trusting the network to figure out how best to achieve them. In other words, the leadership defines what, in terms of strategic outcomes at a portfolio level, and how, within the principles and controls of the organization, is left for the teams to determine. They determine the steer for the business, what should be focused on and how it will be measured, whether that be strategic goals or operational needs. The three things assigned to teams from leaders are boundaries, constraints, and investment. The first defines the team's perimeter of responsibility, the second clarifies the impacts to focus on, and the third determines the team's capacity—how much they can do. It is then up to the operational and tactical teams to define how to achieve the objectives using the cascaded measures as guidance.

At the tactical level, there is a focus on outcomes that will lead to the desired business impacts. The role of an outcome team defines the critical link between desired impacts and the required changes in behavior that product teams need to contribute to. If there is a single team that can achieve an outcome, then that team will take the role of outcome owners, as shown in Figure 7.44.

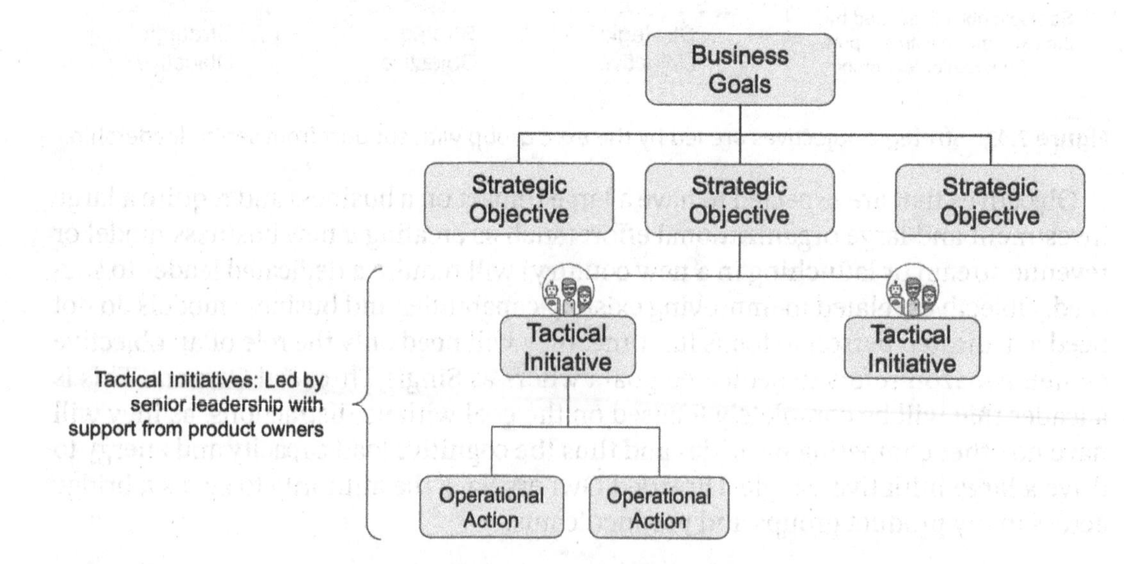

Figure 7.44: Tactical initiatives are led by senior leadership and product group managers.

Operational Level: Delivering the Outcomes

At the operational level, product delivery teams work with outcomes owners on how to improve their capability area to change user behavior. They also must balance this with business-as-usual activities because there are no support teams to hand over the day-to-day running and maintenance of a product. As shown in Figure 7.45, product teams are accountable for determining the best way to improve the capability; they own the process to achieve a desired outcome. At this level teams who have worked closely with outcomes owners determine the output that is most likely to achieve the outcome (where output is the improvement or creation of a business capability).

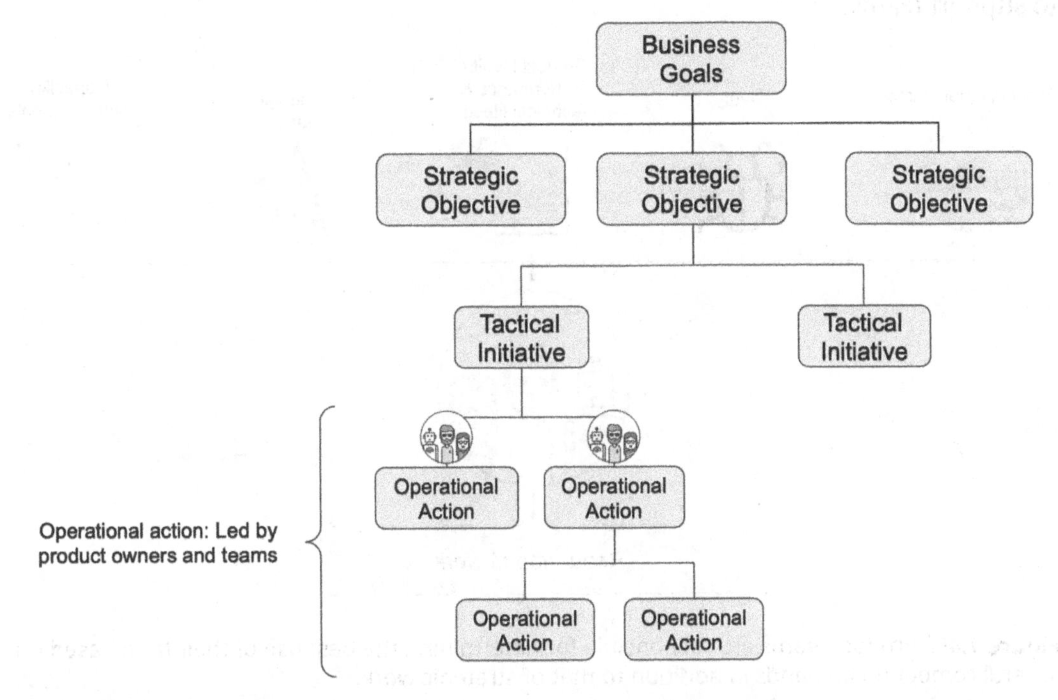

Figure 7.45: Operational action is lead by product owners and teams.

You Build It, You Run It

As shown in Figure 7.46 product teams are responsible for determining the best use of their time based on several competing demands in addition to that of strategic work. There are no separate maintenance or support teams so product teams will need to manage BAU needs. BAU items can include bugs, technical upgrades, updates to maintain current capability tracked as product health metrics, or changes for compliance needs.

In addition, they will need to support other product teams where there is a dependency on them. Not only must product teams balance the needs of strategic vs. BAU improvements to their capability, but they must also ensure the uptime, performance and all the other non-functional requirements. Product teams are responsible for the idea, execution and run time upkeep of what they build. In other words, they build it and they run it. The job of the leadership is to ensure that product owners understand higher level organizational goals and therefore have the context to make decisions on what they focus on; however, there will be times when there is a lack of clarity in the relative value of two or more competing priorities and help is required to decide. This is where leaders can use their power of command and control as an escalation point to support teams.

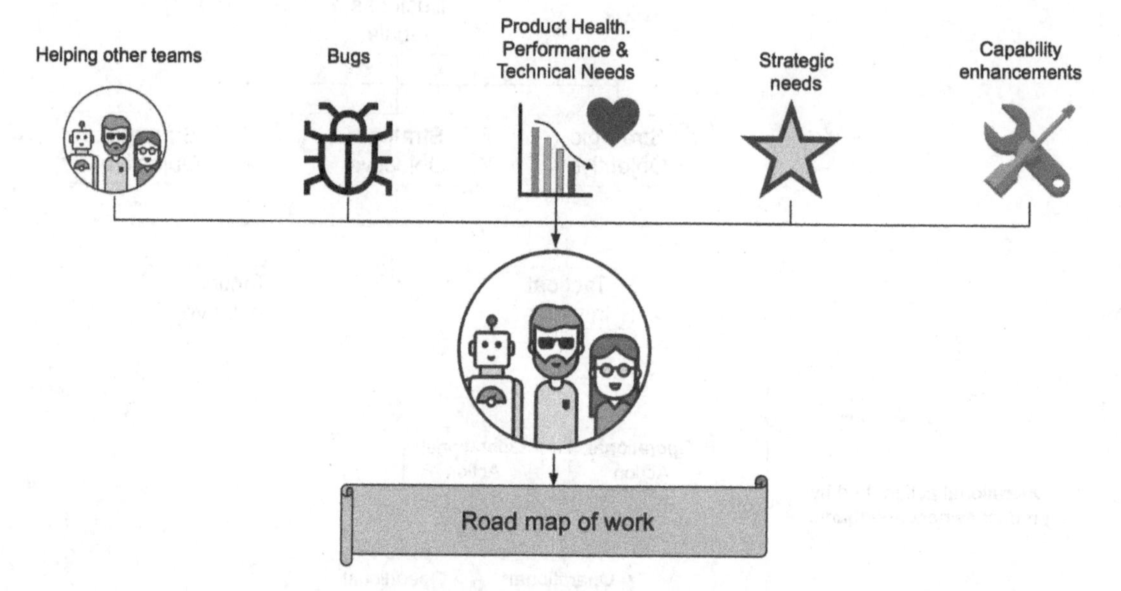

Figure 7.46: Product teams are responsible for determining the best use of their time based on several competing demands in addition to that of strategic work.

Trusting People to Make Decisions

In Chapter 3, "How to Change the System," I introduced the three elements of intrinsic motivation as laid out in Daniel Pink's book *Drive, Canongate Books; (5 July 2018).* Intrinsic motivation is vital when working in complex and complicated problem domains to determine the best path to solve problems. As a reminder, the three elements are purpose, mastery, and autonomy. Purpose is provided through leadership direction and defining a perimeter of accountability, defining outcomes to achieve, and showing how they align to the North Star or business vision. Mastery is achieved by managing the amount of cognitive load and domain complexity people must deal with for teams to develop deep expertise in a problem domain and have enough capacity to focus on improving problem solving. Last, we have autonomy, which is achieved by giving people the authority to make decisions and determine the best path to solve a problem.

As shown in Figure 7.47, in the traditional IT mode of operating as an order taker, there is a focus on compliance to meeting the requirements as prescribed. This required a low level of trust as development teams were effectively delivering to an agreed upon specification acting like a contract. Invariably the solution—often based on a hypothesis—was not going to produce the desired outcome. However, in a compliance mode, the IT delivery team merely delivered based on the requirements given. Since many problems exist in the complex and complicated problem spaces and require intrinsic motivation and the need to sense and adapt, persistent product teams need to actively contribute to achieving an outcome, understanding what is needed rather than what is asked for, collaborating with business peers, and sharing responsibility for delivering an outcome of value. This requires the ability to make decisions, and trust is fundamental to enable that.

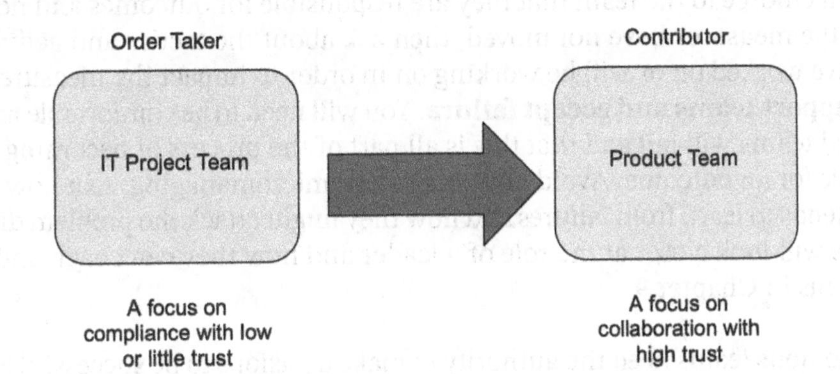

Figure 7.47: Moving from project teams to product teams

Each team should have the authority to make decisions on how best to solve problems and deliver value, whether that is determining how best to pursue goals, impacts, and outcomes or the actual delivery of an output. However, we cannot simply delegate authority to a team. Delegation is about tasks; authority is about making decisions. For a team to have authority, they need to be empowered to be able to make choices for themselves, and to have empowerment, teams need to be trusted.

It may be hard for people from more traditional IT modes of operation to switch from a compliance mindset to a collaboration mindset. Equally, it can be hard for leaders to place trust in their teams, but without trust and the empowerment to make decisions for themselves; teams will not be able to solve complex or complicated problems, and they will default to a compliance mode of working and wait to be told what to do by people far away from the actual problem. To build trust and support the move to a more collaborative method of working, leaders can focus on guardrails, reviews, and support.

- **Clarify the guardrails.** Leaders should add constraints to teams to limit team authority and autonomy to act as guardrails, much like the guardrails parents use to help children on a bowling lane. The first constraint is clearly defining the boundaries of a team's accountability. The second constraint is to communicate clear objectives with meaningful measurements, in other words, articulate the outcomes you want, not the solutions you need. These constraints will define the area of authority the teams can operate within when defining solutions to problems.
- **Review results and outcome, not activity.** When regularly catching up with a team, focus primarily on how they are performing against the objective measure rather than seeking to understand all the activity the team has undertaken. This will enforce to the team that they are responsible for outcomes and not output. If the measures have not moved, then ask about the tactics and activities they have worked on or will be working on in order to impact the measure.
- **Support teams and accept failure.** You will need to be comfortable and accept that teams will fail and that this is all part of the process of becoming accountable for an outcome. Avoid jumping in and micromanaging. Ask how the team intends to learn from failures and how they might attack the problem differently. We will look closer at the role of a leader and how they can coach and support teams in Chapter 9.

Autonomous teams need the authority to make decisions to be successful and move toward business contributors rather than order takers. This can only be achieved through trust and collaboration, not only between teams but also between the levels of an organization. An empowered team is the result of trust from leadership, coaching and support, and the clarity of their areas of influence. Autonomous teams are grown through the change in mental models led by leaders' actions and behaviors. With autonomy comes accountability, but what teams are accountable for differs at different levels.

Project and Program Managers

Most things are products and owned by the product teams. The only exception to team autonomy is when there are outcomes that require the coordination of work across multiple teams. No matter how hard we try, it is unrealistic to think that teams can be independent for every outcome the business requires. Indeed, often multiple business capabilities need to be modified, requiring many different product teams to collaborate to achieve outcome, as shown in Figure 7.48. This is typically done with the role of an outcome owner, whose role it is to facilitate the collaboration and manage the dependencies of multiple teams. Although we want product teams to own a solution end to end, there is often a need to use a leader that sits outside the team to assist in solving problems, acting as a bridge across organizational boundaries and managing cross-team dependencies.

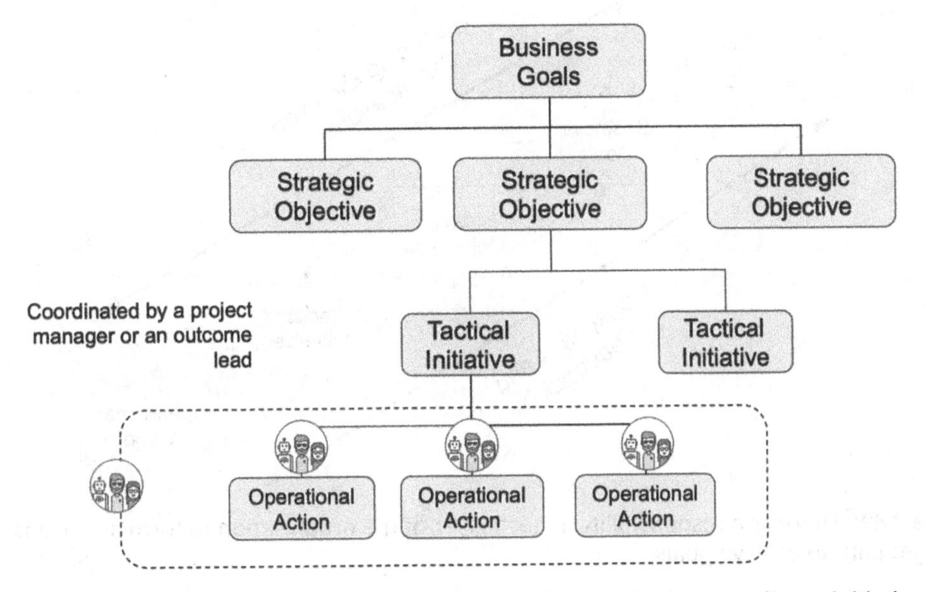

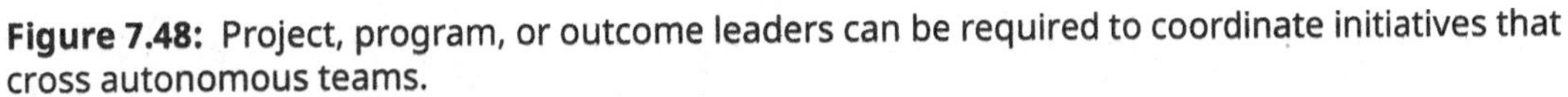

Figure 7.48: Project, program, or outcome leaders can be required to coordinate initiatives that cross autonomous teams.

Everyone Is Responsible for Enterprise Value

There is a collective responsibility at all levels of the organization to contribute to discovering opportunities, overcoming constraints, and driving business value. As shown in Figure 7.49, there is a need to communicate and cascade direction and measures throughout the layers of accountability in the organization and a responsibility at each layer to form a consensus on the right path to achieve goals. There is a responsibility for teams closest to the customer to give feedback to those who set direction and goals, the big bets, and assumptions to validate or correct any hypotheses. This heightened focus on collaboration and contribution when it comes to work differs from the traditional top-down and command-and-control approach to setting objectives. This requires those at an operational level to have intrinsic motivation to solve problems rather than simply be order takers with no accountability for outcomes. We will look at this closer in Chapter 12, when we look at the catchball consensus approach to strategy deployment that Hoshin Kanri prescribes. Empowerment and trust coupled with the clarity of direction encourages a more collaborative organization that is greater than the sum of its parts and is in sync when it comes to adapting and changing.

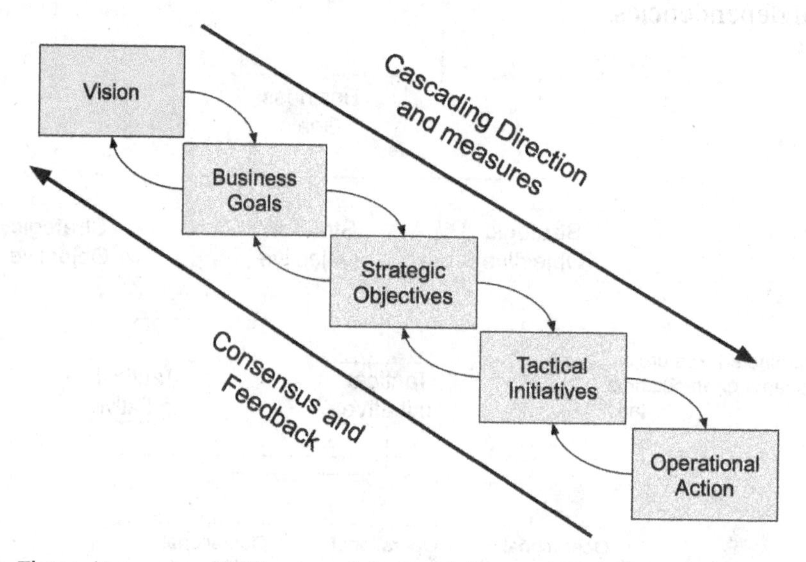

Figure 7.49: There is a responsibility at each layer of the organisation to form a consensus on the right path to achieve goals.

Performance: Monitoring Value

When it comes to reviewing and monitoring the performance of in-flight investments, we should ask ourselves two principal questions:

- Is the work meeting the value as defined by the measurements?
- Should we continue to invest in this work or pivot to a new opportunity?

But how do we monitor value for in-flight projects given that we can have projects at two ends of a spectrum, namely value driven and plan driven? The answer, as you will find in all the areas of our operating model, is to apply the appropriate method based on context. For teams working on complex problems, also known as value driven, you need to review progress against outcome targets on a regular basis and approve funds in a rolling manner. Teams need to have a rough plan before starting and some ideas and operational actions of how they will attack the problem. Then it is a case of measuring value delivered. On the other end of the spectrum is governance for projects in simpler and more obvious contexts, or plan driven, that may require COTS software or adherence to new legislation. This could involve large investments and people commitments for enterprise software integration and configuration. This is to say that these projects can't be achieved in an iterative manner. It's more that there are known unknowns and there is a definitive boundary to scope.

There is nothing wrong with a plan-driven approach. Plans can be delivered iteratively. A plan approach is not the same as a waterfall approach. The difference between a planned approach and a value approach is that there is a known correlation between cause and effect, which means we can have a fixed scope, or at least a very good idea upfront. For example, we would follow a plan-driven approach for moving a warehouse capability from in-house to a third party. The interfaces between our organization and the third party are knowable and so are the many scenarios of stock movements. This is not to say that we won't adapt along the way, deliver value early, or discover issues; it is more that we have a very clear idea of what we need to do upfront.

For value-led projects such as increasing e-commerce conversion, we don't know what will make customers buy more products online. Is it better content? User reviews? UX changes? Price changes? Product configuration? We just don't know, so we need to make a series of small experiments to prove our assumptions and hypothesis on the quest to deliver value. Table 7.7 shows the difference between a plan-led and value-led approach to work and when to use them. Notice that they both focus on delivering value early and often and that their goals are to deliver the desired business outcomes. The difference, highlighted in Figure 7.50, is that in a plan-driven approach, we know the scope, whereas the scope is variable in a value-driven approach. Figure 7.51 shows the different performance methods overlaid on the evolution phases of a Wardley Map.

Table 7.7: Value- vs. plan-driven work.

	Business Outcome	
	Value Driven	**Plan Driven**
Problem Context	Complex, unknown unknowns.	Obvious, predictable, with known unknowns.
Scope	Emergent.	Fixed.
How we measure	Value delivered.	Progression on the plan.
Goal	Deliver value.	Deliver the plan to deliver value. We can't achieve the outcome without completing the scope.

Continues

Table 7.7: Value- vs. plan-driven work. *(continued)*

	Business Outcome	
	Value Driven	**Plan Driven**
How we know when we are done	Enough value delivered, or decision taken to focus elsewhere.	Enough features compete to unlock value.
How to plan	Continuous planning.	Upfront planning.
How to govern	Control through adapting to feedback and adjusting plans.	Control through progress against the plan.
Example	Encourage holiday makers to sign up for insurance before they begin their holiday.	Move warehouse to a 3PL.

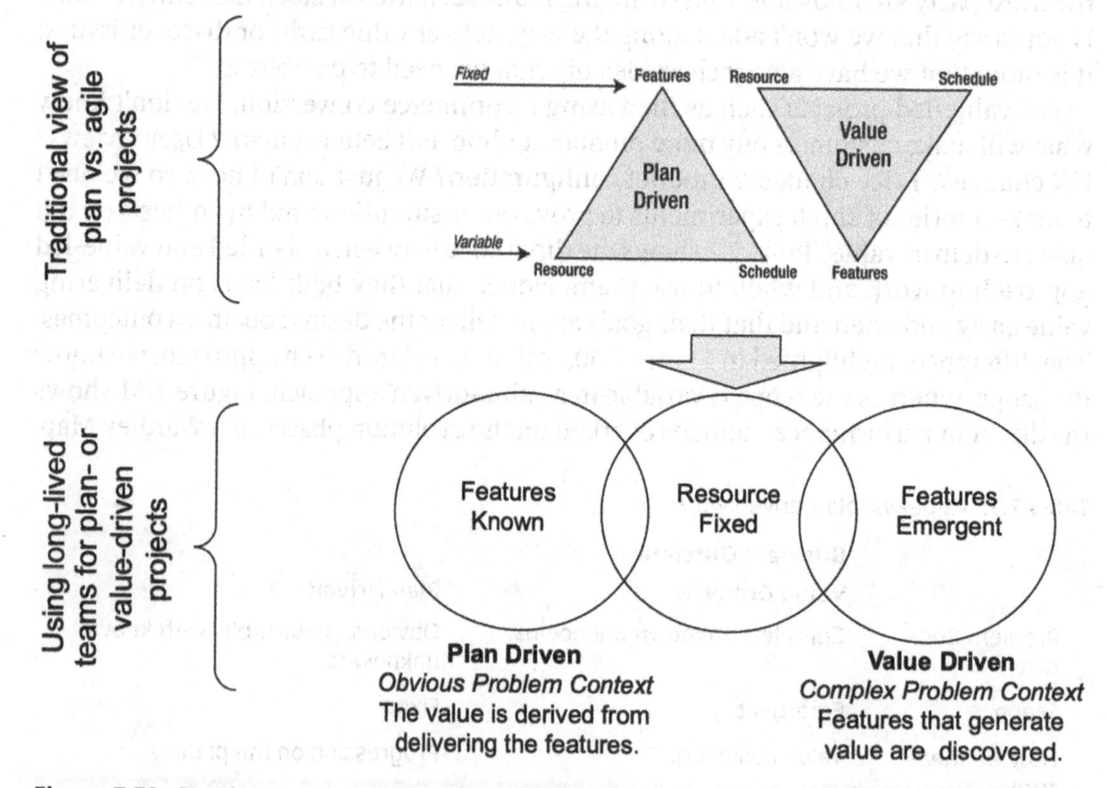

Figure 7.50: Fixed resource capacity can work on a plan driven, known scope, or value driven, unknown scope, work.

The choice between a value-driven and a plan-driven approach is not as polar as described here. In larger initiatives, you may find that areas of the project have characteristics of a plan-driven approach whereas other stages or areas are more value driven. For example, it would make sense to follow a plan-driven approach and a systems integrator's best practice to implement a basic contact center solution with known features. However, once this is integrated, we could then choose to follow a value-driven approach to drive conversion, experimenting and testing hypotheses on what drives buying behavior. Agile extremists may argue that a plan-driven approach is poor, but it is more accurate to say that it is only poor when used in the wrong context. Therefore, always use the most appropriate approach based on context, and if your context changes, your problem moves from complex to obvious, then adapt your approach. Chapter 6 took a closer look at ways of implementing both value- and plan-driven projects.

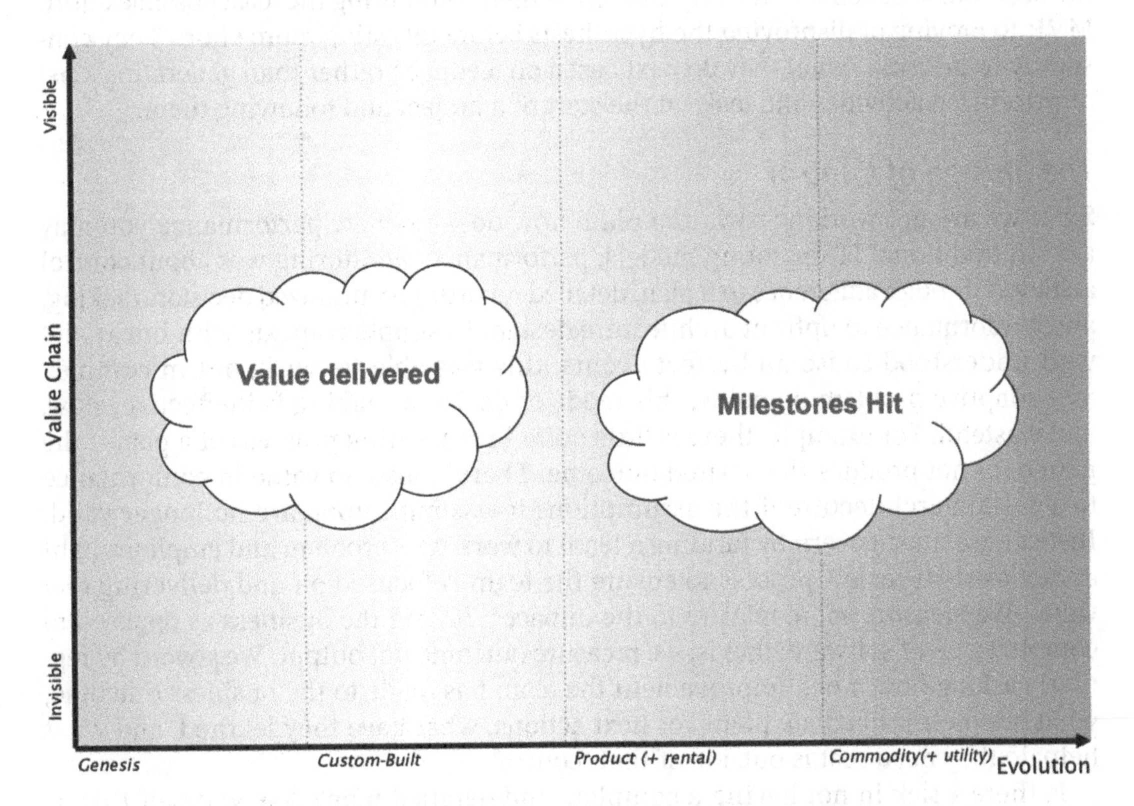

Figure 7.51: Wardley Map showing how the evolution of an area can affect how we review performance

Value-Driven Projects: Govern for Value, Not for Predictability

The whole purpose of pursuing an agile way of tackling complex problems is to frequently adapt based on changes in the context, empirical evidence, new opportunities, and new constraints rather than religiously following a plan. The agile way of working was born to manage uncertainty; its practices enable impediments, incorrect assumptions, and new opportunities to come to the surface early and often to be dealt with by addressing and/or pivoting. Embracing an agile way of working doesn't guarantee success. We need to be comfortable with a level of uncertainty when dealing with complex problem domains because by their very nature they are highly unpredictable. By adopting short-term planning, and through sensing and adapting, we can be more dynamic and less rigid on how we solve problems. A continuous planning process puts value on team learning, which is the essential ingredient to managing complexity. Only through a focus on learning more about the problem domain can we ever hope to derive a solution for it. Teams use data and experiments to determine the best course of action—the requirement—then committing the least possible effort, MVP, to proving or disproving the hypothesis before investing more effort. They constantly re-prioritize what they do next based on feedback rather than generating a list of prioritized activities and tasks at the start of a project and following them.

The Illusion of Control

So, if we are not working toward a plan, how do we review performance you may ask. In traditional IT operating models, performance monitoring was about control achieved through adherence to a plan, detailed reporting, centralized decision-making, and conformance to upfront architecture design. In simple contexts with linear and well-understood cause-and-effect events, this was able to work, but in complex and adaptive problem domains, this mode of decision-making is ineffective, slow, and wasteful. For example, there is little point in measuring progress of a plan if the plan does not produce the desired outcome. There is also no value in conformance to a certain architecture if the assumptions it was built upon are no longer valid. Instead, we must govern by funding a team to work on a problem and employ a light and consistent review process to ensure the team is focused on and delivering real value. We measure value relative to the impact it has on the business as opposed to completeness of software; that is, we measure outcome not output. We govern by regularly asking how much improvement the team has made to the business outcome, what are their immediate plans for next actions, what have they learned, and what help do they need that is outside of their control.

Is there a risk in not having a complete and detailed plan? Ask yourself this: is there not more risk in making assumptions and committing to one plan of action at the very time you know the least about a problem, the start of a project? There is an inherent risk in signing off on a detailed business case riddled with assumptions and ROI projections of an unpredictable future that can change for any number of reasons. There's further risk that once a project is underway, a leader can no longer influence it if it is failing. We can remove this risk by funding capacity to work on a complex problem and regularly reviewing if we are generating value. The shift is to change from making decisions at a coarse level, which is fraught with risk, to decision-making at a more granular level, where mistakes and incorrect assumptions can be corrected at pace. Leaders can still retain control and influence but at a more granular and frequent level through short feedback cycles where the team demonstrates progress and value delivered. At this stage leaders can continue to invest, correct strategic steer, or stop investment.

Simply put, you cannot be agile while rigidly following a plan. If you have a perfect plan and the solution to a problem, then you don't have a complex problem, therefore there is no need to use an agile approach. In cases like this, go with a plan-led approach as it is more suited and more effective. If you are rigidly following a plan in a complex problem domain, it means you cannot be learning. If you are not learning, you are not adapting and correcting your course. Following a fixed plan and being agile are completely at odds with each other. Agile isn't about not planning; it's about constantly planning and replanning to reduce risk—using short delivery cycles with fast feedback so that teams can adapt, pivot, and correct the course.

Value Review

The leadership team should trust delivery teams to self-organize and determine the best course of action to achieve desired business outcomes. However, the teams need to be transparent on what they are doing and ensure that there are fast feedback loops so everyone is informed, aligned, and can contribute and collaborate on work that contributes to outcomes. We don't need detailed plans for value-driven work, but we do need to ensure we have a view on what money is being spent on. Leadership teams need a lightweight overview of progress and learnings to ensure the team is focused in the right areas to continue to fund efforts.

Leaders should collaborate and contribute during value reviews, using their unique position to add narrative, context, and ideas to the discussion of how a business outcome could be achieved and how to further progress against strategic goals. They should reinforce the purpose of the work and how it contributes to the organization's goals. This contrasts with the traditional view of a project review meeting anchored around control and one-way reporting. The leadership should actively support teams learning and taking small, calculated risks in the form of experiments to discover where best to invest. This servant leadership approach is designed to create a safe environment for people to fail, or better articulated as an environment for people to learn without fear. This is what is required to foster a culture of creativity, favoring pace over precision, adaptivity, ownership, and innovation. We will look closer at leadership and culture in Chapter 9.

The value review process, based on continuous delivery, planning, and reviewing, should be candid and transparent, with leaders attacking the problem and not the team. It should focus on three questions:

- What have we learned and achieved?
- What new bets or ideas do we have to invest in?
- Should we continue to invest against looking at other opportunities and needs?

What Have We Learned and Achieved? The first discussion topic at a review is to understand what has been achieved. What feedback have we received that tells us we are on the right or wrong track. What new information have we learned about the problem? How does this affect our working assumptions on solutions? We also need to understand what constraints the team faced that were out of their control to mitigate. Here are some example questions to ask during the review discussion:

- How are you progressing toward the desired outcome? How are you progressing on tactics? What bets have you proved?
- Does the business outcome still make sense? Or have we discovered something that changes where we should invest?
- What obstacles, blockers, or holdups do you face that you cannot resolve yourself? What dependencies do you have?
- What lessons have you learned? What insights have you discovered through the feedback? Were your assumptions correct? What would you do differently? What would you do more of?

What New Bets or Ideas Do We Have to Invest In? The second discussion topic at the review is what plans the team has based on the feedback and learning. Whereas the first discussion topic is about the team feeding back and leaders listening, in this discussion we should expect leaders to contribute and collaborate on insight and ideas based on the analysis of the feedback. Here are some typical questions to ask:

- How is the road map evolving? What new ideas do you have?
- How can we help you? Do you need help, or funding for software?
- What are you planning to do next? What are you hoping to learn? What does success look like?
- What data and evidence has led you to focus on this next?
- How will you measure success? What metrics will you use? How quick can we tell if the assumptions are correct?

Should We continue to Invest against Looking at Other Opportunities and Needs? During funding and feedback meetings, the leadership group determines whether to continue to fund the outcome or pivot to a new goal. Maybe we have made as much progress on a goal as we would like, or maybe the context has changed and a new goal is of more importance? Or maybe our original hypothesis on the goal was based on an incorrect assumption, and we know we are on the wrong path? There are often cases where work toward an outcome, after initially making good progress and delivering value, begins to plateau, and we start to reach the point of diminishing returns. Because of the frequent cadence, we can recognize this situation and make decisions to move the team's focus to the next business outcome.

When deciding on whether we should continue to invest in a team working on an outcome or switch focus, we must take into consideration how much value there is in the team continuing. We can make this view based on what the team has delivered, what they have learned, plus what ideas and road map items they have planned. We can then compare this with the next business outcome in the team's backlog and compare the relative priority and value.

Plan-Driven Projects: Govern for Adherence to a Plan

With regard to plan-driven projects such as investing in enterprise software, the governance model should behave differently. As previously discussed, supporting or commodity capabilities may be better served with off-the-shelf software. This investment may require more stringent review due to the large upfront capital investment, akin to a sure bet. It should be easier to estimate a more realistic upfront cost as well as leveraging best practices from partners to conduct the integration or simply paying setup fees and a monthly rental cost. For these investments, there needs to be due diligence made to ensure we are choosing the right vendor and partner that align to our technical vision and enterprise IT principles. Buying COTS enterprise software still requires a formal sign-off process as this is a big upfront investment and commitment to a plan. Even for operational and commodity projects, we should look for measures of value. For example, a new HR system could be measured in terms of employee satisfaction or security in terms of risk reduction.

When it comes to reviewing performance and progress of a plan-driven outcome, there is more focus on the activities and tasks, especially if value cannot be returned until all tasks are complete. However, there is also a need to ensure our plan is still valid and whether we have discovered new scope or new work that we did not plan for. Here are some typical questions to ask in plan-driven reviews:

- How many tasks are completed?
- What tasks are remaining?
- Has anything changed that means the plan is no longer valid?
- What new tasks have been discovered?
- What dependencies do you have? What help do you need that is out of your control?

Summary

In this chapter, we covered how we can adapt our approach to governance to complement the ways of working rather than fighting against them. We looked at ways to ensure we are doing the right things and that we are doing the things right.

- **Alignment:** Perhaps the most important principle of governance is ensuring alignment. It doesn't matter how well you execute if what you are executing is the wrong thing. To make alignment explicit and visual, we looked at a goal tree and showed how we can link strategic objectives with tactical initiatives and operational action using a case of both direction and measure.
- **Demand:** To manage demand at all levels of the organization, we looked at the lightweight process known as kanban. Kanban ensures that all in-progress and queued work is visible and that we move to a pull versus push method of work to limit work in progress.
- **Investment:** The conventional IT budgeting process is based around the concepts of transient teams, fixed plans, fixed costs, and fixed scope. For anything other than simple problems with obvious solutions this simply does not work. If we are to organize persistent teams around products and focus on outcomes for the reasons laid out in Chapter 5, then the funding process must complement this structure rather than fight against it. Instead of funding for projects in complex contexts, we can fund the capacity of product teams based on strategic need and the strength of a high-level road map. These teams can then have the freedom to determine the best use of the funds to achieve desired outcomes.
- **Prioritization:** Traditional prioritization requires precision to make decisions, which means heavy upfront planning and long approval processes. Given that problems are increasingly based in a complex context and that the environments in which organizations operate are highly volatile, working this way is far too slow. Instead, we need to favor speed over precision using a combination of available data, relative over absolute value, and human judgement (based on persistent long-lived teams with deep knowledge in the domain). By using speed, we can gather real-world feedback quickly to confirm or reject our hypotheses and then take the next appropriate action.
- **Measurement:** Work should focus primarily on customer value before business benefit and be measured in leading measures. The purpose of the work and the measures should cascade down the organization to provide connection and alignment from the tactical level to the strategic and business goals.
- **Decision Rights:** Power and decision-making should be distributed to those closest to the problem and the customer to enable teams to adapt quickly to

feedback. Decision-making can be empowered to people by trusting them. Decision rights are the same across the team, but accountability will change at the three broad levels:

- The IT leadership team shapes the IT strategy determining what will be bought and what will be built. They collaborate with other execs to determine the strategic objectives; they prioritize strategic versus BAU needs and investment in the capacity of teams relative to their importance to business success.
- At a tactical level, there is a focus on outcomes, the glue between achieving a desired impact and the improvement needed in one or more business capabilities. At this level there is also a need to coordinate and orchestrate when more than one team is involved.
- At the operational level, the product delivery team determines how best to use its time to improve or create a capability to achieve outcomes. Product teams are also accountable for balancing strategic outcomes with keeping the lights on in their area of responsibility as product teams think it, build it, and run it.

- **Performance:** The way we review the performance of work is dependent on how we execute the work. Obvious problem domains with known scope favor a more plan-driven approach, whereas nondeterministic projects in complex problem domains benefit with a value-driven approach. Whichever method, we should have a regular cadence of review and planning in order to ensure we are focused on the right direction.

8 How We Source and Manage Talent

What if we train them and they leave? What if we don't and they stay?
—Unknown

Without continual growth and progress, such words as improvement, achievement, *and* success *have no meaning.*
—Benjamin Franklin.

We are only leading when we are centered on making others better.
—Frances Frei

The global increase in demand for IT professionals has been exacerbated with what is termed by Anthony Klotz, a professor of management at Mays Business School at Texas A&M University, as the Great Resignation. Klotz predicted that the COVID-19 pandemic would lead to a mass exodus, as people considered their growth and careers, work conditions and benefits. This prediction was found to be correct. Due to a combination of uncompetitive wages, lack of career progression, appreciation, and flexible or hybrid working, all industries have seen a high movement of workers. Those that are not resigning are quietly quitting—a term used by Peter Drucker for employees that are disengaged with their workplace but continue to stay. This has been a wakeup call for many IT leaders, and the catalyst to reevaluate their employee value proposition to improve both talent recruitment and retention, which is critical to enable IT capabilities.

In this chapter we will look at the options for managing the talent needed to enable the IT capabilities required for business success. Depending on the nature of the capabilities, you may choose to supply it with in-house employees, partner with a third party, or outsource the capability altogether. This will form your sourcing strategy.

For the capabilities that will be retained in house, you will need a plan to manage your talent. We will look at how to take a human-centered approach to recruiting, developing, and retaining talent, one that is influenced by the lean methodology and Daniel Pink's intrinsic motivation principles as covered in Chapter 3, "How to Change the System." Talent management is fundamental for digital business and is likely to be one of the most important things you focus on as an IT leader. In the new hybrid working environment, you are no longer in competition with local businesses for essential talent; you are in competition for talent with companies from around the world.

Sourcing Strategy

Not all capabilities need to be provided by in-house employees. Your first step is to determine your sourcing strategy for talent based on capability needs. There are broadly four types of sourcing:

In House Teams are hired as full-time employees of the business, and they manage solutions themselves. The organization is responsible for hiring, developing, and retaining talent. This approach involves a high level of investment in terms of time, money, and human resources.

Outsource Entire or parts of projects or services are provided by a third party. The organization is still, however, accountable for the overall service delivery.

Team Augmentation In this model we supplement in-house teams with virtual resources to increase the capacity of the team. This is a flavor of outsourcing, where a third party will provide their developers on a time and materials basis.

Managed Service The entire service is outsourced, including the responsibility for maintenance, backups, security, and future enhancements. The third party manages the talent behind the service.

If those are the options we have, then how should we approach sourcing? Broadly speaking, for capabilities that are highly evolved and therefore ubiquitous, we can favor looking at outsourcing or a managed service. These capabilities, while they may be strategically important, are less critical to be delivered by in-house staff. For example, in many organizations, the IT service desk is often outsourced due to the ubiquitous nature of this capability. However, in an organization that requires a service desk to deal with specific and specialized equipment rather than generic laptops or machines, this may be unique and require an in-house team.

At the other end of the spectrum for unique capabilities, those that are differentiating and core to our business, we cannot afford to outsource the knowledge and expertise. As these are critical to the business for gaining competitive advantage, we can focus on in-house teams or augmenting in-house teams with a third party to deliver. Examples include innovation teams or data and analysis engineers required to provide MI and AI big data needs that drive business decisions.

Recruiting

Recruitment is an essential component of the talent strategy. Development and retention are of course important, but they are secondary concerns if you don't attract the right people in the first instance.

Be Clear on Your Value Proposition

Beyond being competitive on pay and benefits, we need to ensure that the opportunity we are offering appeals to what motivates knowledge workers. As covered in Chapter 3, Daniel Pink states there are three factors that help intrinsic motivation—namely autonomy, mastery, and purpose. We need to address all three of the points when recruiting. We will constantly come back to these points throughout the chapter.

Autonomy Be explicit on the decision rights and the responsibilities of the role or the team that you are recruiting for. The decision rights should include ways of working and the working environment (e.g., flexible work arrangements) as well as the work itself. You should also clearly state where the direction of work comes from and any guiding principles that need to be followed. This includes the organization's approach to failure and therefore how psychologically safe the environment is.

Mastery Be clear on the opportunities for growth and state the personal development support that you offer. How will you invest in their development, coaching, and mentorship? What are the career or growth opportunities to take on more responsibilities? Detail the nature of the work opportunities, such as the technology that they will be exposed to. What about the challenges and nature of the problems they need to solve? You need to be explicit on the experience and skills that will be gained from taking on this role.

Purpose Articulate how the role contributes to the organization's business strategy and therefore why it's important. People want to know how they fit in as part of the wider team and how they will contribute to overall business success. Show the problems and challenges that this role needs to solve in context so that there is an understanding of how they will contribute to the business reaching its aspirations. Explain the mission, vision, and strategy of the business as well as how IT strategy contributes to success.

Hire for Attitude as Well as Aptitude

To support an organization's need to exploit the advantage of a successful business model and the exploration of new models, you will need to tackle obvious, complicated, complex, and chaotic problems. A diverse set of problems requires a diverse set of people. Not just in terms of their skills or aptitude (what they can do) but also in terms of their attitude (how they approach doing it). As the Cynefin decision framework illustrates, different problem domains require different approaches and therefore different types of mindsets, and to some extent different personalities, to work. The approaches range from ones that are focused on driving efficiency, predictability and keeping the lights on to ones that are focused on creative and innovative thinking, challenging problems, developing new value, and disrupting existing norms. We need people with personalities at both ends of this spectrum to support a complex and adaptive organization: smart creative types and pure specialists.

The Need for T-shaped as Well as I-shaped People

The T-shaped skill concept is a metaphor to describe the abilities of people. As shown in Figure 8.1, the vertical bar of the letter *T* represents the depth of skills and expertise in a particular field, and the horizontal bar represents the breadth of knowledge and experience.

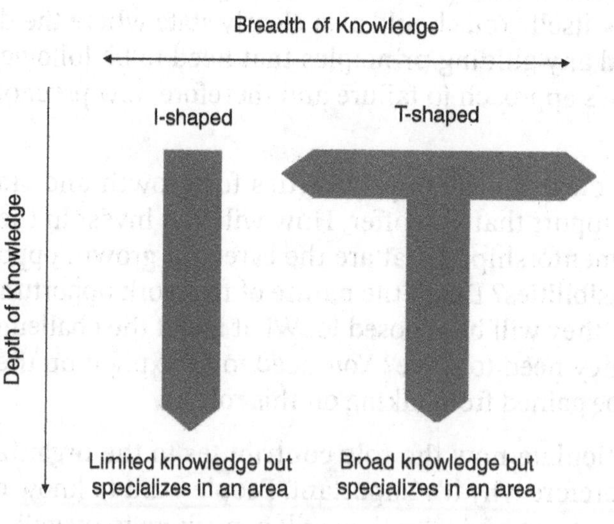

Figure 8.1: T-shaped vs. I-shaped people

At the extremes there are people with two types of ability:

- **T-shaped people** have a deep expertise in one area where they typically excel and a wide but shallow knowledge in other fields. The broadness of the knowledge and expertise enables effective collaboration across disciplines as well as adaptable thinking. T-shaped people have empathy and appreciation for other ways of working and thinking, allowing them to adopt new ideas and approaches with ease. This makes them innovative, smart, and creative. They can combine their technical knowledge, the art of the possible, with complex opportunities or constraints thrown at them to deliver real step change.
- **I-shaped people** are pure specialists that have a great depth of expertise in a single field but with little breadth of knowledge and/or experience across others. Their expertise is typically deeper than that of a T-shaped person who shares the same specialism, and they excel when focused on tasks that require their expertise alone. However, they may have difficulty with being in a cross-functional team or thinking outside of the box as they are focused only on how they can apply their preexisting skill set. At the extreme, they have little or no understanding, perspective, or empathy for other domains of work, which is essential for effective cross-functional collaboration.

It would be wrong to think that we should strive to make all people T-shaped; the reality is that you need different people for different situational contexts. As shown in Figure 8.1, there is a need for T-shaped people to manage complex problems, as with a startup. In these problem domains, as well as deep expertise, there is also a need to be creative and curious, to experiment and empathize with customer problems—there is a need for knowledge and experience in a variety of techniques. As we begin to get clarity on a problem and more certain on its structure, this is where I-shaped people excel and are more effective than T-shaped people. In addition, leadership and managerial positions are a good fit for T-shaped people. They need a shallow but broad understanding of many areas, need to empathize and collaborate well. They do not need deep technical skill, instead expertise in problem solving, facilitating, and critical thinking is where they need to excel.

Explorers, Villagers, and Town Planners

The concept of T-shaped and I-shaped people may seem a somewhat oversimplification, but it does highlight the concept of having the right people in the right place at the right time. Researcher Simon Wardley has taken the concept of attitudes and aptitudes further and described a set of archetypes that match the human characteristics needed for each product, service, or component at its unique stage of evolution. The underlying premise behind it is that he fundamentally disagrees that an organization needs one culture. He argues that there are three cultures, namely Explorers, Villages, and Town Planners, that need to be supported.

Explorers (Pioneers) They are the Explorers (originally known as Pioneers), the ideas people, the ones who conceive new business ideas. They will often fail as they experiment with innovative ways and new concepts to solve problems.

Villages (Settlers) Explorers are typically ill-equipped or have no enthusiasm to extract the full value of their ideas by productionizing and generating sustainable revenue through ongoing efficiencies. This is the domain of the Village (originally known as Settlers). The Village can turn prototypes and incomplete ideas into commercialized products.

Town Planners Town Planners can industrialize components and maximize the economies of scale through standardization and simplification.

No one role is more important than another; all three are critical to the effective running of a business. The attributes of each persona are tuned to meet the needs of the component based on its stage of evolution. Figure 8.2 shows these roles mapped on a Wardley Map. Explorers typically operate with components that are in their genesis or are custom built: Villagers when a component is a product and Town Planners when a component is a commodity or a utility.

Hire for Diversity in Thinking

In *The Diversity Bonus: How Great Teams Pay off in the Knowledge Economy*, Princeton University Press (3 Oct. 2017) author Scott Page advocates the creation of diverse teams to deliver greater impacts. Using his own experience, psychology, and research, the author believes that getting a diverse mix within the workforce results in a "Diversity Bonus." He argues that a diverse team of people will have a greater impact when tackling complex problems versus a team in which members all think and work in the same way. This is due to a greater hive mind based on a variety of experiences and different ways of approaching problems. In my own experience, I have also found this to be the case. Innovative solutions are often the result of vastly different viewpoints and approaches to solve them. Therefore, ensure you are hiring people that can bring diverse points of view, experiences, and ways of thinking to your teams.

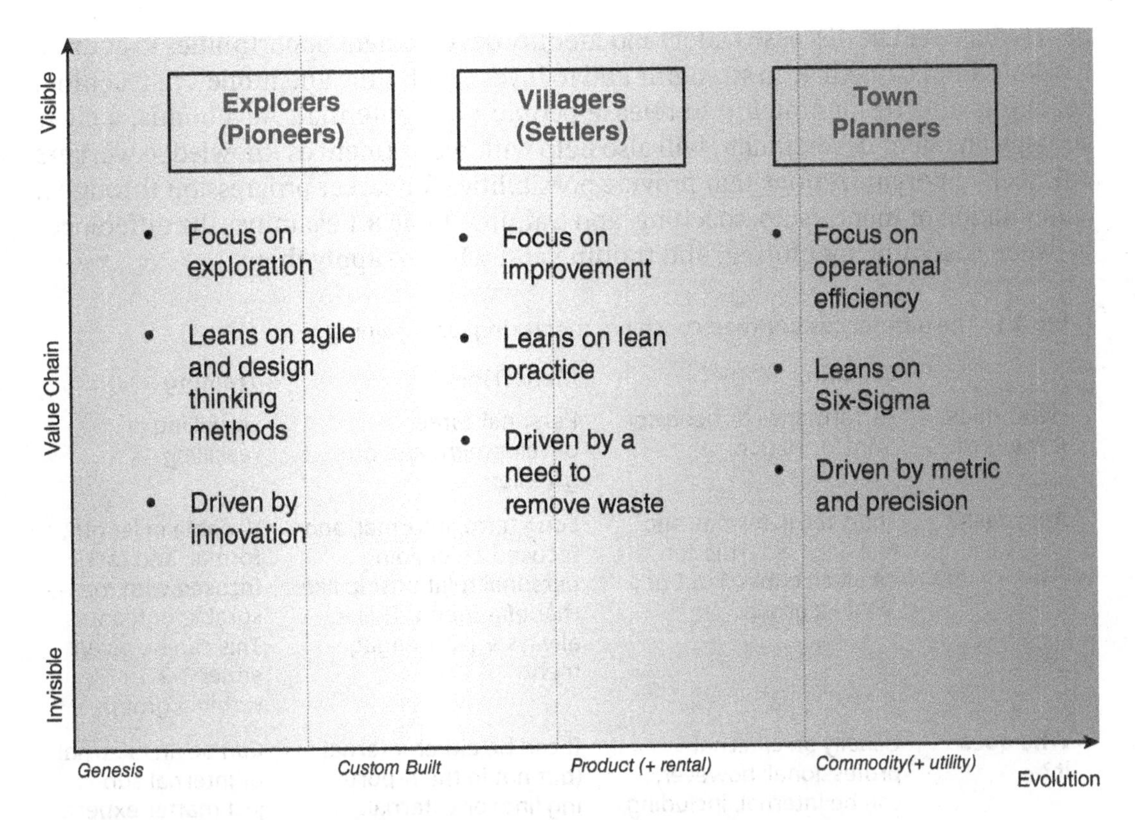

Figure 8.2: Simon Wardley's Explorers, Villagers, and Town Planners

Developing

Developing people within your organization is essential due to the cost and time of hiring against a highly competitive landscape. This is not just a case of sending employees on an annual training course. We need to create the perfect environment in which talent development can excel.

Invest in Mastery through Coaching, Training, and Mentoring

Developing talent and quenching people's thirst for mastery is not something that should be left to chance. The most important responsibility of every line manager is to develop the people they manage. This means as a minimum a manager should hold weekly 1-2-1s with the people that report to them to offer support on the tasks they are undertaking. However, what is needed is a proactive effort to jointly craft a career plan (which will be discussed later) and identify development opportunities to address weakness and capitalize on strengths and to invest, both time and money, in teaching, coaching, and/or mentorship to release people's full potential. As a bonus, a clear strategy on talent development will also help with recruitment as knowledge workers will seek out organizations that provide possibilities for career progression through a combination of mentorship, coaching, and training. Table 8.1 examines the differences between coaching, mentoring, and training and when to apply them.

Table 8.1: The differences between coaching, mentoring, and training

	Coaching	Mentoring	Training
What does it focus on?	Performance or behavior transformation.	Personal career development and progression.	Upskilling or reskilling.
Approach?	Short term, formal, and task-focused. This can be given either 1-2-1 or within a group.	Long term, informal, and focused on building a personal relationship like that of a friend. This is always a 1-2-1 engagement.	Variable in length, formal, and task-focused with measurable outcomes. This can be given either 1-2-1 or within a group.
Who does it?	Usually an external professional; however, can be internal, including within the reporting line. Delivered by the authority or expert in their field.	It can be either internal (but not in the reporting line) or external. Someone with deep experience in the area that the person wishes to progress within.	Can be an external or internal subject matter expert. Can be within the reporting line. Typically, this is outsourced, especially when dealing with new skills.
Examples on when to use?	You want to improve how someone communicates and presents in large groups. A coach would work with the employee to improve confidence in this area and encourage practicing and honing this skill.	You want to develop someone you manage to take on a bigger role. In this case you may suggest to them a mentor who has held the position to share their knowledge, skills, and/or experience.	You want to move to a cloud host provider. You give your team training to become AWS solution architects.

Develop A Growth Mindset

People's opinion of themselves in relation to their ability to learn can have a significant impact on their development progression. Assumptions about their own ability to develop and grow are directly influenced by a person's mindset. A mindset is the series of beliefs people hold about themselves; it's how people make sense of the world and how they perceive themselves. At the extremes, there are two types of mindsets:

- People with a **growth mindset** have a belief that through the right approach, hard work, support, and perseverance they can develop their skills and talents and achieve anything. People with a growth mindset view challenges as an opportunity to develop their capabilities and expand their experience. They appreciate that complex problems require trial and error and see failure as a necessary part of the process to gain deeper knowledge of a problem area.
- People with a **fixed mindset** have the belief that abilities and potential are predetermined—they are either good at something or not. They are more risk averse, avoiding new challenges and situations they are not familiar with that put them out of their comfort zone due to a fear of failure and appearing dumb. This means that they are less adaptable and creative; instead they favor tried and tested ways of working.

Promoting a growth mindset requires a systems thinking approach. This can be achieved through a number of ways:

Change people's perspective on failure, normalize struggle, and embrace imperfection. To encourage a growth mindset, you need to ensure people accept and are comfortable that mistakes are part of the learning process. As a manager, you need to provide physiological safety. This is achieved by changing how you speak and act when dealing with failure. Instead of criticizing, use curiosity to ask questions like, "What am I not seeing here," or, "What may have been wrong with your assumptions?" or "What have you learned from this failure and how will it help?"

Encourage a sense of curiosity by challenging people to learn something new. Providing opportunities for people to get them used to getting out of their comfort zone can help them gain skills in learning how to learn. Start with something small; try a new framework, technique, or practice. Get them to present the experience and knowledge they gained to a small group.

Promote team-based learning. Instead of developing skills as an individual, we can focus on group learning. A team-focused approach provides a supportive and collaborative environment, helping with relationship building, building trust, and accelerating learning.

Articulate challenges as growth opportunities. Explain the learning benefits of tackling new challenges, regardless of whether they succeed or fail to overcome them.

Don't give answers; offer strategies to overcome challenges. It's easy, and often quicker, to give people the answer, especially if you already know it. However, to cultivate a growth mindset, you need to teach people how to learn and the patterns, practices, and principles that can be applied to tackle new challenges.

Take time to review and celebrate progress. Help people reflect on how they have grown and the experiences gained through both successes and failures. Praise the effort, attitude, and the individual development over the result. Taking time to reflect on past experiences and drawing lessons can help prepare people for future challenges.

Reduce Cognitive Load

One of the reasons people are "quietly quitting" is that they are taking ownership of their work-life balance by preventing high levels of stress and burnout. If people feel overloaded, they will become unengaged and check out. If they don't have the space to focus and develop, they will look for other opportunities for a better life balance and one that gives them space to learn and grow. To retain talent, we need to ensure we don't constantly overload people. People will not be able to develop if they have no time and space to think. There are ways managers and leaders can help create the environment where people can achieve mastery:

Limiting the size of the software system with which any given team is expected to work As covered in Chapter 5, "How We Are Organized," we need to base team size around domain and software complexity. As complexity grows, in both software and domain, we need to distil teams to manage cognitive load.

Simplifying communication by defining explicit team relationships and interactions As detailed in *Team Topologies*, Team Topologies: Organizing Business and Technology Teams for Fast Flow, IT Revolution Press; (7 Sept. 2019) by Matthew Skelton and Manuel Pais, if we don't think about how information should flow and create explicit interactions between teams, we leave it up to the teams to figure out, which adds to cognitive load. As observed by American organizational theorist Russell Ackoff, problems that arise in organizations "are almost always the product of interactions of parts, never the action of a single part."

Limiting work in progress As covered in Chapter 7, "How We Govern," we can use Kanban boards to visualize load and create policies to limit the work in progress. By following the practices of Kanban, teams can manage their own workloads, and reduce load, by pulling in work rather than having it pushed to them.

Removing distractions and barriers See Chapter 9, "How We Lead," to see how leaders and managers can remove obstacles for teams. By avoiding distractions such as unnecessary meetings and reports and blockers such as the bureaucratic governance process, we can avoid the interruptions and allow teams to focus.

Supporting teams with platform teams Another pattern in *team topologies* is that of a platform team to reduce the cognitive load of product teams. The purpose of the platform team is to build technology that simplifies common components and thus accelerates delivery for other teams.

Increasing wellness offerings The pandemic highlighted that employees are under enormous pressure both inside and outside of work. To reduce cognitive load, employers have increasingly started to support employees with health matters, both mentally and physically. This reduces the load and can help employees focus when at work.

Retaining

After the investment of attracting and developing talent, it makes sense to do everything to retain them. While it will always be important, we need to look beyond simply rewarding people with higher salaries and more bonuses. Knowledge workers need to satisfy their need for mastery and autonomy to be retained.

Create a Flexible Environment

Flexible work environments are defined by guiding principles rather than strict rules. This gives a level of freedom, or autonomy, for employees, but with still enough control from the employer to ensure the business is not negatively impacted. Research from ManpowerGroup Solutions' Global Candidate Preferences Survey has shown that, compared to individuals in a more traditional working environment, flexible workers had better levels of job satisfaction, dedication, and motivation and are more willing to increase discretionary effort. People surveyed reference an improved work-life balance, increased concentration with fewer interruptions, more time for family and friends, and reduced costs and commute time. To create a flexible working environment, we should allow employees a choice on where and when they work and to an extent what they work on.

Where In the early days of the pandemic, companies had to adopt new hybrid working practices out of necessity. This necessity has now become the norm and the expected way of working. Therefore, a workplace needs to be able to accommodate a hybrid manner of operating, supporting both remote and face-to-face ways of working.

When By creating more flexibility on when people work, companies can help support a better work-life balance. Offering core hours or principles around when teams should be online or in the office can allow flexibility around how employees structure their day. This enables people to allow time for personal appointments that may occur during the typical working day, such as children's school events and other life admin.

What As well as giving people flexibility on "when" and "where" they work, we can, to a certain extent, be flexible on "what they work on." To find the right mix of people and attitudes (Explorers, Villages, and Town Planners) for projects and teams, we need only to build around the volunteers, inviting rather than inflicting change. Allowing people to choose the team that suits their skills and experience or career path rather than assuming who we think is best placed will lead to more intrinsically motivated employees, and those who feel they have skin in the game.

Create a Career Path

People are looking for jobs with progression and the ability to acquire new skills, knowledge, and experience. This is their need for mastery and purpose. However, many engineers are reluctant to become line managers, and even if they aspire to climb the corporate ladder, they may face obstacles in advancing their careers until a higher-ranking colleague moves on or up. It's important for employees to have opportunities for recognized professional growth that don't necessarily require moving up the organizational hierarchy. Therefore, as well as career growth we also need to provide career paths that focus on professional growth and transformation through career development. In addition, we need to think about more lateral, or dare I say agile, career development, a career lattice that allows for horizontal, and diagonal movement within a business. Leaders should not underestimate the importance of mapping out a career path in relation to retaining talent.

Work with talent to define a career progression based on one of three high-level guiding paths:

Career Growth This is the traditional career path, one that is based on climbing the career ladder. Achieved through promotion, and often requires taking on more responsibilities and people. This is easy to quantify as you can measure progress against the corporate ladder.

Career Development For people who don't want to change what they are doing, do not yet know what path they wish to take, or would prefer to become an expert and excel in a particular field, we can focus on career development.

Career development is an ongoing process of personal transformation that involves improving skills, knowledge, and experience. These improvements prepare people for career growth if, and when they are ready. This is qualitative, and based on competency, expertise, and new skills. However, this can be quantified; for example, you can apply competency or proficiency levels such as moving from junior to senior to principal developer.

Career Lattice The most talented employees often have skills that can be applied elsewhere, and they thrive when faced with new challenges. Some people love new challenges; for these you can offer the opportunities to tackle problems across the business rather than in their own department or team. This helps people to form a T-shaped profile by expanding their experience, helping both their personal and career development.

Ensure a Continuous Talent Development

Agile and lean philosophies focus heavily on continuous learning from short cycles, fast feedback, and practical experiments to constantly improve what they do and how they do it. This theme of a continuous feedback and improvement cycle applies to people development as well as business outcome endeavors. The management and continuous improvement procedures at Toyota were documented by researcher Mike Rother in his 2009 book *Toyota Kata: Managing People for Improvement, Adaptiveness and Superior Results*, McGraw Hill (16 Oct. 2009). He observed two high-level practices that Toyota employed to develop staff, the Coaching Kata, and the Improvement Kata. The Improvement Kata is a four-step practice focused on incremental improvements. Table 8.2 shows the Improvement Kata and how we can apply this directly to personal development.

Table 8.2: Using the continuous Improvement Kata for talent development

Improvement Kata Steps	Kata Applied to Talent Development
Get the direction or challenge: Be clear on the end goal or vision and the sense of direction.	Be clear on the direction of your career path. Are you targeting a new position, or to become an expert in your field, or are you looking at new problem domains to challenge yourself?
Grasp the current condition: Clearly and factually define the current situation by determining where you are now.	Understand the skills, competency, experience, or knowledge gaps that you have. Be aware of your weaknesses.

continues

(continued)

Improvement Kata Steps	Kata Applied to Talent Development
Establish the next target condition: Choose an interim objective and set a date as your next step toward the goal. Use this to advance current knowledge and capabilities.	Target the skills gap you want to focus on.
Conduct experiments: Perform a series of minor experiments (changes) methodically and scientifically to get to the next target condition.	Work toward closing the gap through training, coaching, or mentoring or through management help. A person measures their achievement during the experiments, talks about it with a coach, and learns from the experience.

The Coaching Kata is a method that a manager can use to guide a person through the Improvement Kata. It is composed of five coaching questions to help the manager coach the employee toward taking ownership for their personal growth and learning. Table 8.3 shows how we can translate these questions to be more specific about talent development. The purpose of the Coaching Kata is to change the mindset of personal development. Instead of a manager telling the employee what to do, they act as more of a servant leader and ask searching questions of the employee such as "Where do you want your career to go, what motivates you, what interests you? "What skills are you lacking?" "What gap do you want to focus on first?" "What support do you need?" "How will you know you have achieved the skill?" By introducing the Improvement Kata to employees and following the Coaching Kata, we can encourage people to develop their personal development path and have a sense of ownership around their career progression. By employing both the Improvement and Coaching Kata practices to talent development, we can improve our retention of people through a continuous focus on their personal development.

Table 8.3: Using the coaching Kata for talent development

Coaching Kata Question	Coaching Kata Applied to Talent Development
What is the target condition?	Where do you see your career heading? Do you want to manage people or lead in a field? Do you broaden your skills or double down in an area?
What is the actual condition now?	How would you rate your capability against your target role?
What obstacles do you think are preventing you from reaching the target condition? Which one are you addressing now?	What skills do you need to focus on? What gaps do you have? What are the biggest blockers keeping you from performing to your fullest potential?

Coaching Kata Question	Coaching Kata Applied to Talent Development
What is your next step? What do you expect?	How will you bridge the gap? How can I better support you? What resources can I or the company provide that would help you excel further in your role?
When can we go and see what we have learned from taking that step?	How will you measure progress? How will you know you have achieved something or improved in capability?

Summary

You will not be able to hire for every position in your organization, and neither should you. Running everything in-house and managing it is expensive, has a huge opportunity cost, and the ability to fulfil talent in the current competitive market would be extremely difficult to say the least. Your sourcing strategy should determine where you really need to hire and where you can outsource. This should be based on the strategic importance of the capabilities you are enabling and how evolved and ubiquitous they are. At a high level, you should favor in-house talent for core and unique capabilities and outsourcing for generic and supporting capabilities. For the talent we choose to source in-house, we need to focus time and effort on recruitment, development, and retention.

When it comes to recruitment, we need to think about a diversity in mindset. As Simon Wardley has highlighted, there is not one culture but three—Explorers, Villagers, and Town Planners. Team composition within these cultures requires the consideration of both aptitude and attitude. Aptitude refers to the specific skill sets such as development, data integration, user experience, or business analysis. And aptitude refers to the mindsets, cultures, and ways of working. This need for diversity is because different capabilities that IT will build will be in different stages of evolution and will require vastly different approaches as highlighted in the Cynefin framework. Delivering a new product and trying to change customer behavior will require very different methods and techniques versus the integration of a new ERP to consolidate and improve existing processes. As part of attracting these people, outside of a competitive compensation and benefits package, we need to be clear on how they will be developed and the career paths that we will support them with to retain them. Our ability to retain people will largely depend on how they are developed and what you can offer to address the areas of mastery, autonomy, and purpose.

The need for mastery is a key tenet of the knowledge worker. Therefore, learning and development should be a core part of your talent strategy. This is not a case of sending people on an annual training course. You need to craft a career path based on people's desires and support them on that journey, whether it is about growth, development, or a move across the business. Managers need to offer the right support, whether that be training, coaching, or mentorship, as well as providing the space and time for people to learn. The ability to develop talent and enable people to reach their full potential is a key capability of any organization and one that can help retain as well as attract new employees.

A lack of opportunities for progression is often cited by employees as a key reason for leaving jobs. To ensure we retain talent, we need to give people a sense of purpose; we need to focus on their growth and continued improvement. Spending time working with people and co-developing a career path that is personal to them can help with talent commitment to the organization. This shouldn't be a once-a-year annual performance tick boxing session. People will look for new opportunities if there isn't a clear path for growth, be that in terms of promotion, training, new responsibility, expertise, or experience. One of the key ways in which organizations can help grow a continuous focus on personal development is by coaching people in the Improvement Kata. This provides an ingrained learning culture through embedding the practice of continuous personal goal setting and reflective thinking, leading to an increased self-awareness and ownership of their own mastery and purpose.

How We Lead

Anyone who has never made a mistake has never tried anything new.
—*Albert Einstein*

A leader's job is not to fix disengaged people.

A leader's job is to fix the environment that results in disengaged people.
—*David Marquet*

Leadership is not about being in charge. Leadership is about taking care of those in your charge.
—*Simon Sinek*

In the era of the knowledge worker, people are your number one asset. However, good people are scarce. Attracting and retaining people is an essential ingredient in the IT leader's operating model. Key to a motivated workforce is the ability for leaders to create an environment where people can achieve their full potential. Leaders need to ensure people have clarity about the desired business outcomes, have the support to become masters of their craft, and are empowered to determine the best course of action to deliver impactful business outcomes. As covered in Chapter 5, "How We Are Organized," to design an organizational model that delivers, leaders need to invest time in understanding the social and business networks at play in the organization as well as the technology boundaries—namely the sociotechnical architecture. To keep structures working, leaders need to constantly support the people within them, ensuring that impediments are removed and collaboration is unhindered. You don't get this by barking orders at people. Instead of trying to use command-and-control techniques, IT leaders need to change their behavior toward a more people-centric mode of operating. Leaders should invest time in developing people and culture. This is achieved through working on the system, ensuring the environment supports an adaptable manner of working, while leaving the workers to handle the work within it.

In this chapter we will look at the core behaviors leaders must exhibit to build successful teams. By changing leadership behavior, we can influence people's belief system and therefore change culture. Leading by example, trusting, and coaching over commanding and controlling, in combination with visibly changing our ways of working to embrace agile and lean values, we can inspire our teams to adopt the same shift in their mental models, which will lead to an improvement in the patterns of behavior and in turn more positive business outcomes. Fundamentally, you will learn that to build a culture of collaboration and innovation, behaviors from the top must change first.

Adopting New Leadership Behaviors

So, if the job of IT leaders is not to command and control, then what is? How can leaders be ultimately accountable for the work when we have given over responsibility to autonomous teams? In a nutshell, leaders need to lead, not hand-hold—we need to influence the direction of work by setting up the right kind of constraints and the right environment for agility to thrive. We need to focus on the big picture, allow teams to find working practices, remove blockers, and communicate the strategic intent to align those closest to the work and really in control. If we think about it, were leaders ever in direct control of work, or was it just the illusion of control manifested through bureaucratic governance project processes and a focus on standardizing delivery and development practices, all fueled by an underlying lack of trust?

To succeed in a VUCA (volatility, uncertainty, complexity, and ambiguity) environment, we need our workforce to be agile. An agile working environment is drastically different from a traditional workplace, and so by definition the role of the leader is also completely different. We need to change the way leadership teams communicate strategy, objectives, and business goals; how failure is treated; how decisions are made; and how work is worked. We need to turn the relationship between leaders and knowledge workers on its head, as shown in Figure 9.1, from a position of control to a position of support and empowerment.

Good leaders know that in today's rapidly evolving digital world, they must develop successful teams to get results. People on the ground have more domain knowledge than leaders, both from a technical point of view and from the view of specific subdomains of the business that they are working within. They understand constraints and have the skills to capitalize on opportunity and they can do so at pace if enabled.

As software craftsman Guido Dechamps states, "It's not the CEO, but the developers' knowledge that goes into production." Therefore, it is more important than ever before to seek a more inclusive leadership style to support teams and empower them to innovate and collaborate to deliver on the desired outcomes. To achieve this, leaders need to set the right conditions that inspire close collaboration, rapid learning and experimenting, empowerment, autonomy, and the alignment on what is strategically important to the enterprise. To do this, leaders need to lead from a position of service and do everything in their power to develop the flow of work and how work works throughout the organization. Leaders should focus on teamwork, adopting a "we succeed together, and we fail together" mentality. They should show empathy and humility and actively listen to and care about the problems of teams to improve loyalty and motivation. Remember, agility is a team game, or more specifically, the interactions of people in teams, and a leader needs to know their place in that team and how best to contribute to the team succeeding.

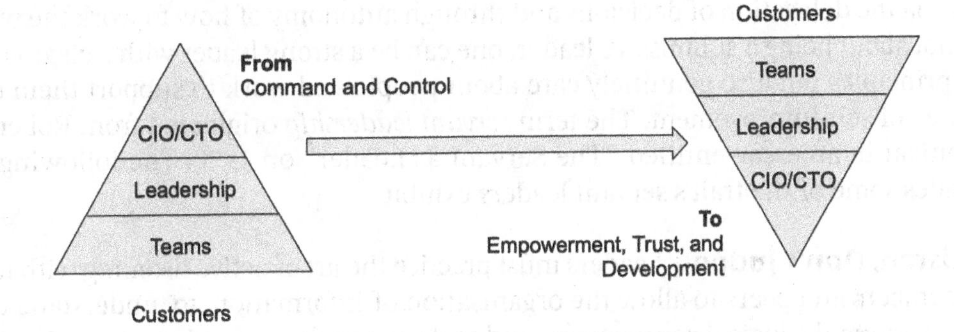

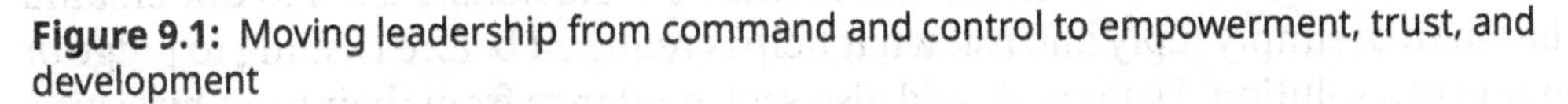

Figure 9.1: Moving leadership from command and control to empowerment, trust, and development

The culture of an organization is a direct reflection of the behaviors of its leaders. We can't dictate culture. However, by leaders exhibiting the behaviors that we will cover in this chapter, we stand a better chance of creating the necessary conditions for a culture based on collaboration, safety, transparency, self-organization, and continuous learning to emerge. We will cover the following new leadership behaviors in this chapter:

- **Embracing Servant Leadership:** A longer-term approach to leadership focused on the needs of others before considering your own
- **Instilling Intrinsic Motivation:** The ability to make people feel incentivized to complete a task simply because they find it interesting or enjoyable

- **Encouraging Growth and Development:** Discovering the potential in people to improve performance, adapt to solve problems, and further their careers
- **Focusing on Improving the System:** Managing the system of work, the environment people work in rather than the people

Embracing Servant Leadership

While there is no single leadership style that is suitable for all contexts that a leader will face, the philosophy of a "servant leader" is particularly suited for complex contexts with a high level of uncertainty that require creativity and high levels of unscripted collaboration. Embracing agility is about more than technology. It is about people. People are at the heart of transformational change, not technology. The core concept around the servant leadership philosophy is that the primary goal of a leader should be to serve others and to make people around you perform as highly effective as possible as opposed to focusing only on self-interests. It is about sharing power with your team via the delegation of decisions and through autonomy of how to work the work. It is not about being a submissive leader; one can be a strong leader with a clear vision and principles but also genuinely care about people and work to support them on a journey of self-improvement. The term *servant leadership* originated from Robert K. Greenleaf in an essay entitled "The Servant as Leader" on 1970. The following list includes some of the trait's servant leaders exhibit:

Listen, Don't Judge Leaders must practice the art of active listening with team members and peers to allow the organisation of information, to understand context and make logical connections, and to draw conclusions. Listen to understand rather than to simply reply and ask what help is required before rushing to judge or jumping to a solution. Leaders should also seek guidance from their team by asking open questions and intently listening to feedback on how to become a better leader.

Understand and Empathize In some respects, technology is the easy part. After all, the computer will only do what you tell it to do (well most of the time). Collaborating with and understanding different people with different backgrounds, perspectives, and different ways of working is hard but crucial for innovation and creativity to flourish. A servant leader views individuals as people, not as resources or workers. They care about people's lives and not just work performance. They take the default view that all people have good intentions and want to do a good job. They ensure people are supported both emotionally and mentally.

Respect One of the most important principles of lean is to treat people with respect. It is crucial to be open and transparent, to listen and treat people in a professional manner, whether it is the intern or the chairman of the board. Hearing out others and taking on their opinions will encourage people to take a more proactive role in work. Respect is even more important when you are disagreeing with people. To show that you have empathy with other people's point of view while also being able to be assertive and disagree without being argumentative is an important behavior to develop.

Mentor and Develop People, not the technology, transform a company. One of the most important responsibilities of a leader is to grow their team and enable them to achieve more than they thought possible. A commitment to supporting the growth of each member of your team through continuous learning and achieving mastery is essential for a workforce that must adapt quickly. Leaders must also look to grow and develop the leaders of tomorrow. They must give up power via delegation of decision-making, provide opportunities for greater autonomy, and offer opportunities and encouragement for others to lead, all the while offering support.

Connect People and Build Community It's not just people that can transform a company; it's the relationship between people. Leaders should focus on facilitating collaboration not just within their own teams but across the enterprise. Build relationships with teams and other leaders across value streams to remove boundaries and flatten an organization to improve the flow of work. Leaders must facilitate the involvement of all levels and break down boundaries when problem solving. Create a sense of unified purpose and community by creating social networks and the spirit of family.

Show Humility, Self-Awareness, and Authenticity However much time you will spend in front of the computer, it will dwarf the time you spend working and talking to others. Show respect, patience, and humility when working with your peers and business counterparts. Understand different personality types and how different types interact in collaborative sessions. Work out how people learn—are they big picture people or detail people? Perfect the art of active listening and show empathy. These are very important traits to get on in this world let alone in the domain of IT. Problem solving is a team sport and a servant leader seeks out contributions, feedback, and opinions. Getting help is not a sign of weakness but is a sign of an inclusive leader and is essential for improving oneself to become an effective leader. Share success and take accountability for failures; own up to your mistakes and seek the lessons from them.

Influence and Persuasion Servant leaders don't manage through command and control. They are not dictators; they are great influencers who can use persuasion with their team and peers to guide people to understand the strategic intent.

When a lead builds consensus over compliance, people will retain autonomy, which will in turn assist with motivation.

Inspire By creating an emotional connection between people and work, leaders can enhance innovation and commitment.

Long-Term Thinker A servant leader keeps an eye on the future while managing the day to day. They need to constantly visualize and communicate the big picture so that they can make trade-offs to avoid short-term wins for long-term problems and to influence the team toward the strategic vision.

Table 9.1 shows the change in leaders' behavior and attitude when they put people first.

Table 9.1: The difference between a traditional and servant leader

Traditional Leader	Servant Leader
Makes all the decisions.	Empowers people to make decisions.
Focus on self.	Focus on team and others.
One-way communication.	Two-way communication, feedback.
Assigns blame to person and punishes individual mistakes.	Assigns blame to system, takes learnings from failure to improve the system.
People are a company resource.	People are human beings and should be treated with respect.
Innovation and creativity is for the select few.	Innovation and creativity is the responsibility of all.
Tells people what output is required and ensures they deliver it to spec.	Tells people the outcome desired. Empowers and supports them to determine how to achieve it.
Motivates people through money and promotion.	Pays people enough but motivates them with purpose, autonomy, and mastery.
Shares information on a need-to-know basis. Treats Information as power.	Transparent; shares the big picture and wider context behind strategy and decisions.
Gives orders.	Asks questions and listens.
Works independently.	Collaborates with others.
Builds hierarchies.	Flattens hierarchies.
Does what's best for themselves and the department.	Does what's best for the organization and the wider system.

Instilling Intrinsic Motivation

Working in complex problem domains requires unscripted collaboration and intrinsic motivation to solve problems. As you have read in Chapter 3, "How to Change the System," intrinsic motivation is driven through a combination of purpose, mastery, and autonomy. Therefore, IT leaders must create an environment in which teams have clarity on the purpose of their role and the desired outcomes, the ability to improve their skills, and the autonomy to work in a manner that gives them the best chance of achieving the outcomes.

Clarify Purpose and Ensure Alignment

For work to be fulfilling, people need to understand the meaning or the why behind what they are being asked to do. Trying to dictate solutions to problems that are complex will not motivate people to contribute and look for better ways to achieve the outcomes, or to pivot and adapt if hypotheses are incorrect. Without clarity on purpose, people can become disengaged, demotivated, and uninvested in their work. People have a desire to do something that has meaning and is important, which goes beyond the need to simply make profits. With a greater level of understanding and context behind a problem, people will feel more engaged. This engagement will lead to increased motivation, which can improve focus, productivity, creativity, and innovation, all of which will contribute to the happiness and loyalty of people.

The role of a leader is to give people clarity by creating a common understanding and alignment on purpose: how the desired outcomes contribute to hitting goals, how the required changes in behavior lead to the outcomes, and therefore why the work is important. As well as clarity on purpose, leaders need to provide clarity on expectations and reporting progress. You cannot give people freedom and autonomy without meaningful constraints, such as clearly stating your expected outcomes and behavior as well as the framework you will operate in. Leaders need to keep the purpose simple and focused and take all opportunities to repeat it and communicate to people. In Part 3, "Strategy to Execution," we will look at a number of techniques to link what people are doing to the overall strategy that can be employed to help alignment and clarify the why behind strategic initiatives.

As shown in Figure 9.2, delegation to teams is only possible through their close alignment to strategic objectives. To ensure teams are on the same page, leaders should communicate the strategic vision and goals of the organization, along with the ever-changing business context, to teams tasked with delivering outcomes. This should be done constantly and consistently to help influence and nudge teams in the correct direction. It is vitally important for leaders to provide an inspirational guiding vision that motivates and inspires teams rather than course correction only, especially when trying to foster a culture of autonomy and where micromanagement is not useful.

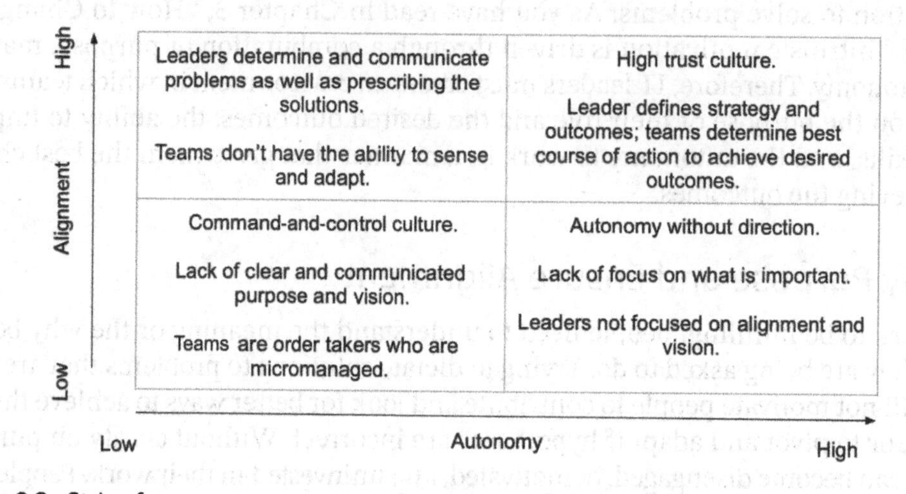

Figure 9.2: Strive for autonomous teams that are completely aligned.

Alignment is a foundation for team autonomy and decentralization of decision-making. By allowing decisions to be made by the teams doing the work, we can reduce the time wasted on getting sign-off and explaining context to someone outside the team that sits higher in the hierarchy. Through consistently sharing information to everyone in the organization, we can ensure cross-departmental teams work more effectively together on the aligned goals. Leadership in an agile and lean environment is about steering and setting direction on which strategic outcomes are important to the business and being clear on the reasons why.

Through clear, consistent, and regular communication of the strategic objectives, everyone should understand exactly why they are tasked with delivering business outcomes and have a clear line of sight on how what they are working on contributes to business success. Teams understanding why their work is important is a key factor to improving motivation and to avoiding people thinking that they are just another cog in a machine. A leader should focus only on orchestration and delegating responsibility. This will encourage new leaders to emerge as teams organize themselves on how they wish to tackle problems. In terms of directional overseeing, leaders should favor a lighter touch governance process to review progress toward goals and ensure alignment to the strategic priorities, intervening only if the team strays off course or if there is a change to the strategic context that teams need to be made aware of.

Empower People Through Trust

One of the factors that can cause delays to throughput for teams is waiting for decisions to be made. Teams need autonomy to move at pace: if we are hiring smart people, we need to trust them and let them get on with it rather than telling them exactly what to do. Leaders need to ensure teams work at pace, and to do this they need faster decision-making, which is achieved by giving teams the autonomy to make day-to-day trade offs on their own, with the knowledge on when to seek further authority for decisions that may have an impact outside of their area of work. It is not the job of leaders, IT or not, to dictate how problems are solved and to make all decisions, especially if they are not close to the problem at hand. Leaders need to set direction and tell people what to do and allow them the time and space to determine how to do it. Eli Goldratt, author of the *Theory of Constraints* (North River Press, 1999), puts it very well: "The minute you supply a person with the answers. . .you block them, once and for all, from the opportunity of inventing those same answers for themselves. . .what you want is action to be taken, then you must refrain from giving the answers." Leaders should ensure teams are self-sufficient, know who to ask for help, make decisions among a cross-collaborative department, and know when to seek approval for bigger decisions.

To help people become empowered, we should lead them by posing questions rather than giving out commands. General George S. Patton Jr. famously told leaders that when working with people, "Tell them what to do, and they will surprise you with their ingenuity." Rather than giving orders, leaders should point out opportunities and guide teams with open-ended questions. That gives people the confidence to explore a problem and come up with innovative solutions that may not be apparent to leaders and other execs who are higher in the chain of command and therefore less knowledgeable on the subdomain compared to the teams responsible for delivery and who understand the constraints and opportunities at a far greater level of detail. This inclusive servant style of leadership helps to foster new leaders and empower teams to own a problem and drive end- to-end accountability in the development of ideas through to production. Innovative problem solving emerges from autonomous and self-organizing teams. By empowering teams and giving them responsibility, they can experiment and evolve solutions that may have not been apparent to leaders.

By empowering others, leaders can focus more on the macro rather than trying to manage the micro. IT leaders should be focusing their efforts as much on understanding business constraints and coauthoring strategy as they do with overseeing development. While leaders should empower teams to be responsible for driving outcomes, leaders must not be absolved of any responsibility themselves. They must accept failure as learning, just as readily as they accept success, and ensure that while they don't make the decisions themselves, they regularly check in to ensure teams are aligned to strategic direction and to correct any course.

Just as it is a wasted effort to try to predict and plan in a complex problem domain, so too is it to try to control and standardize on how the work is worked within one. Teams aligned on strategic outcomes and working in complex problem domains need time to self-organize and determine the best way to work the work. Leaders need to let the ways of working emerge by creating an environment in which teams can determine the best way to organize themselves. A focus on continuous improvement will lead to an emergent way of working that fits the context they are working within. In other words, leaders need to give teams time to form before they can storm, only interfering to provide strategic alignment and to remove impediments outside the control of the team. Once they have set the strategic direction and articulated the goals, leaders should guide the team toward self-organization by stepping back and allowing working practices to emerge.

When organizing and aligning efforts around business objectives in complex problem domains that have clearly desired outcomes but lack detail and clarity on how to achieve them, it is important for leaders to hand over responsibility to teams so that they move further up the business value chain to determine tactics for themselves. Aligning effort and investment around outcomes, as opposed to specific projects, requires leaders to give responsibility to the teams to own the end-to-end process of discovery and delivery. Leaders must set the strategic direction and explain the outcome desired and the business context that lies behind it (i.e., the "what,") and then allow the teams to work within agreed boundaries on the "how."

The single biggest reason organizations fail at agile transformations is not trusting the team to determine how best to work. There is no point in hiring smart people to solve complex problems if you are only going to try to control them by dictating solutions. This is why the command-and-control management model doesn't work in complex problem domains; it assumes that your employees can't be trusted. Trust is an essential ingredient that enables the empowerment of a team and is a core value of both lean and agile philosophies. Command-and-control organizations treat people as resources; agile organizations value people as partners.

Routine simple problems that require simple solutions don't require high levels of trust as work can be explicitly defined upfront. However, for work in complex and unpredictable contexts, trust is vital. In this age of the skilled knowledge worker, it makes no sense, when dealing in complex problem domains, to dictate to smart professionals on what to do just because a leader happens to sit higher on the organization chart. To engage clever individuals, leaders must set up the right environmental conditions (namely, autonomy and trust), communicate the strategic direction, and support the team's ability to solve problems through removing ambiguity, removing any impediments, and providing anything the team needs to enable progress. Leaders should focus on building relationships on trust, both between the leadership and teams and within the teams themselves. An environment and working relationships void of trust will not foster the type of collaborative, innovative, and proactive working required for complex problem domains.

Develop Mastery in People

IT leaders can develop people to their fullest potential by expanding decision-making and leadership skills. Through increasing the number of people that can act as leaders, we can increase the speed at which the organization can learn and adapt. Simply put, leaders need to create leaders. In his book *Turn the Ship Around!* Penguin (8 Oct. 2015), US submarine commander L. David Marquet argues that you should move the authority to the information rather than moving information to the authority. What Marquet means is that those closer to a problem should have the authority to make decisions rather than passing information up a hierarchy. This is a simple but very important point and is at the core of trust and empowerment that is required to succeed in complex problems. To give authority, we need to develop leadership ability.

The role of a modern leader is to provide opportunities for people to lead. Developing leadership capability allows the organization to grow without the hindrance of centralized authority. To develop people, we need to give them opportunities. This means we need to tolerate potential failures. How we deal with mistakes and risk taking will go a long way toward the development of people. By supporting people rather than berating them, we can help them learn from mistakes and thus improve knowledge.

Encouraging Growth and Development

As covered in Chapter 8, "How We Source and Manage Talent," one of the most important jobs of a leader is to recruit, develop, and retain talent. This is achieved through inspiring an adaptive approach to problem solving and encouraging a growth mindset through a focus on continuous learning and development, all with a safe and supportive environment.

Adopt a Curious and Adaptive Approach to Problem Solving

The dictionary definition of *adaptability* is "an ability or willingness to change in order to suit different conditions." Fostering a culture that embraces change, learns from experience, and continuously improves to deliver business outcomes is a difficult task. Leaders can't dictate an organizational culture of flexibility coupled with a readiness to adapt. They can, however, encourage it through their behaviors when faced with competing priorities and ever-changing problems. By demonstrating a flexible, risk-tolerant, and collaborative way of problem solving and the ability to adapt their thinking based on the problem context, leaders can inspire their team to think outside of the box. Culture is not something you do to people, it's an output and experiences of the system that you create. By setting the right conditions—that is, by leading with principles that value empiricism, learning, and continued improvement—you will create the right environment for people to adapt and rise to the challenge of complex problems.

By utilizing such practices as systems thinking, we can adopt new styles of thinking, enabling complex problems to be tackled in innovative ways. We can look beyond how things have always been done, challenge status quos and allow ourselves to see the bigger picture, and give ourselves time to play with a problem before jumping to a solution. We should be comfortable with ambiguity and explore underlying patterns and structures that may contribute to events we can see before jumping to quick solutions. Leaders can open their minds and approaches to problems by challenging the way they think, their assumptions, and their unconscious bias. We can more readily adapt, improvise, and overcome challenges by looking at problems from a different perspective. Deliberately practicing critical and systems thinking strengthens our problem-solving muscles, makes us more aware of how we approach challenges and how others do. This all contributes to being able to alter our perspective to adapt to new context and new information. Successful leaders thrive in complex and uncertain problems by being able to think differently and helping their teams to think differently.

Not only must leaders be open to the way they approach problems they must also have a curiosity to seek out new opportunities and uncover constraints to anticipate future needs, thus being better prepared to adapt to them. By being curious and seeking to understand all aspects of the business, IT leaders can be the catalyst to exploiting new opportunities and removing constraints through challenging the status quo. Curiosity in conjunction with an open mind can spark new ideas and lead to new opportunities. We should ask open questions, observe how value streams work, consider all options, and listen to those closest to the work in each business area to uncover overlooked constraints or opportunities. However, an adaptable open mindset is something you need to develop and constantly invest in. Leaders, like the organization, should be in a state of continuous self-development to prevent becoming tied to a particular way of working or thinking.

Focus on Continuous Learning and Development

Being able to adapt and innovate depends on the skills and the abilities of a diverse workforce. Leaders need to provide opportunities for people to learn and become masters at what they do in terms of technical practices, soft skills, and the problem spaces they are working within. As the demand for knowledge workers and the reliance on technically savvy and commercially minded people ever increases, we need to retain and grow from within. As the environments we work within and the problems we need to solve are constantly changing, we need to also adapt and change how we operate. Leaders need to instill the Kaizen mindset (see Chapter 2, "Philosophies for a New System"), the focus on continuous improvement, within their teams, and develop a working culture that promotes investment in people in order that they can improve their knowledge, ways of working, and problem-solving capabilities.

Only through a consistent, deliberate, and continuous focus on upskilling our workforce, giving them time to improve their knowledge of the business domain, mastering new technology, and trailing new ways of working, will we be able to meet the needs of a complex and adaptive business environment. Agile is all about fast learning through fast feedback. To compete and win in today's rapidly moving world, organizations that foster a learning culture and constantly strive to improve everything they do will have a much stronger chance of succeeding.

Provide a Safe Environment

The core concept of managing uncertainty in complex and unpredictable problem domains is to constantly experiment, learn, and adapt. However, not all the experiments will be successful and worthy of continued investment. The team will often fail while determining what will work and what won't, but these failures will allow the team to adjust to what they focus on and invest in next. To innovate, experiment, and learn through trial and error, teams need psychological safety. If there is a fear that mistakes will be punished, teams won't be empowered to make decisions and will wait to be told what to do. Ask yourself, If an organization will not tolerate failure, how do you think it will be able to adapt when encountering unpredictable challenges? Leaders need a tolerance for failure and to communicate to teams that it's okay to fail fast, or rather learn fast from failures so that they can quickly pivot to correct their course. The adaptability needed to manage in uncertain problem contexts comes from the results of experimentation. By creating a safe environment and through regular coaching, a leader can set the conditions that will encourage creative problem solving and a learning culture that is able to sense and adapt to the changes in the business context and changes to their understanding of the problem space. As shown in Figure 9.3, psychological safety is a basic need. Psychological safety is ensuring that people can work in an environment that can give open and honest feedback without assigning blame and singling out individuals for punishment. It's about creating a culture that encourages people to admit mistakes and the whole team to learn from them.

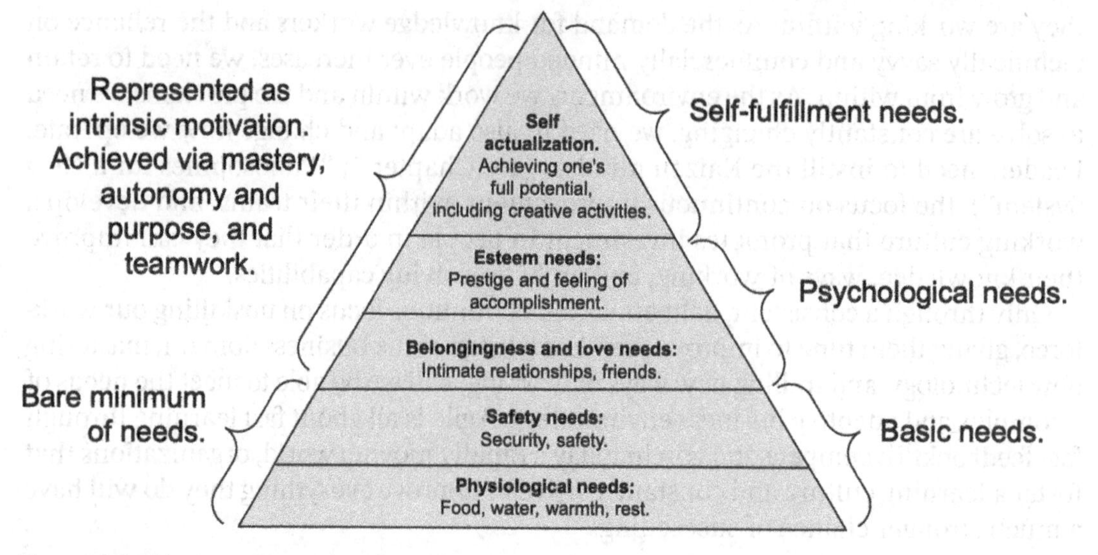

Figure 9.3: Maslow's hierarchy of needs

Failure is inevitable when dealing with the unpredictability and uncertainty of working in complex problem domains. Therefore, we should not try to prevent failure, as making and learning from mistakes is the process that increases understanding. What we need to do is look at the causes of failure and try to minimize them. By leveraging systems thinking, we can challenge ourselves to think about the patterns and structures that are in place that could have contributed to a failure and look to see if they can be altered rather than trying to quickly assign blame to people. It's imperative to foster a culture of trying new things and experimenting with new solutions to complex problems. This tolerance for failure will empower teams to become more innovative and autonomous, allowing them to achieve mastery of their business and technical domains. You can have a hypothesis on a solution, but only by putting the theory into practice will you prove it true or false. This is not a culture about embracing failure; it's a culture of embracing learning at pace.

Focusing on Improving the System

As covered in Chapter 3, the system does what it is designed for, or in other words, you get the organization you design. Therefore, leaders need to work *on* the system rather than simply *in* it. We need to exhibit the behaviors we want to see in our teams and our peers. We should utilize our position and command and control to remove impediments such as bureaucracy and other obstacles that are outside the control of our teams. We should show empathy for both customers and employees by going to the shop floor to see the reality for ourselves, to understand both opportunities and problems. As well as team setup, we should also observe the relationships between teams and departments, ensuring a collaborative organization focused on customer value regardless of how many teams and boundary lines are crossed in the effort to deliver it.

Lead by Example

Restricting agile and lean thinking to the development of software and the IT department will have only limited success on impacting business outcomes. Applying agile thinking to IT development but not changing the leadership and management structure holistically will result in a conflict in values and lead to frustration on both sides of the fence. As you have read, to succeed in a VUCA business environment, agility is a business necessity rather than being just an IT mode of operating. Remember, agile and lean are a set of principles and a value-driven way of thinking that focus on collaboration and pace, even though they may have been born from software development and manufacturing, respectively. These principles are both key to succeeding, and therefore it's vital that they are adopted beyond the domain of IT delivery. If the rest of the organization is not fully engaged with the underlying goals of the principles and values of agile, it will be difficult to improve outcomes. IT leaders need to spread the agility mindset throughout the organization, as shown in Figure 9.4, and ensure other members of the leadership team are supportive, empower their teams to experiment, and tolerate failure in order to find the best way of working to adapt to the reality of the modern business context.

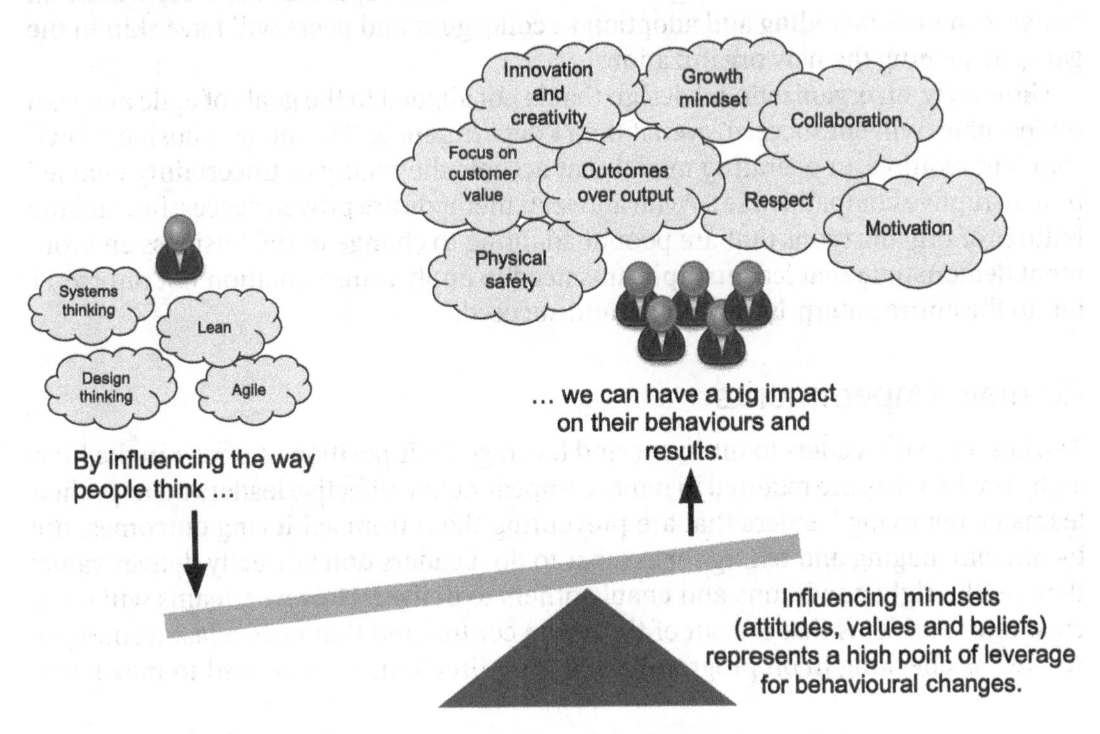

Figure 9.4: Influence your peers by changing beliefs, values, and attitudes.

However, the adoption of agile and lean principles is a challenge due to the fundamental shift in values people need to take on compared to traditional ways of working. By employing systems thinking, we can set up the conditions that will enable transformation to occur. Therefore it is essential that CIOs educate other business leaders and influence their mental models to create an environment that will focus on continuous improvement and the evolution to an agile operating model that works for them. We must engage and inspire our peers through all forms of communication, such as storytelling, visual aids, visits to other businesses, industry white papers, experiments, and minimum viable products (MVPs) to demonstrate what the future of working the business strategy looks like.

To succeed in a business world largely defined by digital capabilities, everyone needs to speak the language of technology and understand lean and agile practices and the art of the possible. The IT leader's role is to act as a guide for their colleagues and peers on this journey of technological enlightenment. Just as how IT folk need to learn about the terminology of the business domain that they are working within, so too must other executives understand and speak fluently about technical practices and ways of working.

IT leaders need to learn alongside business leaders. After all, there is no right or wrong way to embrace agility, and it's important to fit any change to the context of the business rather than opting to cargo cult another's successful operating model. By actively working and learning with other execs and departments, there will be an increase in understanding and adoption as colleagues and peers will have skin in the game of shaping the new organization.

Ultimately, an organizational design that is not aligned to the goals of agile and lean philosophies will not succeed, even if the IT department is. The modern business environment requires an operating model that accepts the reality of uncertainty enabled by a disruptive competitive set. Agile and lean methods are proven successful, and the failures of organizations that are poor at adapting to change in the business environment demonstrate that leadership teams need to apply transformation not only to IT but to the entire enterprise to survive and succeed.

Remove Impediments

The best time for leaders to intervene and leverage their position of power in the hierarchy is when they are required to remove impediments. Effective leaders manage their teams by removing barriers that are preventing them from achieving outcomes, not by micromanaging and telling them what to do. Leaders don't directly deliver value; they set the right conditions and enable others to deliver. However, teams will often encounter problems that are out of their own control and that need a fast resolution. As well as checking in on progress and setting direction, leaders need to make time to understand anything that is causing a team a problem, such as legacy policies preventing them from using third-party services or software, unnecessary bureaucratic red tape, lack of funds, and so on. This includes removing barriers. Continuously challenge and remove wasteful and non-value-add activities. Often the system itself is the reason for blocking or delays. IT leaders are in a unique position to change the system. Fundamentally, a leader adds true value by setting direction, creating the right environment, and removing any obstacles that prevent a team from delivering a fast throughput of value.

Go and See the Work

As you may have gathered, the key skills that a modern leader should focus on are supporting the team through coaching and empowerment as well as developing, aligning, and communicating the organization's strategic focus and desired outcomes. However, a leader cannot do these things if they are too far removed from the realities of customer needs and team problems. Leaders need to have a firm grasp on what is going on at the shop floor. This is less about trying to assert control and more about having a clear understanding of what is happening, what the challenges are from both your team and the internal and external customers. What are the underlying patterns causing the problems? By having a firm understanding of what is happening, you are better placed to improve the system.

Even though we have talked about giving teams the autonomy to get the job done, you are ultimately accountable for results. Therefore, it is imperative to understand constraints as well as opportunities in the system. It is also vital to ensure the environment is conducive to delivering business impacts at pace. The key point is to avoid becoming too far removed from the purpose of the business and the front-line problems of your customers and teams. You must actively go to where the work is to gain a greater level of understanding to improve the system with feedback from your people close to the work. A constant flow of feedback is vital for leaders to improve their organizations.

However, it is important to show action when taking on feedback rather than simply paying lip service to your team. You must purposely be focused on actions to improve the system of work. Feedback is an important source for new ideas. Often it is the people who are closest to a problem that have innovative and creative ideas about how to solve it. Therefore, be open to and encourage feedback from all areas of the organization. Take time to listen and understand people's ideas and points of view, and ensure you explain the reasons if they are not pursued. This will ensure a constant flow of feedback up the hierarchy from people with knowledge of the problem domain and a detailed understanding of the technical capabilities available.

Break Down Silos

A customer journey will pass through many business processes and departments and utilize multiple business capabilities. However, customers will expect a seamless experience regardless of the differing ownership and accountabilities within the organization. As a leader, it is your responsibility to set an example by facilitating the collaboration across teams and departments, actively identifying and removing blockers with your leadership peers that hinder cross-capability work. Simply put, complex and adapting systems require collaboration to develop and evolve. Reducing the pain of hand-offs and coordination will improve collaborative discussions, discovery, and innovation and lead to better designed systems.

Leaders need to break down silos that get in the way of collaboration and a joined-up customer experience, ensuring all departments and capability teams are aligned on the purpose of their tasks and how they contribute to the desired outcome. Leaders should work to form a community based on meaningful respect, trust, and transparency and providing the space, guidance, and any facilitation and coordination to ensure all teams are working toward the same goal.

Summary

Agile literature focuses mainly on a team's ability to manage development efforts in complex problem domains, with no mention of the role of an IT leader. So how does the IT leader add value? Simply put, it is to set a clear direction with context, support the team's effort to deliver outcomes, fix any issues with the system, and for the rest of the time, get out of the way. IT leaders can still flex their muscles, but they should use them to remove impediments, facilitate collaboration, and educate peers all toward the goal of creating a better working environment for people.

In today's fast-paced and adaptive world, the leader is no longer the single authority on everything. Nor should they be. In complex adaptive systems, each person in the value chain needs to be involved in the decision-making on problem solving and opportunity discovery. We must move away from seeing development teams as siloed project delivery machines that we hand detailed requirements to and move to a more fluid model with the ability to pivot, collaborate, and think for yourself, which is of far greater value in complex problem domains. Therefore, supporting the team and creating the right environment to keep people motivated and aligned, autonomous, and empowered is essential. There is a scarcity of talent, and leaders must do their utmost to retain and develop people and create an environment to instill intrinsic motivation, creativity, and innovation. To achieve this, leaders should embrace the following behaviors:

Embracing Servant Leadership A longer-term approach to leadership focused on the needs of others before considering your own

Instilling Intrinsic Motivation The ability to make people feel incentivized to complete a task simply because they find it interesting or enjoyable

Encouraging Growth and Development Discovering the potential in people to improve performance, adapt to solve problems, and further their careers

Focusing on Improving the System Managing the system of work, the environment people work in rather than the people

Agile leaders are humble, respectful, empathetic, and genuinely care for the people they work with. Their belief system and culture, just like the philosophies of lean and agile, are formed around respect for people. All the behaviors that we have covered in this chapter are formed around developing people in your team. As Jack Welch, the CEO of General Electric, famously said, "Before you are a leader, success is all about growing yourself. When you become a leader, success is all about growing others."

III Strategy to Execution

Understanding Your Business

The goal of a business is to create a customer.

—Pete Drucker

Competitive strategy is about being different. It means deliberately choosing a different set of activities to deliver a unique mix of value.

—Michael Porter

It is not enough to just do your best or work hard; You must know what to work on.

—W. Edwards Deming

To become more strategic and offer real value, IT leaders need to move away from the role of order takers and toward the roles of business leaders and cocreators. To achieve this, you need a deeper understanding of your business domain in addition to your technical domain. In short, you need to know your business inside and out. To help exploit the current business, you need to have a detailed understanding of how the business operates as well as the constraints and waste that is in the system. To help explore new opportunities, you need to have a good handle on the environment—the business context—that your business operates in so you can understand how technology can be leveraged to capitalize on a new opportunity or mitigate the impact of a new challenge brought on by a competitor, a supplier, or a change in the wider context such as a piece of new legislation or a geopolitical event.

As laid out in Chapter 1, "Why We Need to Change The System," technology is playing an increasingly important role in the shape and success of an organization. As an IT leader, it is imperative to have a fundamental understanding of the components that make up the business architecture of your organization so that you can communicate more effectively with your colleagues. Through spending time with your peers and investing time at the gemba (the place where value is created) to understand constraints, challenges, and opportunities, you will gain a much deeper knowledge of the business and be in a far better position to contribute than if you simply sit back and wait to be told what to do. In addition to understanding the inside of the business, you will need to be able to look from the outside in, speak to real customers, listen to customer contacts in sales or customer services, and read customer reviews and feedback as well as form relationships with key suppliers and partners to your organization. Only through a deep understanding of your business domain will you be able to build trust, influence your peers, and leverage the art of the possible that technology offers to produce impactful business outcomes and move you from an IT leader to a business leader.

In this chapter, we will look at the frameworks, patterns, and models that will help you map what your business does—the *business model*. How your business does what it does—the *operating model*. The factors that influence the model—the *business context*, and finally, the *strategy*—what your organization is doing to succeed and win in its chosen market. You will see that the business model, operating model, and the strategy are all intrinsically linked. A change in one will have a knock-on effect to the others. With this fundamental understanding and empathy for the rest of the business, you will be in a strong position to contribute to business strategy.

Business Anatomy

The anatomy of a business can be distilled into five components. As shown in Figure 10.1, it includes a purpose, a business model, an operating model, a business context, and a business strategy.

Before anything else you need to have a reason for being, a mission, or a place in the world. This is your *business purpose*. This is your motivation. A business model and operating model describe how your business runs; think of these as your business's system. They show how the various pieces of the organization fit together to produce value. An organization's *business model* describes the high-level means by which it creates value. It is composed of the *value proposition*—what value it generates for its customers, the target customers, markets, and channels, and the *revenue*—how value for customers is turned into value for the business. The *operating model* describes how an organization does the work, that is, how the business is run. It details the organizational design, partners, locations, decision rights, culture, capabilities, and business processes that deliver the business strategy and deliver the value proposition of the business model.

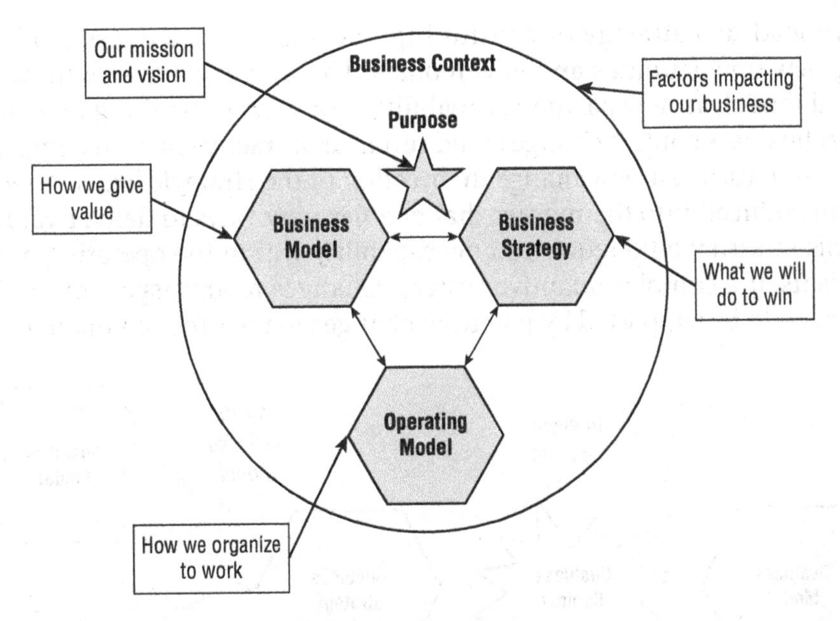

Figure 10.1: The five components of business anatomy

All businesses operate in an environment with a vast number of factors impacting them. Forces such as customer expectations, suppliers, competitors, and so on can disrupt an organization's business model or create opportunities for growth. We refer to this as the *business context*. External political, environmental, economic, social, legal, and technical factors can present both opportunities and threats. Internally, constraints and insights can also provide opportunity. No business is an island and understanding the environmental facts that can influence your business gives you the situational awareness which is crucial to adapting and surviving.

No business can stay still, and all have goals and aspirations. To stay competitive, you need a strategy, a list of choices detailing how you are going to do better by being different. Strategy is heavily influenced by the business context and the competitive forces at play. The *business strategy* clarifies the company's goals and the direction it will head towards to win. The business strategy is the set of choices that explicitly lay out where the business will play, how it will succeed, and what capabilities it needs to do so. The business strategy explains the choices that the business will follow, where to and where not to invest. A strategy can also result in a change to both the business and operating model, often due to changes in the business context.

The components that make up the architecture of a business are intrinsically linked—a change in one can have an impact on the others. As shown in Figure 10.2, changes in the business context can have an impact on an organization's business, which can lead to a strategy of continuing to exploit the business model and/or exploring new opportunities and new models. This can also have an impact on the operational model, as new partners, capabilities, or ways of working may need to be adopted. A business context change could reveal an attractive new customer segment to target that in turn causes a change in direction of the strategic focus. A technology could be introduced into the market that enables new ways to deliver value to customers; this in turn would require a new capability within the operating model. The business is itself a complex adaptive system. Changes to any aspect of the business anatomy need to be supported by potential changes to the other components.

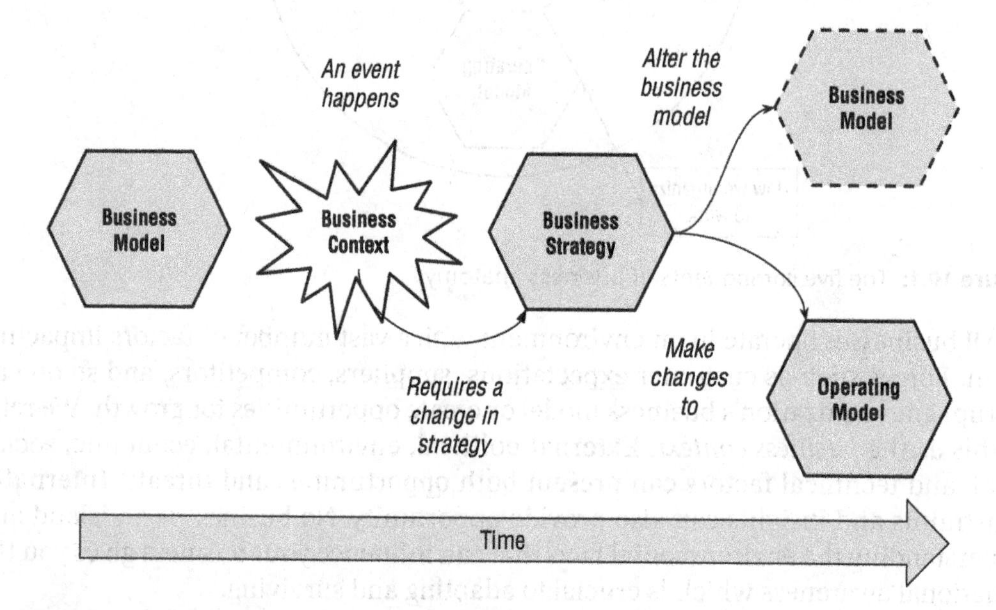

Figure 10.2: How business context changes can change strategy, business, and operating models

As an example, Table 10.1 shows the various ways in which businesses have adapted and changed during the COVID-19 pandemic.

Table 10.1: How businesses adapted during the pandemic

Industry	Impact of COVID-19 pandemic	New Strategy or Model
Restaurants	Restaurants closed due to restrictions, only available for takeout. Key food items in short supply due to panic buying and people reluctant to go into busy supermarkets due to risk of catching COVID-19.	Restaurants have strong supply chains and offer customers basic fresh food items, eggs, flour milk, etc. for delivery along with takeout meals using online platforms such as Uber Eats or setting up Shopify sites.
Hoteliers	No business travel or holiday bookings due to corporate work from home and government no travel advice.	Hotels offer day rate rooms as quiet areas for remote workers working from home without adequate space.
Airlines	Airlines cancel scheduled flights due to government travel advice and country restrictions.	Airlines offer cargo-only flights and transport goods on the empty passenger planes.

Why IT Leaders Need to Understand the Anatomy of a Business

To contribute to business improvement, IT leaders must understand how the business operates. As W. Edwards Demming said, "Management of a system requires knowledge of the interrelationships between all of the components within the system and of everybody that works in it." The purpose of this chapter is to help you understand the architecture of a business. The various frameworks, models, and patterns are presented so you can start to distill a business into its important components. It is not my intention for you to document all the various areas of your business using the frameworks, models, and patterns in this chapter; instead, think of them as tools for you to effectively capture and communicate important aspects of your business as you need them. You can use them to frame conversations with your peers and the CEO, you can use them to capture thoughts and ideas in a structured manner, and you can use them to communicate effectively and concisely with your team. Understanding the anatomy of a business in this way is the starting point for digital innovation and IT strategic contribution that we will visit in Chapter 11, "IT Strategic Contribution," where then you will formally document your technology strategy. As an IT leader, you are also a business leader. You need to understand the lexicon of the business anatomy so that you can converse with the CEO and the board in a powerful and effective language about the business, where it is, how it operates, where it wants to go, and how it will get there. As an IT leader, you will of course have a deep expertise in technology, but you must also have a broad understanding of every aspect of your business. We will now look at each of the five components of the business anatomy and learn about the tools and models that can assist in capturing and analyzing each area. But first we must start with the "why," the purpose of your organization, its mission and reason for being.

Purpose: Starting with Why and Understanding Your North Star

What is the purpose of your business? What is your organization's motivation, beyond making money for shareholders? Why does your company exist? What is your ultimate North Star that aligns your people and creates clarity for customers and potential buyers of your business? We are in an age now where businesses need to be more than about making money; businesses need to have a conscience and be socially and environmentally focused with a purpose beyond profit.

To be explicit about purpose and aid motivation, we can look to the business mission and vision statements to guide us.

Mission This is your reason for existence. This is ultimately what problem you are trying to solve and the value you are offering. This is what drives your company and what is fundamentally core to your business. It should concisely express what you do, how you do it, and who you do it for. This is your overall purpose and what should motivate your team.

Vision This is where you want to get to and what you want to become. It is your aspirational vision of how the organization looks in the future. This is what gives you direction and purpose. It should inspire your people and customers alike. It should cover such things as the problem you are trying to solve, the change you are seeking, and your dreams of the future.

The combination of the mission and value statements helps to define your organization and give it an identity. When done right, with thought and consideration, they form your North Star, giving purpose beyond generating shareholder value. As discussed in Chapter 3, "How to Change the System," purpose is essential for intrinsic motivation in knowledge work. Purpose helps to satisfy the desire of people needing to be part of something bigger and to contribute to something worthwhile. Purpose can help people find meaning in what they do beyond the need to work to pay the bills. A compelling vision of the future state helps the organization to think big and focus on meaningful change to achieve the end state rather than simply looking at hitting the next quarter's sales targets, albeit this is necessary as well. A well-articulated purpose can help with external as well as internal relationships, helping to connect with customers on an emotional level for loyalty, trust, and support and with suppliers for allegiance and cocreation. Clear purpose ultimately makes sure we never lose sight of what is important.

The Business Model: The System of Capturing Value

A business model describes how the organization creates and delivers value to its customers. It defines the market, channels, customers, and segments, the products and services offered, and it defines how it will turn value for customers into financial rewards. To describe the business model—and later in the chapter the operating model—we can leverage the Business Model Canvas. The Business Model Canvas, by Alexander Osterwalder, is composed of nine "building blocks" that describe a business. As shown in Figure 10.3, the Business Model Canvas covers components of the business model as well as the operating model. In this and the following sections, we will focus on the components that relate to the business model and will revisit the canvas when we look at the operating model component of the business anatomy later in this chapter.

We can use the components of the Business Model Canvas, namely value propositions, revenue streams, customer relationships and channels, to map the elements of the business model:

What Value Do We Offer? This section represents the products or services that you are offering to meet a customer need.

How Do We Make Money? This section describes the revenue streams business model—how the business makes money.

How Do We Reach and Interact with Our Customers? What relationships will you form with customers and what channels will you engage with them?

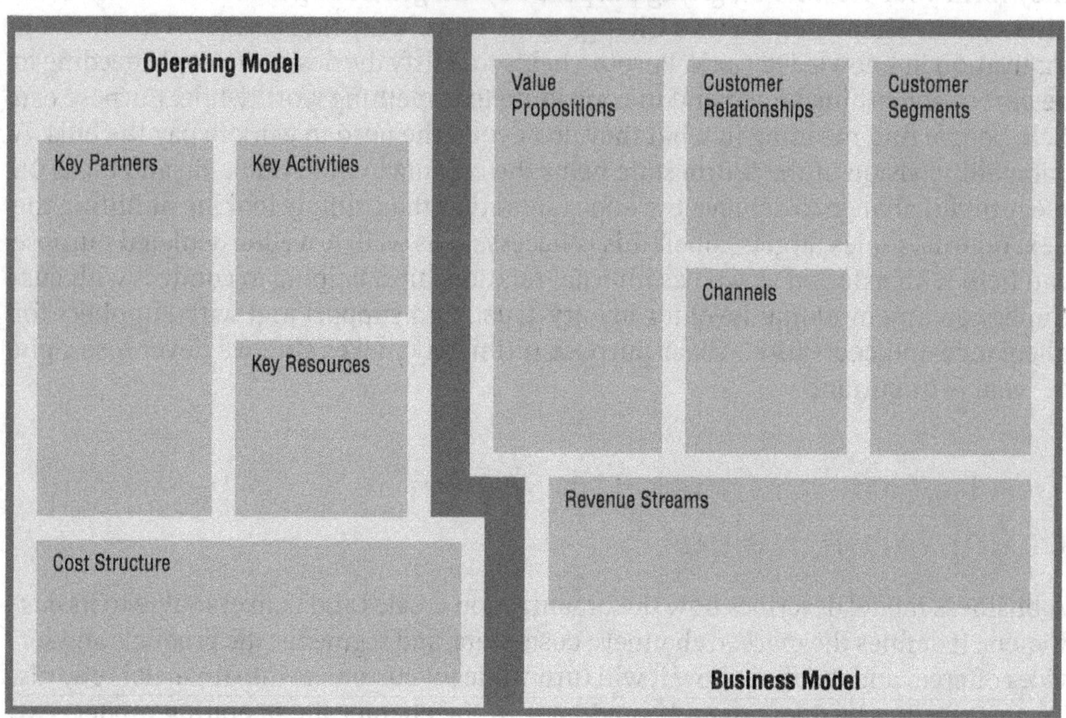

Figure 10.3: The Business Model Canvas covers both the operating and business model.
Source: The Business Model Canvas / Strategyzer AG / CC BY SA-3.0

The combination of these elements, can be specific to each customer segment we are targeting:

> **Who Are Our Customers?** This section covers the target customer segments, and the combination of value propositions, revenue streams, channels and the relationships which are applicable to them.

The Business Model Canvas is a great framework for capturing the business and operating models of existing organizations. However, this model will evolve or completely change based on the direction of the business strategy and changes in the business context. For startups or organizations looking to leverage the business model canvas for new products and services, there needs to be supporting material on why these customer segments are important and why the value proposition is defendable from competitors. This is typically where a strategy can complement.

What Value Do We Offer?

Forming the foundation of any business and thereby business model is a value proposition. It is the range of products and services that you offer to customers to meet a need or a desire. Value propositions come in various shapes and sizes; the following is a list of some popular ways to provide value:

Newness A product or service that didn't exist before. Think electric cars.

Performance Increased performance of an existing product or service. Think fuel-efficient cars, a faster Internet connection, a faster mobile phone.

Customization Products and services tailored to the needs of a specific customer segment. Think travel agents and designers.

"Getting the Job Done" Helping to deliver value to a customer. Think third-party logistics partners and delivery carriers that help complete the customer journey and get product to their doors.

Design The look and feel of a product or service, typically related to fashion brands and consumer goods.

Brand/Status The value comes from a customer being associated with the brand itself. Think endorsed and fashionable brands.

Price Offering products and services at the lowest possible price or even for free.

Cost Reduction Managing costs on a customer's behalf. Software as a service (SaaS), infrastructure as a service (IaaS), and platform as a service (PaaS) now alleviate customers from having to host and manage their software and hardware estates.

Risk Reduction Insurance policies, extended warranties, and service-level agreements (SLAs) reduce the risk for customers in the event of an issue.

Accessibility Allow customers to invest in products and services that were previously out of their reach, such as contributing to hedge funds, shares, and investment portfolios.

Convenience/Usability Aggregates for insurance and all manner of products and services allow customers to find the best deals without having to investigate individual suppliers' offers.

How Do We Make Money?

There are many ways a company can generate income from their value offering. These can also differ between the collection of product and services as well as from each customer segment. Here are some typical methods that organizations employ to generate revenue:

Asset Sale Perhaps the most traditional of all revenue streams is selling physical goods. Think retailers such as Amazon and supermarkets selling goods to customers.

Usage Fee Revenue is generated from the use of a service or product. Think Amazon Web Services (AWS) usage fees for the platform or courier usage per parcel.

Subscription Fee This represents access to a continuous service. Think Netflix for video content, Amazon Prime for free delivery, Apple Music and Spotify for audio content, and newspaper content behind paywalls, which all offer access for a monthly or annual subscription.

Lending/Leasing/Renting Granting access to a product or service for a period that is otherwise too expensive to use full time or only required for a specific job. Think renting equipment for building works, leasing a mobile home for a holiday, or hiring a car.

Licensing Charging for permission for the use of protected intellectual property. Think licensing music for advertising, the use of some proprietary technology, or image rights of a celeb or a sports personality.

Brokerage/Commission Fee By acting as an intermediate service between two parties. Think estate agents selling a house, or eBay and Amazon Marketplace.

Advertising Revenue is generated from charging customers for product advertising. Think Google Ads as well as sponsored products on Amazon search result listings.

How Do We Reach and Interact with Our Customers?

To successfully generate value, organizations must first reach customers and interface with them at each step of their buying journey. Your channels represent the touch points your customers will have with your organization, which will form part of their overall experience of your business. Interfacing with customers can occur at any number of channel phases:

Raising Awareness How do we establish the initial relationships with customers and make them aware of our organization and its products and services? Do we use affiliates, social media, recommendations, endorsements, TV, newspapers, and so on?

Evaluation How do our potential customers evaluate our value proposition? How do we demonstrate the value we can give them? Do we offer demos or limited trials whether by time, feature, or volume? Do we explicitly explain how we solve needs and release benefits for our customers? Do we allow previous customers to review our services?

Purchasing How do we enable customers to purchase our products and services? Will they purchase in store or online? Can they self-serve or do they require salespeople to negotiate and tailor terms and agreements?

Delivery How do we deliver our products or services? Can customers buy over the counter, or do they need the product or service delivered and tailored to their requirements?

Post-Sales How do we provide customer support? Do we offer a no quibble return? Do we have SLAs and various levels of support for customers?

An organization can reach customers through its own channels, partner channels, or a combination of both:

Owned Channels Business-to-consumer (B2C) channels owned or operated by the business can be online, in store, or via a sales team.

Partner Channels Business-to-business (B2B) channels, where a partner manages the interaction with a customer, can include distribution, in store, or third-party websites such as marketplaces.

Mix of Channels Often what is core to an organization is the development of a product or service, and so while they will offer this direct to customers (typically online or a flagship store), they also partner for wider distribution. Although brands such as Nike do direct sales, it generates the vast majority of its revenue via partners. Similarly with cruise holidays, suppliers will market and advertise directly and have a portal for selling but will rely on partners to sell the majority of products.

When you have reached your customers, you need to identify the type of relationship you will form with them. Customer relationships are important for the following reasons:

- To attract new customers
- To retain customers
- To increase sales from loyal customers

The types of customer relationships are as follows:

Personal Assistance A pre-/post-sales personal assistance experience is offered to customers. This could be in person, online, or via the phone. Think consultancies like Gartner and Forrester that offer personal assistance for IT and business projects.

Dedicated Personal Assistance The assignment of an employee to a customer or set of customers to handle all their needs. Think account managers that will build and maintain relationships with a small set of customers.

Self-Service The business provides the means for customers to serve their own needs efficiently and effectively. Think blog engines or the Shopify platform.

Automated Services A more personalized version of self-service, with the ability to identify and tailor experiences and preferences for customers. Think content sites, online banks, and modern e-commerce that present content and products based on customers' previous searches, visits, accounts, and purchases.

Communities Creating peer-to-peer communities to allow customer-to-customer as well as customer-to-business interactions. Knowledge and experience can be shared between clients to improve their use of a company's product or service. Many businesses of consumer technologies offer online communities to allow customers to share knowledge.

Cocreation A relationship between a company and its customers that leads directly to the organization's products or services. Think content creators for sites—reviews (various e-commerce sites) and audio/visual content (YouTube).

Who Are Our Customers?

Last, we need to understand what customer segments our company is trying to serve. Our customers can be segmented in many different manners based on need, profitability, location, or demographic to name but a few. However, for each customer segment we are targeting, we must ensure we have tailored our value proposition, our channels, relationships, and even how we generate a return from them. Here are the customer segment types:

Mass Market There is no focus on a particular customer segmentation. Mass market is typically used for an organization wishing to appeal to a customer base with similar needs. Think organizations such as supermarkets or car insurance.

Niche Market The opposite of a mass market. The customer segment is narrow and focused. Think businesses that offer luxury goods to a small set of customers.

Segmented A company adjusts its value proposition, channels, or relationships to segments within an existing customer base. A business could segment its customer base by age, gender, spend, loyalty or by any number of attributes and data points.

Diversify Businesses that serve multiple unrelated customer segments such as B2B and B2C customers. Think Amazon servicing B2C customers on its shopping platform as well as B2B customers on its Marketplace.

Multi-Sided Platform/Market Some organizations require mutually dependent customer segments to make the business model work. Think brokerage organizations such as estate agents that need sellers and buyers. Publishers need content writers as well as readers to balance the business model.

As shown in Figure 10.4, an organization should tailor its value proposition, channels, relationships, and revenue generation depending on each targeted customer segment.

Operating Model: How We Do the Work

As shown in Figure 10.5, the Business Model Canvas covers components of the business model as well as the operating model. In this and the following sections, we will focus on the components that relate to the operating model. The operating model articulates how the work is done to realize the value described within the business model.

The operating model has the following components:

Key Activities The activities directly involved in delivering on the value proposition. The most important things your company does. These will be your unique or core value streams as opposed to other supporting activities such as HR and finance. Key activities can be grouped as problem solving (think consultancies), platform/network (think eBay, Facebook), and production (think Apple, Nike).

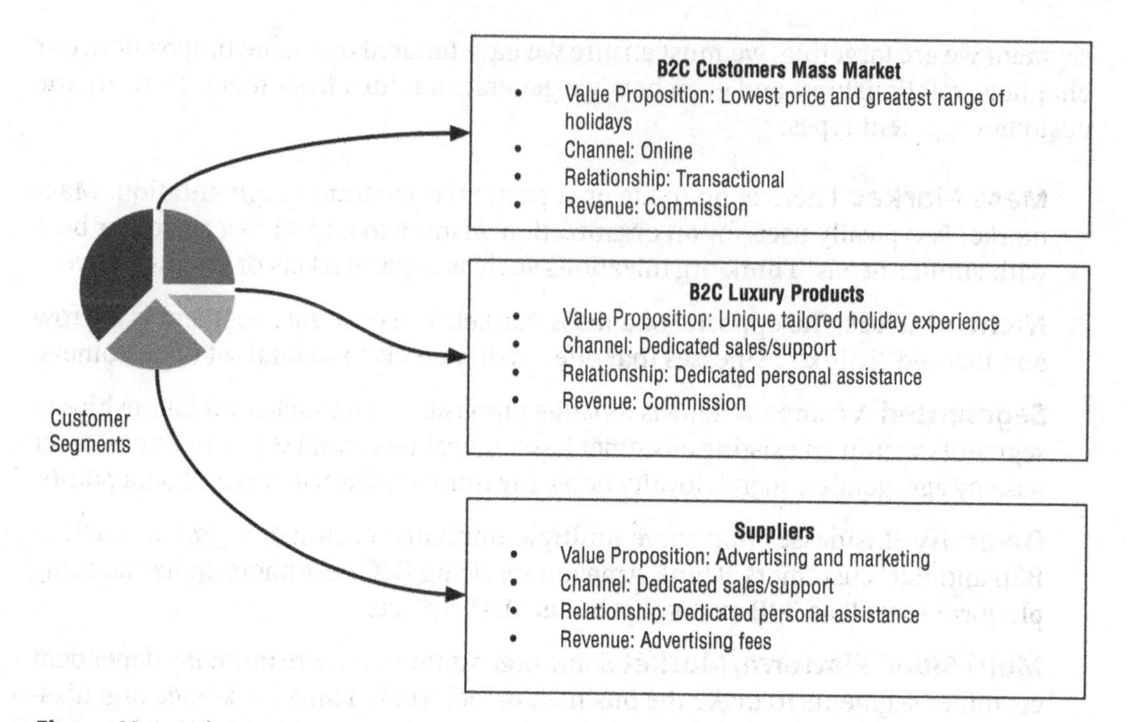

B2C Customers Mass Market
Value Proposition: Lowest price and greatest range of holidays
Channel: Online
Relationship: Transactional
Revenue: Commission
B2C Luxury Products
Value Proposition: Unique tailored holiday experience
Channel: Dedicated sales/support
Relationship: Dedicated personal assistance
Revenue: Commission
Customer Segments
Suppliers
Value Proposition: Advertising and marketing
Channel: Dedicated sales/support
Relationship: Dedicated personal assistance
Revenue: Advertising fees

Figure 10.4: Value propositions, channels, relationships, and how revenue is made can be tailored to a customer segment.

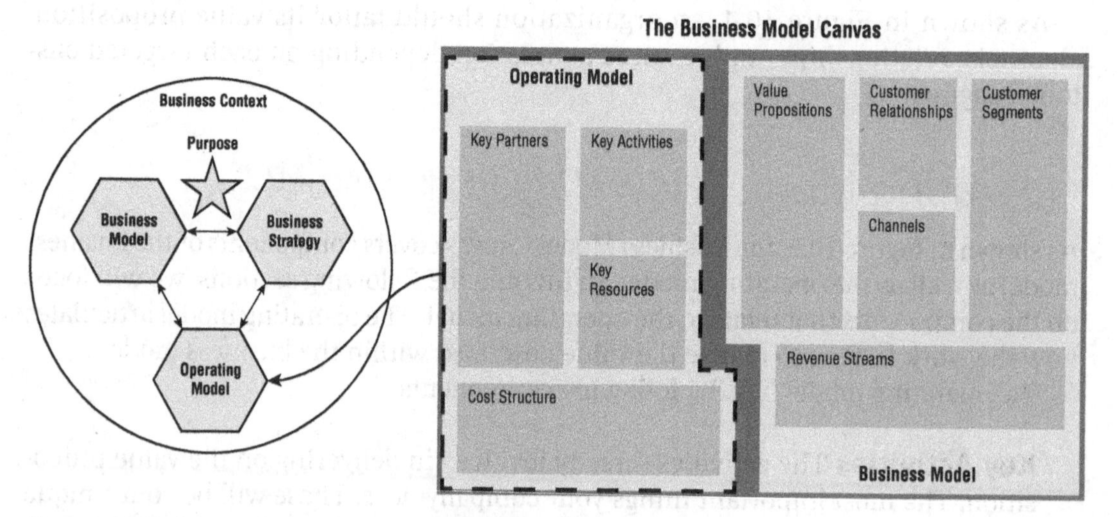

Figure 10.5: Components of the operating model

Source: The Business Model Canvas / Strategyzer AG / CC BY SA-3.0

Key Resources The resources and assets that are necessary to support the value proposition. Key resources can be grouped as physical (think buildings, warehouses, vehicles), intellectual (think brands, copyrights), human (think scientists, experts, highly skilled employees), and financial (think cash, lines of credit, and leasing options).

Key Partners These are the network of the organization's suppliers and partners that contribute to the key activities to deliver on the value proposition. Partnering with other third parties enable your organization to focus on their core activity. The motivations for forming partnerships can be grouped as cost optimization (think a product organization partnering with a company for third-party logistics such as warehouse and home delivery), reduction of risk (think joint ventures or strategic alliances), and the acquisition of skills and capabilities (think licensing part of a product such as an operating system or outsourcing development to a third party).

Cost Structure The key costs that support the key activities, resources, and customer channels. All businesses are keen to reduce costs, but some value propositions are based on a low-cost profile. These business models are cost-driven. Businesses like low-cost airlines and supermarkets can be considered cost-driven. Organizations that aren't competing on cost can be described as value-driven, with a focus on creating unique and innovative products and services. Businesses like Apple and luxury watch makers can be considered value-driven.

In addition to the operating model components of the Business Model Canvas, another good framework that you may wish to use to understand the operating model of your business is the Operating Model Canvas based on the book by Mark Lancelott, shown in Figure 10.6.

We won't look at all the sections of the operating model as many will align with the operating model of IT covered in Part 2, "Designing An Adaptive Operating Model," of this book. For now, we will focus on what is going to be important for IT to understand to add value (see Figure 10.7), namely the businesses key activities. The key activities can be distilled down to:

Value Streams How value is delivered from customer request to receipt

Business Capabilities What is required for the activities in the value stream steps

Business Process, People, Systems How work is done

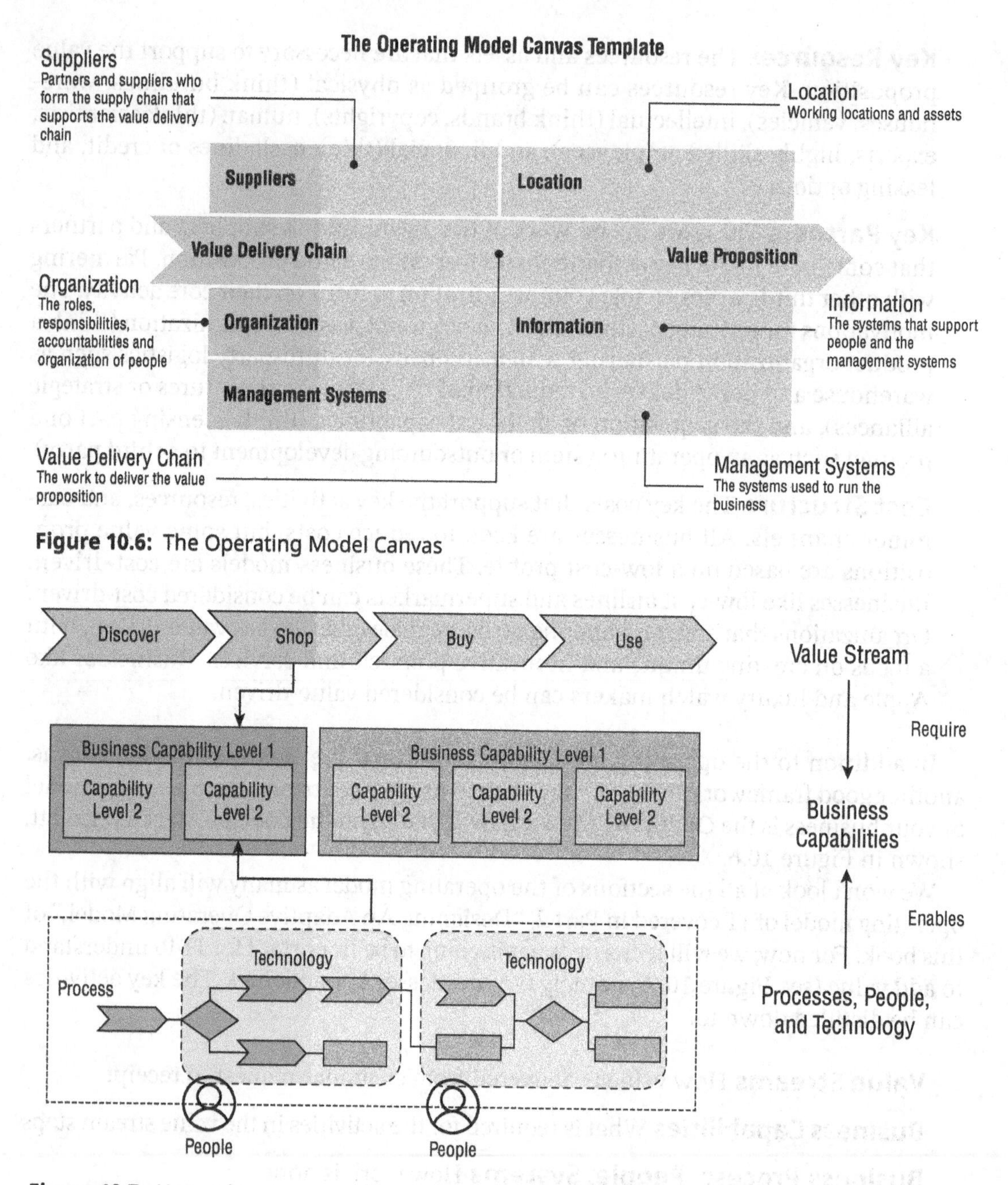

Figure 10.6: The Operating Model Canvas

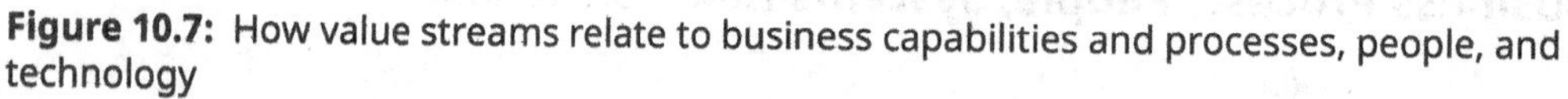

Figure 10.7: How value streams relate to business capabilities and processes, people, and technology

How We Deliver Value: Key Value Streams

A value stream consists of a sequence of activities, whether performed by the organization itself or by a third party, that turn a customer request into something of value. Examples of value streams are "Buy a lottery ticket," "Apply for a mortgage," or as shown in Figure 10.8, "Order takeout food." Value streams are created from an outside-in view, anchoring the initiation event from the perspective of a customer, be that an internal or external one. By mapping the organization's key value streams from a customer's view, we can then map back to what is required from the business to fulfil those activities rather than mapping solely from a business perspective. In other words, outside-in rather than inside-out. To help focus on the customer, we name the value stream in terms of the value proposition we are providing or the job the customer wants to get done. By visualizing the flow of value, we can better understand the steps involved, what activities offer no value, and where to focus investment to poorly performing steps. However, the biggest benefit of thinking in terms of value steams is that we never lose focus of the end-to-end process, which avoids siloed thinking and only localized improvement. In Chapter 6, "How We Work," we looked at a tool called value stream mapping, which provides a structured way to make improvements that optimize the entire process of value delivery.

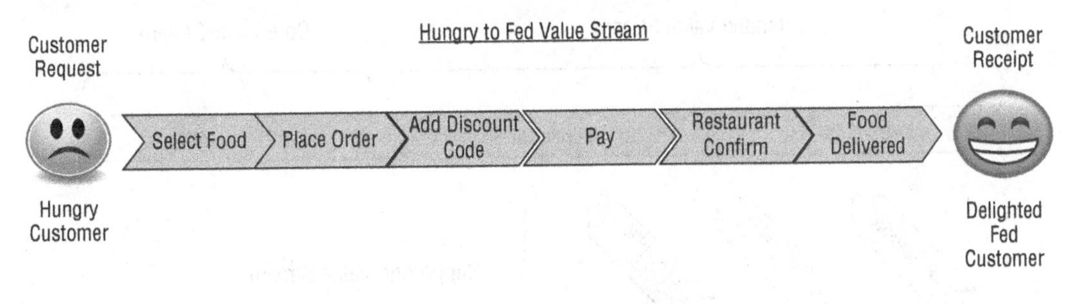

Figure 10.8: An example of a value stream

Value streams are very closely aligned with customer journey maps, also covered in Chapter 6. A customer journey map, shown in Figure 10.9, like a value stream, is designed from the point of view of a customer. Each value stream or activity within it can be related to one or more steps in a customer journey.

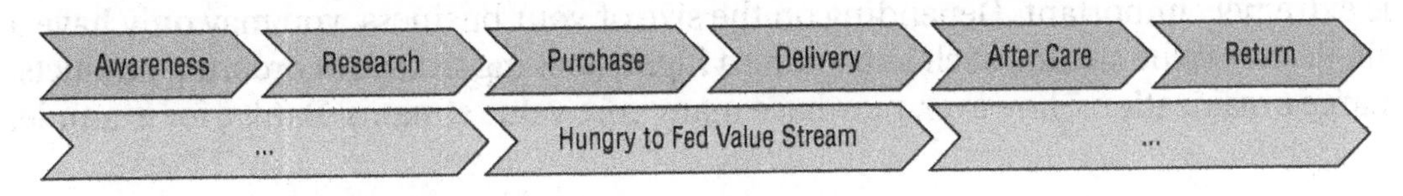

Figure 10.9: How value streams align to customer journeys

Types of Value Stream

There are three main types of value stream, as depicted in Figure 10.10:

Core Value Stream Core value streams link directly to how your organization delivers on its value proposition. These value streams typically involve external customers requesting a product or service. Examples include applying for a loan, bidding on an auction, ordering food, or submitting your tax returns.

Extended Value Stream An extended value stream includes the pre- or post-activities of the core value streams. For example, if your core value stream is to provide the service to book a holiday for a customer, the extended value stream will include providing the customer with a quote and providing post-booking services. Other examples include processing payments, servicing a leased car, and handling returns.

Supporting Value Streams Supporting value streams are typically the internal activities that support business operation. Examples of supporting value streams include recruiting, onboarding and employee performance reviews, budgeting or forecasting, and the development "concept to cash" value stream.

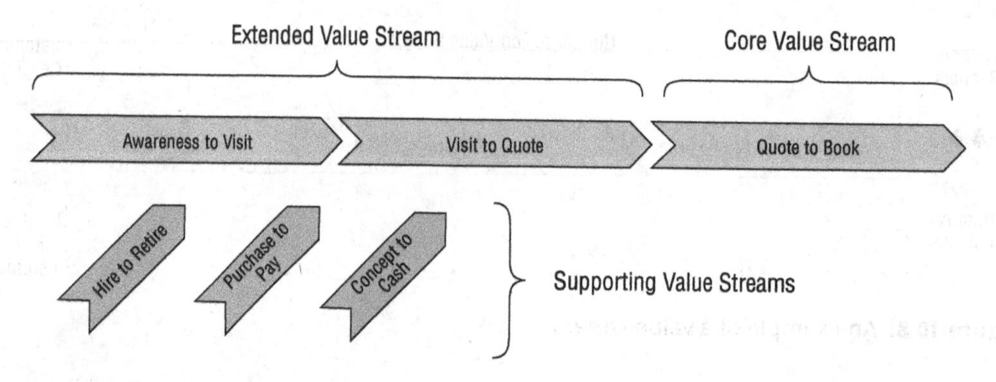

Figure 10.10: Core, extended, and supporting value streams

Whether core, part of an extended, or supporting a value stream, each starts with a request or a trigger and has a series of activities that provide value. The ability to understand each value stream and improve flow by identifying and eliminating waste is extremely important. Depending on the size of your business, you may only have a single core value stream, such as booking a flight, booking tickets, or ordering products. Large organizations, however, may have many core value streams. Banks, for example, offer loans, mortgages, savings accounts, and many other products and services; each one is a key value stream to the business, giving direct value to a customer.

The Value of Thinking in Value Streams and Journeys

As shown in Figure 10.11, value streams are typically highly cross-functional in that the activities from customer request to customer receipt are handled by many different functions of the organization. Traditional improvement initiatives are aimed at a unit or departmental level with little overall optimization of the delivery of value. However, by thinking in terms of value streams or customer journeys, we can start to move toward integrated optimization programs organized around the end-to-end delivery of value. This is a practical step toward taking a systems thinking view of optimizing an organization's value proposition. Table 10.2 highlights the difference between functional thinking and thinking in terms of customer journeys and value streams.

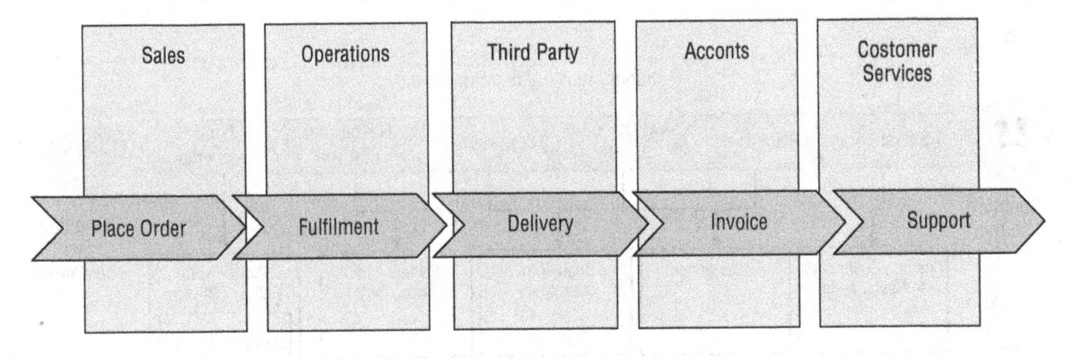

Figure 10.11: Value streams cross business units

Table 10.2: The difference in thinking in silos vs. thinking in value streams

Thinking in Functional Silos	Thinking in Customer Journeys or Value Streams
Focus is on the improvement of a function's activities and processes.	Holistic focus on the customer journey or value stream.
Only incremental gains; larger benefits not achieved due to disconnect and handoffs between functions.	End-to-end improvement in customer experience.
Local optimization: while there may be operational improvement, this can be at the expense of the overall performance and customer experience of the value stream or journey.	System optimization: Focus is on the removal of the biggest constraint of the system; teams focus on the end to end for bigger impact.

What We Need to Do: Business Capabilities

Business capabilities, as shown in Figure 10.12, do the work at each activity step within the value stream. This gives us a link between how value is provided for customers and how the value stream relates to the required internal capabilities. You can think of a business capability as an abstraction of a business function. The capability represents the *what*, whereas the people, process, and technology represent the *how*. A business capability is an extremely important and fundamental building block of the business anatomy. It defines the organization's capacity to successfully perform key business activities. Understanding what business capabilities are required to support key value streams helps to answer the question, What do we need to do to win? It provides the link between the business model and the operating model and, as you will see in Chapter 11, the business strategy and IT's strategic contribution.

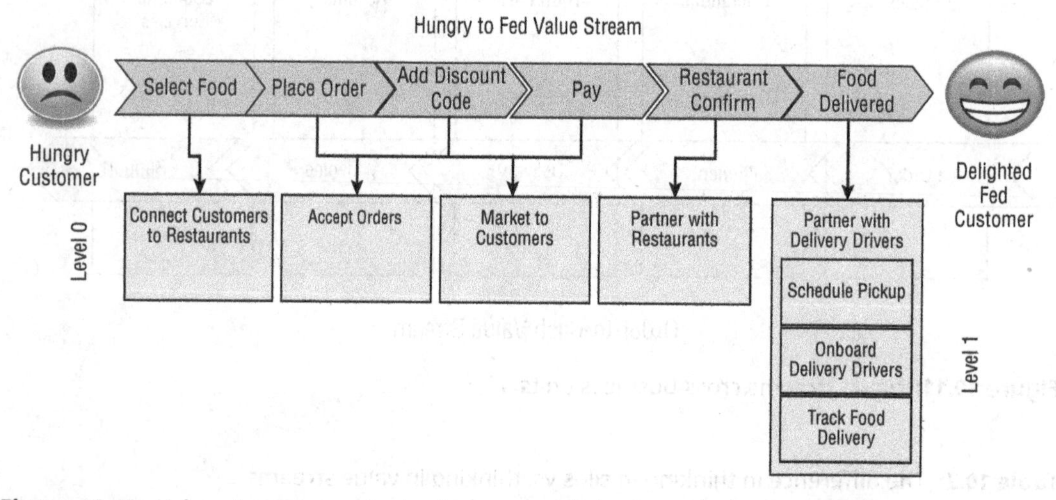

Figure 10.12: Value streams require business capabilities to perform the activities.

How We Do It: People, Process, and Technology

While a business capability is an abstraction of a business function, they are composed and enabled by technology, process, and people. Take the example of the capability of an online food ordering company to "Onboard Delivery Drivers" as shown in Figure 10.13:

People These are the people involved in the delivery of the capabilities. In this example, this could be HR or members of the operations team.

Process Business processes are the steps, tasks, and workflows necessary to get work done. In this example, this could cover the steps to register new delivery drivers and set up contracts and work schedules.

Technology The systems used to capture and store data and automate processes. In this example, there could be some form of CRM to manage all delivery drivers' information and the onboarding process.

All components are required to enable the capability of onboarding delivery drivers.

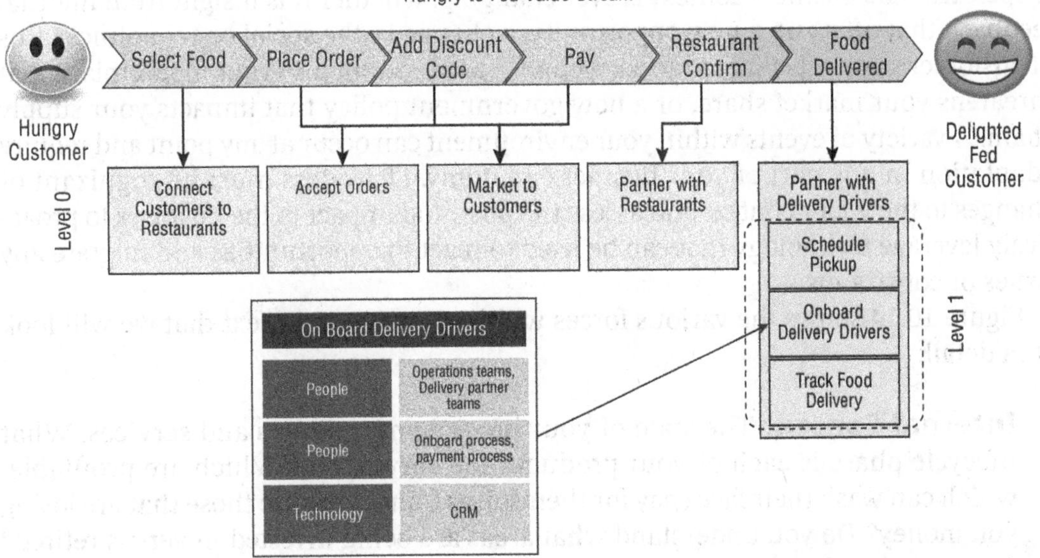

Figure 10.13: Business capabilities are composed of people, process, and technology.

Value streams provide a common understanding of how a company delivers customer value. They orchestrate the delivery of value by utilizing business capabilities. By modeling value streams, we can determine where investment in capabilities is required to improve the value proposition. This can be thought of as the strategic direction, that is, what must happen to improve customer experience. As business capabilities are enabled by technology, along with people and processes, we can identify the management systems that are related to capabilities to determine if they sufficiently support the business. Or indeed if technology can alter or replace entire value streams themselves.

Business Context: Understanding What Can Impact Us

Every business must understand the environment that it operates within, from both an internal and external perspective. This is known as the business context, and analyzing it to determine what factors will influence an organization enables the organization to adapt strategy and business models to changes in the environment in which it operates. The business context is ever changing, whether it is insight from internal feedback that offers up a new opportunity, a change in the social or economic status of customer segments that reduces revenue, a new entrant into your marketplace that threatens your market share, or a new government policy that impacts your supply chain. A variety of events within your environment can occur at any point and require adaptation in any part of your business anatomy. IT leaders must be cognizant of changes to the wider context and assess the potential impact to the business to proactively leverage technology that can be used to react to opportunities and migrate any issues or constraints.

Figure 10.14 shows the various forces within a business context that we will look at in detail:

Internal Context The state of your incumbent products and services. What lifecycle phase is each of your products and services in? Which are profitable, which can wash their face (pay for themselves), and what are those that are losing you money? Do you understand what areas are being invested in versus retired? Do you know the strategic importance of each product, the market size and customer segments?

Microeconomic Elements These represent the industry-specific contextual factors that can impact your business:

Market Demand The understanding of the market demand profile and attractiveness of different customer segments and how this can change your strategy as your business model.

Market Supply Your competitive position in the marketplace. While it's vital to keep an eye on your rivals, you also need to understand how your competitive position can be affected by suppliers, customers, and new entrants.

Macroeconomic Factors The wider factors outside of what is specific to your industry. These represent all the other factors that could impact the business from political to environmental and everything in between.

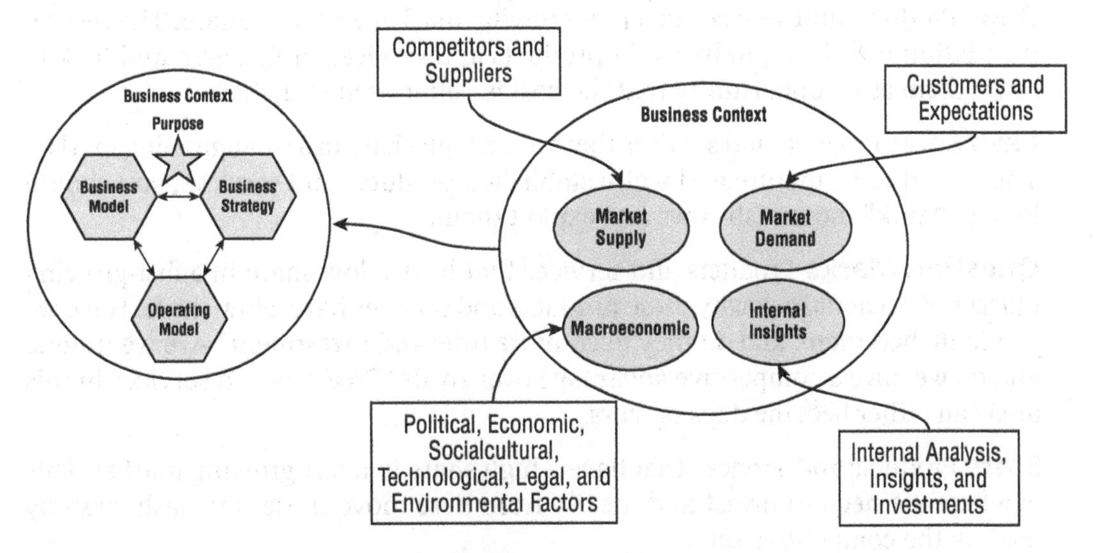

Figure 10.14: The factors in your environment that can affect your business

The business context will heavily influence the direction of your organization's business strategy. However, it is worth noting that strategic decisions are based on a combination of incomplete data and judgement. This is often because we can't get perfect data about our environment. Therefore, it is important to understand that the models that I will present in this discussion to analyze the business context are all subjective. That is not to say that they are not useful, but because they are high-level frameworks for analyzing data, they can only be effective based on the quality of that data. Data on the size of the market, your competitiveness in relation to rivals, global changes, and investment strategies for products will be garnered from industry and trade journals, market research firms, off-the-record chats with your suppliers, competitors, and in some cases your best guesses. In business, as in IT, you will rarely have all the facts; therefore, you need to use experience and judgement to make decisions based on the information that is available to you at any given time.

Internal Context: Portfolio Analysis

A tool that can help to analyze your organization's portfolio of products and services, aka value propositions, is the growth-share matrix. This chart, shown in Figure 10.15, was designed by Bruce Henderson in 1970 for the Boston Consulting Group to assist organizations in analyzing their product lines. It is based on two measurable axes: market share and market growth rate. There are four quadrants in the matrix that are used to map your product offerings:

Dogs Products and services in a low-growing market with low share. The recommendation here is not to invest in products and services in this area and look to retire them as an opportunity that the cost is better used elsewhere.

Cash Cows Products and services that have a high share in a slow-growing market. These tend to be mature and well-established products and services; you should look to "milk" these cash cows and exploit them.

Question Marks Products and services that have a low share in a fast-growing market. You should ask why these products and services have a low share: Have we just launched them, and do they need more time and investment? Are we unique and do we have a competitive advantage over rivals? Products and services in this area can either become dogs or stars.

Stars Products and services that have a high share in a fast-growing market. This is where we need to invest and ideally from here move these into cash cows by leading the competitive set.

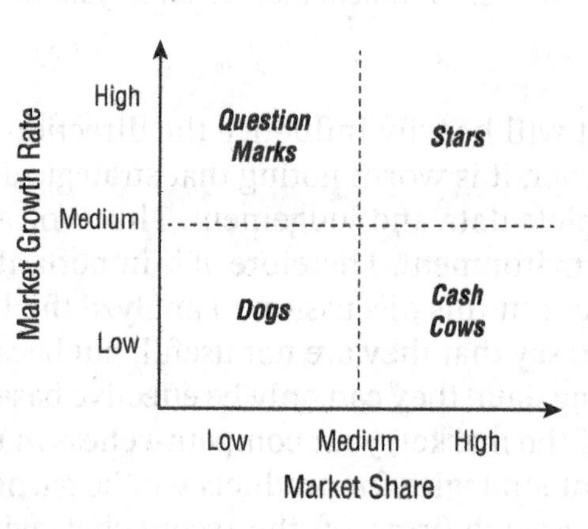

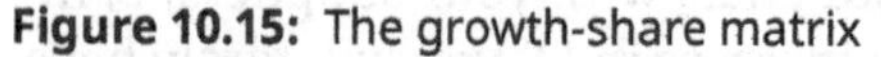
Figure 10.15: The growth-share matrix

You can also map your organization's portfolio of products and services on a product lifecycle model, as shown in Figure 10.16. This can be overlaid with the growth-share matrix to show products and services that need exploring versus exploiting.

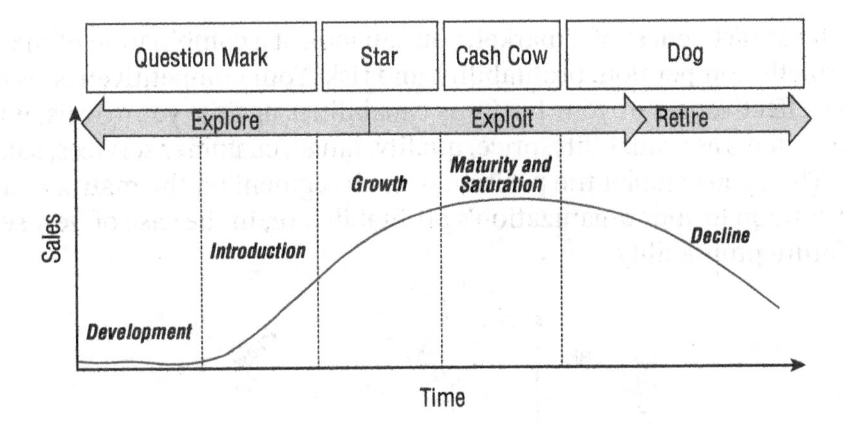

Figure 10.16: The product lifecycle model

Smaller enterprises and startups will only typically have a single product or service, but larger organizations will have many. For example, consider Google's various products and services and where they sit on the growth-share matrix:

Question Marks Google Cloud, Google Shopping

Stars YouTube, Google Assistant, Google Maps

Cash Cows Android, Google Ads, Chrome

Dogs Google News, Google Groups, Google Glass

The position of a product or service on the growth-share matrix and the product lifecycle model will heavily influence the level of investment organizations allocate for improvement.

The Demand: Customers and the Market

In what new market areas should your business compete? What customer segments should you target with your products and services? The attractiveness/advantage matrix, shown in Figure 10.17, can be used to analyze potential new areas to focus on and areas that are under threat in the marketplace. The matrix shows how competitive your organization is, and reveals markets where you should continue to invest, markets where you ought to withdraw, and markets that you may wish to enter. The matrix is based on market attractiveness along with your competitive position. To derive a rating for the attractiveness of a market, you can look at a combination of market size, demand growth, competition, profitability, and risk. Your competitiveness is based on the relative effectiveness of your business capabilities against your rivals, which can include such items as availability, price, quality, range, customer services, sales, scale, and so on. Then you can plot the position of each segment on the matrix and the size of its contribution to your organization's profitability or, in the case of new segments, estimate future profitability.

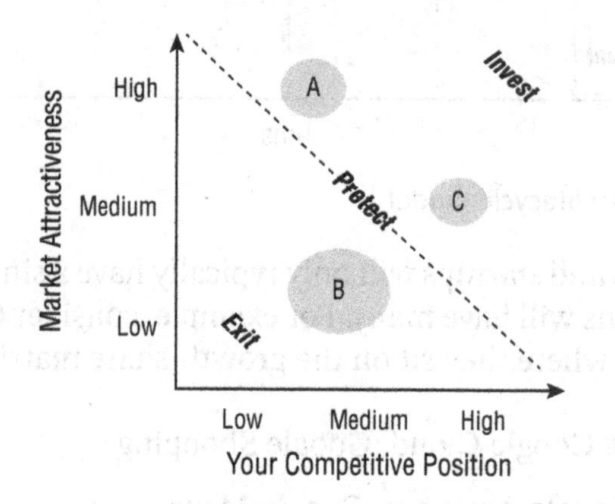

Figure 10.17: The attractiveness/advantage matrix

The Supply: The Competition and Porter's Five Forces

Michael Porter developed the Five Forces model, shown in Figure 10.18, to help organizations gain a better understanding of the competitive environment in which they operate. Porter recognized that while organizations will closely monitor their direct competitors, they often don't look at other factors that can have a material effect on the organization's competitive position. By analyzing the strength and direction of each force on the organization you are better placed to understand your position in the competitive landscape. Based on this analysis you can then shape a business strategy to improve your chances of winning.

The five forces identified by Porter are as follows:

Competitive Rivalry How much rivalry is there in the environment you work within? Do your competitors have deeper pockets than you? Do they have better capabilities, products, and services than you? How aggressive are they? In an environment with few competitors or where you are the only organization offering a particular product or service, you will be in a strong position. However, where there are many businesses vying for the same customer base, you will have intense competition with aggressive tactics to gain market share, especially if the market is not growing. If you are not on top of customer services and your offering does not meet the expectations of customers, they could quickly go elsewhere.

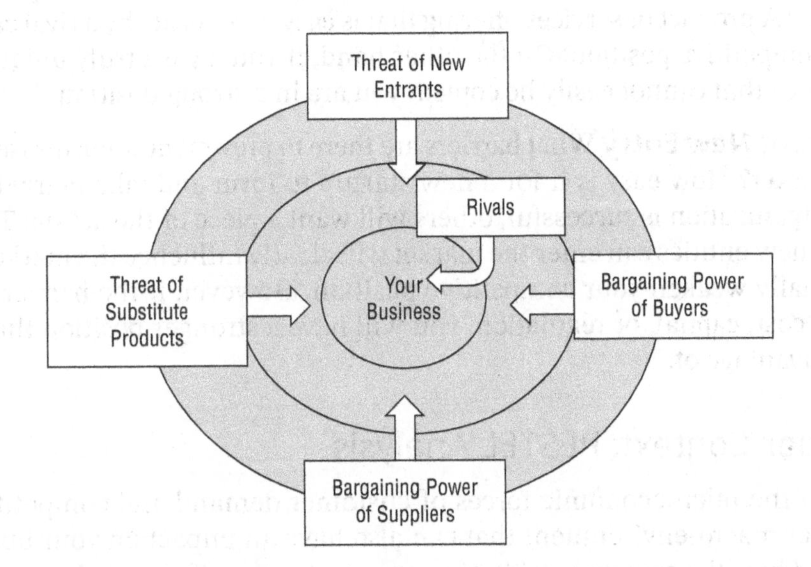

Figure 10.18: Michael Porters Five Forces model

Buyer Power To what extent can your customers influence your prices? Do your products and services differentiate from those of competitors enough to avoid customers being able to play you off against a rival? How costly is it for a buyer to switch services? Are you reliant on a single large buyer or is your revenue spread across many buyers? If you are beholden to a single large buyer or if you have competitors that offer the same products or services, where buyers can easily switch, you will be in a weak position. However, if you have a unique offering and low competition and are not reliant on a few large buyers, you will be in a much stronger position.

Supplier Power How much influence do suppliers have on your business? Do you need them more than they need you? As with buyer power, suppliers can also impact your price and your market share. Are you reliant on a single supplier, or do you have options to source inputs from various supplier markets? How important are you to your suppliers? Do you have a relationship where you can cocreate value? You will be in a stronger position if you have choice when it comes to supply. However, if you are reliant on a single supplier, then the strength lies with them, and they will be able to dictate what they charge you.

Threat of Substitution How easily can your products or services be replaced or superseded by competitors? Can your customers switch to a competitor with ease and without disruption to their experience? Can customers live without your product and service and take the work on themselves, perhaps moving to a manual process? A product or services offering that is easy to replicate by a rival can weaken your competitive position. On the other hand, if you have a truly unique product or service that cannot easily be copied, you are in a strong position.

Threat of New Entry What barriers are there to prevent new businesses entering the market? How easy is it for a new startup to form and take market share? If your organization is successful, others will want a piece of the action. The ease at which new entities can enter the market will clearly influence the market and will potentially weaken your competitive position. However, if the barriers are high due to cost, capital, or regulation, you will have a stronger position that you can take advantage of.

The Wider Context: PESTEL Analysis

Outside of the microeconomic forces of customer demand and competition there is the wider macro-environment that can also have an impact on your business. To understand how the context outside of your industry can affect your business, we can employ PESTEL analysis. PESTEL, as shown in Figure 10.19, is an acronym that stands for Political, Economic, Social, Technological, Environmental, and Legal. It is a useful template to break down the wider impacts that can challenge a business and ensure you are at the very least aware of factors outside of your specific business. It is worth noting that it is important to identify the factors and key drivers that will have a direct impact on your business rather than going deep into detail on each of these areas.

Political Factors What changes are the government putting in place that could affect your organization? How will Britain pulling out of the EU impact trading? What about governments stopping noncritical travel to other countries during the pandemic? How does a change in a political party and a new mandate impact your business?

Economic Factors What factors will have an impact on your business and market demand? How susceptible are you to changes in exchange rate? Are your customer segments sensitive to economic change, are they low income/young families, or are they more financially secure retired/empty nesters? Are there new taxation regulations or laws? Are your customer segments affected by economic changes and will this influence demand prospects?

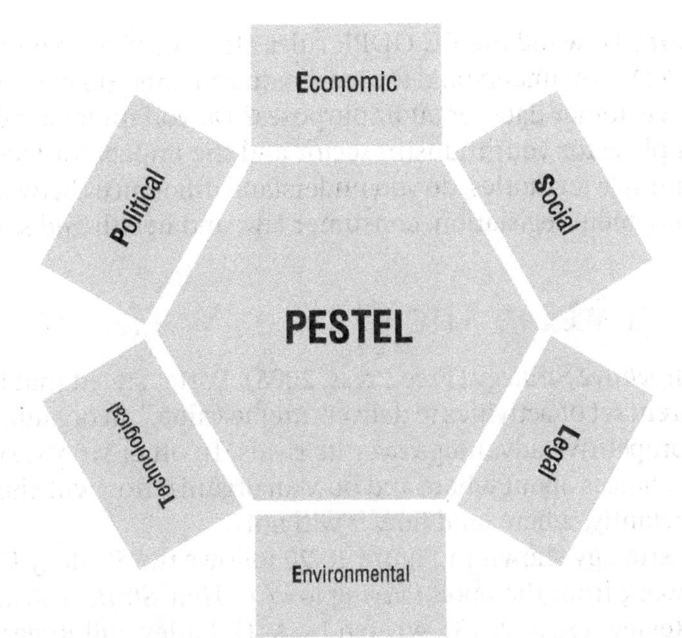

Figure 10.19: The PESTEL model

Social Factors Are your customers' attitudes and expectations changing to how you offer your products and services? Do they want you to support charities? Are your customer segments' lifestyle changes affecting how they communicate with you? Are there any emerging trends you need to be aware of for your customer segments? Are your projects still appealing to the next generation?

Technological Factors This is perhaps one area where you feel you have a good handle on what is happening. However, ensure that you look at how technology is being used in unrelated business. Look at technologies that you are currently not using. Are you keeping up with industry trends? How many white papers have you read on new technology, new methodologies, and ways of working in the development arena? What new cloud services are available to simplify your estate? Try not to become too blinkered on what you are currently working on and take an outside-in approach to understand new trends and advances in technology.

Environmental Factors How much do ecological factors impact your business? How important is this in the minds of your customers? Do they want you to be carbon neutral? There is an expectation for companies to have a conscience and to focus on the responsibility of sustainability. At some point this will more than likely become a legal requirement.

Legal Factors How did the EU GDPR rules affect you? Are you clear on the PCI DSS controls? Do you understand the legal requirements you have on retaining and disposing of customer data for audit purposes? Do you understand the regulatory measures in place for your industry sector and the impact for technology? If you operate in multiple territories, do you understand differences between what is legal, such as employment legislation, consumer law, and health and safety.

Business Strategy: The Choices We Make to Win

In his book *Competitive Strategy* (Free Press, 2008), Porter stated that by "deliberately choosing a different set of activities to deliver unique value," an organization can create a sustainable competitive advantage over its rivals. In other words, strategy is about making explicit choices about where and how an organization will choose to compete and, more importantly, where, and how it will not.

The business strategy shown in Figure 10.20 follows the Strategy Choice Cascade. This is a framework from the book *Playing to Win: How Strategy Really Works* (Harvard Business Review Press, 2013), written by A. G. Lafley and Roger Martin. Lafley and Martin define strategy as "an integrated set of choices that uniquely position a firm in its industry so as to create sustainable advantage and superior value relative to competition." The Strategy Choice Cascade presents strategy as the answer to five interrelated questions.

1. What is your winning aspiration? The purpose of your enterprise, its motivating aspiration.
2. Where will you play? A playing field where you can achieve that aspiration.
3. How will you win? The way you will win on the chosen playing field.
4. What capabilities must be in place? The set and configuration of capabilities required to win in the chosen way.
5. What management systems are required? The systems and measures that enable the capabilities and support the choices.

There is feedback, iteration, and consenus to arrive at an answer to each of the five questions, thereby ensuring a consistent and reinforcing set of answers at all levels of the organization and importantly ensuring that the necessary investments in capabilities, people, and process are in place to act on the strategy.

As mentioned earlier, a good strategy can advance a business and allow the business model to adapt and evolve to changes in the business context, thus ensuring it is not disrupted by others and that it keeps a competitive edge. In other words, a business model is where you end up based on the choices you make in your strategy. As shown in Figure 10.21, the where-to-play and how-to-win choices replace the customer segments and value proposition of the incumbent business model. The capabilities represent a link between the strategic vision and the operational execution. Capabilities are an important concept that we will revisit in Chapter 11 when determining how IT contributes to business strategy success.

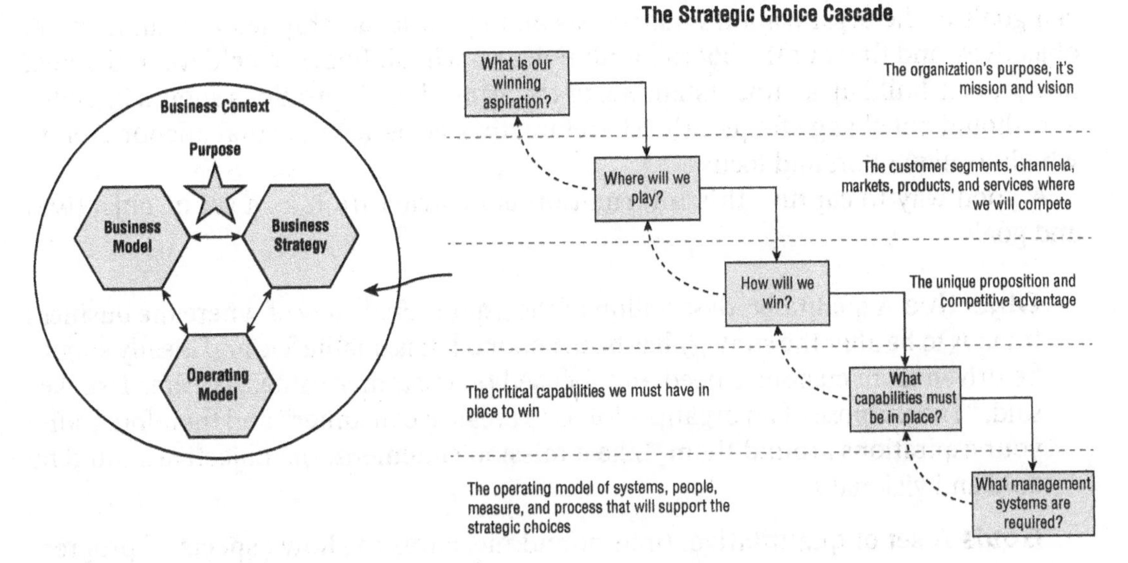

Figure 10.20: The Strategy Choice Cascade

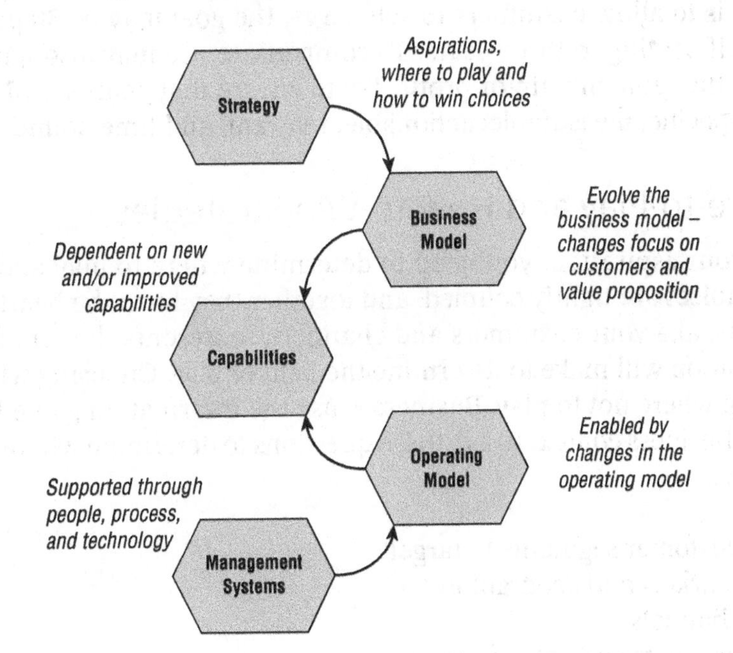

Figure 10.21: How strategic choices cascade throughout a business

Aspirations, Objectives, and Goals

A good strategy starts with an end goal, an aspiration of what success looks like and how it will be measured. An organization needs to have a clear understanding of what the winning aspiration is to set direction. If it's not explicitly stated, look for financial goals in the organization's budget, speak to your leadership team to understand objectives, and find out the board's ambition and what it hopes to achieve in the next few years to build up an understanding of the aspiration. Aspirations should be stable and should not change frequently. Therefore, they act as a contextual anchor around which to align effort and focus.

A good way to capture this information is to structure it as a set of objectives and goals.

Objective A qualitative description of the aspirational view of where the business intends to be after the strategy has been actioned. It is qualitative and ideally should be driven from customer need and focused on customer value. As Peter Drucker said, "The purpose of an organization is to create a consumer" and therefore frame your aspirations around them. Like a mission statement, the objective should be bold and visionary.

Goals A set of quantitative, time-bound measures to show aspects of progress against the objective. An objective can have many compensating goals. If the objective is to allow customers to self-serve, the goal may be 30 percent of customers self-serving in three years. To compensate, we may also have the goal of retaining the same margin or profit. Try to ensure that goals are SMART, in that they are specific, measurable, actionable, relevant, and time bound.

The Where-to-Play and How-to-Win Strategies

To achieve your aspiration, you need to determine where to play and how to win. These two choices are tightly coupled, and together they form the heart of a strategy. Where to play, aka your customers and channels, represents the set of choices that your organization will make to determine the field of play. Choosing where to play is also choosing where not to play. Businesses use the information gained from understanding the business context to ask these questions to determine where the company will compete:

- What customer segments to target?
- What markets and geographies?
- What channels?
- What products and categories?

How to-win, aka your value proposition, defines the choices for winning in the chosen where-to-play field. The organization needs to determine how it will offer value to customers in a unique and sustainable manner that sets it apart from competitors. In other words, what is the organization's competitive advantage that will allow it to win in the chosen where-to-play area? Generically speaking, Porter describes three major ways for how to win: cost leadership, differentiation, and a narrow focus.

Winning on Cost Charge the lowest price or use high margins to invest in creating competitive advantage. Focus on cost reduction as the strategy.

Winning on Differentiation Create a product or service that is perceived by consumers to be of greater value and that differentiates you from your competitors. Focus on brand distinction and delighting consumers as the strategy.

Winning through a Narrow Focus Don't address all of an industry; focus on a segment of it.

Knowledge of the market, both customer needs and competitive set, along with the PESTEL analysis gained from understanding the business context, is vital input in determining where to play and how to win. For how to win, there are some other good frameworks that guide the framing of a winning strategy:

The Value Disciplines Model This can help to clarify how organizations achieve their competitiveness. Whether that is by operational excellence, customer intimacy or product leadership.

The Value Proposition Canvas This can help ensure that a product or service is positioned to meet the needs and expectations of what a customer will value.

The Value Disciplines Model

Michael Treacy and Fred Wiersema's research discovered that organizations who explicitly articulate and focus on competitive advantage and customer value will be more successful than organizations that don't focus and try to be all things to all people. Treacy and Wiersema categorized the three approaches to market differentiation as follows:

Customer Intimacy Businesses that focus on customer intimacy offer tailored products and services to narrow customer profile segments. Their model is based around customer centricity and, due to the cost outlay, building long-term loyalty. This overlaps with Porter's principle of winning through a narrow focus. Bespoke travel agents and consultancy firms are examples of organizations that compete on customer intimacy.

Operational Excellence Operational excellence is about being the best on price and convenience. This strategic approach focuses on driving costs out of the process of product and service production and delivery. This value discipline overlaps with Porter's principle of winning on cost-competitive focus. Supermarkets are examples of organizations that compete on operational excellence.

Product Leadership Product leadership places a focus on creating unique or leading-edge products and services. There is a strategic investment in future innovation and creating consumer demand. This overlaps with Porter's principle of winning on differentiation. Apple and Nike are examples of organizations that are focused on product leadership.

It's important to note, however, that even if a business focuses its efforts in one area, it still needs to have a minimum capability in the other areas. As you can see from Figure 10.22, you can rate your effectiveness in each area as well as your competitors' effectiveness. Then you can see where you are performing or underperforming to determine if a chosen strategic choice of how to win is appropriate and if you require investment in that area. For example, if your capability that you intend to focus on is not mature, are you willing to invest in it? If a competitor has a higher maturity in a particular area, do you want to compete there?

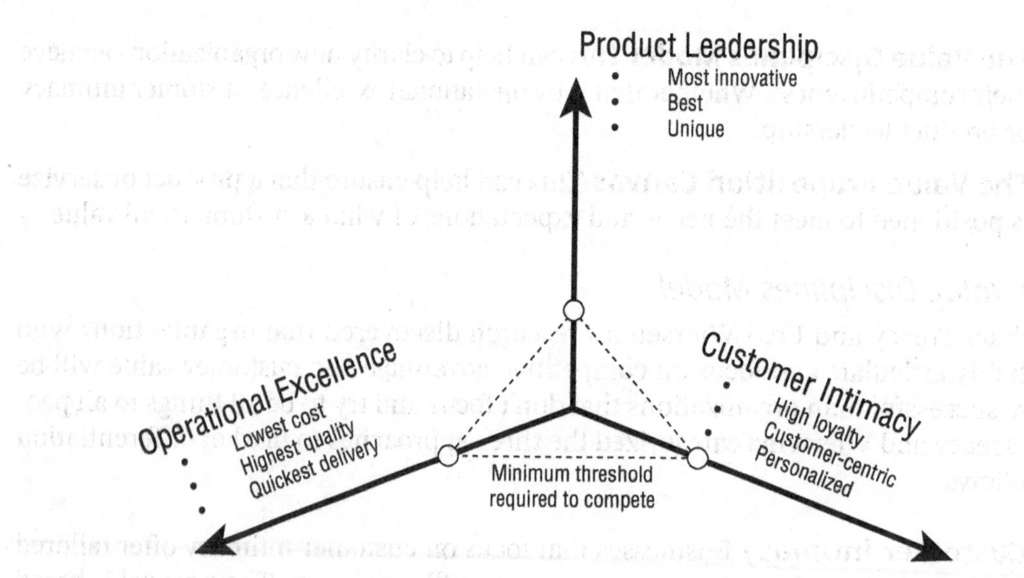

Figure 10.22: The Value Disciplines model

The Value Proposition Canvas

Your organization's value proposition should be derived from the needs of the customer. The same authors of the Business Model Canvas also developed the Value Proposition Canvas, a tool that can help ensure that a product or service proposition is correctly positioned within the market—this aligns nicely with the how-to-win and where-to-play strategic choices. This framework also ensures that an organization anchors any choice on how to win with a true customer need.

As shown in Figure 10.23, the Value Proposition Canvas is based around two components: the customer profile and the company's value proposition.

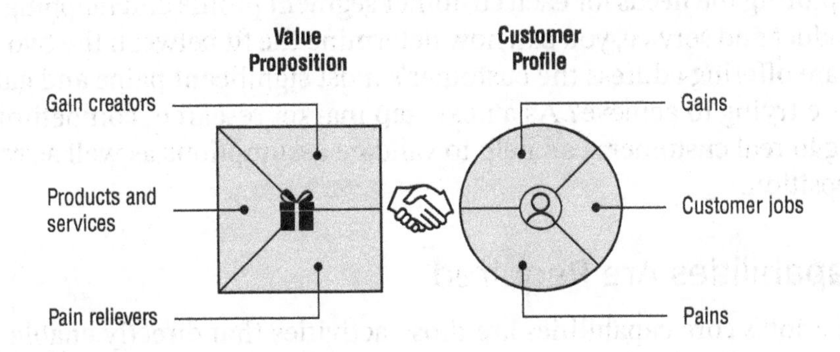

Figure 10.23: The Value Proposition Canvas

Products and services are designed to meet a customer's need: however, often the ability to fulfill this need falls far below customer expectations. To avoid missing the mark, a customer is profiled based on three needs. A profile should be created for the distinct gains, pains, and jobs of each customer segment.

Customer Jobs These are the jobs, tasks, and problems customers are trying to solve, be they functional, social, or emotional. These are the tasks customers wish to satisfy.

Customer Pains These are the current pain points related to trying to achieve the jobs mentioned in the preceding item. These are the negative experiences, emotions, and risks that the customer experiences with the way things are currently done.

Customer Gains These are the expectations, benefits, aspirations, and positive outcomes that would offer value to a customer.

The value proposition to meet the needs of a customer profile, in other words segment, is composed of the following elements:

Products and Services The products and services offered to meet customer needs. This is your value offering to a customer.

Gain Creators How your service creates positive outcomes and offers added value for the customer.

Pain Relievers How your products and services remove customer pain points.

After capturing the needs for each customer segment profile and mapping the value of your product and service, you can now determine the fit between the two. Does the value you are offering address the customer's most significant pains and gains in the jobs they are trying to achieve? As a next-step market research, competitor analysis and talking to real customers can help to validate assumptions as well as refine your value proposition.

What Capabilities Are Required

An organization's core capabilities are those activities that directly enable the how-to-win strategies. Porter visualized the activities in an activity system as shown in Figure 10.24. The activity system shows the capabilities and related activities required to deliver on the strategic choices of where to play and how to win. By understanding what capabilities are essential to support how-to-win strategies, those of high importance that have low maturity, an organization can more effectively prioritize investment. This enables a business to clearly see where it should focus its time and effort to be different and where it can afford to have parity or be good enough.

The capabilities within the activity system need to be created or improved to deliver on the where-to-play and how-to-win choices. Avoid being distracted by the incumbent capabilities and instead work backward from the capabilities that are critical to achieve the where-to-play and how-to-win choices. This will enable you to understand what new capabilities you need, what improvements in existing capabilities you need, and what capabilities are not contributing to the strategic choices. Strategy creation is not a top-down exercise, the Strategy Choice Cascade has feedback and consensus at each level, and therefore the investment required in new capabilities and improvements to existing capabilities needs to be estimated to understand the feasibility of a how-to-win choice. There is no point in taking a strategic path if you have no hope of being able to provide the necessary capabilities to succeed in it. You also need to understand if the capabilities are distinctive and defensible, that is, can a competitor easily gain the same capabilities and void your competitive advantage?

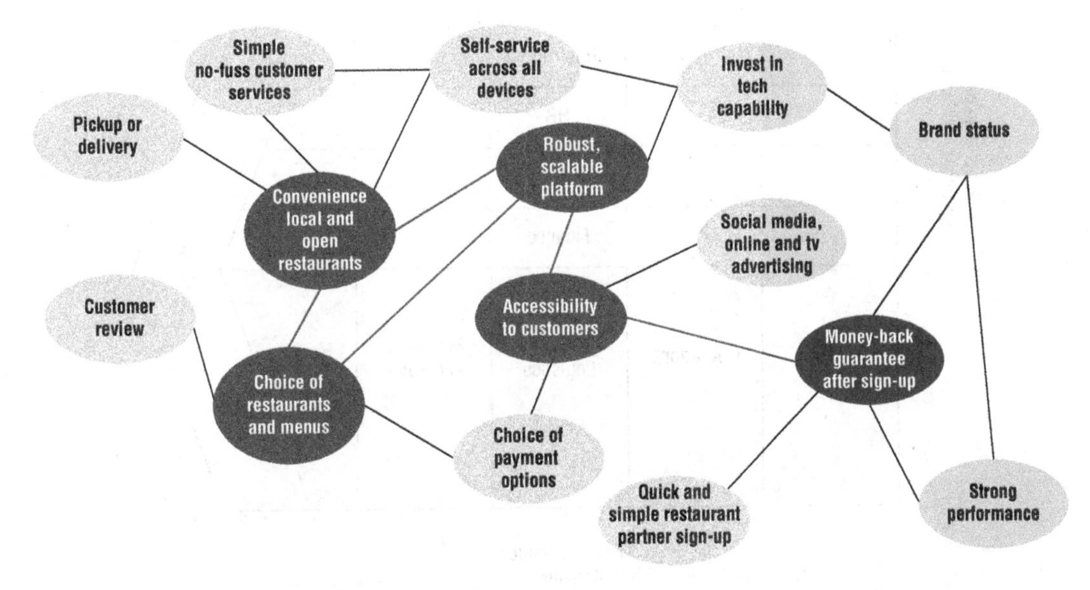

Figure 10.24: Porter's activity system

There are some other good frameworks that help to identify and assess your organization's core and competitive capabilities and enable weak strategic areas to be addressed or strategies changed:

Porter's Value Chain

Value chain analysis based on what an organization does.

Wardley Maps

More recently Simon Wardley has introduced a new method of mapping to determine where an organization should differentiate.

Porter's Value Chain

According to Porter, a company's competitive advantage "stems from the many discrete activities a firm performs in designing, producing, marketing, delivering, and supporting its product. Each of these activities can contribute to a firm's relative cost position and create a basis for differentiation." To better understand the activities and capabilities that support the how-to-win choices, it is useful to distill the system of the organization into a series of value-generating activities referred to as the *value chain*. The value chain model captures the collection of activities that a company performs to create value for its customers. Figure 10.25 shows a generic value chain model that identifies the primary and supportive activities as introduced in Porter's book *Competitive Advantage*.

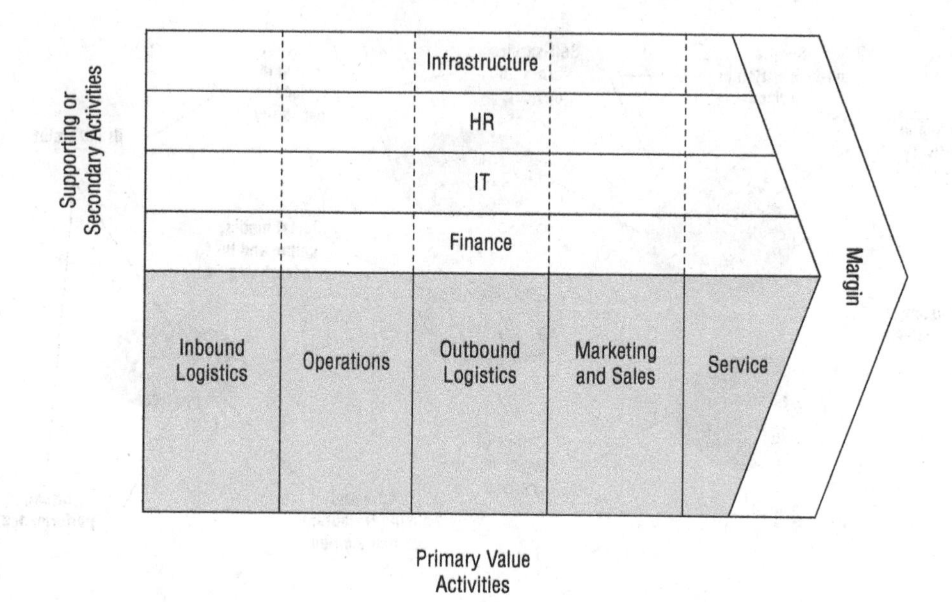

Figure 10.25: Porter's value chain

A value chain can be used to focus on the primary activities that support and enable the strategic how-to-win choice.

Winning on Cost If the how-to-win choice is based on winning on cost, you can identify the cost drivers for each activity and seek to find cost savings. Can automation be used to reduce manual effort and save costs? Can we outsource a commodity activity that can be done cheaper and better by a third party? Can we source raw materials from a cheaper supplier but still retain quality? Can we move manufacturing to a different country?

Winning on Differentiation If the how-to-win choice is competing on a unique offering, be that operational excellence, customer intimacy product, or another, then you need to identify the activities that support that value generation and ensure that you can offer a unique, different, or at least better, capability compared to your competitors.

The value chain has a very different purpose to the value stream as covered in the section "Operating Model: How We Do the Work" earlier in this chapter. Value streams are focused on the activities directly involved in delivering value to an end customer, the areas that add value to a product or service, whereas the value chain covers all the activities that an organization must undertake to support the value proposition. Value chains are useful from a high level of abstraction, to understand where to focus effort and investment to gain a competitive advantage. In *Competitive Advantage*, Porter explains that "the value chain disaggregates a firm into its strategically relevant activities in order to understand the behavior of costs and the existing and potential sources of (competitive) differentiation." Whereas value streams are useful for improving the flow of value to a customer and highlighting issues, constraints, and problems, especially when the flow crosses department silos.

Porter's value chain is a good way to determine which capabilities and activities are key to enabling the competitive advantage. Understanding the links between activities can clarify how capability needs are fulfilled, whether that be build, buy, partner, or outsource. This analysis directly supports the cost or differentiation advantage. However, it has some drawbacks: for one, it lacks an anchor—how do the capabilities relate to a customer need? There is also no concept of position, only the left to right flow of the product or service. This means it's not easy to see what is visible to a customer. Last there is no easy way to map the evolution of an activity, which would make it easier to see which are competitive and which are commodities. That would help to validate the how-to-win choice—for example, if your how-to-win strategy relies on commodity capabilities, will it truly give you a competitive edge? Porter's value chain is a good start to modeling the activities and capabilities of a business; however, Simon Wardley has introduced a more effective way to map using his Wardley maps technique.

Wardley Maps

A Wardley map, introduced in Chapter 2 and shown in Figure 10.26, is a map of the structure of an organization or a service including the components—the value chain—required to serve a need of a customer. The key difference from Porter's activity maps is that the Wardley map adds both an evolution and dependency view on top of the value chain view. This means the map can help us to understand the most appropriate method to use to bridge any capability gaps as well as dependencies between capabilities. This visualization of components, evolution, importance to user, and dependency all anchored to a user's needs aids the strategic conversations on where best to focus business and IT strategic actions as well as the most appropriate method of applying them.

Another key benefit of a Wardley map over a capability map is that we can see the evolution of components and thus movement. The underlying dependencies between components are constantly evolving over time from custom-made to commodity. This gives us situational awareness, which in turn makes it easier to make informed strategic choices on where and how to invest. It is out of the scope of this book, but Wardley maps can be used to map many abstract aspects of business strategy, such as political and cultural, not just the value chain. We will dive deeper into Wardley maps and how they can help with IT strategic contribution in Chapter 12, "Tactical Planning: Deploying Strategy."

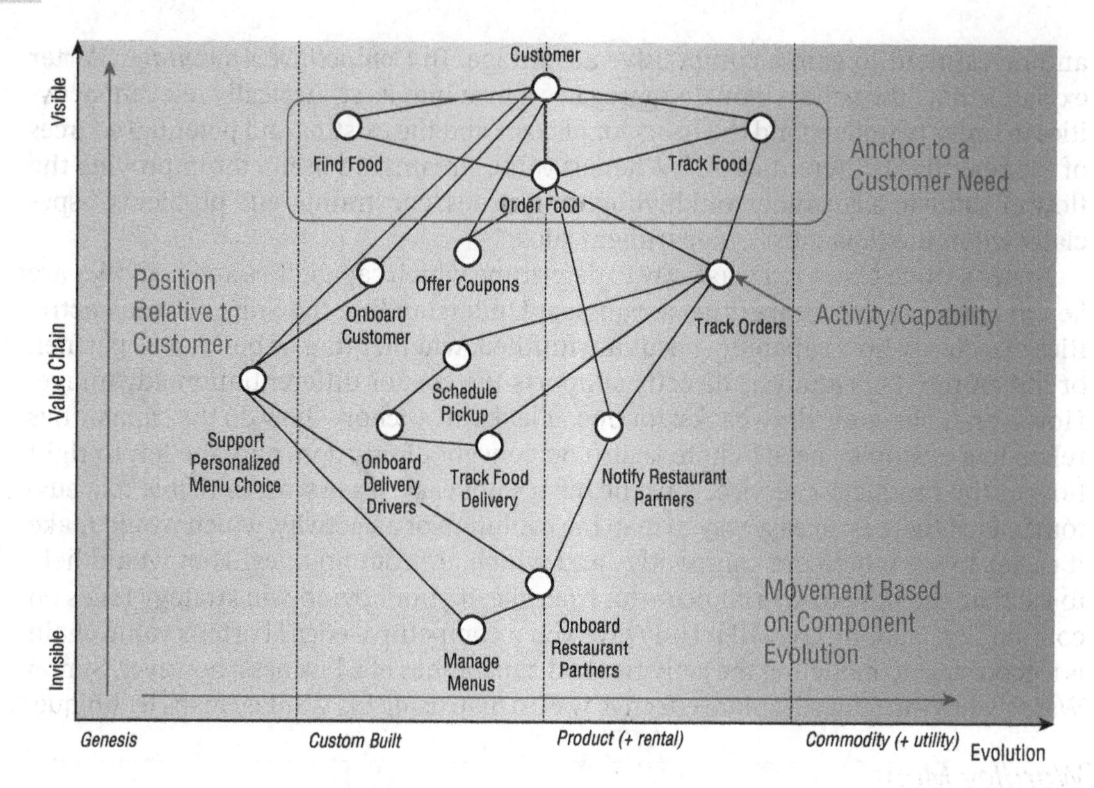

Figure 10.26: An example of a Wardley map

What Improvements to the Management Systems Are Required

The final strategic choice in the cascade focuses on the management systems required to build and maintain the distinctive capabilities that are required by the unique strategic choices. This is perhaps the most important aspect as it ensures effective action throughout the organization and is fundamental to executing the strategy. At the management systems level there is a need to provide the right structure, accountability, and decision rights to support the delivery of the core capabilities. A need to ensure that the most appropriate ways of working, and specific measures are employed. As well as ensuring the necessary required skills, capabilities and talent are in place. In short, the management systems are the operating model that will be created to enable the creation or improvement of capabilities critical to success. Martin says, "The key to great management systems is their individual match with the distinctive capability." In this real sense the structure, or rather operating model, follows the strategy.

Capturing and Communicating Business Strategy

Having a winning strategy is one thing but if it is not understood by the rest of the organization then it will fail to be effectively deployed and executed. When it comes to strategy communication, there are some frameworks that can help distill the strategy to a one-pager for the organization to ingest; these are the OGSM framework and strategy maps.

OGSM as a Framework for Strategy Communication

How do we deploy strategy? Is everyone aware of strategic vision? Can people see how their work contributes to the organization's goals and aspirations. How do we cascade strategy to employees so that they make day-to-day decisions to deliver on the strategic choices? As shown in Figure 10.27, the OGSM (objective, goals, strategies, and measures) is a concise framework that will help clarify and communicate your strategy. It is composed of the following components:

Objective This is your statement of intent and maps to the qualitative part of the strategic aspiration.

Goals This is how you will measure progress toward the aspiration. This maps to the quantitative part of your aspiration. As mentioned earlier. these should be SMART goals.

Strategies These are the choices that have been made for the where-to-play and how-to-win questions. This represents what you will do and focus on to achieve the goals rather than how you will do it.

Measures Measures are the numerical benchmarks used to monitor progress against the chosen strategies to determine if the strategies are working. If the goals are lagging in indicators (profit after three years), then the strategic measures ideally are leading indicators of business outcomes that contribute to the goals.

Strategy Maps as a Framework for Strategy Communication

Strategy maps, as shown in Figure 10.28, are the evolution of the balanced scorecard, developed by Kaplan and Norton, that connect the strategy with the drivers that will produce the business outcomes. A strategy map can help to communicate a strategy in a cohesive and concise manner.

There are four perspectives in a strategy map:

Finance These are the financial measures that typically show the health of a business. This section maps to the quantitative part of your aspiration. As mentioned earlier, these should be SMART goals.

Customer The customer value propositions composed of the choices of where to play and how to win.

Internal Processes This represents how to achieve your financial and customer goals, which map to the capabilities of the Strategy Choices Cascade.

Organizational Capacity This section defines the core competencies and skills, the technologies, and the corporate culture needed to support an organization's strategy. This maps to the management systems of the strategic choice cascade.

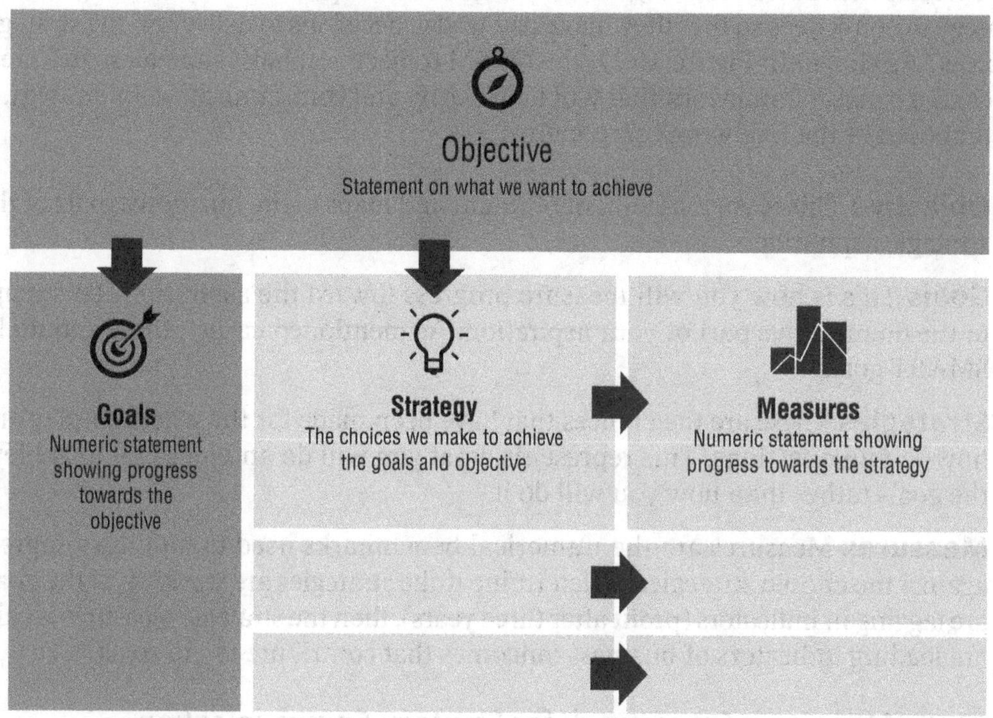

Figure 10.27: The OGSM framework

A strategy map should be read from bottom to top. This tells a story of what actions have a cause and effect on the desired strategic outcomes. By investing in employee skills and knowledge, we can create and improve the internal processes to deliver the capabilities that in turn provide value to customers who in turn provide us with financial rewards.

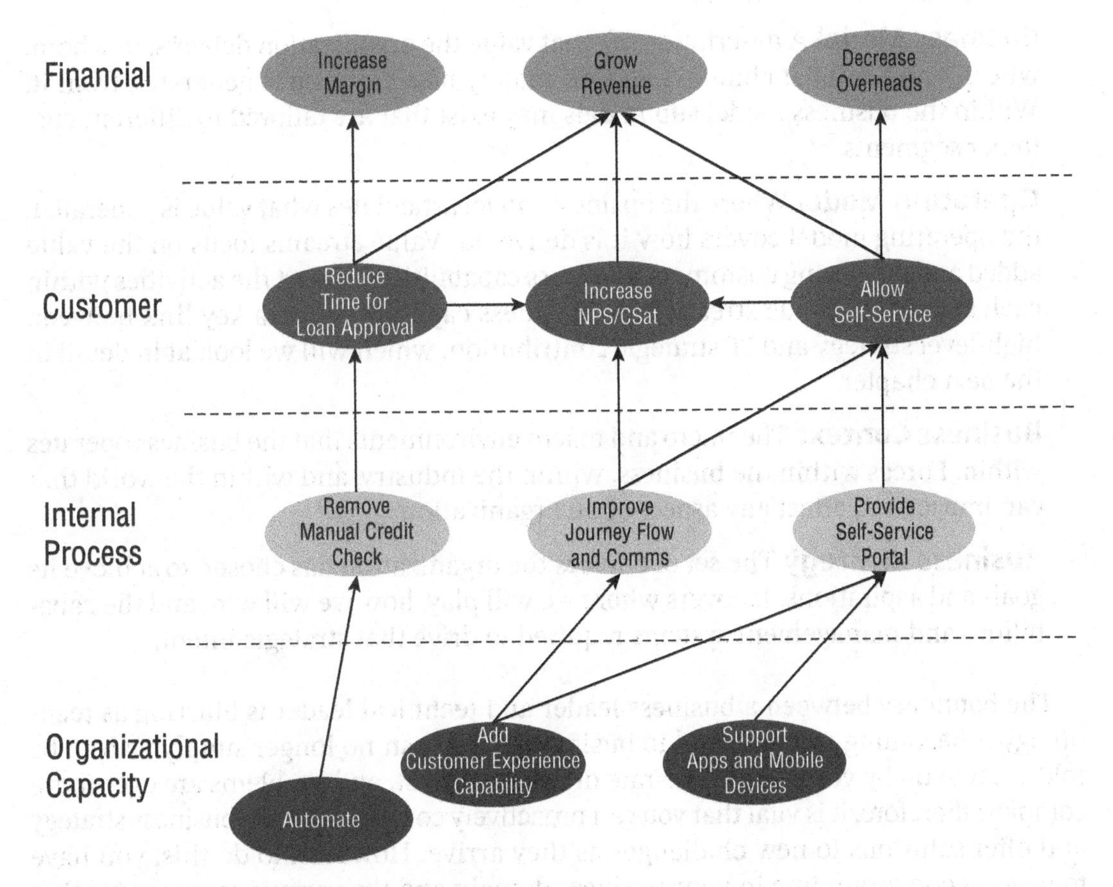

Figure 10.28: An example of a strategy map

Summary

In this chapter, we looked at the various elements that make up the anatomy of the business. We created a model of the business and presented frameworks that you can use to map the various components of the anatomy that enable more effective communication with your peers as well as a deeper understanding of the business. We looked at the following components:

Purpose Why the business exists and where it sees its place in the future. This is important to understand and communicate to give people a sense of meaning and belonging to something bigger.

Business Model A description of what value the organization delivers, to whom, where, and via what channel, and ultimately how revenue is generated from it. Within the business model submodels may exist that are tailored to different customer segments.

Operating Model Where the business model articulates what value is generated, the operating model covers how it is delivered. Value streams focus on the value added and delighting customers. Business capabilities support the activities within each step of the value stream flow. Business capabilities are a key link between high-level strategy and IT strategic contribution, which will we look at in detail in the next chapter.

Business Context The micro and macro environments that the business operates within. Forces within the business, within the industry, and within the world that can impact and affect any aspect of an organization.

Business Strategy The set of choices the organization has chosen to achieve its goals and aspirations. It covers where we will play, how we will win, and the capabilities and management systems required to drive that strategic intent.

The boundary between a business leader and technical leader is blurring as technology is becoming more central to businesses. You can no longer simply wait to be told what to do by your peers. The rate of change is fast, and problems are ever more complex; therefore, it is vital that you can proactively contribute to the business strategy and offer solutions to new challenges as they arrive. However, to do this, you have to have a good grounding in your business domain and the various components that make up your business anatomy. This means taking time out of the technical domain and engaging with the rest of the business. After all, you are a business leader; you just happen to be the custodian of the organization's technical capability. With this knowledge you will be in a far better place to contribute to the organization's strategic ambitions, which is the topic of the Chapter 11.

IT Strategic Contribution

Technology can bring benefits if, and only if, it diminishes a limitation.
—*Eli Goldratt*

If everything is important, then nothing is important.
—*Patrick Lencioni*

Strategy is essentially an intent, rather than a plan.
—*Steven Bungay, author of The Art of Action*

As introduced in Chapter 10, "Understanding Your Business," the business strategy explicitly lays out the choices the organization will take to achieve its business goals and objectives. The organization will require the combined efforts of its major capabilities (IT, marketing, etc.) to work together to agree on a set of strategic initiatives that will deliver on those strategic choices. The contributions to the strategy represent each department's own strategy. However, conceptually there is only one strategy, and the strategic actions of IT, just like the marketing or finance departments, are embedded within it. The IT strategic contribution should not only be aligned to supporting the actions of other departments, but it should also inform the strategy itself, through a process of consensus and feedback based on the technical art of the possible. This is a shift from creating an IT strategy that merely supports the rest of the business to one that is integral and shapes the business model and strategic outlook itself through a focus on digitalization.

There is a myth that you don't need strategy if you have agility. Unsurprisingly, this is simply not the case. Agility cannot replace strategy. Agility is the ability to adapt your path toward a goal. Strategy is about clarifying the goal and giving clear direction and alignment. Without this there won't be a focus on the right things. If you are not focusing on the right things, then it doesn't matter how agile you are at achieving them as it's likely to have little benefit to the success of an organization. As Peter Drucker famously said, "There is surely nothing quite so useless as doing with great efficiency what should not be done at all." Strategy is the choices we make on where to focus effort. It is about where we need to be agile to sense and adapt to exploring new opportunities, where we need to exploit our advantage by being lean and removing waste, and where we need to standardize and outsource when there is no competitive advantage.

Without a strategy you have no framework for making decisions or understanding trade-offs. Without the strategic bigger picture, it is difficult to determine the effective use of the organization's resources and capabilities. Without a strategic context, all initiatives can be deemed worthy of investment if they result in a measurable business improvement. This, however, quickly leads to siloed thinking and local rather than global optimization, or in other words, departmental goals over the business goals. Strategy can be used to determine that which is critical to success and that which is simply an improvement to business performance. This can speed up priority decisions and quickly highlight wasteful or vanity projects. Fundamentally, you need a target; you need a clear set of choices, a strategy for how you intend to contribute to the organization's success. We can then be agile and sense and adapt on the path to achieving that target.

In this chapter we will look at how to create an IT strategy and embed it in the business strategy. You will come to understand why business capabilities are the key to linking the business strategy to IT contribution. We will look at the importance of measuring work in terms of business outcome to ensure the improvements in capability result in positive business impacts that contribute to the strategic objectives. Finally, we will examine two methods for communicating the IT strategic contribution to effectively align the organization to a North Star, ensuring there is a shared purpose across the department.

Linking IT Execution to Business Strategy Using Enterprise Architecture

A *strategy* is the explicit set of choices that an organization makes to succeed in achieving its objectives and goals. It acts as the ultimate North Star, guiding investment and focus. A *plan* is a list of activities that describe how you will get there. In IT strategy deployment we have two levels of planning: tactical and operational. The relationship of IT strategy and tactical and operational planning can be seen in Figure 11.1. Strategy represents the highest level of intent, the city plan if you will. It focuses on the long-term actions IT will focus on to close capability gaps.

Tactical planning takes that intent into a concrete vision that we can drive toward. You can think of this as the district plan. It defines the large steps to implement the strategic actions. Operational planning covers the programs and projects that we can execute to move toward that vision. Think of this as the blueprints that will be used to construct the buildings.

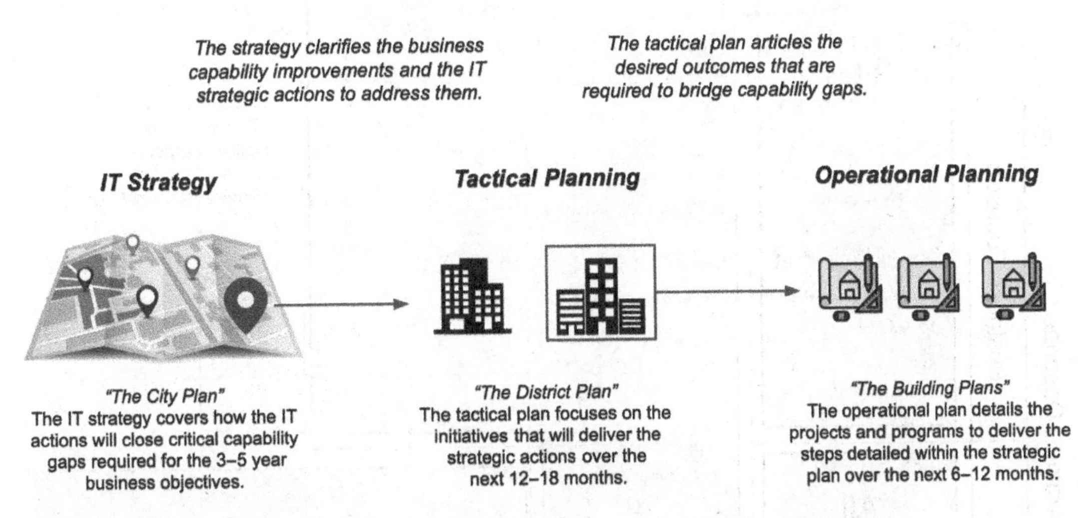

The strategy clarifies the business capability improvements and the IT strategic actions to address them.
The tactical plan articles the desired outcomes that are required to bridge capability gaps.
IT Strategy
Tactical Planning
Operational Planning
"The City Plan"
The IT strategy covers how the IT actions will close critical capability gaps required for the 3–5 year business objectives.
"The District Plan"
The tactical plan focuses on the initiatives that will deliver the strategic actions over the next 12–18 months.
"The Building Plans"
The operational plan details the projects and programs to deliver the steps detailed within the strategic plan over the next 6–12 months.

Figure 11.1: The three levels of IT planning: strategic, tactical, and operational

To guide our strategic, tactical, and operational planning, we can utilize the practice of enterprise architecture. Enterprise architecture (EA) is a blueprint for how IT will be used to contribute to meeting business goals. EA has gotten a bad reputation for being too high-level and impractical to be of any use. This is largely down to the proliferation of the many overly abstract and complicated EA frameworks. However, in *The Practice of Enterprise Architecture: A Modern Approach to Business and IT Alignment* (SK Publishing, 2018), Svyatoslav Kotusev has written a very practical book on the subject based on the reality of EA rather than what people believe it should be. In the book, Kotusev details the CSVLOD (Considerations, Standards, Visions, Landscapes, Outlines and Designs) Model, as shown overlaid with the three levels of strategic planning in Figure 11.2.

We will use this model and the various artifacts associated with each component to build our strategy. Kotusev's book is centered around aligning IT action with business strategy. At its core, EA is nothing other than an alignment exercise, alignment between the business intent and IT action. EA plays two roles in alignment; the first is to convert strategic intent into something actionable, and the second is a bridge from action to technology contribution. As we will cover in Chapter 13, "Operational Planning: Execution, Learning, and Adapting," EA is also a continuous process rather than a one-off exercise.

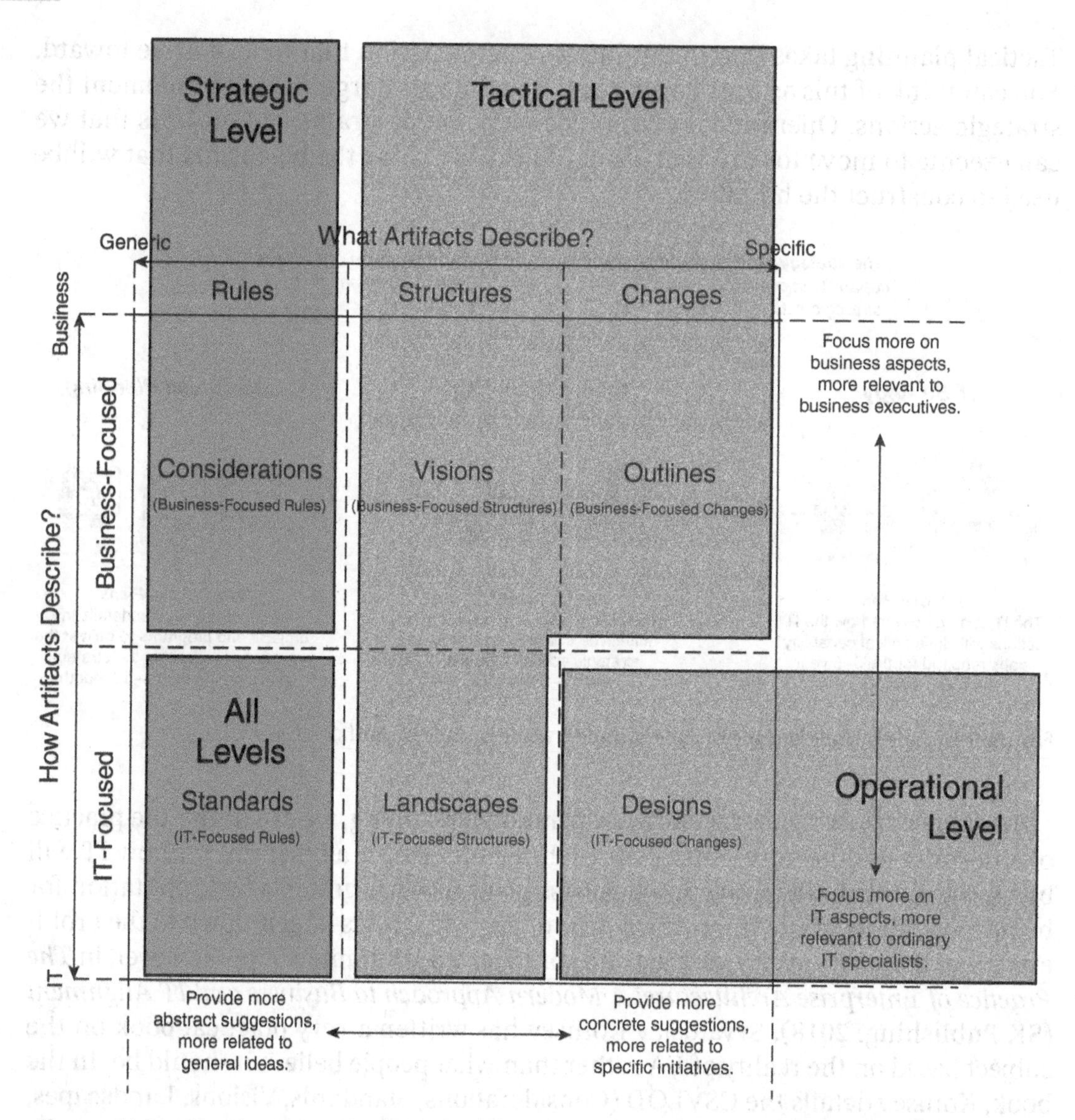

Figure 11.2: How Kotusev's CSVLOD model and its artifacts relate to strategic planning levels.

- **Strategic Level:** Setting out our long-term strategic intent
  - **Considerations** are high-level, clear, and concise statements that describe long-term strategic intent. These can be in the form of guiding principles, policies, or descriptions of actions we will take. They are simple to read and business-focused, applicable to both IT and business. These are the fundamental considerations we need to take into account when planning.
- **Tactical Level:** Building a vision and a bridge
  - **Visions** represent the conceptual view of the organization, detailing the decisions on what IT will deliver to an organization in the long run. They

are typically in the form of target state diagrams or a road map of initiatives. As with considerations, they are business-focused but are more detailed in order to guide and prioritize IT investments.
  - **Landscapes** are the technical artifacts describing the current IT landscape. Typically these are technical diagrams and inventories of IT applications, infrastructure, and interfaces.
  - **Outlines** are the high-level descriptions of IT initiatives required to deliver the vision. Again, these are focused on the benefits to the business but also contain more detailed solution overviews and options.
- **Operational Level:** The executable projects to build the bridge
  - **Designs** are the detailed and actionable solution artifacts used for teams to execute IT projects.
- **Applicable to all levels:**
  - **Standards** describe the high-level technical rules, principles, patterns, and best practices to guide to IT execution.

To help make these concepts more concrete, I will step through each level of the planning hierarchy, showing both inputs from the level above and outputs for the level below to give a flavor of how we cascade and align on intent. In Chapter 12, "Tactical Planning: Deploying Strategy" and Chapter 13, we will also look at the feedback flows to ensure consensus and the ability to adapt the strategic direction based on the results of our actions.

Strategic Level

The business strategy produces a set of choices. These choices will require cross-functional execution to deliver. In this sense the strategy acts as a guide, making explicit the business intent, which makes planning simple. With an explicit strategy it is clear which actions and initiatives make sense and are likely to lead to results against the strategic goals due to the alignment cascade. To clarify the link between business strategy and IT operational execution, we will use a simple business strategy example as shown in Table 11.1.

Table 11.1: An example business strategy and measures

Business Strategy	Strategic Measures
Where to play ■ UK ■ Online ■ B2C **How to win** ■ Focus on operational excellence. ■ Reduce manual processing to avoid head count. ■ Outsource supply chains to lower costs. ■ Drive sales online to reduce call center costs.	■ Overheads as a % of revenue ■ Sales heads as a % of revenue ■ Margin gains

IT strategy, the topic of this chapter, is composed of the set of choices on what actions IT will take, it is *WHAT* you are going to do to contribute to business success. Consider the example shown in Table 11.2 that demonstrates the link from business strategy, using the "Reduce manual processing to avoid head count" strategic objective, to the IT actions needed to mitigate the challenges the business has with achieving it. You will notice that at a strategic level, the actions of IT and other departments are high-level. This is intentional. The strategy is not directly actionable. The purpose of the strategy is to articulate the strategic intent of each department in order to cascade planning alignment. This is what Kotusev refers to as a Consideration.

Table 11.2: An example of how IT actions relate to business strategy

Strategic Objective (Where to Play & How to Win Choices)	Business Challenge	Other Depts' Action	IT Strategic Action (High-Level Intent / Consideration)
Reduce manual processing to avoid head count.	Manual processes requiring a large head count.	Ensure consistency in processes. Exclude the need / outsource.	Automate back-office processes.
	Lack of maturity in handling customer requests.	Provide the information for customer to self-serve.	Provide self-service portal for customers.

Tactical Level

If strategy is the *WHAT*, then tactical planning is the *HOW*. Tactical planning, covered in Chapter 12, defines the road map of strategic initiatives and the investment plan required to fund it. However, to define a plan, we need to have a destination or a vision. From a strategic action, we can determine a target state, in terms of both the technical architecture and the operating model. This is achieved by understanding the current IT landscape and how best to bridge capability gaps. The target state acts as a vision of what we need to have in place to achieve the strategic objectives. Continuing the example from Table 11.2, consider Table 11.3, which shows how both the technical architecture and the operating model will be changed to meet the IT strategic actions.

Table 11.3: An example of the technical architectural and IT operating model consideration to achieve the IT actions

IT Strategic Action (High-Level Intent / Consideration)	Target State Designed to Meet the IT Action	
	IT Architecture Considerations	**IT Operating Model Considerations**
Automate back-office processes.	Leverage Robotic Process Automation (RPA) platform for fulfilment activities. Implement platform for customer services self-service.	■ Sourcing: Partner. ■ WoW: Take a lean approach and remove waste. ■ Org Design: Project team. ■ Sourcing: Partner + in-house team. ■ WoW: Take a lean approach and remove waste. ■ Org Design: Product team.

Tactical planning is the glue between the strategic direction and the operational actions. The tactical plans, made up of initiatives, are the steps required to move from where we are, the ASIS state, to the target, or TOBE, state. Table 11.4 highlights the tactical plans, the steps, required to realize the IT action from Table 11.3.

Table 11.4: An example of how the IT tactical plans link to the IT action

IT Strategic Action (High-Level Intent)	Tactical Plan (Strategic Initiatives)
Automate back-office processes.	■ Automate the handling of customer service requests. ■ Automate the fulfilment of orders.

Operational Level

Operational planning, covered in Chapter 13, is the distillation of tactical plans into executable projects and programs. Table 11.5 lists the programs of work, including the list of projects, for each of the strategic initiatives within the tactical plan. The projects include both technical architecture changes and operating model changes.

Table 11.5: An example of how the strategic initiatives break down into programs and projects

Tactical Plan (Strategic Initiatives)	Operational Plan (Projects and Program)
Automate the handling of customer service requests.	■ Project: Integrate Chatbot. ■ Project: Automate order updates via self-service.
Automate the fulfilment of orders.	■ Program: Automation Program. ■ Project: Automate Seat Allocation. ■ Project: Automate Invoicing. ■ Project: Automate Tickets.

Strategy Nomenclature

The terminology that people use to describe the creation of strategic planning throughout the business and in IT is often used interchangeably or inconsistently, leading to confusion. Table 11.6 is a glossary of all of the terms that are associated with each of the levels of strategic planning to help your orientation.

Table 11.6: Terms used as the various levels of strategy deployment

Flight Level	Terms	Description
Strategic level / long-term view *What we do*	Business Objective & Goals, Vision	Business aspiration and measures.
	Business Strategy & Goals / Strategic Objectives	The Where to Play and How to Win choices.
	Strategic Actions / Department Strategy / Strategic Contribution / Strategic Plays / Strategic Intent / Product Strategy	The high-level actions that each department will contribute to achieving the strategy.
Tactical level / mid-term view *How we do it*	Strategic Initiatives & Measures / Strategic Plans / Tactics / OKRs (objectives)	If strategy is about what you want to achieve, a strategic initiative explains how you can achieve it.
Operational level / short-term view *How we execute it*	Programs	A collection of projects with a common theme, to deliver an initaitive.
	Projects / OKRs (tactics)	Can be mandatory, linked to strategy, run the business, or technical/security.

Creating an IT Strategy

Figure 11.3 shows the inputs, process, and outputs of an IT strategy. Ideally, as an input you will have a well-thought-through, explicit, and clear business strategy communicated to an aligned organization to anchor IT strategic actions. However, the reality is that we don't live in a perfect world, and often you will need to make assumptions about the strategic direction of your business. If you don't have an explicit business strategy you can assume one from what you know about the business aspirations, context, and operating model or from understanding departments' strategic focus and needs. The reason why it's important to understand strategy is so that we can determine the key initiatives, from the many requests we have, that are directly linked to delivering that strategy. With a focus on the critical areas, we can then uncover the obstacles and barriers we have and determine the list of business and technical actions that are required to overcome them, thus ensuring our efforts and actions have the maximum impact. The outputs are the list of IT strategic actions along with the measures, ideally framed in business outcome terms, that will address the challenges, along with a document to be used to effectively communicate and deploy the strategy.

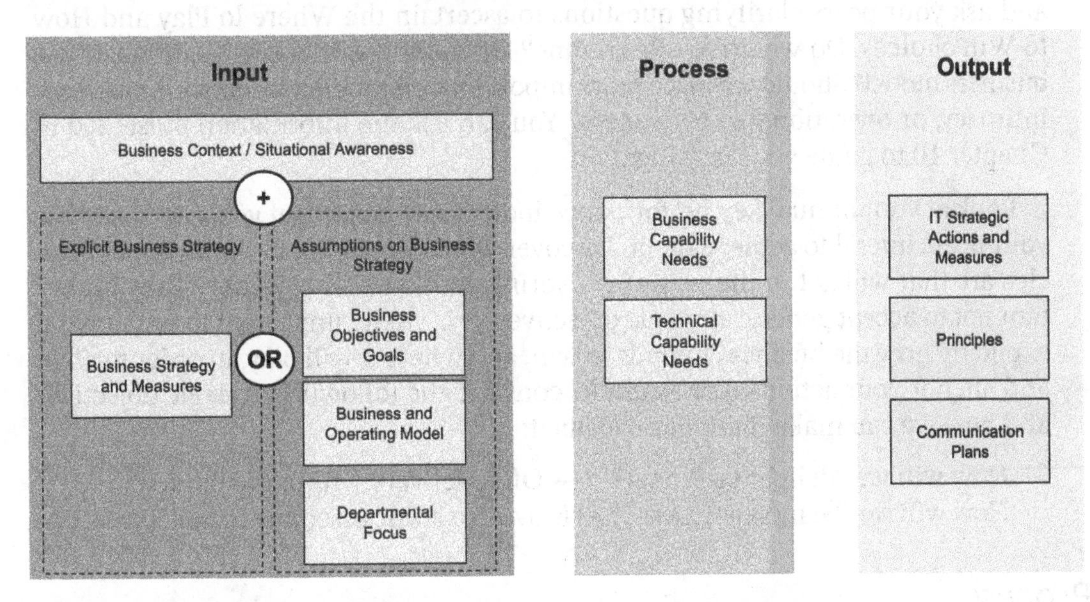

Figure 11.3: The inputs, process, and outputs of creating an IT Strategy

Inputs

The inputs to an IT strategy, which were covered extensively in Chapter 10, are as follows:

The Business Context / Situational Awareness The understanding of the environment in which your business operates, both macro and micro. What is going on in the environment around you? Political? Competitors? What could impact your business? What are the key trends? What threats does the current context pose? What opportunities can you glean from the trading environment? What are the key challenges and barriers with the current mode of operating? Do you have technical risks? What technical obstacles do you have for supporting the business? If you really understand your environment, you will make better decisions and discover more opportunities. As they say, if you don't know where you are, a map won't help.

The Business Strategy or Your Assumptions on Strategy If you don't have an explicit business strategy, then you must make assumptions based on what you know about the business context, business aspirations, business model, and the operating model. You will need to make explicit your assumptions on strategy and ask your peers clarifying questions to ascertain the Where to Play and How to Win choices. Do we drive sales online? Open in new markets? Launch a new business model? Should we place more importance on product leadership, customer intimacy, or operational effectiveness? You can use the information presented in Chapter 10 to guide your investigation.

Look at your annual key performance indicators and budget goals and ask how your peers intend to achieve them. Discover what the biggest constraints or obstacles are that will get in the way of delivering your annual objectives. It is important not to accept generic strategic objectives. It is vitally important to understand explicitly how the business intends to win, and where it will play, in order to align and anchor your actions. For example, consider the following strategic objectives and how we can make them more explicit:

How will we "delight customers"? → Offer delivery within two hours.
How will we "be more efficient"? → Move to a home assembly model like IKEA.

Process

So how do we ensure there is a clear relationship between business success and IT contribution? As shown in Figure 11.4, we need to link IT actions and measurable impacts to removing the barriers that stand in the way of achieving the strategic objectives. You will also notice in Figure 11.4 that in addition to the business strategy, we may also need to overcome any operational essential, aka business as usual (BAU) obstacles or risks that are stopping, or could stop, the running of the business. While large BAU needs may not be the things that help us win, they may be the things that will result in us losing. For example, while customers may not care about our internal financial processes, without the right level of maturity in this area there is a large risk that the business will not be able to operate.

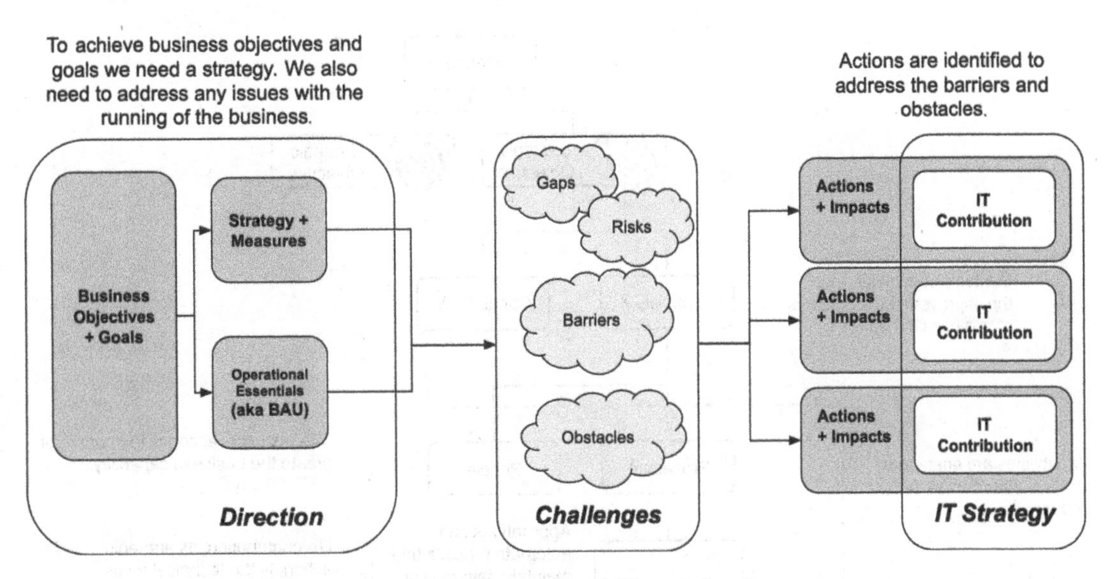

Figure 11.4: The link between business demand and IT contribution

The barriers that prevent us from achieving our strategy or risk the running of the business can be described in terms of lacking maturity in business capability. Business capabilities, as covered in depth in Chapter 10, are the things a business does, not how it does them. They are the abilities required to deliver value. As shown in Figure 11.5, we use business capabilities to link IT action to business strategy:

- Strategic objectives, the Where to Play and How to Win choices, require the improvements or creation of business capabilities.
- Capabilities are enabled by process, people, and technology.
- The strategic actions for the enterprise are to improve or create business capability. IT's contribution, its strategic action, is the technical focus.
- IT can go beyond contribution and use digitalization to completely change *HOW* the capability is achieved, or removing its need entirely rather than improving it. This is a move away from only contributing toward coauthoring business strategy.

To provide the IT strategic contribution, we must identify the actions needed to enhance or acquire business capabilities that are required by the strategic objectives. It is these business capabilities that provide the link between the high-level strategy and IT strategic contribution. What a business does changes very infrequently—it is how the business does what it does that is subject to change, and therefore we can build a strategy around the organization's capabilities.

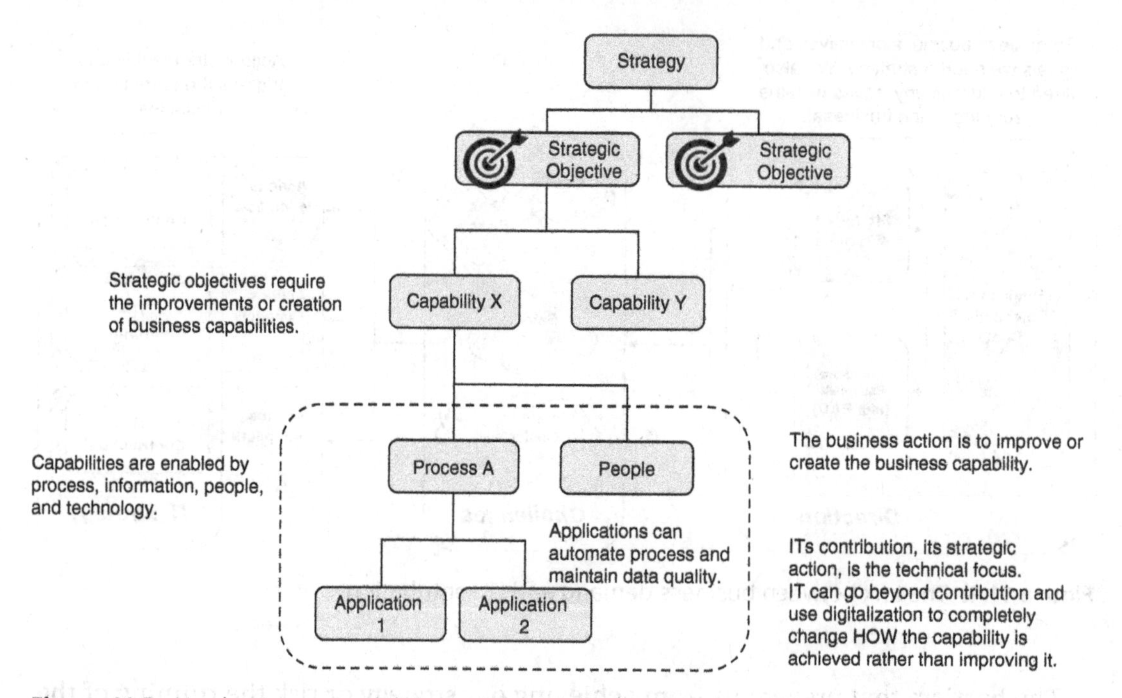

Figure 11.5: Use business capabilities to link IT to business success.

Outputs

The output of the IT strategy consist of the following:

IT Strategic Actions and Measures In his book *The Practice of Enterprise Architecture*, Svyatoslav Kotusev refers to a consideration artefact called Direction Statements. Kotusev describes Direction Statements as "the most action-oriented EA artifacts of all Considerations. While other Considerations merely describe how an organization needs to work or analyze the technology environment, Direction Statements point to a certain direction where an organization needs to go in the future and explain the rationale for this direction. However, they still do not provide any specific details regarding how exactly it should be done. Essentially, Direction Statements only indicate where an entire company needs to go without specifying how." I have been referring to these Direction Statements as strategic actions. These actions will guide the more specific planning decisions at the tactical and operational levels.

The IT actions will bridge the business capability gaps or create new capabilities. These actions should clearly articulate what IT is going to do, in alignment with other business functions, to contribute to business success. The IT strategic actions should focus on the key strategic contributions and therefore should not represent everything you will do. Do not be distracted by noncritical capabilities (i.e., avoid Bikeshedding, aka the Law of Triviality). Don't spend time on relatively unimportant but easy-to-grasp capabilities that you will supply, such as a service desk or collaboration tools, unless these capabilities are critical to the organization and will result in competitive advantage. Instead focus only on what is critical to enable the capability directly linked to business strategic choices.

The strategic actions that you propose need to be measurable. They should not simply show an improvement to a capability, they also need to show how the improvement leads to business impact that contributes to a strategic objective. For example, the capability improvement measure could be quantified as a business outcome of "We will adjust our prices every 20 mins by scraping our top 5 competitors." That would lead to a business impact of "increasing market share by 10% due to price parity" that contributes to a strategic goal of "becoming the market leader."

Principles Another Consideration artifact is Principles. Where Actions, or Direction Statements, are very much action-oriented and focused on the strategic objectives, Principles are more general abstract heuristics or guidelines on how to approach solutions, taking insights of the business context into consideration when articulating design guidance. Actions and Principles complement each other and offer two types of strategic direction setting. As an example, we may have Actions to provide a sales CRM and a content management platform but a Principle to ensure that any solution is multi-language and can support different brands.

A Communication Plan The best strategy in the world is useless if you can't deploy it and align people on why, where, and how to focus their efforts. Communicating your strategy, whether to the board, your boss, your peers, or your team, is as important as the strategy creation itself.

Determining IT Contribution to Addressing BAU Challenges and Achieving the Strategic Objectives

To determine the strategic IT actions, we must first identify the improvement required in, or the creation of, the business capabilities that are key to delivering the business strategy. To assist with this, we can create a business capability model, an example of which is shown in Figure 11.6. A business capability model is an abstraction of business operations, not of what an organization does, but rather it is a model of the capabilities it requires to deliver on its value proposition.

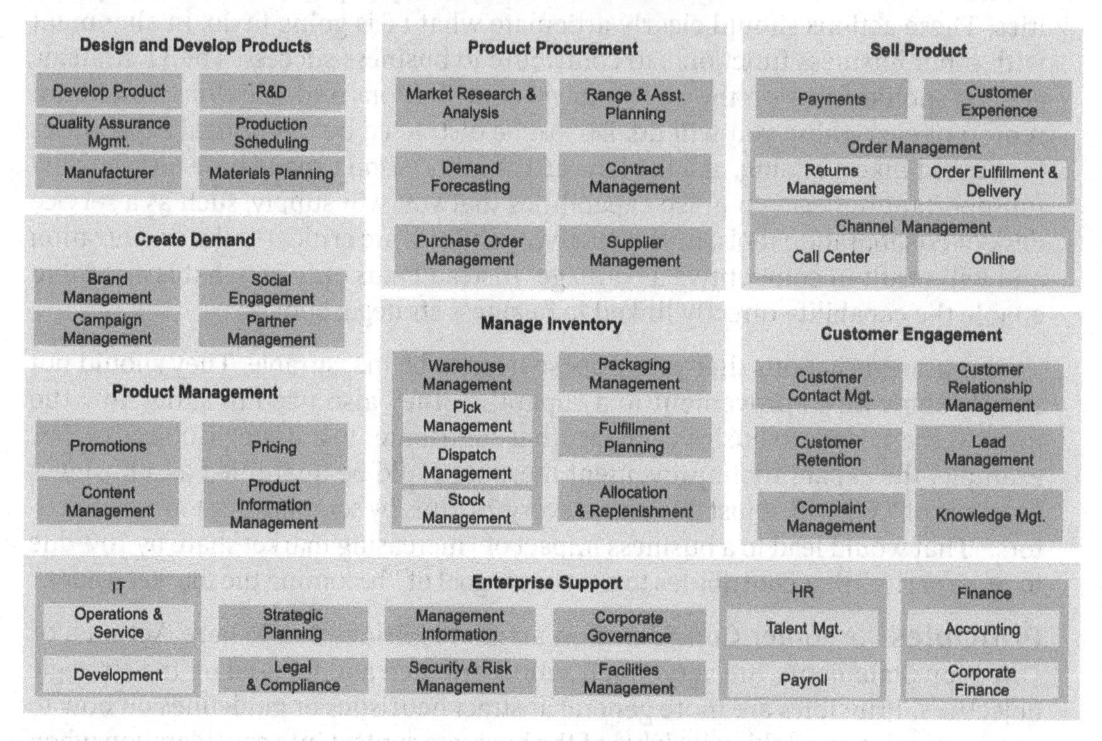

Figure 11.6: An example of an e-commerce business capability model

Modeling business capabilities gives several benefits:

- A business capability can be decomposed into sub-capabilities and or business processes. We can then map these processes to applications to clearly understand the business & IT alignment.
- Business capabilities help to clarify business components, offering a domain language for concepts that span the organizational structure.
- A business capability model can be used as a battle map to highlight areas in need of investment and areas that are critical for business success.
- As business capabilities are inherently stable, we can use them as the basis for organizational design.

The business capability model will be used as an input to designing the technical architecture target state as will be covered in Chapter 12. During the IT strategy creation, we will highlight the capabilities that require investment. This model of capabilities will be used along with the IT strategic actions to determine the best technical solution to bridge the maturity gap based on how evolved the capability is and the technical status of any existing technology that enables it.

There are many articles on the correct and incorrect ways of naming capabilities and the various levels of granularity that you should map to. However, the purpose of modeling the capabilities in the organization in relation to IT strategy is to distil the organization to the appropriate level of granularity that will help you understand the link between business architecture and technical architecture. Therefore, don't lose sight of the end goal by worrying about naming conventions or the definitions of capabilities versus business processes. One tip I will stress though is, where possible, try to avoid modeling the organizational structure and instead focus on capability, the WHAT an organization does, regardless of WHO or HOW it is done. Whatever naming convention you use, the most important point to take away is that you must use terminology that is specific to your organization. Therefore, useful as they are, you should avoid copying reference capability models verbatim. Instead, use them as a guide but agree with your business peers on names that mean something to your organization.

It is important to note that capability modeling cannot be achieved in a vacuum; you must work closely with your peers and their wider teams to understand the capabilities, new or improvements to existing, that are crucial to delivering on the strategic direction. Remember, different parts of the business will need to converge to deliver capabilities that help to realize strategic choices. By visualizing the capabilities that are required by an organization, we can create a shared understanding of how to connect the high-level strategy to execution and where we will focus effort. As with many things, the journey is more important than the destination. While a complete business capability model is a very useful artifact, the real value is in the conversations with your peers on understanding needs and desired outcomes of different areas of your business, which all helps to inform how IT should focus its time and energy for maximum impact. British statistician George Box famously wrote, "All models are wrong, some are useful." His point was that we should not waste time on trying to create a perfect model but instead build something that will help answer the specific question we have.

Using Value Streams to Clarify Capability Improvements Required to Achieve the Strategic Objectives

As shown in Figure 11.7, the best way to anchor business capability discovery to strategic objectives is to understand the related value streams. We covered value streams in depth in Chapter 10. The value stream represents the end-to-end collection of activities that create the value for a customer. When we have mapped the value stream, we can then map the capabilities needed to deliver this value and identify any gaps, barriers, obstacles, or opportunities. This will ensure that we capture only the most important capabilities required to deliver on our strategic objectives rather than modeling everything our business does. Value stream mapping can be broadly achieved using two methods, namely customer journey mapping, which measures the customer's experience, and value stream mapping, a lean method for discovering waste. Both of these techniques were covered in detail in Chapter 6, "How We Work."

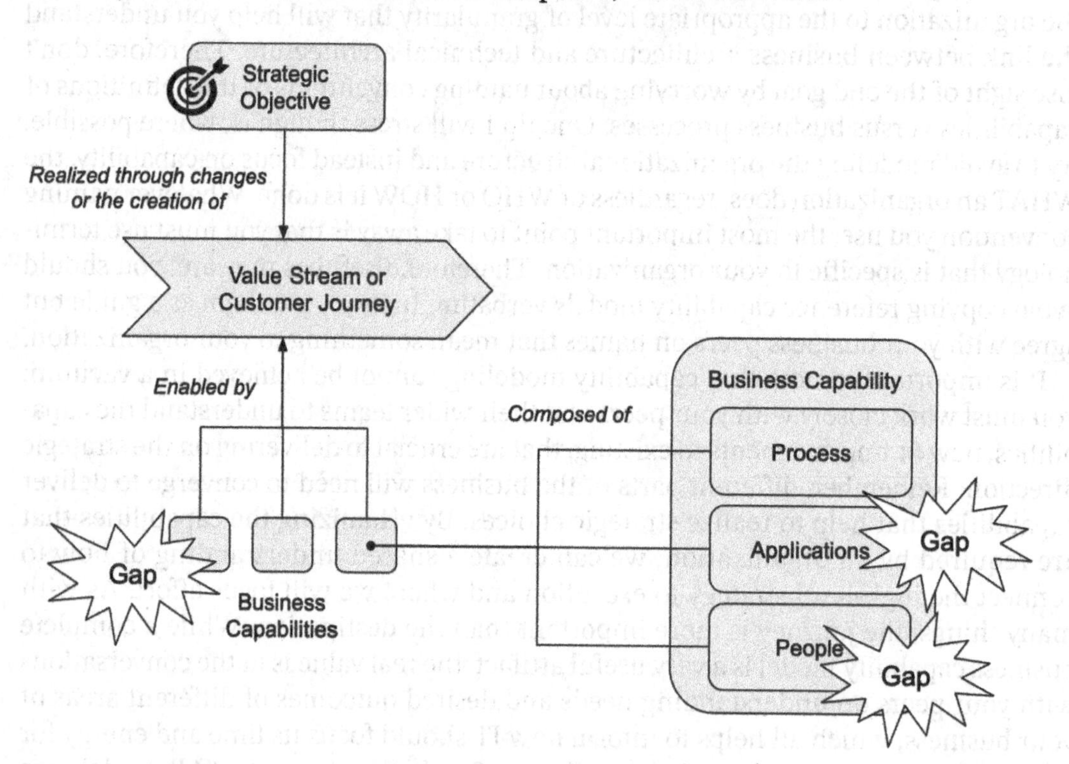

Figure 11.7: Mapping business capabilities from value streams

To demonstrate the process of mapping capabilities, we will use a simple example of a catering marketplace called A Food Company. Think Just Eat or Uber Eats but for corporate functions, with a strategy of driving sales online, expanding into the B2C market, and focusing on operational excellence, as shown in Table 11.7.

Table 11.7: The case study business strategy

"A Food Company" Business Strategy	
Strategy	**Measures**
Where to Play: ■ B2C + private functions ■ B2B + corporate events ■ UK *How to Win:* Move corporate sales online. ■ Improve corporate events online booking conversion. Expand into B2C markets: ■ Acquire and integrate a B2C business. Lean back-office operations: ■ Avoid overheads. ■ Increase caterer sign-ups.	*B2B Growth:* ■ Online revenue 75% *B2C Expansion*: ■ 10% revenue from B2C during 1st year *Lean back office operations*: ■ Overheads % of revenue *Caterers signed up:* ■ Increase by 50% YoY.

Mapping the Business Capabilities

To demonstrate how to map business capabilities, we will use two of the strategic objectives from Table 11.7 to highlight the two high-level methods of discovering business capabilities. For the strategic objective of "Improve corporate events online booking conversion," we will use a customer experience map. For the "Increase caterer sign-ups" strategic initiative, we will use a value stream map.

Strategic Objective: Improve Corporate Events Online Booking Conversion

As shown in Figure 11.8, the strategic objective is about changing customer behavior. Therefore, it makes sense to use a customer journey map to highlight the pain points a customer experiences with our service. The value stream trigger event is the need for catering at an event. The steps in the value stream represent the individual activities that work together to enable a customer to contract a caterer for an event. Once we have mapped the customer experience, we can identify the business capabilities involved and use them to link to action.

Table 11.8 lists a sample of the capabilities required for each step in the value stream along with the pain point a customer or the business is experiencing.

Table 11.8: An example of customer and business pain points mapped to capabilities

Customer Journey Step	Customer / Business Pain Point	Capability Involved
Find a caterer: A customer can choose from street food suppliers, restaurant caterers, and mobile bars based on location and availability.	■ A caterer is not available during searching, only after enquiry. ■ There are no indications of per-person costs. ■ Lack of content on menu options and set-up. ■ No reviews from other customers, so customers nervous about contracting. ■ No clarity on staff, glass hire costs, or availability.	Manage Caterers Caterer Schedule Menu Management Content Management
Submit Enquiry: Once a customer has found a menu they like, they can submit an enquiry about their event and needs. An event coordinate will be assigned at this stage.	■ Customers submit enquiries by phone. There is a lot of back and forth to get the information. ■ Customers are updated with many emails.	Manage Quotes Customer Engagement
Get Quote: The event coordinator will liaise with suppliers and customers to get personalized quotes and finalize details on the event.	■ Event coordinators manually contact caterers. This can take days. Again, there are multiple steps to confirm customer requirements.	Caterer Management Customer Engagement
. . .	. . .	. . .

Strategic Objective: Increase Caterer Sign-Ups

Figure 11.9 shows the example value stream map for the onboarding caterers, which supports the "Caterer sign-ups" strategic objective. The trigger is that a sales agent has made an introduction, and the end result is the ability for the caterer to advertise and accept business on the platform. The steps are the activities to run financial background checks and environmental food agency certification and set up the caterer's account. Similar to the previous objective, we can map the capabilities required and identify any shortcomings that generate waste in the value stream.

Building the Capability Model

After the value streams for each strategic objective need are mapped and the related business capabilities are captured, they can then be deduplicated and organized into a business capability model as shown in Figure 11.10. The capabilities in the model are grouped into a logical structure and, if appropriate, distilled to a lower level of granularity to make explicit key sub-capabilities. This model now represents all the business capabilities that support the strategic outcomes. Some will represent existing capabilities that may need improving and others will reveal that new capabilities are required. Some capabilities will support more than a single value stream.

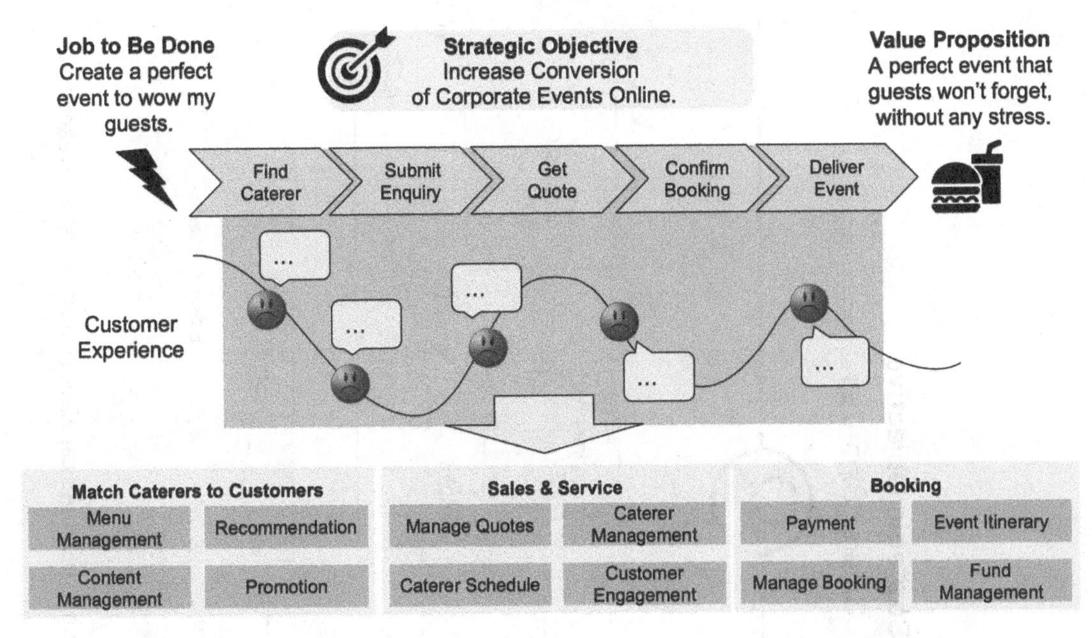

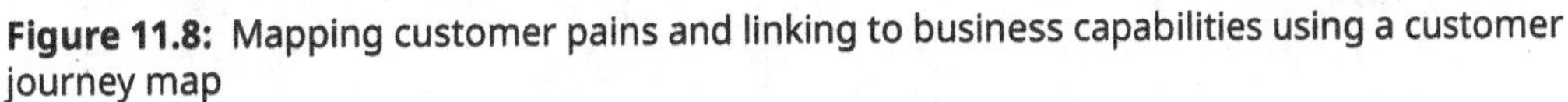

Figure 11.8: Mapping customer pains and linking to business capabilities using a customer journey map

Identifying Barriers and Opportunities Preventing Us from Achieving the Strategic Objectives

The next step is to assess capability improvement needs; we do this by rating the ability for each capability to execute. To determine how mature a capability is, we can see how well it is supported by technology, people, and process. Determining the maturity of a capability is subjective, and it can be more art than science. It is essential to talk to domain experts and key stakeholders to identify any significant capability gaps or performance problems. You can also compare the capability against industry standards, if they are available, or best-in-class systems, which may be in the form of software vendor documents, presentations, or white papers. In terms of using a measurement, you can use something high-level and simple such as Low, Medium, or High, or you can look to use an industry standard benchmark such as the capability maturity model (CMM).

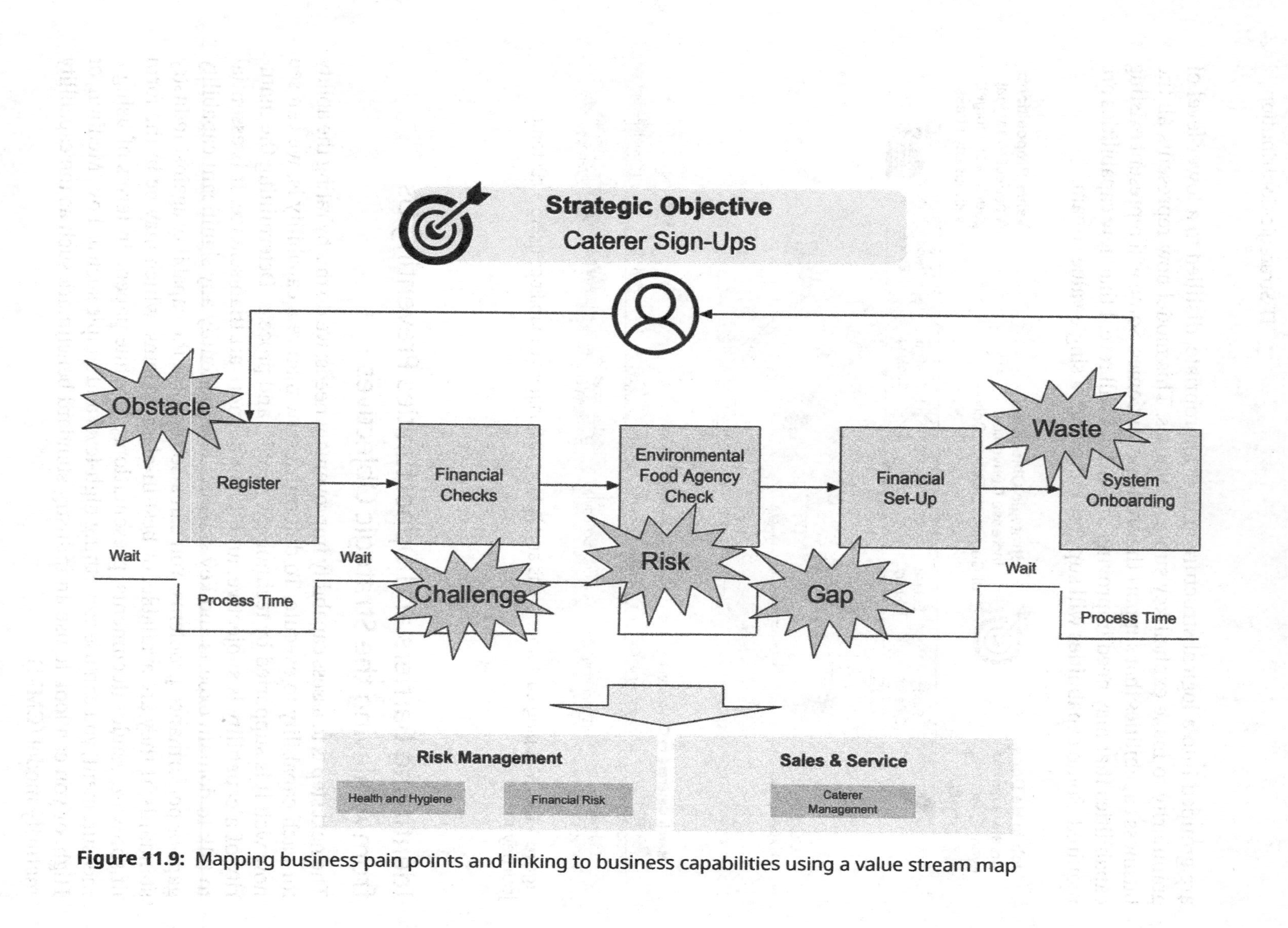

Figure 11.9: Mapping business pain points and linking to business capabilities using a value stream map

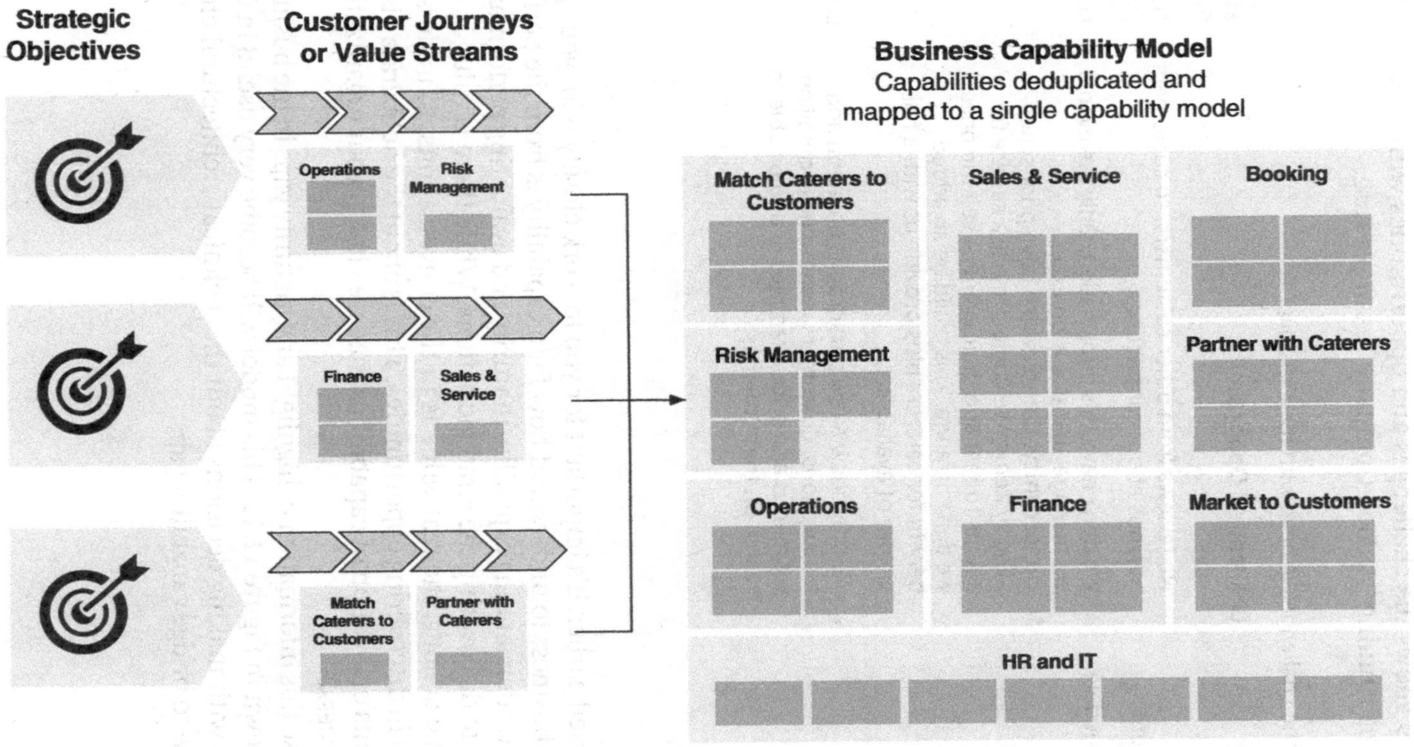

Strategic Objectives

Customer Journeys or Value Streams

Operations | Risk Management

Finance | Sales & Service

Match Caterers to Customers | Partner with Caterers

Business Capability Model
Capabilities deduplicated and mapped to a single capability model

Match Caterers to Customers | Sales & Service | Booking

Risk Management | Partner with Caterers

Operations | Finance | Market to Customers

HR and IT

Figure 11.10: Consolidating mapped business capabilities into the business capability model

Table 11.9 shows the challenges and opportunities with some of the capabilities map from the running "A Food Company" example case.

Table 11.9: The challenge we have with the capability

Capability	Challenge / Opportunity	Maturity Rating
Content Management	■ No reviews from other customers, so customers nervous about contracting.	New Capability Needed
Caterer Schedule	■ We don't have this information. Only the events we manage, and this is on Excel. ■ **Opportunity:** Caterers also have poor scheduling systems. We could provide a service for agents to manage their events, even the ones we don't arrange. This could be an added benefit for caterers plus provide us with a full picture of availability.	Low
Menu Management	■ We only show a few paragraphs on menu and food options. Customers want to have detail on set menus and price examples and to be able to search on these themes.	Low
...	...	

As discussed earlier, it's important for you to work closely with your peers in other parts of the business to understand how far a capability is from the performance that is required. Understanding capability maturity and evaluating performance with your peers is key to discovering the strategic actions. While there is value in documenting your strategic actions, the real value is in the conversations, in understanding constraints, and in discovering opportunities. This is about the deliberate discovery of the improvements or creation of capabilities that are vital to your organization achieving strategic success.

We can use this information to highlight any maturity gaps in the business capability model as shown in Figure 11.11. This model will become very useful in Chapter 12. It will help us with making decisions about the technical architectural choices we will make of how to bridge capability gaps.

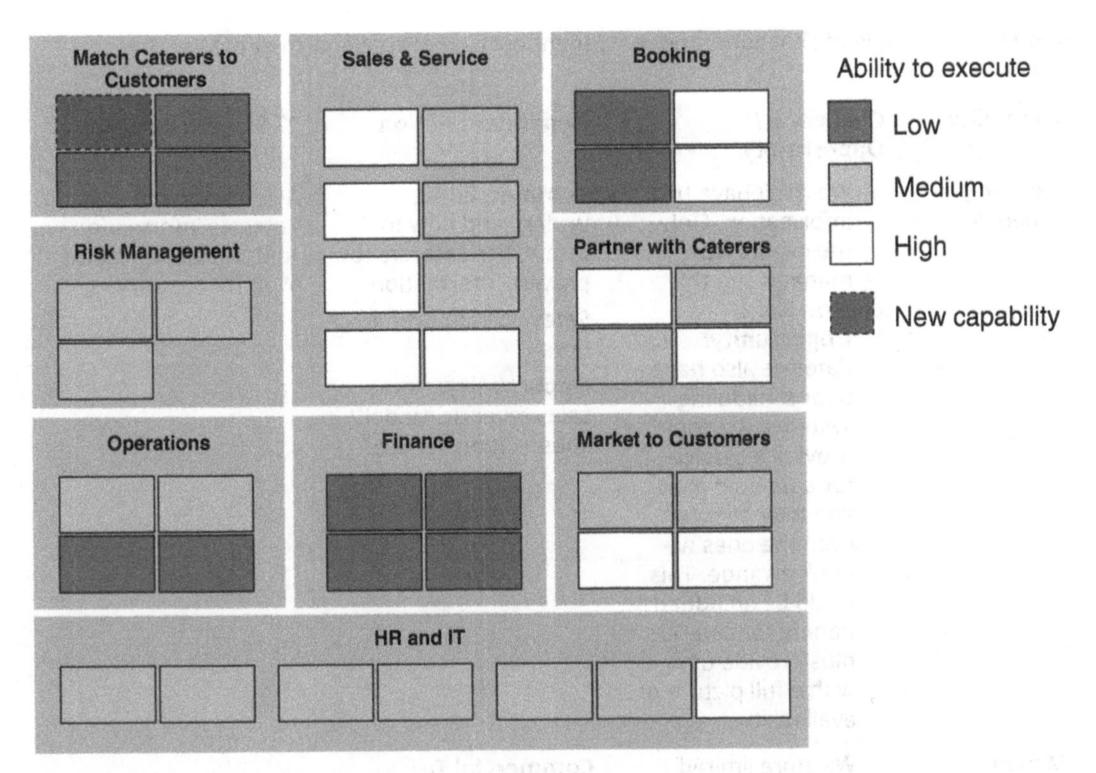

Figure 11.11: Highlighting the maturity games on a business capability model

What Actions Must IT and the Rest of the Business Take to Remove the Obstacles?

What is the gap in maturity? Are we trying to make things better, faster, or cheaper? Whose behavior are we trying to change? Is there a dependency on other departments in terms of an alignment on process or people involved? These are all important questions to help clarify the exact type of improvement that is required in a business capability and therefore the appropriate action. When determining actions, we should ensure what we are proposing is feasible, in terms of both costs and complexity. Although we won't go into the detail of proposing the exact technical solution, we do need to ensure we are able to follow through on the strategic action. There is no point in promising the world if we know it to be too complex or too expensive for our business. Table 11.10 shows the strategic actions, along with the dependency and alignment with other departments, to address capability shortcomings based on the running example from previous sections.

Table 11.10: Example of IT actions, and the dependency on other departments to address a capability barrier

Capability	Challenge / Opportunity	Dependent Action	IT Strategic Action
Caterer Schedule	■ We don't have this information. Only the events we manage and this is on Excel. ■ **Opportunity:** Caterers also have poor scheduling systems. We could provide a service for agents to manage their events, even the ones we don't arrange. This could be an added benefit for caterers plus provide us with a full picture of availability.	**Commercial** To understand how to incentivize caterers to provide information **Sales** To support partners **Opportunity:** New sales revenue model / subscription service	1.1 Provide full caterers information and availability on the Web for customers.
Manage Caterers	■ We store limited information.	**Commercial** To understand how to incentivize caterers to provide information	
Menu Mgt.	■ Only show a few paragraphs on menu and food options. Customers want to have detail on set menu examples and search on these themes.		
...	...	...	...

When you are looking at how technology can contribute to removing a barrier (the HOW we perform the capability), don't simply think how we can make something more efficient; ask how technology can transform the value delivery entirely. In other words, avoid capabilities that "scrape burnt toast." You could improve on the ability to scrape burnt toast to make it more efficient, but wouldn't it be better to focus on the problem of the toast being burnt in the first place? This is exactly where we can challenge the status quo and challenge how we perform the capability rather than simply make what we currently do more efficient. Don't just take requirements verbatim. Don't fix the symptom; fix the root cause. For example, "Do we need a capability to manage the call center better or do we need to focus on the customer self-help capability?" This is where we need to use system thinking to understand real root causes rather than symptoms along with our technical know-how and art of the possible to help digitalize the business—moving us from contributors to digital leaders.

Understanding Operational Essential (BAU) Capability Needs

Outside of the strategic objectives, you need to understand the major operational risks, barriers, and challenges that can damage your business growth. While these are not directly related to the business strategy, they are related to achieving the business goals. Table 11.11 lists some examples of the operational risks with related mitigation actions.

Table 11.11: Examples of large operational risks that require IT strategic action

Operational Risks	Dependent Action	IT Strategic Action
Invoicing is a very manual process; we often miss late payments and don't chase clients for money owed.	**Finance** Consistent process.	Provide accounts payable platform.
We are heavily reliant on demand from PPC search engines.	**Marketing** Invest in non PPC demand generators.	Reduce site latency. Support upper funnel content with blog platform.

Defining Principles to Guide Technical Solutions

Strategic principles guide decision-making on technology choices and implementations at both the tactical and operational levels. Principles are articulated in the following way:

- **Name:** A clear and concise representation of the guideline.
- **Statement:** A brief summary that defines the principle.
- **Rationale:** The reason and justification of why the principle exists and why we want to change behavior.
- **Implications:** Details the tangible actions to take to adhere to the principle.

The rationale for each principle is directly linked to observations from the analysis of the business context and business strategy. This ensures that the principles are explicitly tailored for your organization and will therefore guide decisions that will have a material effect on business success. Without this link we end up with generic statements such as, "We will adhere to data security and compliance laws." While this kind of principle is certainly not controversial, and is good practice, it offers no actionable guidance for decision-making; for example, were we really going to provide a solution that didn't comply with data compliance laws? As with IT strategic actions, we are not looking to boil the ocean with a list of every good piece of advice we can think of. Instead we need to focus on the key principles required to change behavior in order to mitigate or capitalize on a material impact to the business. Therefore, keep the count of principles short and tied to the implications on IT of the key business context changes. A good rule of thumb is that if you can't tie a principle back to an observation in the business context, then you should ask yourself if you need it. The example principles in Table 11.12 are based on the following insights from the business context and business strategy.

- **Technical risk:** Traditionally, we have built everything in house, which means we have a large platform to manage. Increasingly, development resources are focused on noncritical capabilities.
- **New market attractiveness:** We improve market share with strategic partnerships and white label our platform.
- **Business risk:** We are unable to scale the business model as we have many manual processes that require head-count growth in line with revenue growth.

Table 11.12: Principles based on observations of the business context

Principle	Statement	Rationale	Implications
Avoid unnecessary development.	Only build where there is a strategic advantage.	Traditionally, we have built everything in house, which means we have a large platform to manage. Increasingly, development resources are focused on noncritical capabilities.	When looking at a new solution, we should first exhaust a COTS solution before looking at developing a solution ourselves.
Enable the business to operate as a platform.	Enable the business to support a white label revenue stream.	We improve market share with strategic partnerships and white label our platform.	We will design the software systems to support different clients and channels. All systems should be defined by brand and support white labeling.

Principle	Statement	Rationale	Implications
Build for automation and scale.	Ensure any solutions will scale with the business and do not rely on manual processes.	We are unable to scale the business model as we have many manual processes that require head-count growth in line with revenue growth.	Ensure all software systems can manage 10x the growth without adding additional people to manage the process.

Determining Strategic Actions for IT Capability Maturity Improvements

As well as contributing to capabilities for the wider business, the IT department needs to ensure it has the sufficient maturity in its own management capabilities. There are a number of frameworks to model the IT capabilities for the IT function, but one that takes a holistic view of IT is the Open Group's IT4IT Reference Architecture, as shown in Figure 11.12.

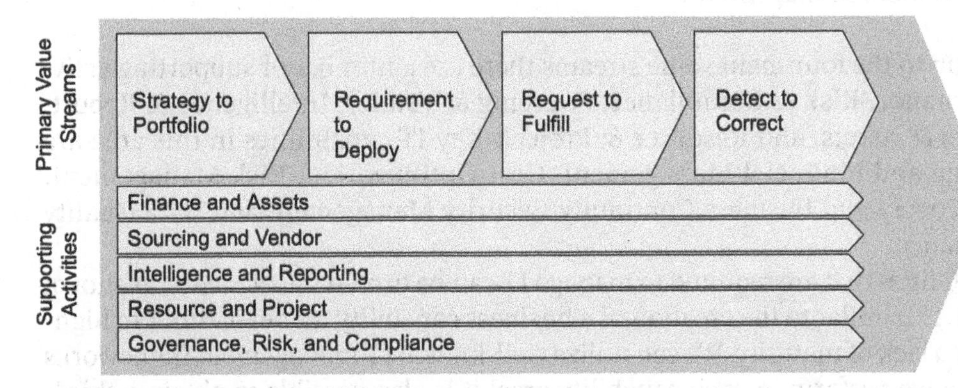

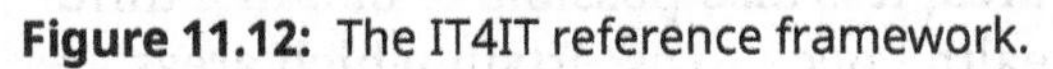
Figure 11.12: The IT4IT reference framework.

The IT function itself needs to be managed in a manner similar to how the rest of the business is managed. The IT department has a value chain, value streams, capabilities, processes, people, and systems just like the business. The IT4IT Reference Architecture is based around the IT value chain and highlights the high-level value streams that generate value for the business. The value chain concept was covered in detail in Chapter 10. Just as with the capabilities that enable the business value chain, we can optimize the activities of the IT value chain in terms of maturity and performance.

The IT4IT Reference Architecture contains the following value streams:

- **Strategy to Portfolio** covers the capabilities required to align business goals with IT efforts. This is also known as the planning phase. Key IT capabilities in this area are Enterprise Architecture, Portfolio Management, Demand Management, and Innovation.
- **Requirement to Deploy** covers the capabilities and activities that deliver the technology to enable business capabilities. This is also known as the building phase. Key IT capabilities in this area are Project Management, Application Development, and Testing and Quality Assurance.
- **Request to Fulfil** covers the capabilities to service requests from the business. This is known as the delivery phase. The key IT capabilities in this area is Service Management.
- **Detect to Correct** covers the capabilities required to detect and correct issues. This is known as the run phase. Key IT capabilities in this area are Availability Management & Capacity Management, Operations Management, and Incident and Problem Management.

In addition to the four main value streams there are a number of supporting activities: Governance, Risk & Compliance, Sourcing & Vendor, Intelligence & Reporting, Finance & Assets, and Resource & Project. Key IT capabilities in this area are IT Budgeting and Financial Management, Cost Optimization, Risk Management, Disaster Recovery and Business Continuity, Security Management and Data Quality and Governance.

The capabilities that are required to manage IT can be turned into a model, as shown in Figure 11.13, similar to the creation of a business capability model in order to highlight gaps or a lack of maturity. We can utilize well-known IT management frameworks to score how we perform in each capability area; it is also possible to obtain a third-party audit as some form of due diligence health check to ascertain which areas are in need of investment and improvement. We looked at some frameworks in Chapter 6 that can help with measuring IT capability maturity in key areas.

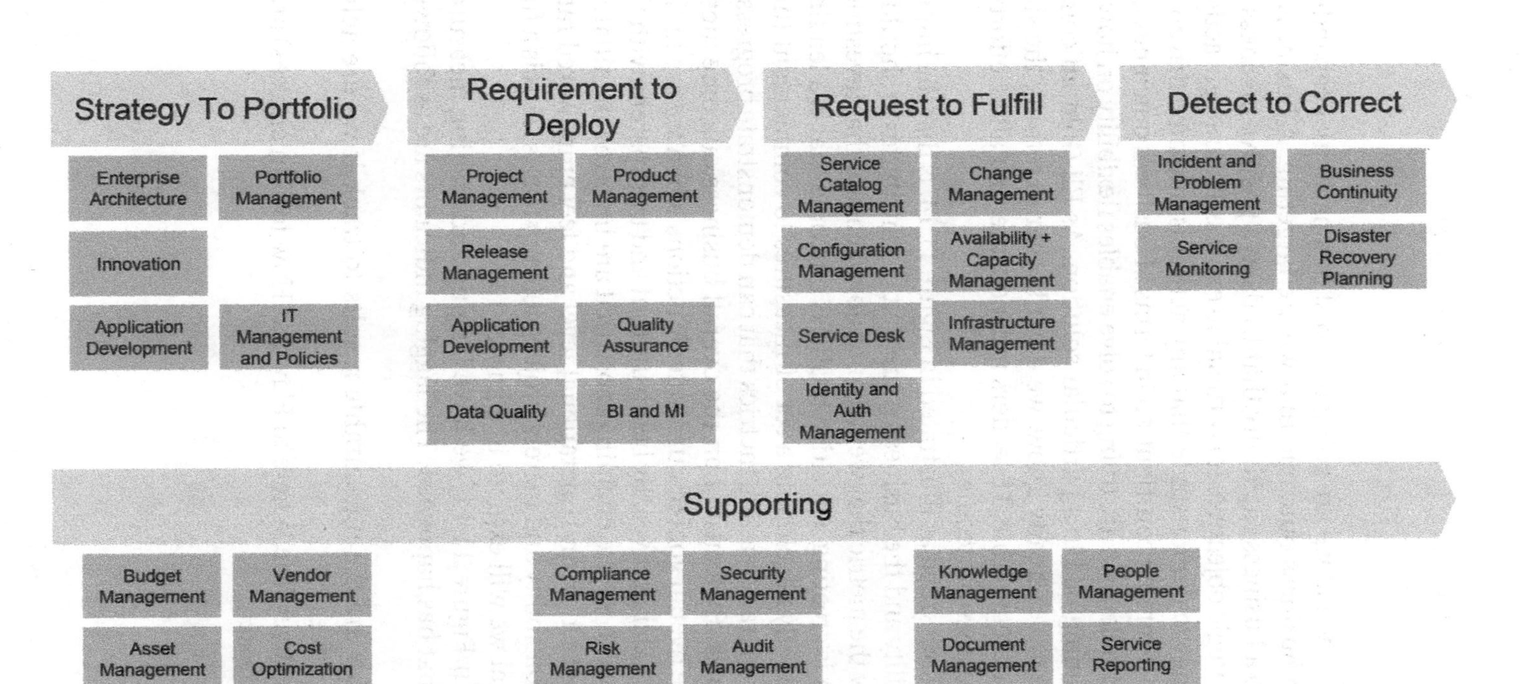

Figure 11.13: The IT capability model based on the IT4IT reference framework

Measuring Contribution in Terms of Business Outcomes

It is no good just to say that we are going to invest in a particular capability. We must have a measurable business outcome that we are targeting. A business outcome is an improvement to a business capability that leads to a positive business impact that contributes to a strategic objective. It is crucial to express IT strategic actions in terms of business outcomes to demonstrate that any action is outcome driven rather than technology-focused. Framing contribution in terms of business outcomes and focusing on the underlying problem we are trying to solve enables flexibility on how it is tackled and avoids us becoming tied to a particular solution. As Eli Goldratt comments, "As long as we think that we already know, we don't bother to rethink the situation." To have a seat at the strategic table, IT leaders must ensure that all IT efforts ultimately result in positive business outcomes.

The key point to stress, and as illustrated in Figure 11.14, is that the improvement in the business capability and the business value that it contributes to should be measurable. We must show the metric that we will monitor to ensure that investment in capability improvement is creating a positive business outcome that is a leading indicator of a business impact. By developing a set of measures to complement the capability improvement, we can track leading metrics that can demonstrate progress toward the overall business goals. By focusing on the overall business outcome metric, we will not lose sight of the overall goal of our strategic actions when we are in the weeds of delivering projects and initiatives. By being upfront and transparent with the outcome we are working toward, we are enabling those that are involved in initiatives to make better decisions at a tactical and operational level and feel empowered rather than to be absolved of responsibility and merely project delivery teams. This is a fundamental and key concept that we will explore in Chapter 13.

As highlighted in Figure 11.14, we have lagging and leading measures. Lagging measures tell us what has happened. The lagging indicators are as follows:

- Business goals are the measurable results to track if we are achieving our business objectives.
- Strategic objectives (the Where to Play and How to Win choices) are measured in terms of impact.

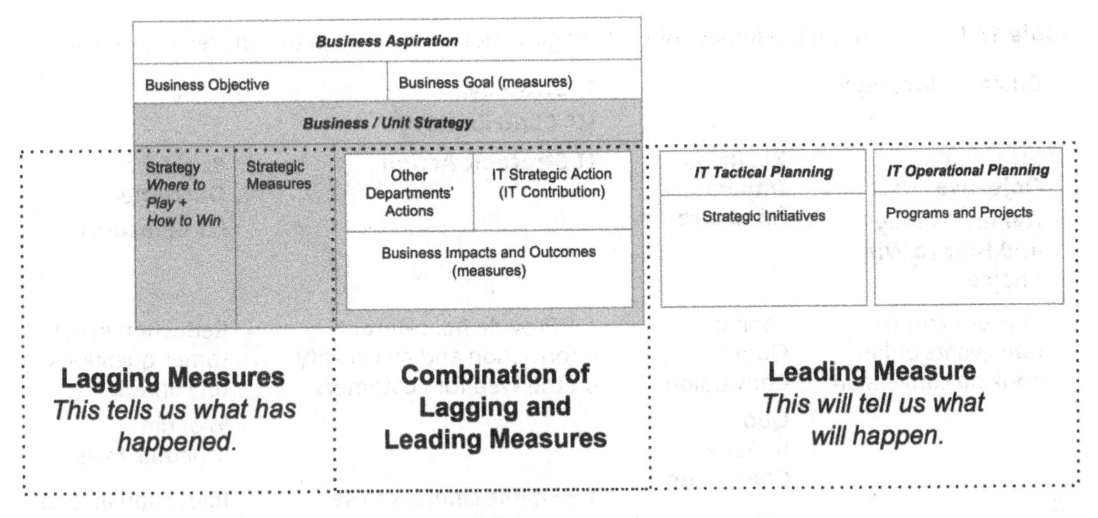

Figure 11.14: The use of lagging vs. leading measures in strategy deployment

The leading indicators tell us what will happen:

- At a strategic contribution level, business outcomes are the changes in behavior that are used to measure the impact of strategic actions.
- More granular or sub-business outcomes are the changes in behavior that are used to measure the strategic initiatives.
- Even more granular outcomes or output are how we measure the operational projects and programs.

To complete our running example, Table 11.13 shows how we will measure the impact of strategic actions. The cascade of measurements enable a clear link between the business strategy and the operational projects and programs. As we move from the strategic level to the operational planning phase, our measures will become more granular. Note that the metrics don't focus on technical measures or adherence to plans or budgets; where possible we should focus on business metrics that any leader in any other department could understand and see the contribution that IT is making to the strategic choices of the business.

Table 11.14 shows how we can use measurements to track the progress of IT capability improvements required to improve IT management that will directly support business success.

Table 11.13: Measuring the impact of IT strategic action in relation to the strategic objectives

Business Strategy		IT Strategy (IT Contribution)	
Strategic Objective (Where to Play and How to Win Choices)	**Business Impact (Measure)**	**IT Strategic Action**	**Business Outcome (IT Measure)**
Improve corporate events online booking conversion.	Look to Quote Conversion Quote to Book Conversion	1.1 Provide full caterers information and availability on the Web for customers.	Reduction in customer questions on content % of time schedule miss
		1.2 Enable caterers to self service quotes and customers to confirm bookings online.	Reduction in lead time from quote to book
Increase caterer sign-ups.	New B2B sign-ups	1.3 Automate the sign-up and approval process.	Reduction in time from B2B introduction to on-site selling
Drive visitors to the site.	. . .	. . ..	..

Table 11.14: How to measure IT capability improvements

IT Capability Weakness	Action and Measurement
Our ability to recover from a loss of data or service has not been tested or documented.	Move from Our Disaster Recovery and Business Continuity position Maturity level 2 to level 4. *Level 2: Established or reactive disaster recovery strategy in place but lack proper capabilities and handled on best-efforts bases.* *Level 4: Managed or proactive disaster recovery managed as program where dedicated team manage/maintain/validate various components of disaster recovery program.*
Our security posture is poor and we have a heightened security risk due to our position as SME.	Become compliant in all of the CIS controls organizational categories: ■ **Reactive Organization:** The organization's ability to respond to an active cyber incident (70%+) ■ **Proactive Organization:** The organization's ability to prevent, monitor, and detect events prior to a cyber incident (70%+) ■ **Expert Organization:** The level at which the organization demonstrates an expert level of cybersecurity practices. (70%+)

IT Capability Weakness	Action and Measurement
Highly manual effort to maintain infrastructure availability and security.	Move to managed services. Reduce the number of systems where we need to patch the operating system.

Communicating IT Strategic Contribution

The purpose of a strategy is to create a shared understanding of the goals the organization has, what they intend to do to achieve them, and perhaps more importantly what they are not going to do. By being transparent on the strategic choice cascade, we can promote an understanding on what is important for business success to everyone in the organization to ensure that all efforts are driven and focused on delivering on what makes a difference—in other words, creating an effective workforce. Therefore it is vital that leaders focus on ways to communicate and engage people in the strategic posture from members of the board to potential new hires and everyone in between. Different scenarios call for different methods of communication. For investment decisions and boards, there is a need for a high-level summary. For due diligence, or your direct reports, a more comprehensive document is required. For communicating to the wider organization, a high-level visual storyboard is sometimes more effective. Either way any communication must create a compelling vision and a clear focus for the direction of effort, clarifying exactly the WHY behind the WHAT.

There are two mediums of communication that you will need to master:

Full Document For offline reading and presentations. A document that will contain detail on all the aspects of a strategy. Typically I keep this high-level and easy to read and to present, so I tend to favor slides.

One Pager Aka the Elevator Pitch. For presentations and strategy refresh, add a storytelling narrative over the top to improve engagement and emotional commitment. Communicates a clear vision and consolidates the strategic choice cascade and IT's contribution with key metrics to measure success.

Full Document

In this section I will present an IT strategy embedded within a business strategy using the example business that has been used throughout the chapter. I will show how the story and typical sections of an IT strategy can be displayed in a slide format. This is by no means an example of a comprehensive strategy document. My hope is that it will help to make some of the components of the strategy discussed earlier in this chapter more tangible as they can be seen in context. By using a slide format, we can force ourselves to be concise and focus on the value of the IT strategic actions and their link to overall business goals and strategic alignment without trying to boil the ocean—remember the strategy is what we will do, not a detailed plan on how we will do it.

The strategy document needs to be concise and consistent; the detail comes in the target state document and the tactical and operational plans. The true value of an explicit strategy is in the conversations and the impact on day-to-day actions. Think of the full document as an aide-mémoire, something you can refer to for new people or to refresh understanding on strategic choices and their implications. It is not the end, merely a useful medium to help communicate our strategic intent and facilitate consensus and alignment.

How Would McKinsey Do It?

For some guidelines on how to structure the strategy, we can leverage the tips and tricks of some of the top management consultants. The Pyramid Principle, as shown in Figure 11.15, is a concept in effective and structured communication created by Barbara Minto during her time at McKinsey & Co during the 1970s. While many people would build their argument from the bottom up, the Pyramid Principle begins with the answer, then summarizes and sequences the arguments for it before supporting those arguments with data. The aim of the Pyramid Principle is to help structure thoughts and arguments concisely to be easily digestible for readers, who are often only interested in the answer or recommendation. We will use this framework to construct the IT strategy deck.

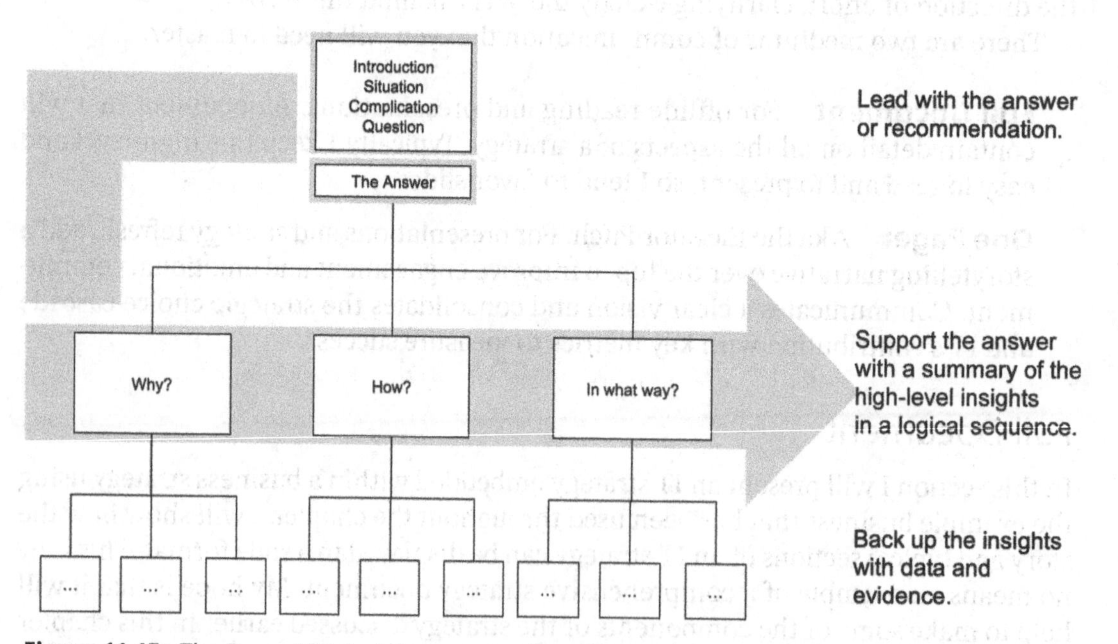

Figure 11.15: The Pyramid Principle structure for effective communication

Because you start with the answer, the introduction is extremely important—especially as many may not get past this point. You must communicate the context and question that you are answering. The introduction can be based on the SCQA framework, which stands for Situation Complication Question Answer:

- **Situation:** The context, containing recognizable and agreed points.
- **Complication:** The problem, or reason why there is need for action.
- **Question:** Poses the question of how the complication or problem can be overcome.
- **Answer:** Your main idea on how to overcome the problem.

The next step is to summarize and sequence your argument. If people want to dig into why this is the answer, then it's important that you structure your argument in a clear and compelling way, a way that ensures readers reach the same conclusion as you have. The first step is to summarize your information by abstracting it into logical groupings that allow readers to reach the same insights, or the "so what's" that you have. The second step is to ensure that there is a logical ordering to these summaries. You can order chronologically if time is a key concept. You can order them in terms of significance, highest impact to lowest. Alternatively, you can choose a structure that has meaning to your business domain. Whatever the sequence, ensure you make the flow simple and easy for the reader or audience to follow. Lastly, you need to support your arguments with data to ensure your summaries are based on indisputable facts and are objectively verifiable to support your argument and opening answer.

When writing out the strategy, you need to clearly show the link between the strategic choices down to the IT strategic actions and how you will measure your contribution impact. Anyone reading the document should reach the same conclusions about what IT should do, and why it is doing it. The document should make it obvious what should be done and how IT will make an impact. Be concise and limit yourself to conveying the key strategic choices, actions, outcomes, and expected impacts to better communicate the message. This will ensure your document is high-level, business-focused, and digestible to readers. Supporting data should be added to the appendix and referenced where needed. Table 11.15 shows how to map the IT strategy to the Pyramid Principle template. The strategy template, shown in Figure 11.16, lays out the typical elements that should be included when documenting IT strategic contributions. However, this template should be used as a starting point; what specifically is in your strategy should depend on your unique context and audience needs.

Table 11.15: Mapping an IT strategy template to the Pyramid Principle

Pyramid Principle Template Section	IT Strategy Section
Introduction, Situation, Complication, and Question	■ Business Context ■ Business Strategy
Answer	Executive Summary on IT Strategic Action
What are the obstacles we must overcome?	The business capability gaps to address
How will IT contribute?	IT strategic actions (IT contribution) and business outcomes measures
What other actions must IT take?	■ The IT capability gaps to address ■ BAU actions

The Introduction, Situation, Complication, and Question

The first part of the document sets out the demand that requires IT and the rest of the organization to take action. This section includes the business context as well as the business strategy or your assumptions on business strategy.

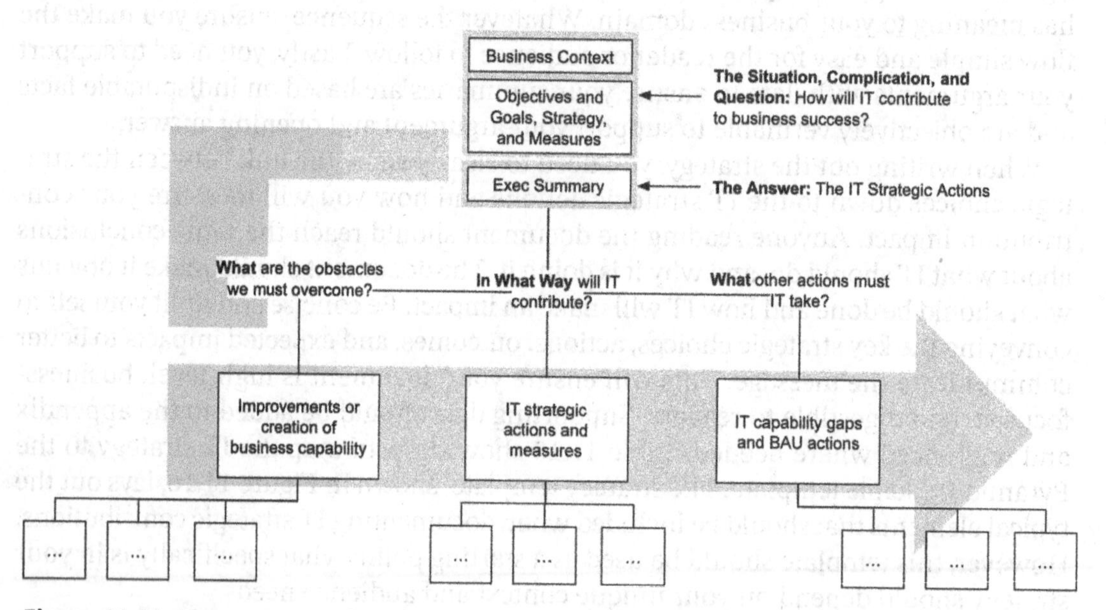

Figure 11.16: The structure of an IT strategy document based on the Pyramid Principle

The Key Factors Affecting Our Business If the strategy is all about the choices and direction of travel, then the business context represents the landscape and the business environment that the strategy exists within. The business context should focus on the things that will impact the organization, be they threats or opportunities, driven from external changes or internal feedback, and how these things—environment, political or technical—will impact the competitive landscape and ultimately why your organization should take notice. These are best represented as scenarios based on what is known and educated assumptions about likely future developments. The business context is the foundations and fundamentals of the climate in which the business operates, which provides the explanation to the strategic choices in the rest of the strategy.

The business context section, as shown in Figure 11.17, will rarely be written down and available as concise as displayed here. Instead, it will be found in talking to peers, understanding what is happening in the wider environment of where the business operates and plays. Think about the relevant political changes that could affect the organization, technology trends, and advancements that could change business models and what the competitive set are doing. Look at different industries to see how external factors are impacting them that could affect you and at how customer attitudes and expectations are changing. Information will be available via industry papers, industry specific conferences, and online white papers and blog articles. This information will be known; however, you will typically need to extract and distil this information to make it explicit. Some of our actions will need to address large BAU needs, it is a good idea to highlight these in the business context to ensure line of sight between IT action and business need.

Key factors that have a material impact on our business

New market Attractiveness	Threat of Substitutes	Competitive Rivalry	Business Risks
Observation: • There is no real offering in the B2C market, which is estimated to be worth £250m a year. So What? • We have the capabilities to exploit to expand offering to B2C. Observation: • The B2B market in Europe is fragmented, with no dominate player. So What? • Early discussions of how we expand into Europe and take advantage of the market.	Observation: • Feedback shows that B2B customers are frustrated by the slow process of dealing with their requests. So What? • Customers are increasingly moving to competitors with digital ways of self-service.	Observation: • Our competitors are already partnering with large brands to gain market share of B2B business. So What? • Our systems can't support white label to offer a partnership model.	Observation: • Inefficiency and manual process. So What? • As we scale, we will need to increase overheads unless we automate processes. Observation: • We are heavily reliant on demand from PPC search engines. So What? • As we look to take market share, we can't afford high PPC costs. Observation: • We have long delays in process payments and mistakes are often made. So What? • We are losing out on revenue.

a Food Company

Figure 11.17: Highlighting key factors that will have a material impact on the business

The Choices of How We Will Win The objectives, goals, strategies (maybe more than one if there are multiple business units) and measures are how the business is going to win. This is the business strategy and the choices based on input from the information in the business context—the organization's vision and mission, the business model, and the current factors impacting the business environment. There should be a clear connection between the insights of the business context and the strategic choices of the business strategy. The strategy can be thought of as "So what?" to the analysis of the business environment—insofar as to say, "Given conclusions to the current environment, this is what we are going to do to win." As covered in Chapter 10, and as shown in Figure 11.18, the OGSM framework clearly and concisely summarizes the organization's strategic position.

You may not always have this level of strategic clarity, hence the need to deep dive and understand your business as covered in Chapter 10. However you will always have some sort of annual budget targets or three-year business goals. Even if you don't have a perfect business strategy document, you must explicitly lay out the assumptions on business strategy that IT strategy is based upon. Explicitly calling out your assumptions on business strategy is important as the strategic actions you arrive upon are neither right nor wrong when considered in isolation; instead, think of them as the logical conclusion based on the business demand. Therefore, if the IT strategy is wrong in the eyes of others, it's probably the assumptions that need to be reexamined.

This slides section should not list every objective that the business intends to pursue over the strategic time period; instead it should be limited to the key objectives that advance the organization toward the business goals and overall objectives. If you find that the number of strategic objectives cannot fit on a single slide, then it is a sign that perhaps this is merely a wish list of items and thus will provide no framework for prioritization or focus during strategy deployment.

The IT Strategic Contribution Executive Summary

The answer, as shown in Figure 11.19, is in the form of an executive summary showing how IT will contribute to business success. Include the business outcome measurements for readers to quickly and clearly understand the business needles IT strategic actions intend to move. People are busy and may not have time to read the full document and join up all the links between context, goals, and action. You need to be able to distill and communicate IT's contribution in a manner that is explicitly clear and meaningful. This one-pager represents your elevator pitch. An elevator pitch is a concise summary that enables any reader to comprehend it in a short period of time. This summary should be able to get across IT's value contribution in the time it typically takes to ride an elevator, hence the name.

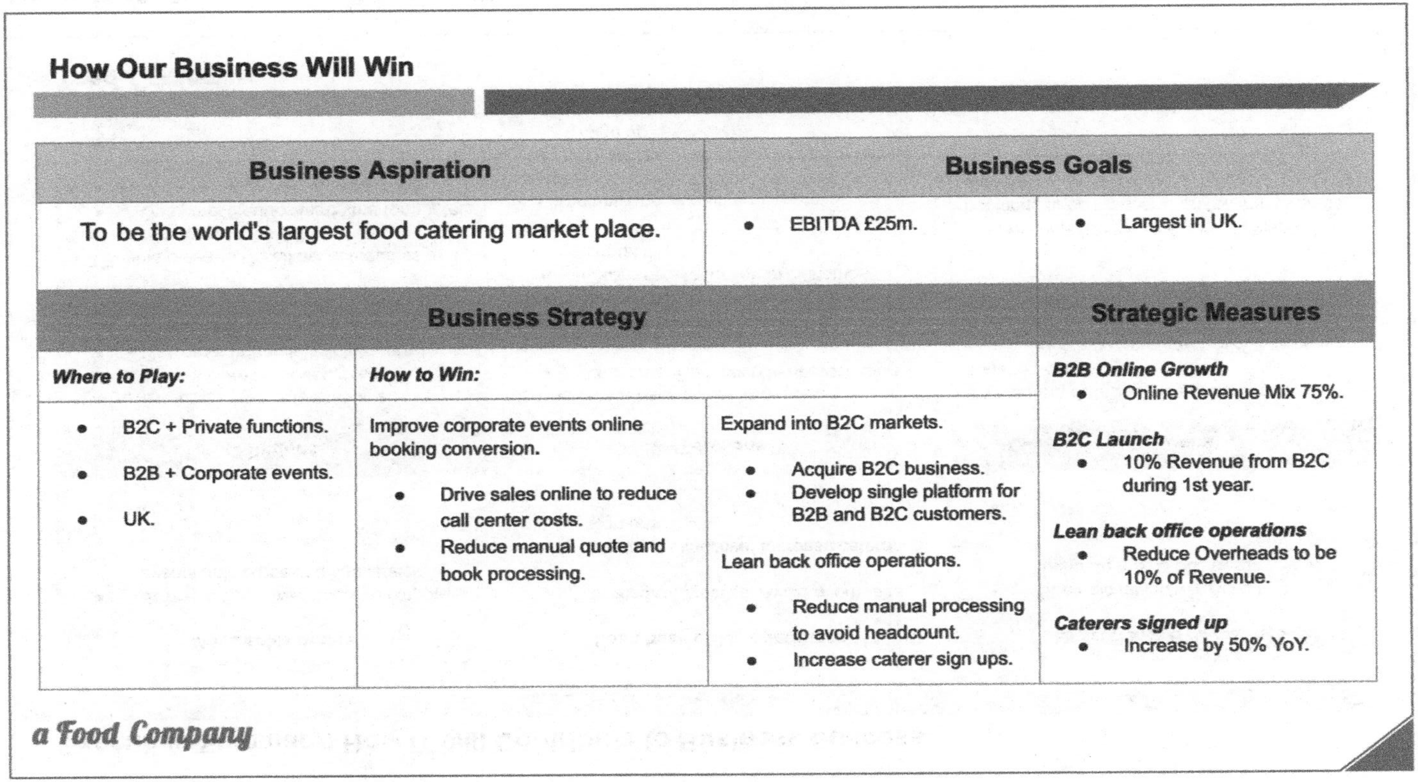

How Our Business Will Win

Business Aspiration		Business Goals	
To be the world's largest food catering market place.		• EBITDA £25m.	• Largest in UK.
Business Strategy			**Strategic Measures**
Where to Play:	***How to Win:***		***B2B Online Growth*** • Online Revenue Mix 75%. ***B2C Launch*** • 10% Revenue from B2C during 1st year. ***Lean back office operations*** • Reduce Overheads to be 10% of Revenue. ***Caterers signed up*** • Increase by 50% YoY.
• B2C + Private functions. • B2B + Corporate events. • UK.	Improve corporate events online booking conversion. • Drive sales online to reduce call center costs. • Reduce manual quote and book processing.	Expand into B2C markets. • Acquire B2C business. • Develop single platform for B2B and B2C customers. Lean back office operations. • Reduce manual processing to avoid headcount. • Increase caterer sign ups.	

a Food Company

Figure 11.18: A business strategy laid out using the OGSM template

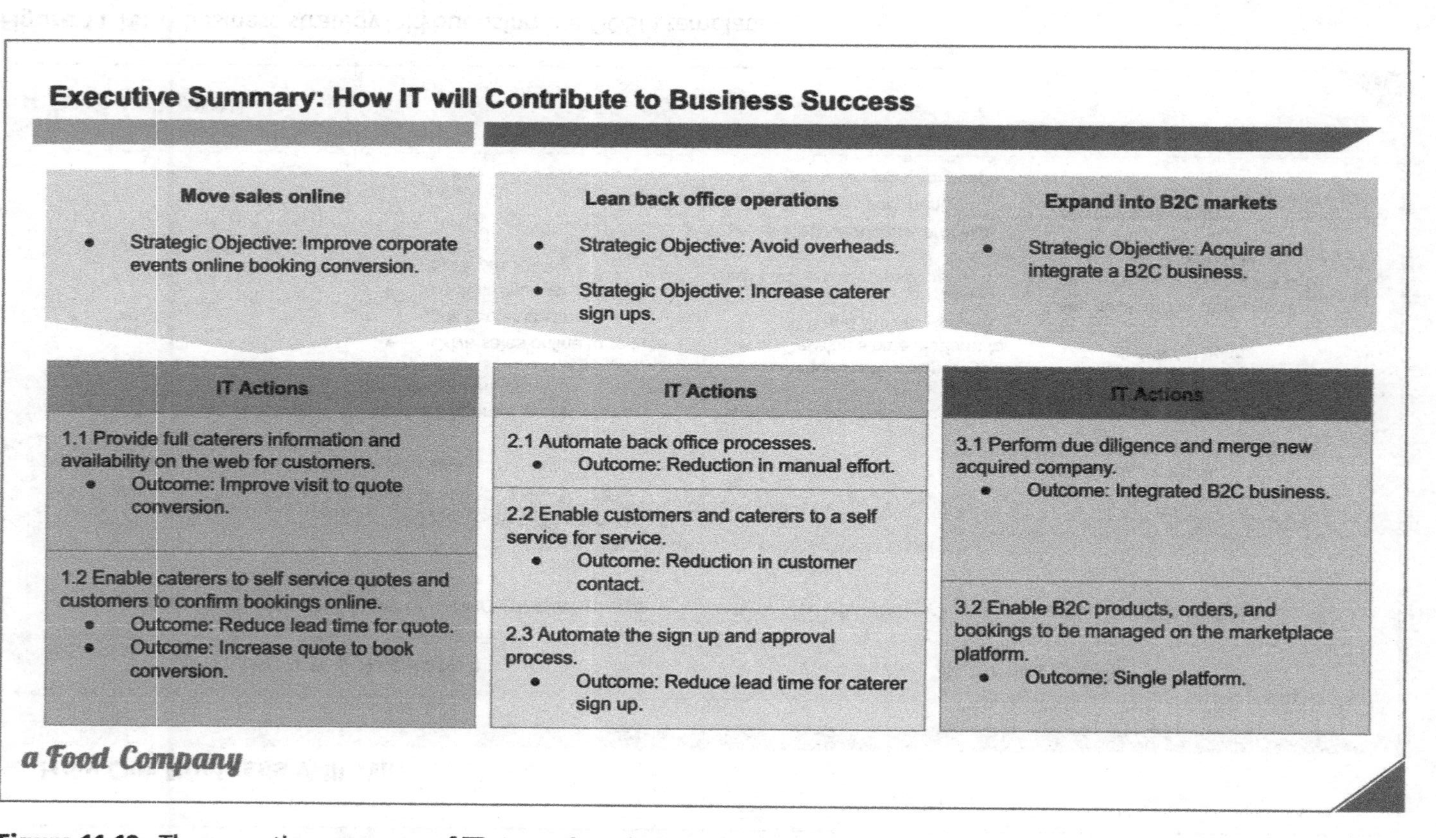

Figure 11.19: The executive summary of IT strategic actions

What Are the Obstacles We Must Overcome?

Now that we have given the answer as to how IT will contribute to business success, we can demonstrate how we have come to that conclusion. As previously mentioned, the capabilities are the link between the strategy and the IT contribution. We have covered how to map the capabilities that are critical to the strategy extensively in this chapter. All this information needs to be distilled to one or two slides maximum for each strategic objective, so focus on the most critical capabilities and what improvements are required. Figure 11.20 shows a sample of mapping capabilities from a customer journey, highlighting the pain points customers experience and that are preventing the strategic objective from being achieved. Figure 11.21 shows a sample of mapping capabilities from a value stream map, highlighting the obstacles and barriers in business performance that are again preventing a strategic objective from being achieved. Notice that we have tagged the business or customer pain point with the business capability. This is important as we can use this to anchor strategic actions in the following slides.

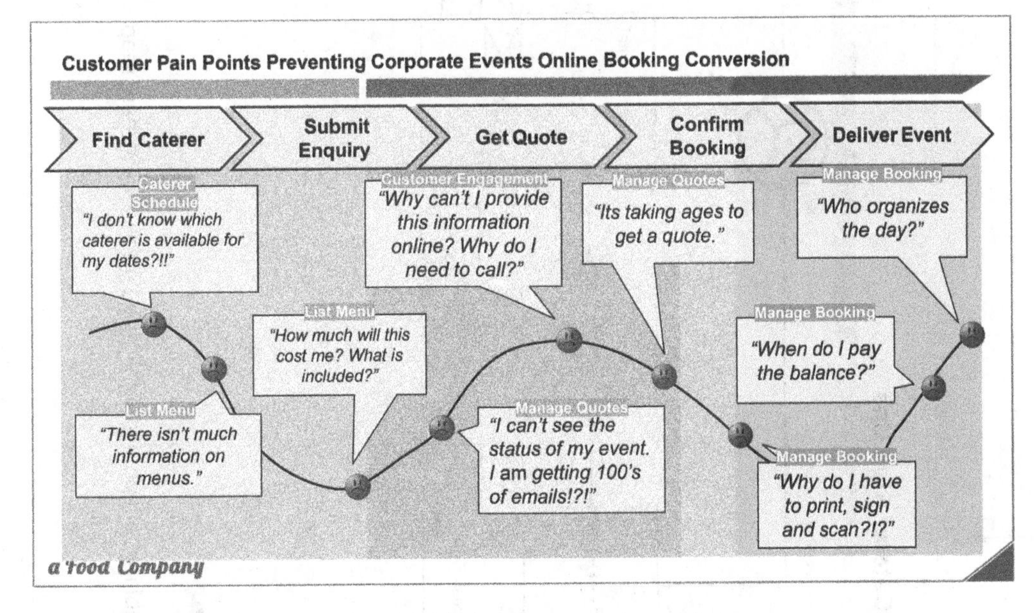

Figure 11.20: Capability pain points highlighted on a customer journey map

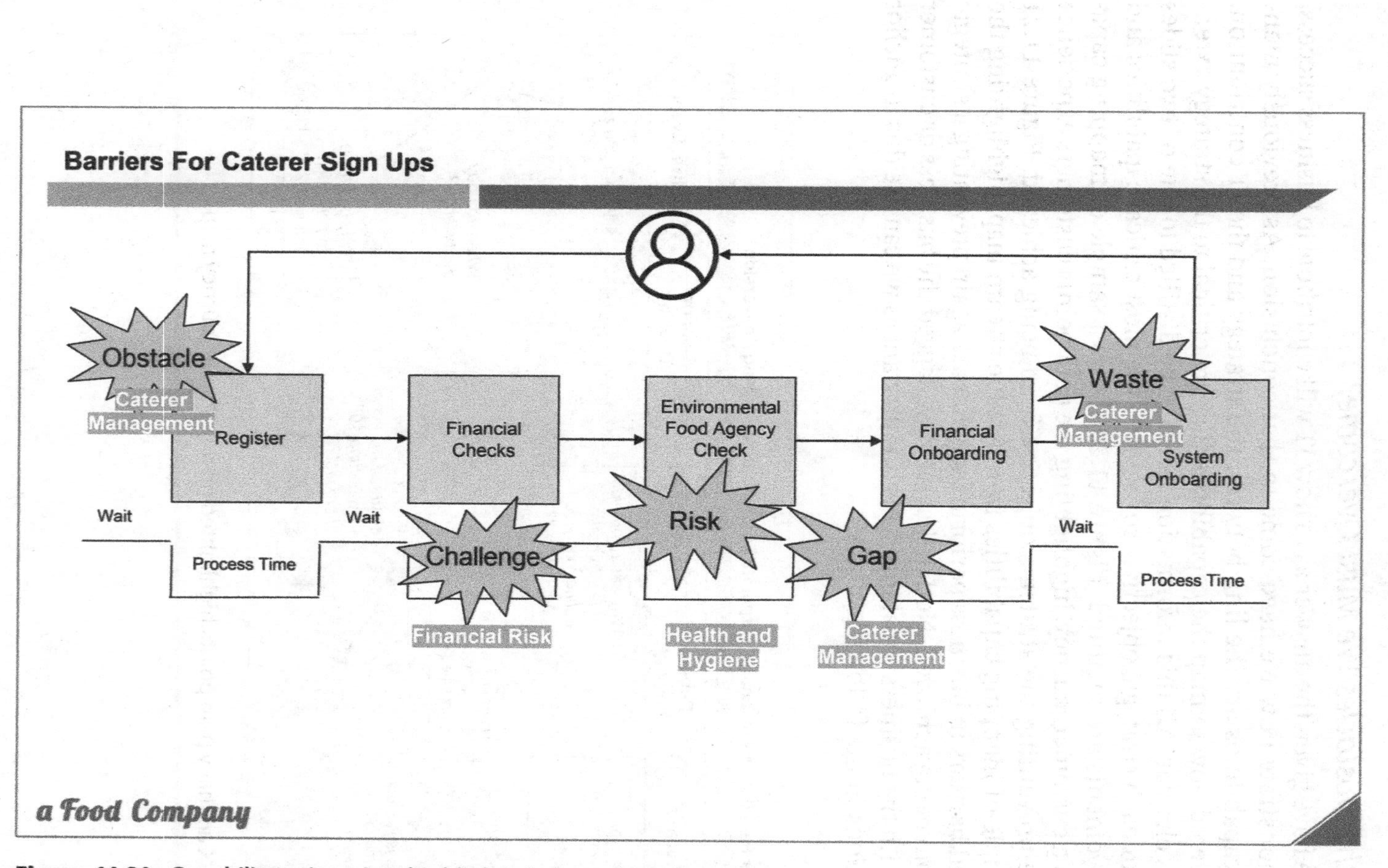

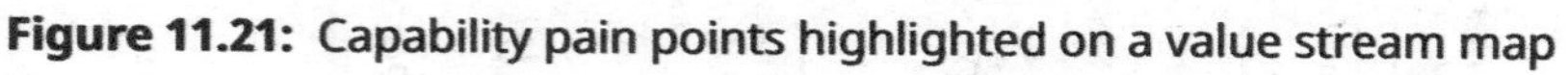

Figure 11.21: Capability pain points highlighted on a value stream map

In What Way Will IT Contribute?

The next section of the document should articulate the contribution IT will make to the creation or improvement of business capabilities, including, most importantly, how you will measure the improvement and how it will contribute to a business impact that is aligned to the business strategy. This is basically the distillation of this chapter into one or two slides.

Figure 11.22 shows how IT will contribute to filling the key capabilities gaps. Remember this is what we will do, not how or in what order. This is the strategy, not the target architecture or tactical and operational plans. This should clearly state what critical investments are required in technology rather than simply listing everything the IT department needs to do to support the organization, such as service desk, etc. Figure 11.22 also shows how we will measure the capabilities improvements and how they in turn will contribute to high-level business impacts. Ideally, we would be able to show the specific improvement in a time-bound manner; in some cases this will be possible, e.g., reducing process time down to zero man hours for automation. However, often we will not, and instead, we should define the metric we will monitor to ensure improvements are having their intended effect. Clearly showing the measure we intend to improve with the IT action reinforces the outcome over output principle. For example, even if a project is on budget, on time, with no errors it will not make an impact unless it produces a positive business outcome. As covered earlier, try to include multiple metrics and ideally a leading measure. This will give you early feedback on the success of your actions and give you an opportunity to change direction if your assumptions were proved to be incorrect.

What Other Actions Will IT Take?

Outside of the actions that IT will contribute to addressing the barriers to achieving the strategic objectives, there are actions to address key BAU risks and key IT capability gaps. Figure 11.23 shows how we link large BAU risks, while Figure 11.24 and Figure 11.25 show how we link IT capability improvements, both of which were highlighted in the business context section.

What Are Our Guiding Principles?

The principles section, shown in Figure 11.26, demonstrates that you have understood the key factors influencing your business and you have codified the guidelines that will guide decision-making and behaviors to address them.

IT Strategic Contribution to improve corporate events online booking conversion

The IT strategic action of...	Will improve the business capability of...	Leading to an outcome of...	Contributing to a business impact of...
1.1 Provide full caterers information and availability on the web for customers.	List Menu Caterer Schedule	% of caterers signed up to provide availability scheduling % of caterers that have 10 or more menu items	Reduction in customer questions on content Look to Quote Conversion
1.2 Enable caterers to self service quotes and customers to confirm bookings online.	Customer Engagement Manage Quotes	Lead Time to supply customers with quotes	Quote to Book Conversion

a Food Company

Figure 11.22: Linking IT strategic actions to improve business capabilities and how the outcomes contribute to business impacts

IT Strategic Contribution To Address Key Business Risks

To address the risk of ...	IT will...	Leading to a business outcome of...
Missing late payments owed money	4.1 Provide accounts payable platform	Reduction of Invoices not collected Visibility of outstanding invoices
Becoming reliant on demand from PPC search engines	4.2 Reduce site latency	Improved core web vitals that will improve our google ranking
	4.3 Provide blog platform	% of visits from blogs

a Food Company

Figure 11.23: Linking IT strategic actions to mitigate business challenges or capitalize on business opportunities

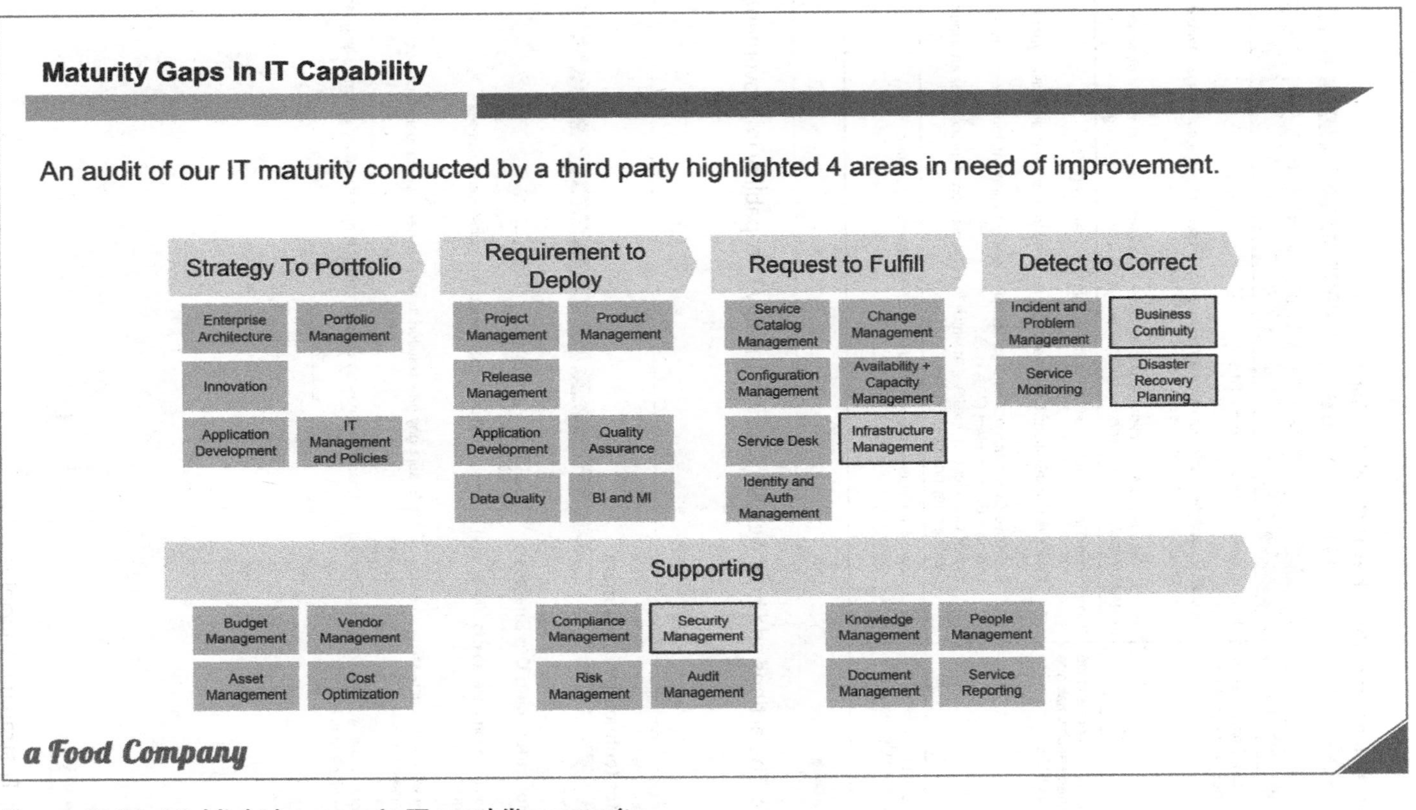

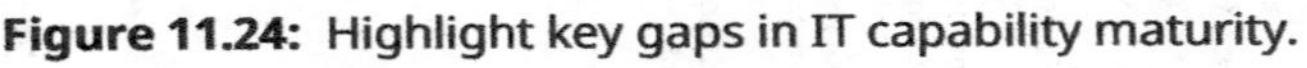
Figure 11.24: Highlight key gaps in IT capability maturity.

IT Strategic Actions To Address Key IT Capability Gaps

Weakness	IT Measure
Our ability to recover from a loss of data or service has not been tested or documented.	Move from Maturity level 2 to level 4. *Level 2: Established or Reactive disaster recovery strategy in-place but lack proper capabilities and handled on best-efforts basis.* *Level 4: Managed or Proactive disaster recovery managed as program where dedicated team manage/maintain/validate various components of disaster recovery program.*
Our security posture is poor and we have a heightened security risk due to our position as SME.	Become compliant in all of the CIS controls organizational categories: 1. Reactive Organization: The organization's ability to respond to an active cyber incident (70%+). 2. Proactive Organization: The organization's ability to prevent, monitor and detect events prior to a cyber incident (70%+). 3. Expert Organization: the level at which the organization demonstrates an expert level of cybersecurity practices (70%+).
Highly manual effort to maintain infrastructure availability and security.	The number of systems where we need to patch the operating system.

a Food Company

Figure 11.25: The actions and how we will measure the key IT capability improvements

Guidance for technology investment

Key Factor Influencing Our Business	Guiding Principle
Our systems can't support white label to offer a partnership model.	Ensure all technology supports a multi-tenant business.
As we scale we will need to increase overheads unless we automate processes.	Ensure any new solution avoids the need for manual processes.
Early discussions of how we expand into Europe and take advantage of the market.	Ensure any new system can support multi-currency and multi-language.

a Food Company

Figure 11.26: IT strategic principles

The One-Pager

The full document should be enough for everyone to understand what IT's contribution is and why the actions link to business success. However, the full document can be long and take time to digest, and for the message to stick, the message needs to be repeated. To communicate effectively, we need a concise message, and dare I say catchy medium to have at hand at town halls, meetings, and reviews in order to reenforce the strategy. By being consistent and repeating IT's strategic position, you can quickly build alignment. I find that you can build a simple one-pager based on the executive summary from the full strategy document. However, you can go a step further and make it more graphical in order for it to be memorable, as shown in Figure 11.27. A simple clean message visually communicated can be very effective at spreading your story and vision.

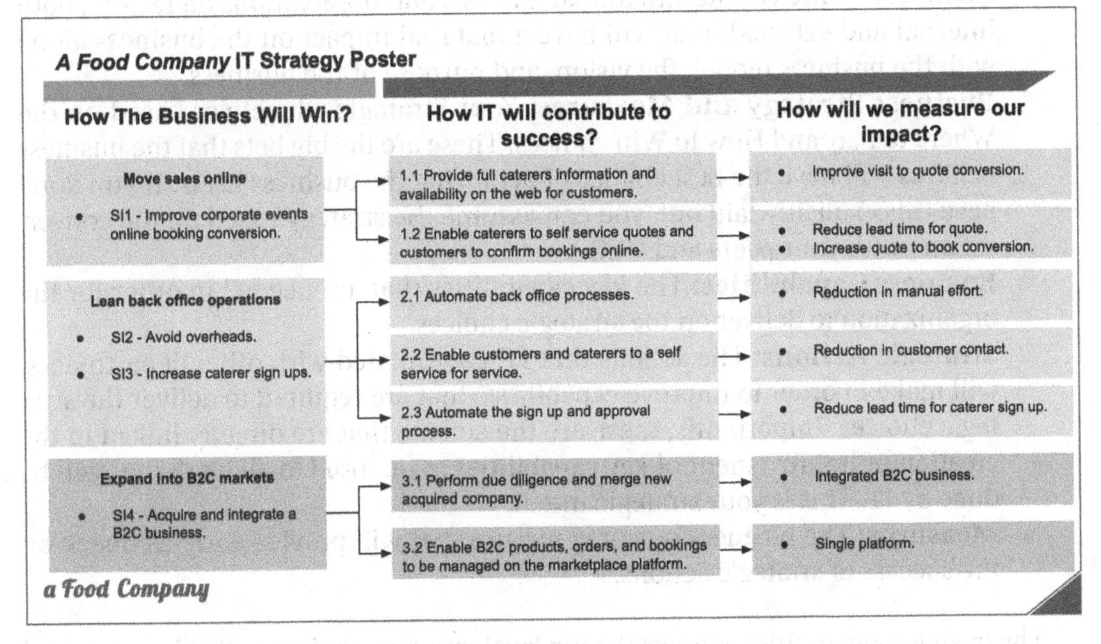

Figure 11.27: A one-page poster to communicate IT strategy

Summary

Ideally, for an organization there should only really be a single strategy, the business strategy. Within this single strategy there should be an aligned contribution from all of the major departments, including the strategic actions that IT intends to make. The words *strategy* and *plan* are often used interchangeably; however, they are not the same thing and they represent two distinct concepts. The IT strategy represents WHAT contribution IT will make in order to achieve the strategic objectives, whereas the tactical and operational plans reveal HOW this strategy will be delivered, detailing the resources, priority, and budgets required. We will cover the tactical and operations plans when we take a look at deploying the strategy in Chapter 12 and Chapter 13. The importance of an explicit strategy cannot be overlooked; even an environment of mature agile and adaptable capability needs direction of travel, a purpose and vision people can align around. The strategy is a key artefact to help make priority decisions and understand trade-offs. The strategy is about explicitly what is the right thing to do, and why.

The anatomy of a strategy, including the embedded IT strategic actions, is typically composed of the following elements:

- **Business Context:** The situational analysis and observations on factors, both internal and external, that will have a material impact on the business along with the business model, the vision, and purpose of the business.
- **Business Strategy and Measures:** Your Strategic Objectives based on the Where to Play and How to Win choices. These are the big bets that the business believes will have the best chance of achieving the business goals. If you don't have this explicitly laid out, you can assume the strategy based on the conversations with your peers and CEO.
- **Business Capabilities:** The key capabilities that are needed in order for the organization to deliver on the strategic choices.
- **Strategic Actions:** The actions that IT, coordinated with other departments, will make in order to improve capabilities that are required to deliver the strategic choices. Importantly, these are the actions that are directly linked to the creation or improvement of key capabilities as opposed to all work that will be done by IT. This is your strategic intent.
- **Measures:** The business outcome metrics that will provide early feedback on the success of strategic actions.

The business capabilities connect the top level-strategic choices to the departmental strategic actions. Therefore, they are key to understanding how to be effective and the areas in which to invest. Business capability modeling comes from the enterprise architecture community and enables a visual map to be built and highlighted to promote collaboration and facilitate the discovery of critical capabilities. Business capability modeling should be anchored to the strategic objectives and needs of users and is dependent on collaboration and alignment with peers on the consensus of what is required to deliver on the strategic choices. The value is in the conversations and discovery versus the output of a document or following a particular framework.

Everything we do must be anchored to delivering business outcomes that contribute to strategic impacts. It is very easy to recommend actions that improve capabilities; however, in order to be strategic we must be able to measure the improvement in a capability in terms of the value to the organization. We define a business outcome as an improvement in a capability that leads to a measurable business value being produced that is aligned with the strategy of the business. In other words, we must clearly show that any strategic contribution to improve a capability can be measured, using a business metric, and in turn that improvement contributes to overall business value. For example a strategic action to reduce processing time for a particular function from 1 hour to 1 minute does represent an improvement, but this may not be of any real value. If the reduction of processing does not remove a higher-level constraint, then we can not say that we are delivering true business value; we are only delivering an improvement.

This is the essence of strategy, ensuring that improvements remove a true constraint or reveal a new opportunity to produce business value. By explicitly showing how a measurable improvement in capability contributes to a strategic objective, we can clearly show a link between what strategically matters to the business and what is nothing more than local optimization. Furthermore, by quantifying investments in terms of business outcome, we can focus on the problem, the desired outcome, rather than be carried away with our technology output. This enables us to adapt and respond if our iniaitives to achieve them don't have the desired impact, and hence why leading as well as lagging indicators are important to help course correction.

Tactical Planning: Deploying Strategy

I have always found that plans are useless, but planning is indispensable.
—*Dwight D. Eisenhower*

Strategy without tactics is the slowest route to victory. Tactics without strategy is the noise before defeat.
—*Sun Tzu,* The Art of War

While we must acknowledge emergence in design and system development, a little planning can avoid much waste.
—*James O. Coplien*

Strategy, as covered in Chapter 11, "IT Strategic Contribution," defines the high-level guidance on what a business and IT need to do to achieve the organization's long-term goals and aspirations. However, the strategic actions only give us direction; they are not immediately actionable. To understand exactly what solutions we will invest in and the most appropriate way of operating, we need to understand the tactics required to deliver on the strategic direction. Tactical planning is the process that turns strategic intent into IT initiatives that in turn can be distilled into programs and projects to be executed at the operational level. You can think of tactical planning as the bridge between high-level strategic action and low-level operational execution.

In this chapter, we will begin with a consideration of what is fundamental to tactical planning. That is, we will consider the need for a strong technical vision, how planning horizons are affected by our unique business context, and the need to constantly adapt and refine plans in relation to the changing business environment. To deploy strategy and align the organization at both the tactical and operational levels, we will look at a strategic deployment policy known as Hoshin Kanri. Hoshin Kanri will enable us to cascade strategic intent to all levels, ensuring that the entire organization has a line of sight between what they are doing and the strategic goals. Hoshin Kanri also introduces the idea that from the strategic down to the operational level, there is a need for the ability to feed back to form a consensus on direction and alignment on planning efforts. We will explore how to feed back and adapt at both strategic and tactical as well as operational levels in detail in Chapter 13, "Operational Planning: Execution, Learning, and Adapting."

Finally, we will cover the process of creating a tactical plan. Understanding that while it is impossible to plan a full three- to five-year strategy in detail, we do need to understand the feasibility, even at a high level of our ambitions, as well as the concrete initiatives that we will fund in the short and medium term. The process itself will cover the creation of a target architecture and target operating model to act as a guiding vision to fulfill capability gaps required to achieve strategic objectives. We will look at how to define the portfolio of tactical initiatives, turning the business needs into concrete goals and understanding high-level solutions, effort, and costs based on the target architectural and operating model suggestions. Last, to produce a tactical road map, we will need to prioritize initiatives based on business benefit along with a balance against investment target allocations, dependency resolution, and the needs to support operational essentials as well as technical optimization and rationalization.

Planning Considerations

All businesses need a financial plan, which typically spans three to five years and covers what it can achieve in terms of financial goals and aspirations along with a strategy of how to achieve it. Depending on the context of your organization, this could be for a bank loan, for the board looking for a return on their investment, or for potential new investors looking to acquire your business. The business goals represent the long-term view on what it can achieve based on the analysis of market conditions, capabilities, and strategy. In terms of how we achieve the strategic and business goals, we need a plan. These plans, as well as validating the feasibility of how we are going to achieve aspirations, support long-term investment decisions such as transformation or consolidation programs or large-scale infrastructure moves.

The process of creating a tactical plan will differ widely for each business based on the unique context that they operate within. However, there are three high-level planning principles that are universally applicable that we must consider when deploying a strategy, namely:

- Creating a strong vision to guide planning
- Embracing flexible planning horizons
- Building in the ability to adapt both plans and strategy

Creating a Strong Vision to Guide Our Planning

As the guardians of technology, it is our responsibility to ensure we have a cost-effective, reliable, scalable, and easy-to-manage technical estate that supports the needs of the business. However, without strong guidance we can end up with an IT landscape full of duplication, high complexity, low integrity, poor reliability, and technical debt as well as both quality and performance challenges. This can be exacerbated in the world of distributed and autonomous product teams where there is danger of siloed thinking and the risk of a "not invented here" mindset, which is the tendency to avoid using, sharing, or buying existing solutions. To avoid this trap and ensure there is a coherence for the IT architecture, we need an aligned technical vision. This technical vision will help us during planning, clarifying where we standardize and provide platforms or shared services, how we integrate and share data, and where we need to build and where we will buy.

Designing an architecture upfront is great for identifying the high-level structure of the IT landscape to ensure there is a cohesion and integrity to the architecture. However, at a software architecture level, there is often a need for a more emergent approach as the high-level design hits reality, especially when the problem domain is complex. The 11th principle of the agile manifesto even states that "the best architectures, requirements, and designs emerge from self-organizing teams." The SAFe framework refers to these two architectural approaches as intentional architecture and emergent design:

- **Intentional Architecture:** This represents the high-level enterprise and solution architecture. Purposeful, deliberate, and planned to ensure the whole system is designed with integrity and coherence.
- **Emergent Design:** Software architecture designed iteratively and guided by standard and principles.

Our technical vision needs to encompass both the intentional architecture and emergent design. The vision for the intentional architecture manifests itself as a set of target architecture suggestions to guide planning efforts at an enterprise architecture level (strategic) and high-level design documents at a solution architecture (tactical) level, as shown in Figure 12.1. Neither of these are detailed enough to serve as a blueprint to build a solution. Instead, they are designed to influence the overall architecture, guiding solution design and determining the system integration points and boundaries, how systems will be consolidated, and any important nonfunctional requirements.

To ensure there is governance at the software architecture (operational) level to support an emergent design, the guidance is in the form of a set of technical standards and principles. These standards can help guide teams to build evolutionary and incremental architectures in a manner that gives autonomy to teams but within boundaries. While we want teams to have autonomy, there is little value in each team choosing its own cloud providers. Therefore, there is a need for governance, albeit led by consensus rather than top down. It is important to point out that Figure 12.1 features feedback loops so that emergent design, the reality, can inform the intentional architecture.

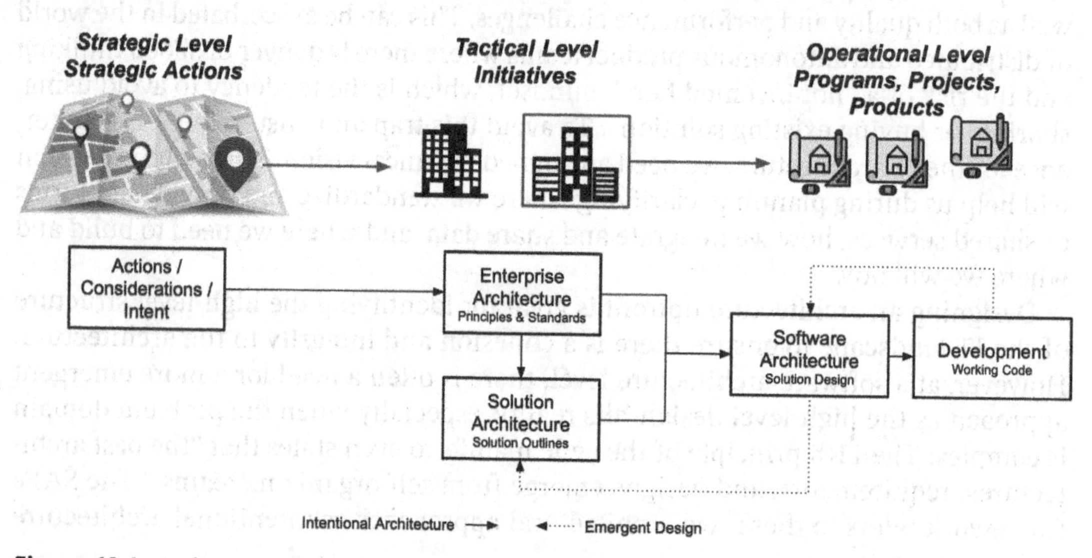

Figure 12.1: Architecture abstraction from strategic to operational

Embracing Flexible Planning Horizons

Because of the environmental uncertainty that modern organizations operate within, plans that support three- to five-year business goals and aspirations can only possibly offer an idea of what might be feasible from a time, complexity, and financial point of view. Because of the speculative nature of long-term business road maps, we must ask ourselves how far we should focus our tactical plans into the future. While it makes no sense and will certainly be costly to meticulously plan everything for such a long period, it would be foolhardy to not plan at all and make it up as we go. So, the question is what time frame should we aim for? As with many things in IT, it depends and will be influenced by your unique business context. The factors that will influence your planning horizon include, but are not limited to, the stability of the environment, the maturity of the organization, the needs of different parts of the organization, and the investment strategy of the owners.

If we examine one of these factors (namely, the planning needs of different parts of the organizations), we can consider two ends of the spectrum. At one end we will have capability needs that we are able to plan far into the future. For example, while many things will change, many things will not. Jeff Bezos says,

> *I very frequently get the question: "What's going to change in the next 10 years?" And that is a very interesting question; it's a very common one. I almost never get the question: "What's not going to change in the next 10 years?" And I submit to you that that second question is the more important of the two—because you can build a business strategy around the things that are stable in time. . . . When you have something that you know is true, even over the long term, you can afford to put a lot of energy into it.*

For example, as part of an M&A (mergers and acquisitions) process, there may be a need to consolidate warehouses to optimize costs. This is a certainty that we can make plans around that span beyond a 12-month time frame.

In contrast, some areas of the business will be more dynamic and therefore it makes sense to focus on a shorter horizon with only high-level themes of what we might do beyond the short term. For example, organizations with a digital offering may need to react quickly to competitors' movements and releases or capitalize on new technologies being realized such as generative AI to stay relevant for their customers.

While there is no right or wrong way to plan, you do have to plan. Planning is not the enemy of agility. Planning is essential; the agility comes from your ability to change those plans when the information you based those plans on changes. At the very least you should have a plan that gives at least some indication of what you aim to invest in over the next 12 months. The level of detail within these plans will vary based on the business problem at hand. Plans must include indicative costs, effort, and benefits. While it is of course difficult to give exact figures for these attributes, we do need to start to understand these at a high level so we can get a feel for the tactical initiatives; at this level a range of estimates +/- 25 percent is typically okay. It's important we do this consistently so that we can understand initiatives relative to each other to assist in prioritizing.

Building in the Ability to Adapt Both Plans and Strategy

As you have read, trying to predict the future is impossible for complex contexts. Plans with Gantt charts that map out the next three years are nothing more than wishful thinking. Planning for three years ahead in anything other than a stable environment is foolhardy and a waste of time and resources. And is there such a thing as a stable environment? Can any business nowadays say that it operates in an environment that is impervious to disruption? But paradoxically we do have things that we can be certain about—the basics, such as security, large consolidation, cost optimization, and automation. However, things do change, and we must accept that and have a framework that deals with it.

To have an effective planning process, we need to be able to adapt when new information challenges our assumptions, whether they're based on our high-level strategies or the outcomes of our operational actions. Our strategy and plans are not static; they capture thinking at a given point in time and therefore are context sensitive. As the context changes, so must the strategic outlook. Strategy should be thought of as the by-product of an organization that is continuously learning how to reach its vision by constantly improving, sharing experience and knowledge from feedback loops, experimenting, and innovating. The system of strategic thinking, tactical planning, and operational execution needs to be flexible enough to adapt when the context changes through external events or internal experience—the feedback loops. Just as complex projects are tackled in iterations, so too must strategic planning be tackled. Therefore, our planning framework needs to support feedback and the ability to adjust the strategic direction and investment as the context changes around us.

Following Hoshin Kanri to Deploy Strategy

To support a lean and agile strategic planning process we will look at the principles and practices of Hoshin Kanri. Hoshin Kanri, or Policy Deployment, is a strategic deployment process that ensures strategic goals are communicated clearly throughout an organization and drive action at every level. Business goals, strategic objectives, and measures are mandated by leaders in a top-down manner; the responsibility of delivery and execution of those goals is left with the levels of management and operational employees below. Hoshin Kanri provides clear and consistent communication on direction and strategic intent without dictating the actions at each level of the organization. As shown in Figure 12.2, this clarity aligns each level of the organization, enabling a clear link between the strategic intent and the operational execution.

There are two main benefits of employing Hoshin Kanri. First is the removal of waste from miscommunication, a lack of focus, or focusing on the wrong areas that are of little or no value to the overall goals of a business. Second, it supports intrinsic motivation by being clear on purpose and allowing for autonomy. The purpose and vision of the organization is communicated along with the context and the choices, or the why, behind its strategies. Responsibility is delegated, and thus autonomy, for each level to determine the best way to fulfil the high-level strategic-level goals. As discussed in Chapter 2, "Philosophies for a New System," people have intrinsic motivation to focus efforts and drive results when they have clarity of purpose, an understanding of why a particular strategy is important, and the autonomy to determine the work that has the best chance of achieving a goal.

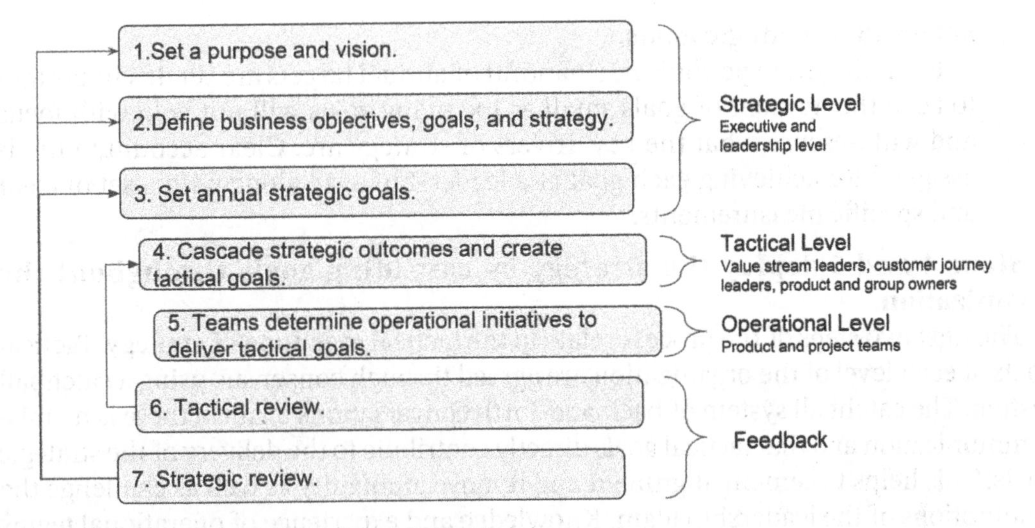

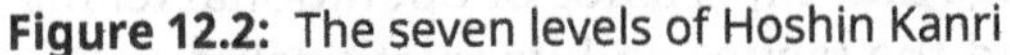
Figure 12.2: The seven levels of Hoshin Kanri

The process of Hoshin Kanri consists of the following seven steps:

Steps 1–3 Set and communicate the North Star, long-term goals, and the strategy.

The first set of steps in the Hoshin Kanri process focus on alignment and direction. It is about communicating the vision, goals, and strategic choices of the organization, often referred to as the organization's True North or North Star. Through consistent and constant communication, people will understand the purpose of the organization and the context behind the chosen strategic path, which contributes to intrinsic motivation when executing on the strategy.

1. **Set a purpose and vision.**
   Explicitly define the purpose and vision of a company and communicate to the entire organization. Everyone should be aligned on where the organization wants to be in the future and understand why the company exists and the value it provides to customers.
2. **Define business objectives, goals, and strategy.**
   The business objectives and goals are the qualitative and quantitative measures that show progress against the vision and mission; typically they span many years. The strategic choices define the areas that will be focused on that represent a competitive edge and will enable the business to win by achieving the high-level business objectives and goals.

3. **Set annual strategic goals.**

   From the strategic choices, a handful of annual targets are set. It's important to keep the number of goals small as too many goals will not help with focus and will obscure what the key drivers of strategy are. Clear accountability is assigned for achieving each goal at a leadership level along with a set of clear and specific measurements.

Steps 4 and 5 deploy the strategy by cascading goals throughout the organization.

The next two steps of the process relate to the tactical execution of strategy. Tactical goals at each level of the organization are agreed through consensus using a catchball system. The catchball system of back-and-forth conversations ensures there is no miscommunication and that tactical goals directly contribute to the delivery of the strategic goals. This helps to cement alignment and remove ambiguity as well as challenge the assumptions of the leadership team. Knowledge and experience of operational teams is used to ensure that any assumptions around the tactics and its relationship to the overall goal are correct.

4. **Cascade strategic outcomes and create tactical goals.**

   The annual goals and measures are deployed to all levels of the organization where tactical goals are created. Meaningful measures are employed that clearly indicate progress against an outcome that rolls up to impact the strategic goals rather than showing progress against an activity or a plan.
5. **Teams determine operational initiatives to deliver tactical goals.**

   Teams plan and execute the programs and projects that will deliver the tactical initiatives. They use fast feedback cycles as detailed in the next step to ensure that they are working on the right initiatives to drive progress of tactical goals. Generating tactics is not a one-off exercise; teams constantly look at new and better ways to deliver the goals as likely many of their assumptions and hypotheses on initiatives will be incorrect as they gain more experience.

Steps 6 and 7 feedback loops to review, check assumptions, and correct course.

To improve progress toward goals, there is a flow of information back up the cascade in the form of regular progress checkpoints. These feedback loops provide the opportunity to change or adjust tactics and the investment capacity of them. The strategic planning process of Hoshin Kanri is built upon flexibility, adaptability, and feedback from all levels of the organization. In essence, this is very similar to the PDSA (plan-do-study-act) process as covered in Chapter 6, "How we Work," just applied to the strategic planning process.

6. **Monthly goals (tactical) review.**
   The monthly reviews ensure tactical work is in line with strategic goals.
7. **Yearly goals (strategic) review.**
   Annually there is a review to show progress against the strategic goals. However, most companies would review their strategic goals on at least a quarterly basis.

Hoshin Kanri is a top-down approach for strategy deployment, but it focuses on the cascade and translation of strategic intent rather than the cascade and alignment to goals at each level. In this way, you can think of it as a framework to manage direction and alignment. As strategy transcends the levels of an organization, the goals and objectives from the level above are not transcribed directly into the objectives for teams at the level below. The implementation of goals at lower levels is left to the management and operational level, business units, and delivery teams, those that are close to the work, or Gemba. Strategy and outcome, not specific tasks and output, are communicated. In the simplest of terms, Hoshin Kanri is a framework for strategic alignment based on three important principles:

- Consensus and collaboration on targets and strategies at each level of the organization
- Autonomy to translate strategy into a team's own objectives and goals
- Honest and open feedback loops at all levels and the ability to change or adjust plans

The first principle is to ensure consensus on strategy and agreement on goals. Using bidirectional goal setting ensures we have realistic targets, which is critical for ownership, motivation, and team commitment. A process known as catchball is used to ensure information flows up as well as down the company. The word *catchball* evokes the exercise of throwing ideas back and forward between the levels to gain a shared understanding of the right thing to do and thus align. The catchball process, shown in Figure 12.3, uses frequent cadence to ensure the opinions of those closest to the work flow up to management and leadership levels as well as contextual information, the why, following down. As teams from a lower level can feed back up to a higher level on the assumptions made, we should expect this process to challenge higher-level goals and strategies, which can result in changes to high-level target outcomes and focus, an example of a learning organization in action.

The second principle is to ensure teams remain autonomous but aligned. Strategy rather than goals and targets is cascaded and translated at each level of the organization. This creates alignment and allows autonomous teams to interpret the translation of the strategic intent into their own goals and objectives. The third principle is feedback loops, namely by using the plan, do, study, and act feedback method. We will look at feedback and how to adjust in Chapter 13.

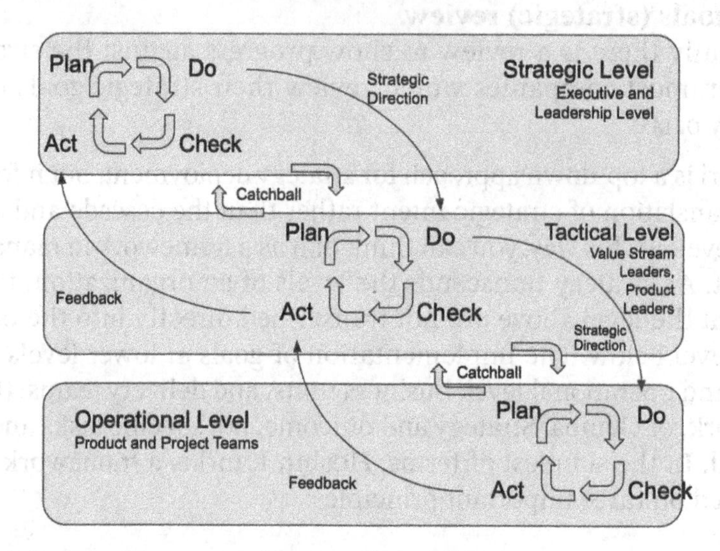

Figure 12.3: Feedback and consensus for strategic cascade

Creating a Tactical Plan

The process of tactical planning, as illustrated in Figure 12.4, takes the business strategy and needs of the various functions of an organization as input. From this point the process is to identify the business capabilities that require investment, understand how the IT landscape needs to change to address capability gaps, define the initiatives to address those gaps, and then prioritize the initiatives guided by the organization's investment strategy. The output of this process is a tactical IT road map that should be refined on a regular basis to consider changes in the micro and macro environment, aka the business context.

Inputs

The main inputs for the tactical planning process are the business and IT strategies as covered in detail in Chapters 10, "Understanding Your Business," and Chapter 11, along with individual business unit or department strategies. They cover the fundamental needs that will shape the IT tactical plan, as follows:

- **The business model:** The operating model, value streams and business capabilities.

- **Business context:** The macro- and microenvironment factors that will have a material impact on the business.
- **Aspirations:** Business strategy. Articulating the desired future course of action for business and IT.
- **Forecast:** Three- to five-year financial projections.
- **Functional strategies such as marketing strategy, finance, product strategy.**
- **Operational essential needs.**

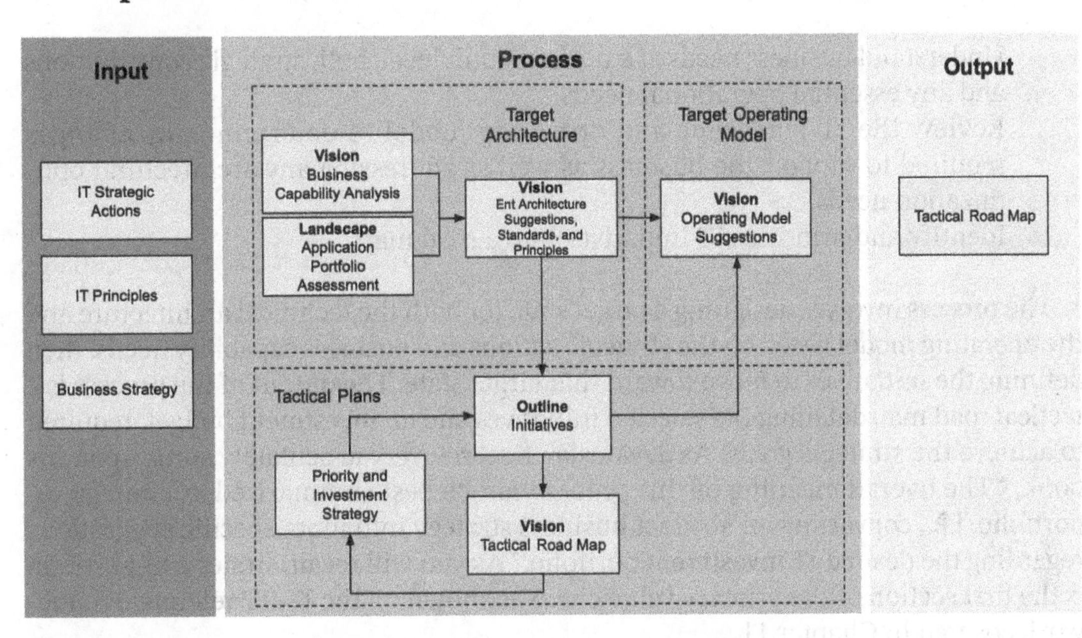

Figure 12.4: The inputs, process, and outputs of creating a tactical plan

This information isn't often in a concise and clear state. It is typically captured through informal discussions with the exec group, reviewing board reports, keenly observing the competition, listening to the opportunities and constraints discussed at trading meetings, understanding the factors that make up the financial budgets, and digging into the high-level communications shared at company town halls. Regardless of how the information is structured or presented, an understanding of the business and departmental needs is vital for effective IT tactical planning and often requires IT leaders to be proactive in obtaining it.

Process

The process of tactical planning is essentially about gathering facts and making decisions. As introduced in Chapter 11, we will again utilize the CSVLOD model from Kotusev's book *The Practice of Enterprise Architecture.* Figure 12.5 lays out the flow of a common approach to tactical planning, what Svyatoslav Kotusev refers to as decision paths. I say common because different organizations will have different ways of operating, and different scenarios require more or less process. What I present here is the core artifacts and decision paths that have worked effectively for me in the past. As highlighted in Figure 12.5, we will use several artifacts covering the standards, visions, landscapes, and outline components of the CSVLOD model to support our planning decisions. Three processes are as follows:

1. Understand business needs at a business unit level, both strategic contributions and any essential operational needs.
2. Review the IT landscape and operating model to determine any changes required to support the business as well as addressing any architectural optimization needs.
3. Identify and prioritize IT initiatives into a road map.

The process involves designing a target state for both the technical architecture and the operating model based on the strategic actions and business capability needs, then defining the initiatives to move toward that target state. The output of the process is a tactical road map detailing the selected initiatives and an investment budget, required to achieve the strategic goals. As Svyatoslav Kotusev very articulately sums up in his book, "The overall meaning of this process can be best summarised as strategy-to-portfolio, i.e., converting an abstract business strategy into more specific suggestions regarding the desired IT investment portfolio." As you will recall, strategy-to-portfolio is the first section of the primary value chain as shown on the IT4IT reference framework covered in Chapter 11.

Clarify the Business Needs

The first step is to reaffirm the business areas that need focus and investment based on the guidance of the IT strategic actions. The business strategy requires changes to or the creation of value streams, customer journeys, and or business capabilities to achieve the strategic goals. These business needs are the basis of the IT tactical plan along with any operational essential needs and the higher-level direction from the IT strategy. As shown in Figure 12.6, this information can be captured in the form of a heat-mapped business capability model to build consensus and alignment on the areas the business deems critical to success.

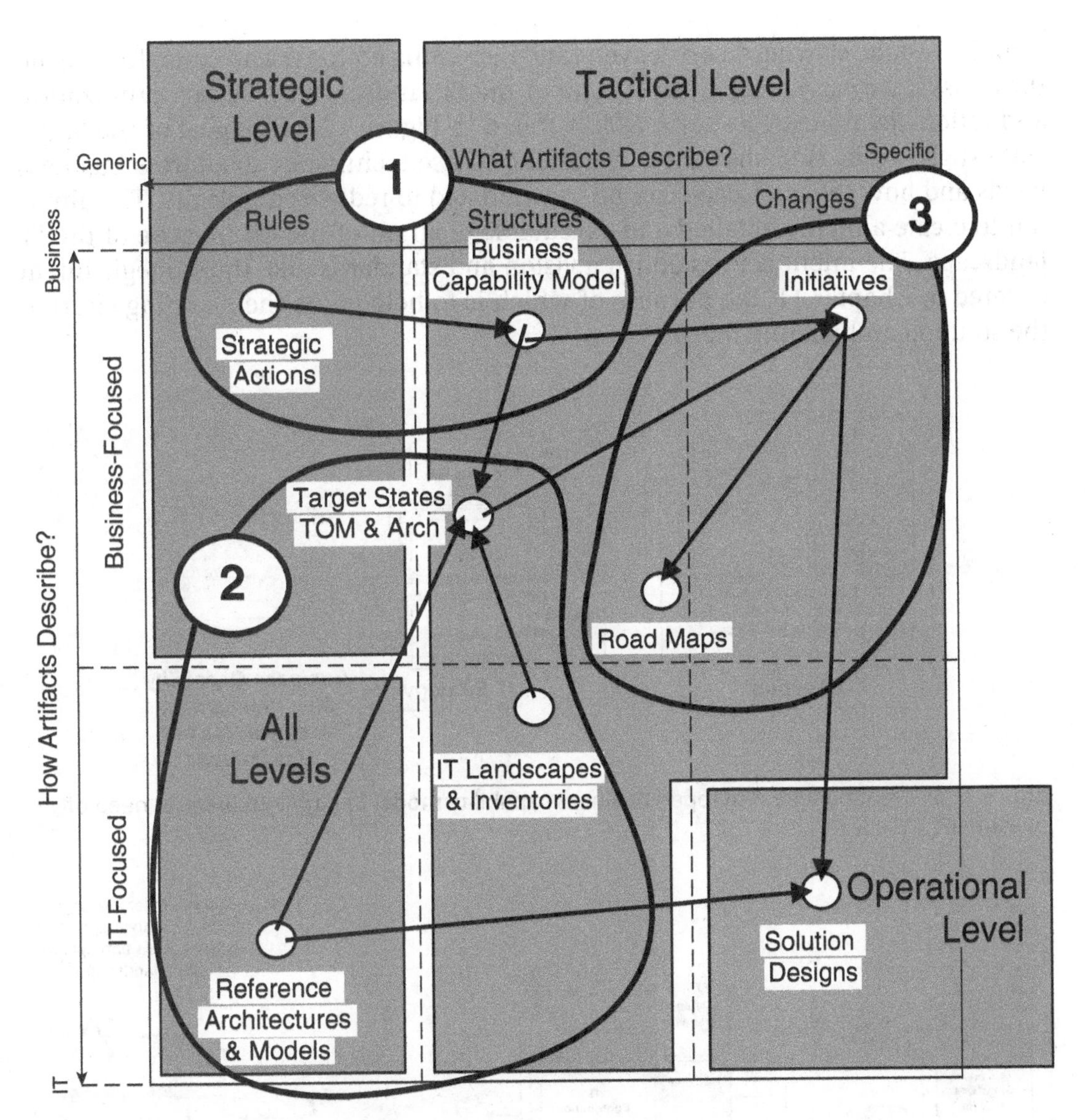

Figure 12.5: The three main processes of tactical planning

Review the Technology Landscape

Before we dive straight into solutions to address the gaps of business, we need to review the IT landscape. Without due care and attention, the constant development and evolution of the IT landscape can become an intangible mess, complicating the technical estate, slowing down delivery, and increasing both risk and costs. To mitigate these risks, we can create a target state of the IT landscape based on optimization and rationalization suggestions, as illustrated in Figure 12.7. The goal of the target state suggestions is to show how we intend to use technology to address business needs and how the landscape can be consolidated to reduce complexity. The aim is not to create a highly detailed and executable diagram of the future state of the IT landscape. The intent is to establish a vision and standards, like the strategic intent covered in Chapter 11, the purpose of which is to help guide the planning effort at the solution architecture level.

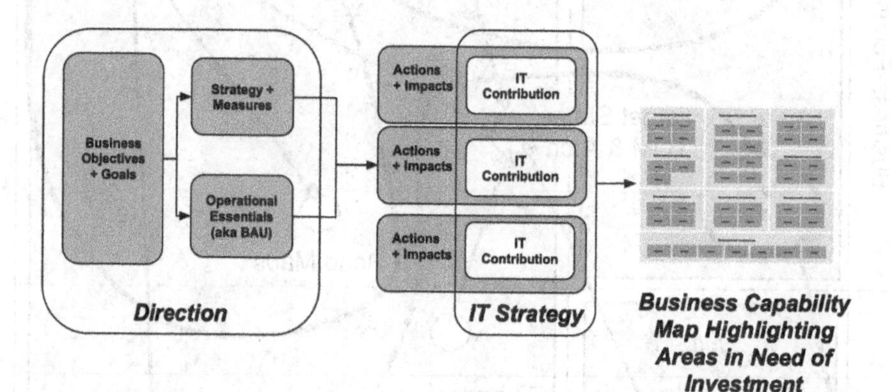

Figure 12.6: Creating a heat-mapped business capability model to highlight areas in need of investment

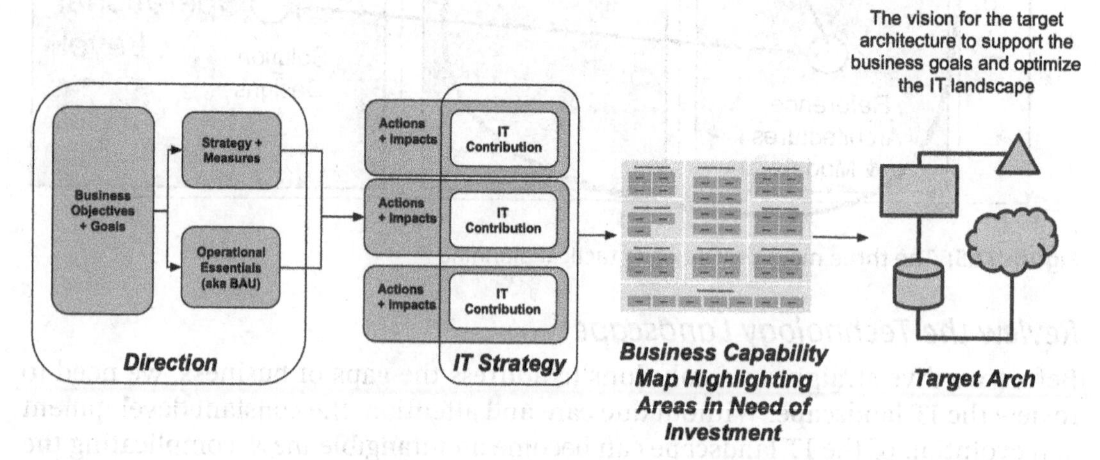

Figure 12.7: Define the target architecture to address business capability gaps.

Review the IT Operating Model

Once the target architecture is defined, we can then understand the high-level impact to the operating model, as illustrated in Figure 12.8. Do we need to change our ways of working? Do we need to partner with a third party? Do we need to hire for new roles or train our existing teams? Do we need to create new product teams? How does the architecture choices affect the team structure? How does it affect funding and measurement? Just as there is a need to understand the end state for the technical architecture, we also need to understand it for the operating model and how we will do the work to create it.

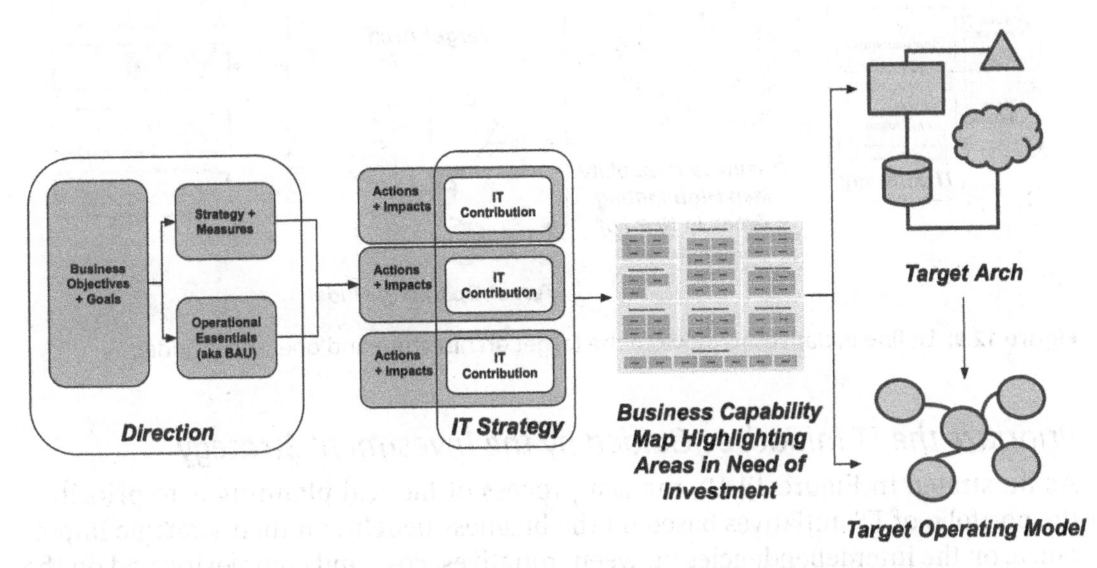

Figure 12.8: Define the target operating model to support bridging technical gaps.

Identify the IT initiatives Needed to Address Business Needs

With clarity on the business needs, along with the vision of the high-level ideas of how to meet them and optimize the IT landscape, we can begin to explore IT initiatives as illustrated in Figure 12.9. The IT initiatives will define a high-level solution architecture and rough estimates on the relative effort, cost, and benefit in order to help with prioritizing and funding. The IT initiatives will address the gaps and shortcoming of the technical contribution of business capabilities. They can be small projects such as evolutions to existing systems, or they can be large programs of work consisting of multiple projects introducing new products and services or transforming large parts of the business. At this stage, we are explicitly talking about high-level tactical initiatives rather than low-level operational projects. The tactical initiatives bridge the gap between the abstract strategic actions and the low-level operational projects and programs, ensuring there is a link between operation action and strategic intent.

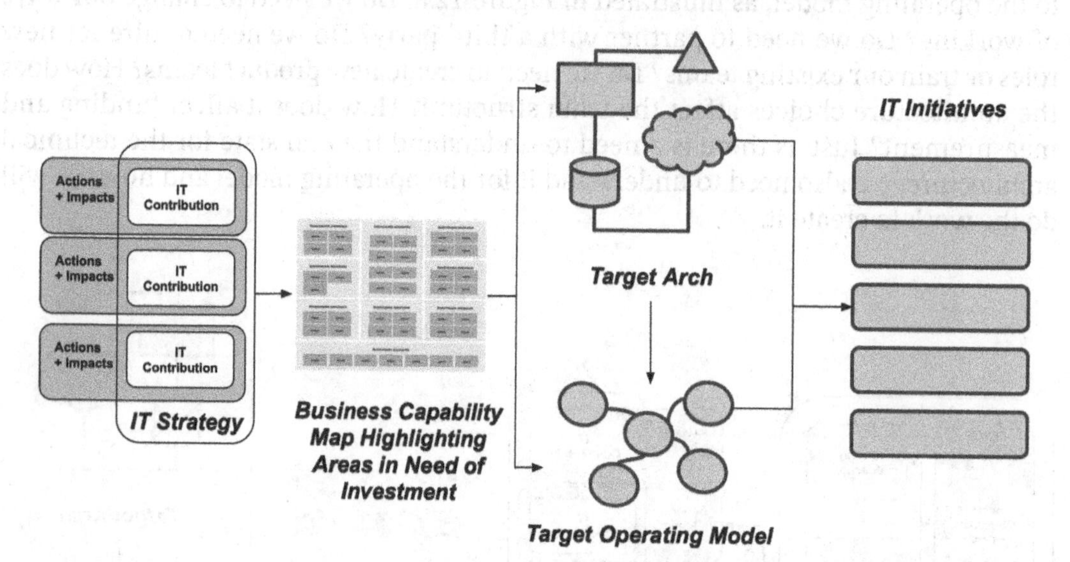

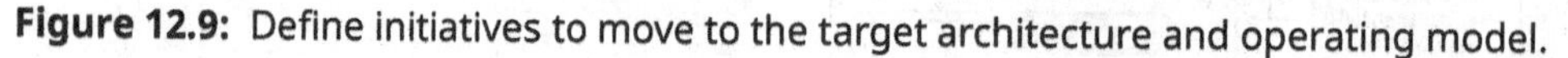

Figure 12.9: Define initiatives to move to the target architecture and operating model.

Prioritize the IT Initiatives Guided by the Investment Strategy

As illustrated in Figure 12.10, the last process of tactical planning is to prioritize the portfolio of IT initiatives based on the business benefit; on their strategic importance; on the interdependencies between initiatives, cost, and complexity; and on the investment strategy. Once prioritized and team capacity funded, the initiatives can be put into a road map. This is often a process that requires consensus, alignment, and, dare I say, a little bit of politics to navigate between the various functions of the business because there is often more demand than IT has the capacity or funds to supply.

Output

The output of the tactical planning process is a document detailing the road map of initiatives for an agreed period, typically at least the next 12 months, along with an investment plan. As shown in Figure 12.11, the document will also include the supporting evidence as to why the initiatives and their phasing has been chosen.

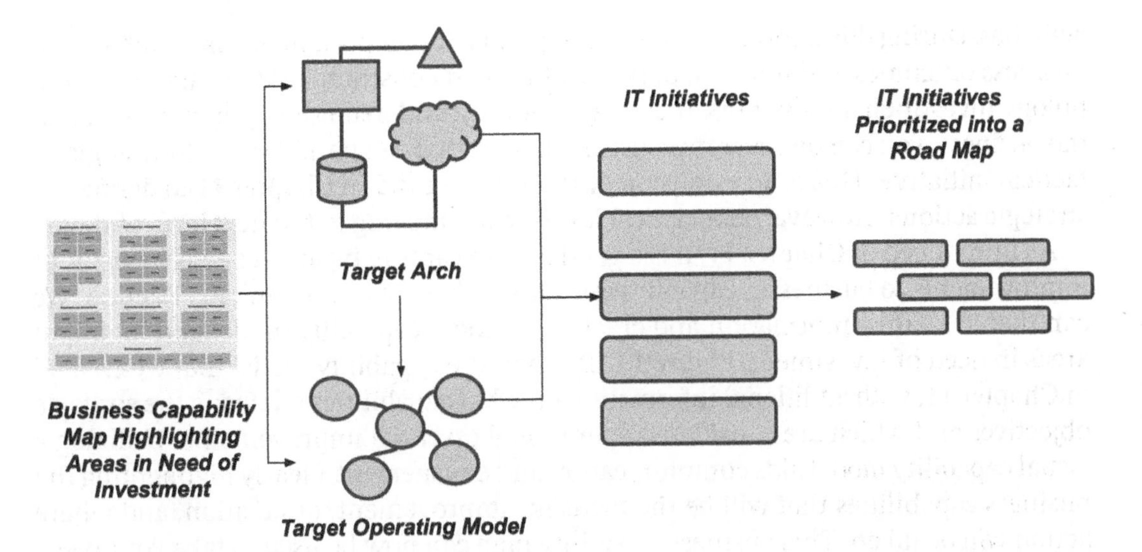

Figure 12.10: Prioritize initiatives into a tactical road map.

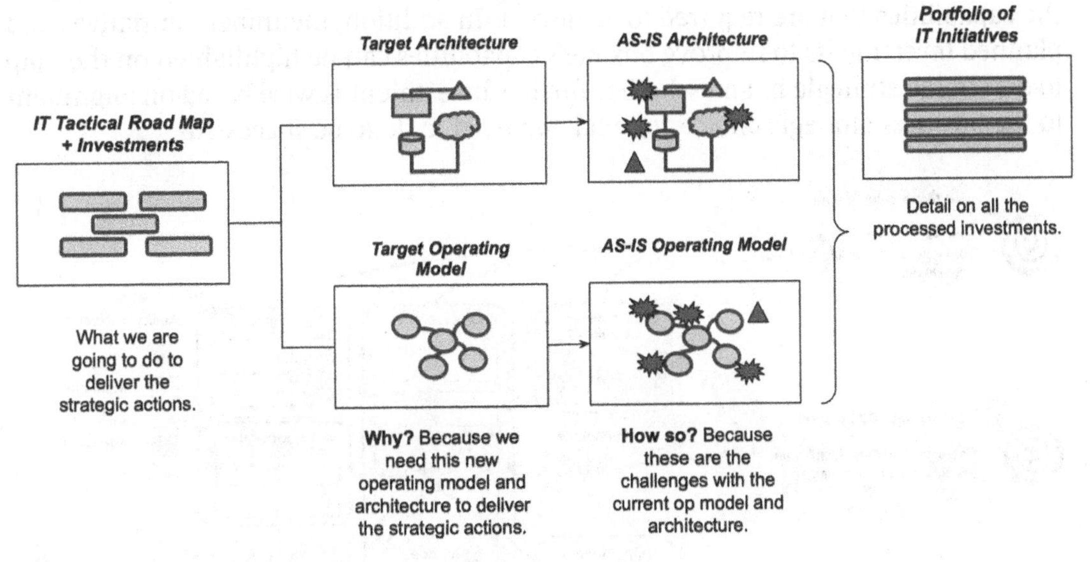

Figure 12.11: The output of the tactical planning process

Clarify the Business Needs: Where Do We Need to Focus Our Investment?

IT tactical planning begins with a consideration of the strategies and operational essential needs of the rest of the organization. In many cases, IT leaders take the initiative, often by necessity, in engaging the rest of the organization in planning activities. During this information gathering period, it is important to talk about desired business outcomes and impacts, opportunities, and constraints rather than the technology investments they think they may need. By understanding plans from across the business, you can discover reoccurring themes that can be addressed by a single IT tactical initiative. This is an extension of the work we did in Chapter 11 to define the strategic actions. However, at a tactical level we are moving to the next level of detail.

As introduced in Chapter 11, it is a good idea to capture business needs in terms of improvements to businesses capabilities as this makes it easier to link to IT action. We can then take this information and create a business capability model and highlight areas in need of investment. Figure 12.12 shows the capability model that we created in Chapter 11, with additional information on what capabilities related to the strategic objectives and which are considered operational essential improvements. Utilizing a visual capability model aids communication and engagement, clearly highlighting the business capabilities that will be the focus for improvement, or creation, and where action will be taken. This business capability map can now be used to take your peers and teams through where you intend to focus and how. This exercise provides the link between the areas of IT strategic action and the organization's strategy via the glue of the capabilities that are required to achieve it. In addition, incumbent initiatives and planned investments to improve business capabilities can be highlighted on the map to determine strategic fit and whether further investment is wise based on alignment to the business strategic choices of what we need to do to be successful.

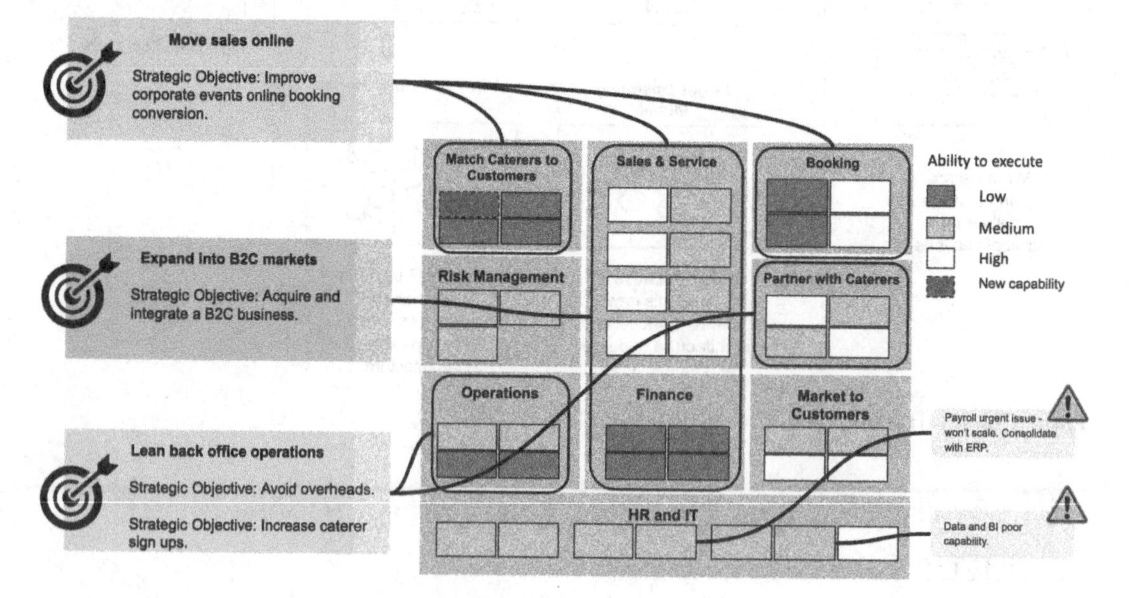

Figure 12.12: A business capability model highlighted to show areas in need of investment

Review the Technology Landscape: What Do We Need to Optimize?

The target architecture is the master vision for the technology landscape, like a city plan. It's the choices on *what* technology we intend to put in place to deliver IT actions needed to improve or create business capabilities. It's important to note that we are not looking to create a detailed technical landscape. At the strategic level, we are concerned about conceptual accuracy rather than architectural precision. Remember, the purpose of the target architecture is to guide tactical planning. It is not a detailed blueprint that can be given to a team to execute.

Although it is high level, it provides our best hypothesis on what IT solutions, platforms, and infrastructure is required to meet the business capabilities. It will be used to influence the choices of initiative solutions and allow us to take a holistic view of the IT landscape to optimize and rationalize our technical estate. The target architecture can support both the intentional architecture and emergent design as it is in the form of both standards and principles as well as explicit target architectural suggestions. As shown in Figure 12.13, the target architecture can cover all elements of the landscape, from the application to data, infrastructure, and security architectures.

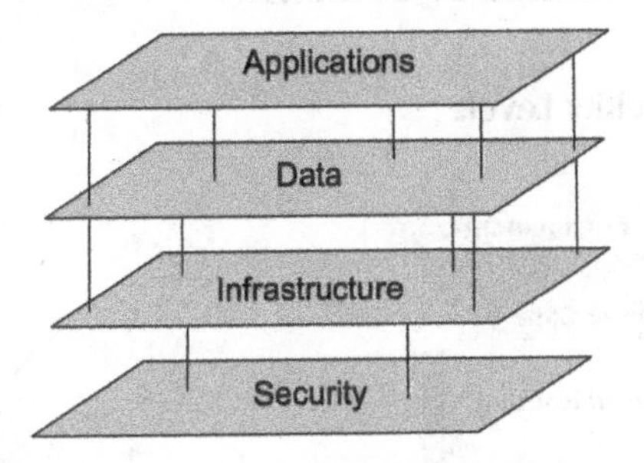

Figure 12.13: Technical architecture layers to review

The process of creating a target state begins with an examination of the existing structure of the IT landscape and the identification of the applications and technology that are supporting the capabilities in need of investment. We then analyze the estate and ask how well the technology supports the business; what the technical condition is; whether there is duplication, complexity, and risk; or if there are cost optimizations. From this fact-finding mission we can then analyze the estate to determine if we should replace or tolerate, invest or remove. To help with addressing technical gaps, we can categorize the evolution and strategic importance of each capability. This enables us to understand where it makes sense to build and where to buy. To ensure we design a target state in the optimal way and provide guidance to teams, we can utilize reference architectures and industry best practice information to guide the target technical architecture from languages to platforms, integration strategies, and vendor applications solution references. The result of this process is several target state suggestions that can be used to inform and guide the solutions for the initiatives.

Capturing and Assessing the Application Landscape

To understand how existing technology is used to support the business, we can map applications to capabilities and processes. This understanding and analysis provides the information needed to optimize the IT estate and address business needs. Figure 12.14 shows how a business capability can be decomposed into sub-capabilities and processes; this can then be mapped to an application. The level of granularity required will depend on the area you are mapping and how complex the capability is. Remember, we are not looking for architectural precision here; we are looking to understand the relationship between capabilities and technology to make strategic decisions. There is unlikely to be a 1-2-1 mapping between your applications and business capabilities. You will find that some applications support many business capabilities, and some business capabilities aren't supported by systems at all, and instead the only system is a spreadsheet. Mapping the IT systems to the business capabilities that they support helps to bridge the gap between the business and IT domains.

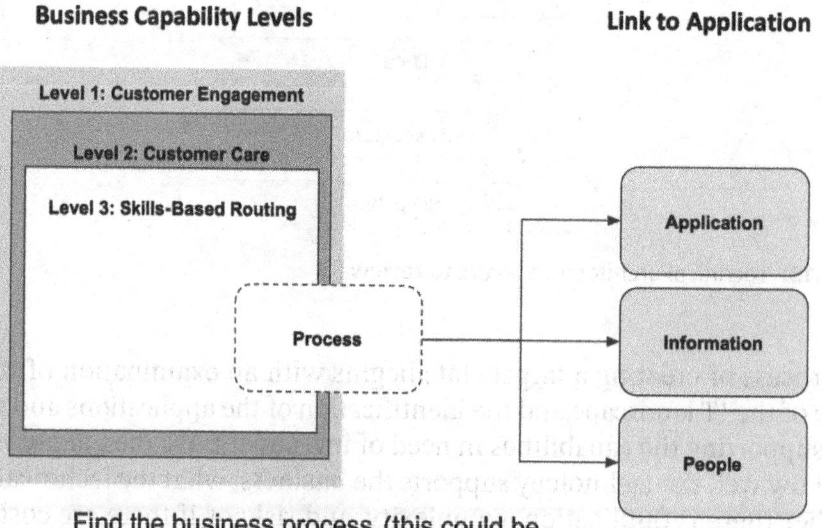

Figure 12.14: Mapping applications to business capabilities

We can capture and assess the landscape using two complementary techniques: technical diagrams and inventories. The combination of these two methods will allow us to understand the IT landscape and answer the following questions:

- What are the applications, databases, and infrastructure that support business capabilities?
- How are the technical components connected, how does data flow, and where are the dependencies?
- What is the technical condition of each component, how much does it cost, and how well does it support the business area?

Technical drawings, as shown in Figure 12.15, visualize how the IT components interact to support various business capabilities and processes. This can help to highlight areas of complexity and duplication. There is no right or wrong to document the landscape; free-form diagrams are just as effective as more formal notation. I personally favor Simon Brown's C4 model, but Archimate is also very popular among enterprise architects. It's okay to document the landscape progressively—for example, focusing on the technical architecture that is related to the business areas in need of investment or areas that are known to be problematic. Ideally, we want to map everything in to have a holistic end-to-end view, but this may not always be possible in a large estate.

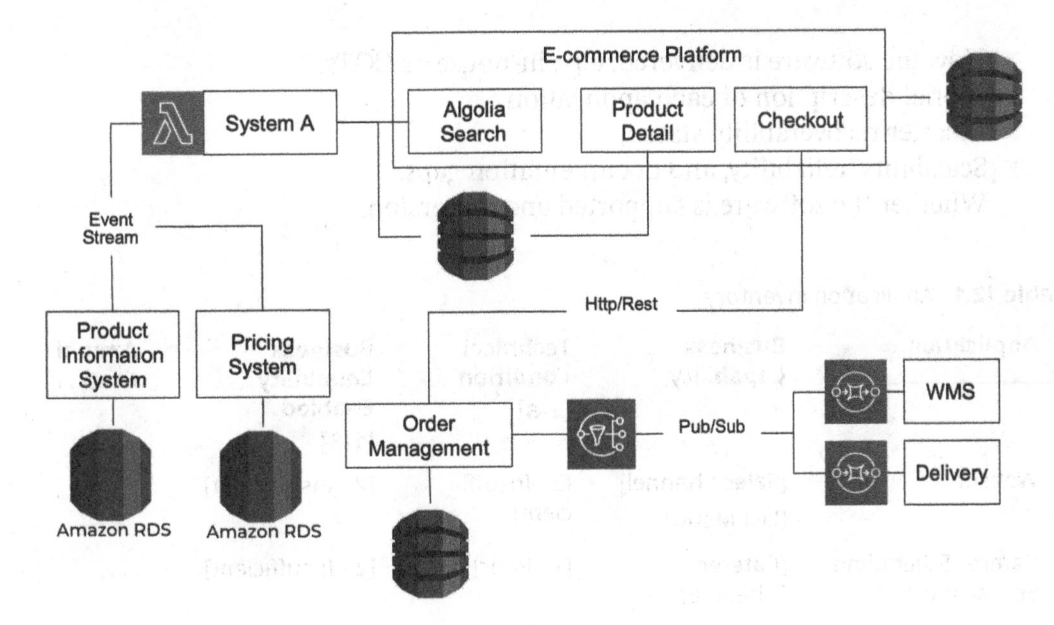

Figure 12.15: Example of architecture diagram

Application inventories focus on the properties of technical components. This information is typically captured in a spreadsheet, as shown in Table 12.1. Once the applications have been captured and mapped to business capabilities, we must evaluate them to determine if they are fit for purpose. We can evaluate and score applications on the following attributes:

Technical Condition: An assessment of the technical condition and system quality of the application. How expandable is the system and how easy is it to modify? Is it a big ball of mud or is it well architected? Does it use modern frameworks and languages? Is it able to scale to meet business demands? What are the vulnerabilities of the application? Are there challenges with regulatory compliance, data sensitivity, security, supportability, capacity, and availability constraints or IT skills availability?

Business Capability Enabled: How effective is the application at supporting the business capability?

Annual Cost: This represents the running cost of the application for a year. This should include hosting, any licenses and the man hours spent on maintenance, and break/fix or enhancements.

Other columns could include,:

- How the software is delivered, e.g., in-house or COTs.
- A brief description of each application.
- Disaster recoverability status.
- Scalability, reliability, and documentation gaps.
- Whether the software is supported and its version.

Table 12.1: Application inventory

Application	Business Capability	Technical Condition [1–5]	Business Capability Enabled [1–5]	Annual Cost...
Website	[Sales Channel] [List Menu]	[2 - Insufficient]	[2 - Insufficient]	...
Caterer Scheduling Spreadsheet	[Caterer Schedule]	[1 - Poor]	[2 - Insufficient]	...

Application	Business Capability	Technical Condition [1-5]	Business Capability Enabled [1-5]	Annual Cost...
CatererQuoter	[Manage Quotes] [Customer Engagement]	[1 - Poor]	[3 - Adequate]	
Caterer Sign-Up Spreadsheet	[Customer Registration]	[1 - Poor]	[2 - Insufficient]	...
Oracle Netsuite	[Finance]	[5 - Perfect]	[4 - Appropriate]	...
...	...	...	...	...

Once the landscape has been captured through diagrams and inventories, we can analyze the information to answer these questions:

- Which IT components are in a good technical condition?
- How well do applications support business capability?
- Where do we need to invest and improve?
- Which IT applications are legacy and should be replaced or modernized?
- What duplicate technology can be consolidated?
- What systems should we tolerate?
- Where do we need to reduce costs?
- What do we need to retire and decommission?
- Where is the risk?

As shown in Figure 12.16, the result of the application analysis can be mapped on a matrix. On the y-axis we have technical conditions from [1 - Poor] to [5 - Perfect]. The x-axis is business value. The size of the circle represents the estimated total cost of ownership. This can then provide life cycle advice for each of them based on the data points, and therefore determine the strategy of investment:

- **Tolerate/Reevaluate:** Tolerate applications that have a good technical value but a low business value. There is little to gain for investing here unless there is a strong business case.
- **Invest/Maintain:** For applications that are strategically important to the business and have a high technical value, we can continue to invest and maintain the levels of performance.

- **Replace/Modernize:** Replace applications that have high business value but have low technical value. Focus on reducing cost and risk.
- **Retire/Consolidate:** Eliminate and retire applications that no longer provide sufficient business value and that have a low technical value. These applications could be causing pain so they may need to be prioritized.

Table 12.2 also shows the result of the analysis against the application inventory.

Table 12.2: Application Analysis.

Application	Technical Condition [1-5]	Business Capability Enabled [1-5]	Application Analysis
Website	[2 - Insufficient]	[2 - Insufficient]	Replace/Modernize
Caterer Scheduling Spreadsheet	[1 - Poor]	[2 - Insufficient]	Replace/Modernize
CatererQuoter	[1 - Poor]	[3 - Adequate]	Replace/Modernize
Caterer Sign-Up Spreadsheet	[1 - Poor]	[2 - Insufficient]	Replace/Modernize
Oracle Netsuite	[5 - Perfect]	[4 - Appropriate]	Invest/Maintain
...	...	...	...

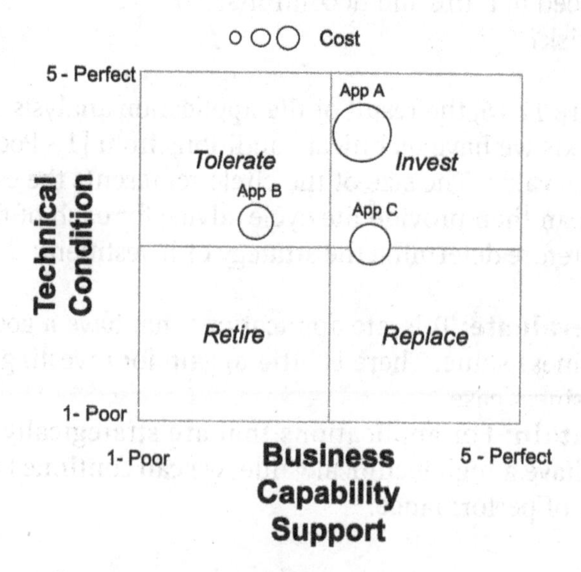

Figure 12.16: Plotting applications on a matrix to determine life cycle decisions

This process gives us the information required to understand how to optimize and consolidate the landscape. If required, we can overlay the results onto the business capability. This important link helps to visualize how well systems contribute to the maturity of a capability. The map acts as a heat map, showing where technical investment is needed; it can also clearly highlight where there are gaps altogether, something that can be missed if we were to map applications first rather than business capabilities first. This illustrates why the process of business capability modeling is so important.

Reviewing the Security, Data, and Infrastructure Layers

In the previous section we investigated in detail about how to gather facts and make decisions at the application level. In much the same way, we also need to capture data and review the security, data, and infrastructure layers. In addition to the data, it is helpful to gather information on data flows and the interfaces between applications. Table 12.3 lists approaches for gathering this information and how to analyze it.

Table 12.3: Approaches to review security, data, infrastructure, and interfaces

Technical Layer	Approach
Security	■ Review the security posture against a framework such as CIS (Center for Internet Security) or ISO 27001. ■ Review any compliance needs such as GDPR or PCI.
Data	■ Capture where data is stored. Is it backed up, who has access to it, does it contain personal or sensitive information? ■ Capture information on data quality and governance.
Infrastructure	■ Catalog each infrastructure component (databases, servers, networks), including both on-premises and in the cloud. Ask will the infrastructure scale, are systems backed up, are systems patched, etc. ■ Diagram to reveal complexity and duplication.
Interfaces	■ Catalog each interface between applications (on-premises and in the cloud). ■ Diagram how data moves across the business to reveal complexity, duplication, and other issues.

How Do We Fill the Gaps: Should We Build or Buy?

Now that we have a good handle on what capabilities require investment, we need to consider any strategic design principles to guide our solution choices. To help us understand whether to build or buy, we can analyze the business capability model once more. Not all capabilities are of equal importance in terms of difference between your organization and your competitors. This is important because you need to understand where a capability needs to be good and where it needs to be good enough; that is, equal to that of a competitor. Broadly speaking, capabilities can be separated into three buckets: competitive capabilities, supporting capabilities, and commodity capabilities.

- **Competitive** (aka Defining / Innovate / Core): These capabilities define what is unique in your organization and thus represent your competitive advantage. These capabilities are core to your success and can differentiate you from others in your market.
- **Supporting** (aka Shared): While these capabilities do not define what your organization does, they support the delivery of what your competitive capabilities do. The capability to market to customers is a good example of a supporting capability. Many organizations have this capability, but it does not define you. However, if your organization is defined by communication and targeted e-mails on limited time offers, like Groupon or Wowcher, the capability to market may be core and need to be sophisticated to set you apart from the pack. What is core to one business may well be considered supporting to another.
- **Commodity** (aka Enabling / Parity / Generic): These are the capabilities that you would expect to find in many organizations. They do not define a business, but the business can't operate without them. Because these capabilities aren't core, having world-class maturity here won't gain any competitive edge. These capabilities need to be good enough and thus not candidates for investment and focus. HR, financial accounting, and payroll capabilities are typical commodity capabilities.

The categorization of capabilities can then be mapped to our model as shown in Figure 12.17.

Figure 12.18 shows a broad high-level mapping of which IT method of closing the gap is most appropriate for each capability depending on its criticality to the organization. Broadly speaking, we should look to develop competitive capabilities in-house because by their very nature, they need to be differentiated from the competitive set; buying an off-the-shelf solution would allow others to gain capability parity. That is not to say we can't leverage off-the-shelf components in delivering our competitive capabilities. We don't have to reinvent the wheel; we can build upon it. Typically, for supporting capabilities we can look for leading software and utilize best practice patterns. For commodity capabilities we really need to align to industry best practices and get something vanilla out of the box without customizing or even outsourcing this capability altogether.

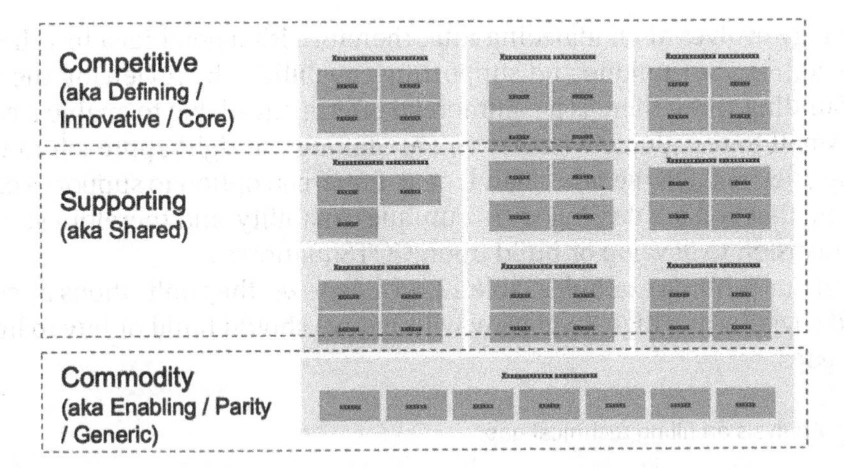

Figure 12.17: Business capability categories

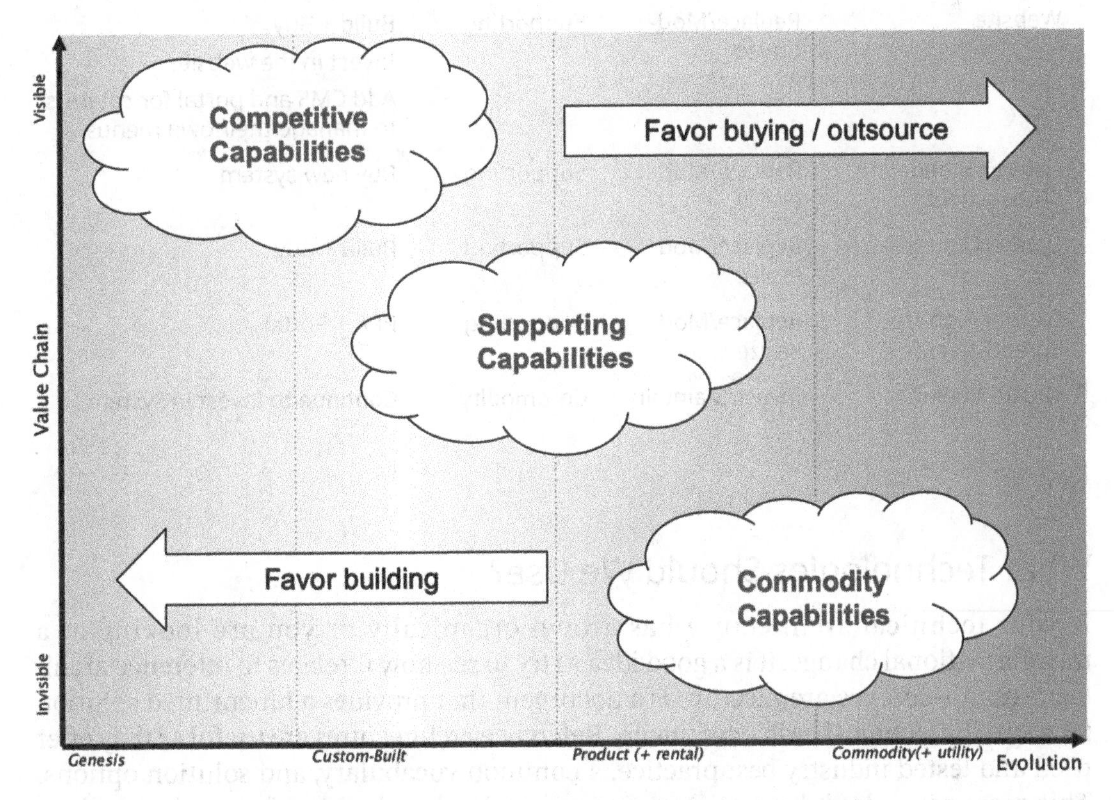

Figure 12.18: Choosing a method when bridging technical gaps based on the capability grouping

Technology evolves at an alarming rate; therefore it's a good idea to validate how you are enabling your unique and supporting capabilities by reviewing the software market. Building when there is a commercial offering available to enable a capability is a code smell to reevaluate whether you are taking the right approach to the build versus buy question. On the other hand, no commercial option to support a capability can indicate that you are dealing with a unique capability, and therefore you can validate the decision to develop or build upon COTS platforms.

Based on our running example, Table 12.4, categorizes the applications by capability group and suggests at a high-level view whether we should build or buy to bridge the technical gaps.

Table 12.4: Analysis on filling technical gaps

Application	Application Analysis	Capability Group	Suggestion on How to Bridge Gap
Website	Replace/Modernize	Supporting	Build + Buy Invest in the website Add CMS and portal for caterers to manage their own menus
Caterer Scheduling Spreadsheet	Replace/Modernize	Supporting	Buy new system
CatererQuoter	Replace/Modernize	Supporting	Build + Buy
Caterer Sign-Up Spreadsheet	Replace/Modernize	Supporting	RPA + Portal
Oracle Netsuite	Invest/Maintain	Commodity	Continue to invest in system
. . .	. . .	. . .	. . .

What Technologies Should We Use?

If your technical architecture has grown organically or you are looking at a transformational change, it is a good idea to try to see how it relates to reference architectures. A reference architecture is a document that provides a blueprinted solution for a specific technical problem domain. Reference architectures are useful as they offer tried and tested industry best practice, a common vocabulary, and solution options. This means you don't have to start from scratch when looking for a solution for a particular problem space. For example, the IT4IT framework as covered in Chapter 11 is an information reference architecture for a holistic view of IT capability; ITIL and

COBIT are examples of two others. Solution vendors will have reference architectures to show you how their product fits into a typical environment such as Oracle, Adobe, and SAP. AWS has many system structure reference architectures as part of its AWS Well-Architected framework, including a Security Services Architecture and DevOps pipeline. The reference architecture for a Modern data architecture is shown in Figure 12.19.

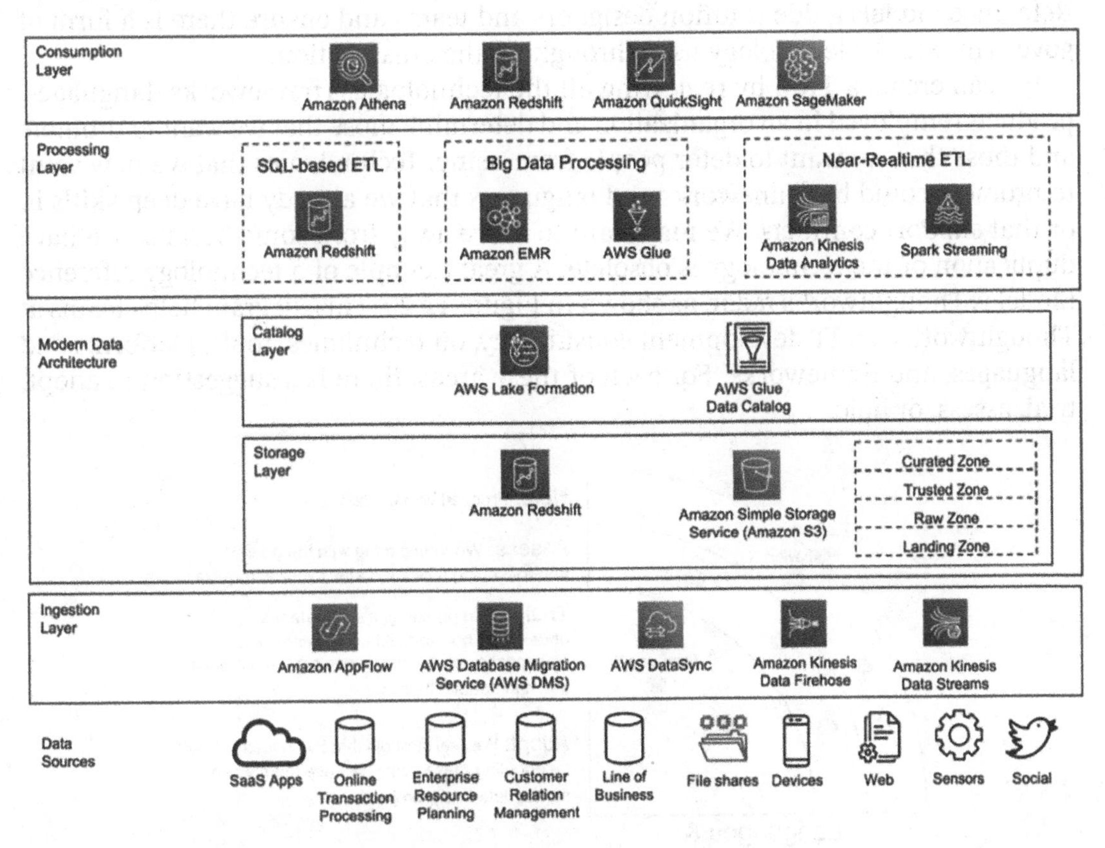

Figure 12.19: AWS Modern data architecture reference architecture

How Do We Guide Build Efforts and Emergent Design?

As previously mentioned, not all architecture can be defined up front. Some of the landscape will evolve, particularly complex problems, through an emergent design. Even with intentional architecture, we are only defining the high-level solution design, boundaries and interfaces. Teams will still need to determine the structure of the actual software architecture. To guide teams when creating the solution designs, we can utilize technology reference models (TRMs). Where a reference architecture can provide a template to a common problem domain based on best practice, reference models provide the recommended configuration of frameworks, languages, software, tools, and ways of working that are appropriate to your organization. In other words, reference architectures are organization-agnostic templates, whereas TRMs define the standards, principles, and lists of approved software specific to your organization. Reference models guide solution designers and teams and ensure there is a form of governance to the technology used throughout the organization.

We can create a TRM by reviewing all the technologies (frameworks, languages, products) employed in an organization and determine those that we want to promote and those that we want to deter people from using. Technologies that we may want to promote could be frameworks and languages that we already have deep skills in or that support contracts. We may want to move away from some because we have duplication or the technology is obsolete. A great example of a technology reference model is Thoughtworks radar, as shown in Figure 12.20. This diagram helps teams at Thoughtworks, an IT development consultancy, on techniques, tools, platforms and languages, and frameworks. For each of these areas, there is a suggestion to adopt, trial, assess, or hold.

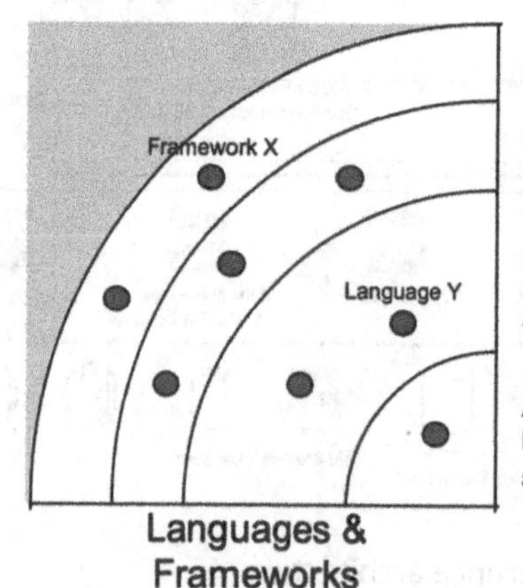
Framework X
Language Y
Languages &
Frameworks
Hold: Proceed with caution.
Assess: Worth exploring with the goal of understanding how it will affect your enterprise.
Trial: Worth pursuing. It's important to understand how to build up this capability. Enterprises can try this technology on a project that can handle the risk.
Adopt: We feel strongly that the industry should be adopting these items. We use them when appropriate in our projects.

Figure 12.20: Thoughtworks technical radar

The Target State Suggestions

As shown in Figure 12.21, to communicate the target architecture suggestions, we can provide high-level diagrams, annotated to show how, and importantly why, a change in the IT landscape is required to either address a business need or optimize the architecture. It's important to stress that we aren't looking for architectural accuracy; instead, this suggestion represents architectural intent. The target architecture suggestions are a communication and alignment device. The suggestions will be used to guide both tactical initiatives and operational solution designs. These suggestions are a form of architectural governance, helping to shape the IT estate toward something that is coherent and fit for purpose. In terms of actual notation, anything that is useful in communicating intent is effective. In my experience, this is typically in the form of slides or simple diagram tools rather than complex enterprise architecture platforms.

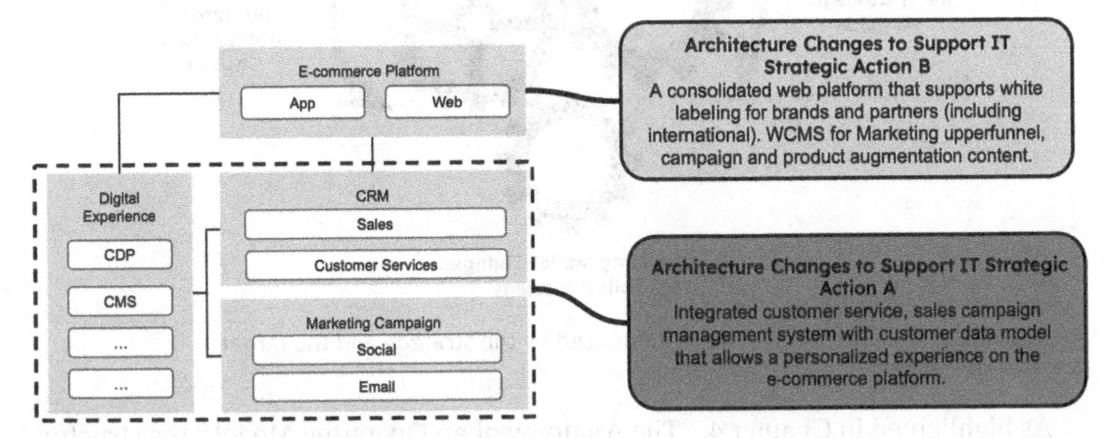

Figure 12.21: An example of a target state suggestion

Review the IT Operating Model: What Do We Need to Change?

As Alfred Chandler said, "Structure follows strategy." The operating model, covered in Part 2, "Designing an Adaptive Operating Model," is intrinsically linked to and exists only to deliver the strategy. Changing the operating model must be made in conjunction with the vision of the target architecture; without doing so, change will lead to an operating model inconsistent and ineffective with the strategy it is designed to deliver. As illustrated in Figure 12.22, the target architecture, influenced by the current landscape and strategic actions, has the biggest influence on the target operating model. Behaviors, ways of working, and the organizational structure, when misaligned with the strategic intent, will impact IT's ability to deliver the desired business outcomes. Like the strategy, the operating model design should not be viewed as something that is fixed or static once deployed. It will need to evolve to meet the changing needs of a digital business, to exploit technology advances, and to respond to other changes in the business context.

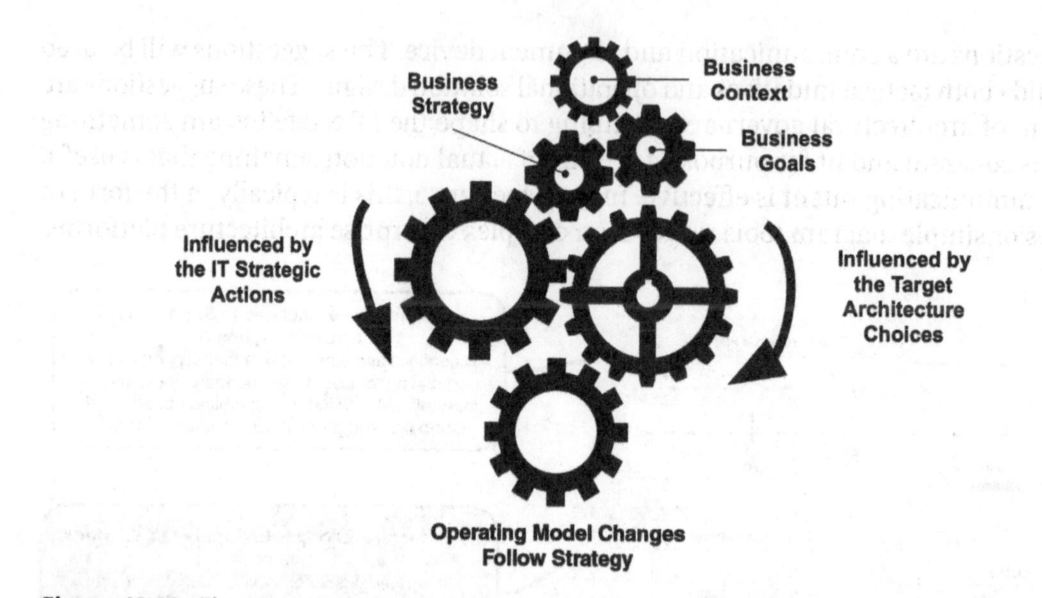

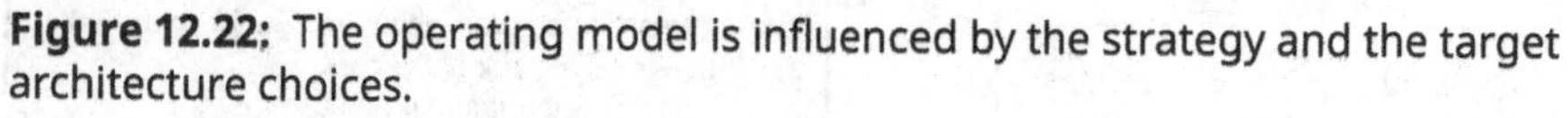

Figure 12.22: The operating model is influenced by the strategy and the target architecture choices.

As highlighted in Chapter 4, "The Anatomy of an Operating Model," the structure and organizational design of IT may be highly visible, but it is only a single component of an operating model. Simply imitating the organizational design of another company, such as Spotify, without making the necessary changes to the other integrated components will have little behavioral or cultural impact. We need to review the effectiveness of IT using a framework such as COBIT or IT4IT. This ensures that we gather facts about all areas of IT that may need to be improved to address strategic actions and related target architecture suggestions. Table 12.5 highlights examples of core IT capabilities that have a gap and how we might suggest addressing it.

Table 12.5: IT operating model capability gaps

IT Capability Gap (Risk, security, complexity, performance, cost)	Target Operating Model Suggestion
[Product Management] ***1.1 Provide full caterer's information on the web for customers.*** Teams currently deliver projects and are transient in nature. To support this strategic action and the customer digital journeys, we need to continually and iteratively improve customer's digital experience.	■ Hire an experienced product manager. ■ Change topology of the team to form a product team around the customer online journey. ■ Invest in product management skills and expertise.

<table>
<tr><th>IT Capability Gap
(Risk, security, complexity, performance, cost)</th><th>Target Operating Model Suggestion</th></tr>
<tr><td>[Application Development]
1.3 Automate the sign-up and approval process.
We don't have the capability in-house to support this strategic action. In-house development is limited and will take time to upskill in new technologies such as a RPA (robotic process automation).</td><td>■ As this area is a supporting capability, partner with a team to provide an augment team who has expertise in RPA so we can free up in-house devs to focus on areas of differentiation.
■ Hire a product manager to manage this area.</td></tr>
<tr><td>[Security Management]
■ We have no security strategy or plan.
■ We are too small, however, to warrant a full-time CISO.
■ No pen test.
■ No vulnerability.
■ No security strategy or plan.</td><td>■ Partner with a virtual CISO.
■ Align to a security framework to quantify security posture.
■ Admin to achieve compliance in all areas.</td></tr>
<tr><td>[Service Desk]
■ We will be unable to manage a 24/7 service desk when opening a new office in the US.
■ Supplying kits outside of the UK will be challenging.</td><td>■ Outsource services desk in the US.</td></tr>
<tr><td>[Infrastructure Management]
■ We don't have the skills to manage all the hardware.
■ We don't have any specific database. No skills.
■ The environment is not being managed.</td><td>■ Partner with expertise to move to the cloud and leverage managed services.
■ Invest in training and upskill team in cloud infrastructure.</td></tr>
<tr><td>. . .</td><td>. . .</td></tr>
</table>

How Wardley Maps Can Help Inform Target Architecture and Operating Model Choices

The target architecture diagrams presented earlier in the chapter are useful, but they have their limitations in terms of communication and clarity. If you take the example of a target application architecture diagram presented earlier, as shown in Figure 12.23, how easy is it to determine if a solution space should be built rather than bought? If we were to move the position of the solution spaces, does that tell us anything different? Where is the point of reference to the strategic outcome that we originally anchored the creation of the business capability to? How does it help us with understanding any changes required to the operating model? To answer these questions and to truly make strategic decisions you need position and movement, which is where Wardley maps excel.

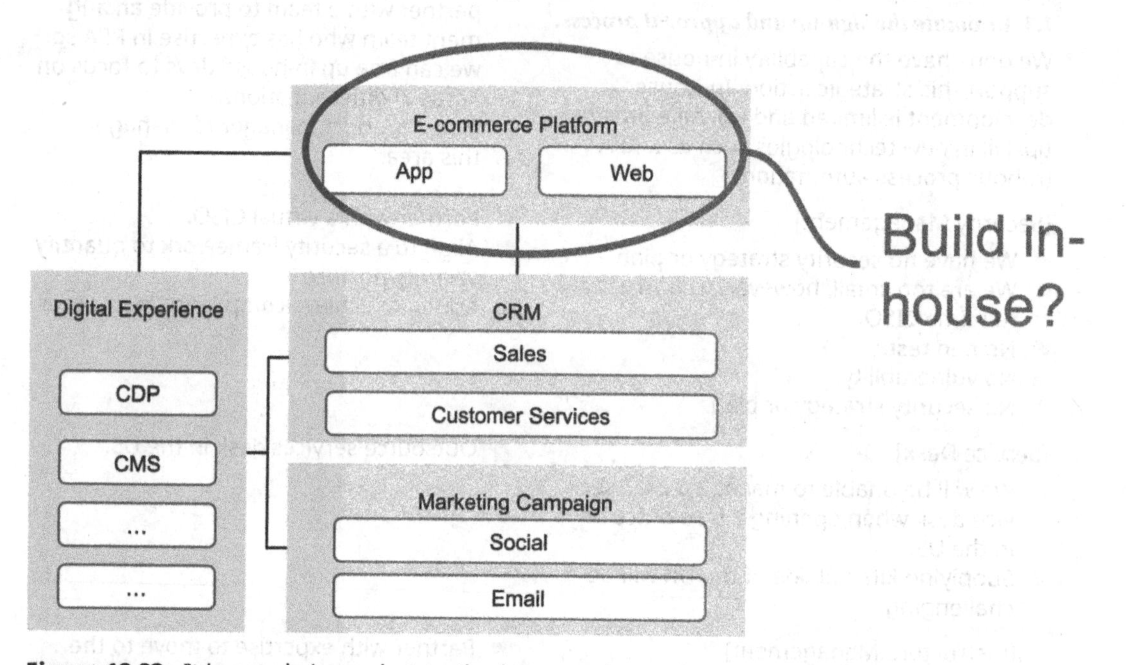

Figure 12.23: It is not obvious what method to use when using an application architecture diagram or the TOM implications.

If you compare the target architecture diagram to a Wardley map, as shown in Figure 12.24 and introduced in Chapter 2, we can have a much more informed conversation on how to bridge the gap in capability maturity using a Wardley map. The position of the customer relationship management (CRM) solution on the architecture diagram does not help us to understand the most appropriate method to support the capability. However, the position of the CRM component in the Wardley map reveals how evolved this component is. As CRM is highly evolved, we should initially look to see if we can leverage an off-the-shelf component to meet our needs rather than immediately trying to build something. Just as with the target architecture approach, we use Wardley maps to make a model that represents the facts of the strategic situation so we can make sense of it and make the most appropriate decisions.

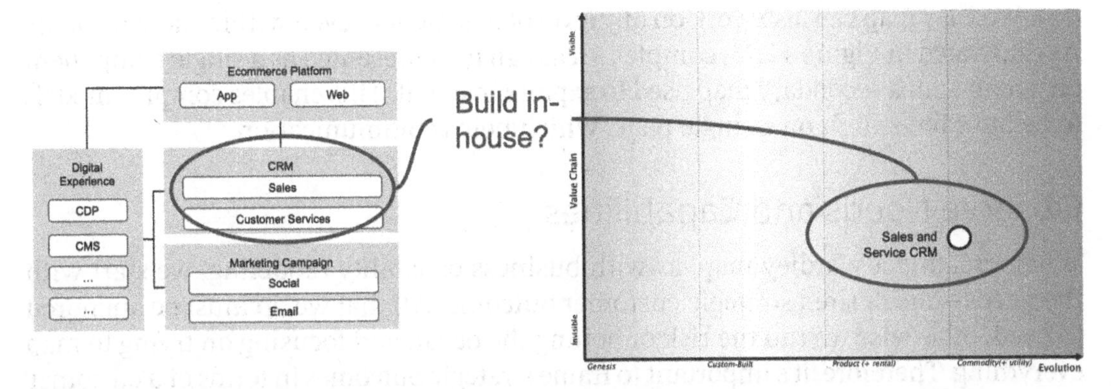

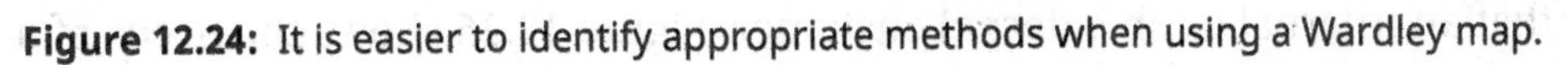
Figure 12.24: It is easier to identify appropriate methods when using a Wardley map.

As illustrated in Figure 12.25, because Wardley maps deal with the entire value chain, we can model all dependent components. This includes, but is not limited to, activities, people, data, knowledge, practice, and technology. This allows us to see links between customers and need (value) and capability and technology in a single view.

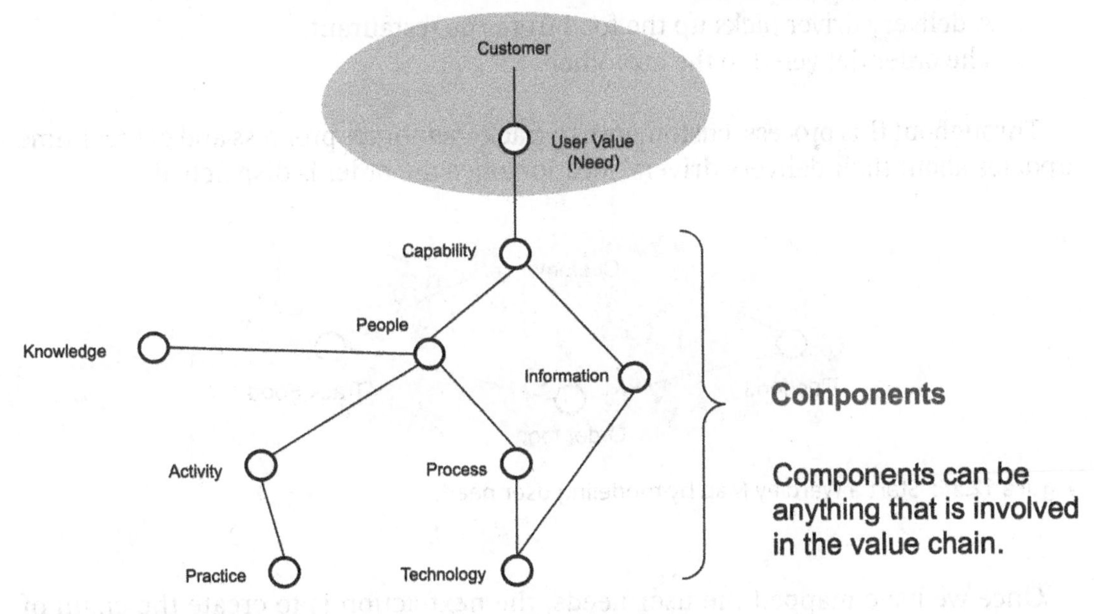

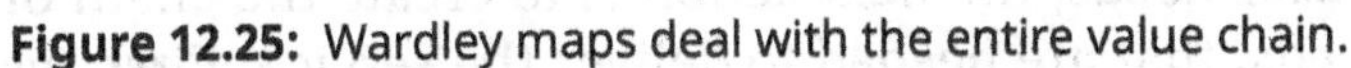
Figure 12.25: Wardley maps deal with the entire value chain.

A Wardley map can also work on any level of abstraction, even within the same map. As illustrated in Figure 12.25, complex areas can be represented as a single component on a map with a secondary map used to explore in detail. This enables complex models to be built but still fit on a single page, which helps communication.

Observe Needs and Capabilities

When creating a Wardley map, as with business capability modeling, we start with the needs—the desired strategic customer outcome. All that we do must be anchored to need, otherwise we run the risk of boiling the ocean and focusing on trying to map everything. Therefore, it's important to frame strategic outcomes in terms of a customer need, be that an internal or an external customer. Figure 12.26 shows the needs of a customer who wants to get food delivered similar to UberEats or JustEat.

The value stream steps are as follows:

1. A customer searches on the website to find food available in their area.
2. A customer places and pays for an order.
3. The restaurant accepts and prepares the order. This lets the customer know their order is being prepared.
4. A delivery driver picks up the food from the restaurant.
5. The order delivered to the customer.

Throughout this process, customers can track their order progress and get real-time updates about their delivery drivers' location once the order is dispatched.

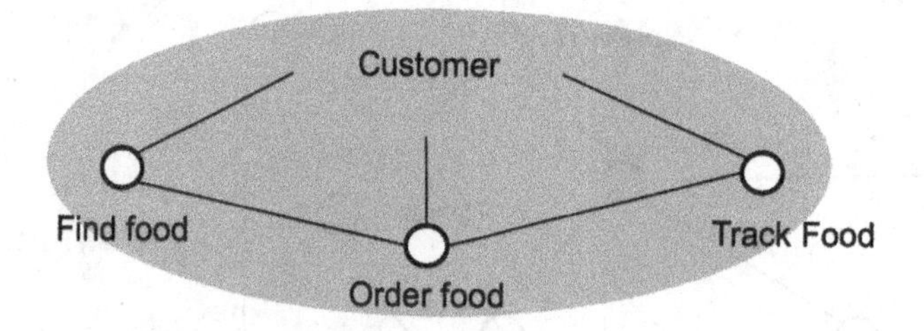

Figure 12.26: Start a Wardley Map by modeling user need.

Once we have mapped the user needs, the next action is to create the chain of dependencies, or capability needs, that are needed to deliver the service. First, we map the capability dependencies classified by the value each component has to the customer—in that the components closer, or more visible, to a customer are more valuable than those further away. Figure 12.27 shows the dependency chain to support the value stream. Unlike in the capability model, we can now see the dependencies between the components.

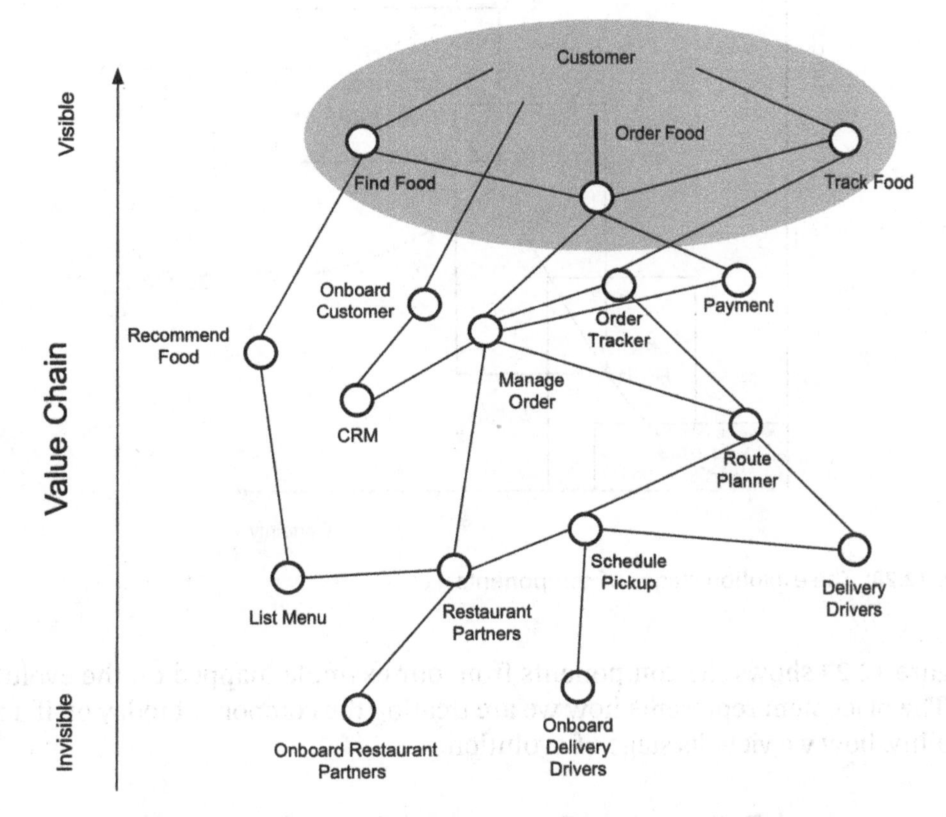

Figure 12.27: Mapping the dependencies between capabilities on a Wardley map

Next, we organize the capability dependencies by the maturity of each component. Just as in the capability modeling process discussed earlier, capabilities can range from the competitive to the commodity. To determine where a capability is on the evolutionary spectrum, we need to agree on how common the component is. How evolved is the capability? Is it unique to our business or do our competitors also have it, or is it ubiquitous in many businesses? Can we buy the capability, or better still, is it available as a service or must we build it? As shown in Figure 12.28, the y-axis represents ubiquity and the x-axis represents certainty; for example, commodity components like water and electricity have a high level of ubiquity and a high level of certainty in the provision of them. Components evolve through these stages over their lifetime driven by climatic patterns such as competition. The various characteristics of the evolution of a capability, which will enable you to map the position of that capability and understand its movement, was covered in Chapter 2.

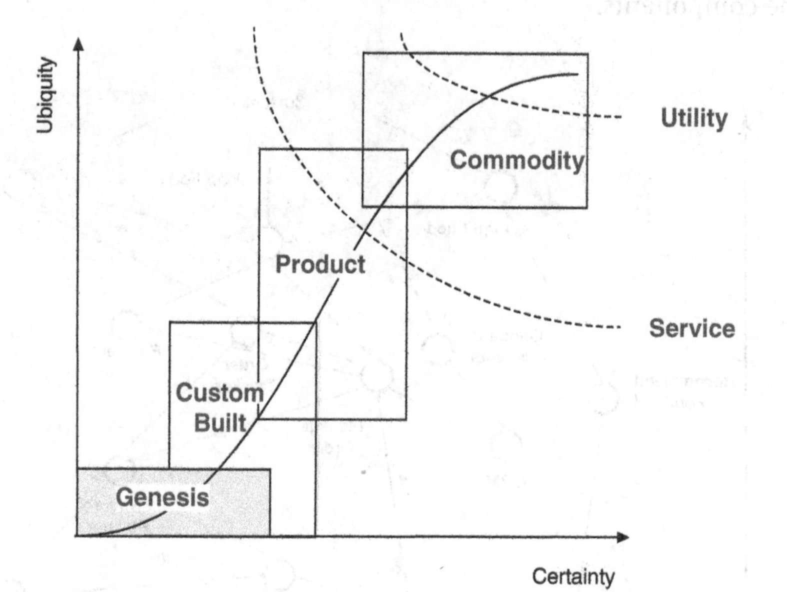

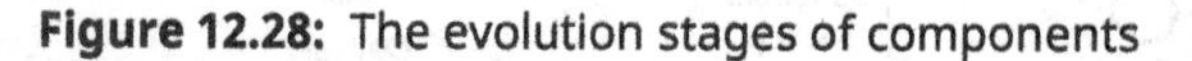
Figure 12.28: The evolution stages of components

Figure 12.29 shows the components from our example mapped on the evolution axis. The placement represents how we are treating the component today or, if a new capability, how we view its stage of evolution.

Orientate and Determine Focus and Direction

After we have repeated the process for each strategic outcome, we can then overlay the different maps to create an aggregated view, as shown in Figure 12.30.

We can then challenge the aggregated map and remove duplication, bias, and areas of inertia. The aggregated map addresses these issues in the following ways:

Inertia Maps can highlight capabilities that have evolved from custom build to commodity and force us to address them if we should continue to invest in a bespoke solution rather than aligning to an out-of-the box solution to reduce cost and allow effort to be focused on areas that are truly still in the genesis or custom build phase.

Duplication In large enterprises, there can be duplication in how a capability that is required by more than a single area is provided. By visualizing this, we can see where and how the same capability is delivered and move to consolidate.

Bias Each business area will deem to have unique needs and require a different solution to fulfil a capability need even though they may serve the same purpose. The maps visualize and challenge this bias to reduce complexity and ideally offer a capability that can accommodate all needs, or certainly discuss this as an option.

Communication Central to the Wardley maps is the visual nature that enables clear and effective communication. They can provide a significant amount of value within a single page map via a common language that is easy to understand for both technical and nontechnical people. By laying out our bias, duplication, and inertia to adapt to evolving capabilities, we can have an informed discussion about how to strategically invest to support business goals.

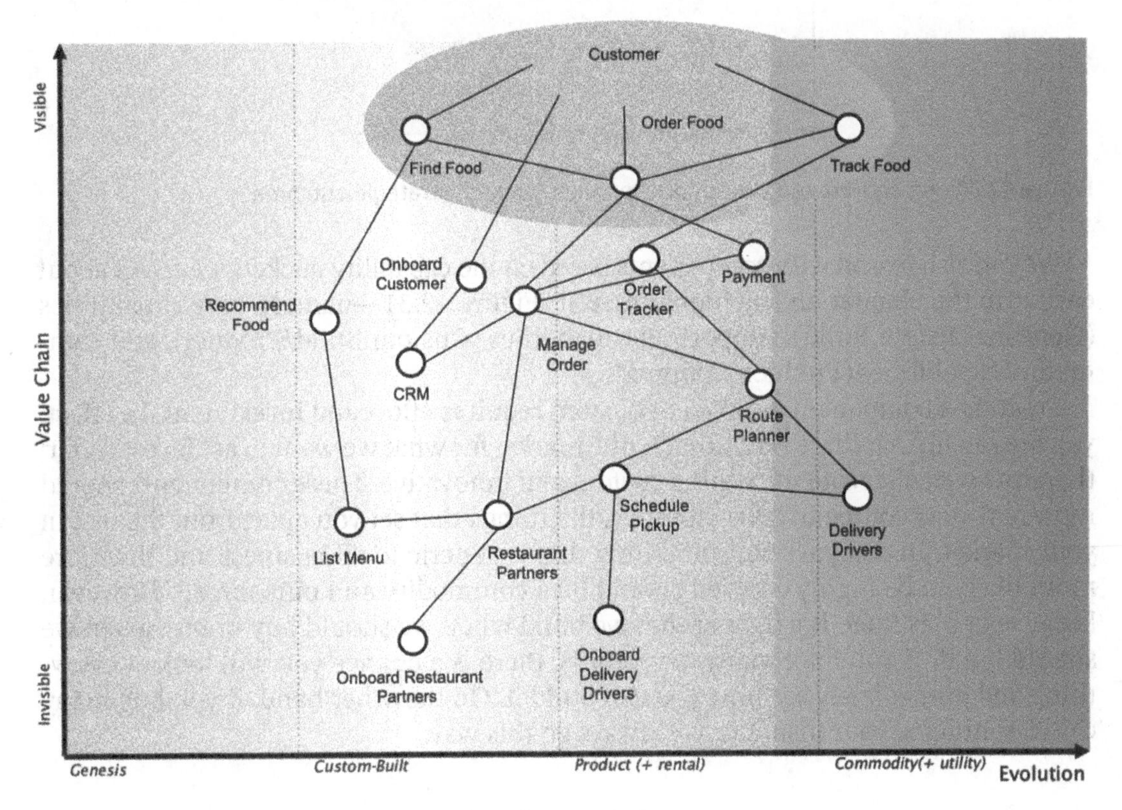

Figure 12.29: Placing capabilities based on their evolution

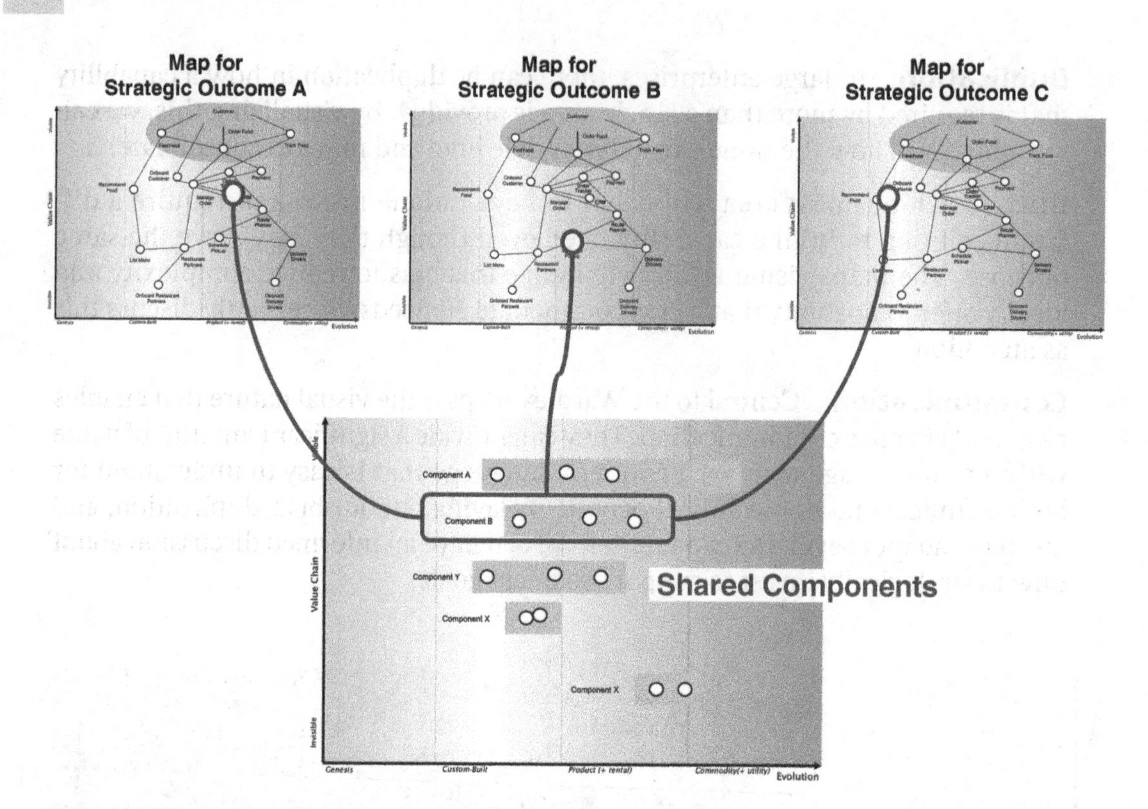

Figure 12.30: An aggregated view of all the maps for each strategic outcome

We can also reason with components based on the capability buckets we spoke about earlier in the chapter and as highlighted in Figure 12.31—namely, core capabilities (Genesis/Custom Build), supporting capabilities (Custom Build/Product), and commodity capabilities (Product/Commodity).

Treating a component as novel or custom requires effort and investment. In effect, you are saying that there is nothing in the market for what we want to achieve or offer; therefore we will create something custom and innovative. These components should represent your competitive capabilities, the things that set you apart from the rest in your market. In comparison, something that is generic to all business and therefore more likely to be highly evolved should be a commodity and outsourced. However, because of bias there is a danger that we build when we should buy or buy when we should build. If you have many developers, there is a chance you will look at everything and immediately see how you can build it. On the other hand, if you buy many COTS solutions, your instinct may always go this way.

Therefore, when working with your business peers when you are mapping business components, you need to challenge their placement and ask if this is the best way to deal with them? For example, you can ask the following questions:

- If a component is in the Genesis stage, ask, "Can this be achieved in a more standard way?" or "Who else does this?"
- If we are treating something as Custom, ask, "Should we be innovating here?" or "Is there a simpler way to do this?"
- If something is in the stage of Product, ask, "Can we negotiate our license?" or "Is there a newer version or a better version from another vendor?"
- Are we aware of what a commodity is in our microenvironment? When is the last time we attended a technical expo for our industry to understand how vendor offerings have evolved?
- How many components are we building? If we are building many, this can be expensive and suggests we are not focused on what is truly unique to us.

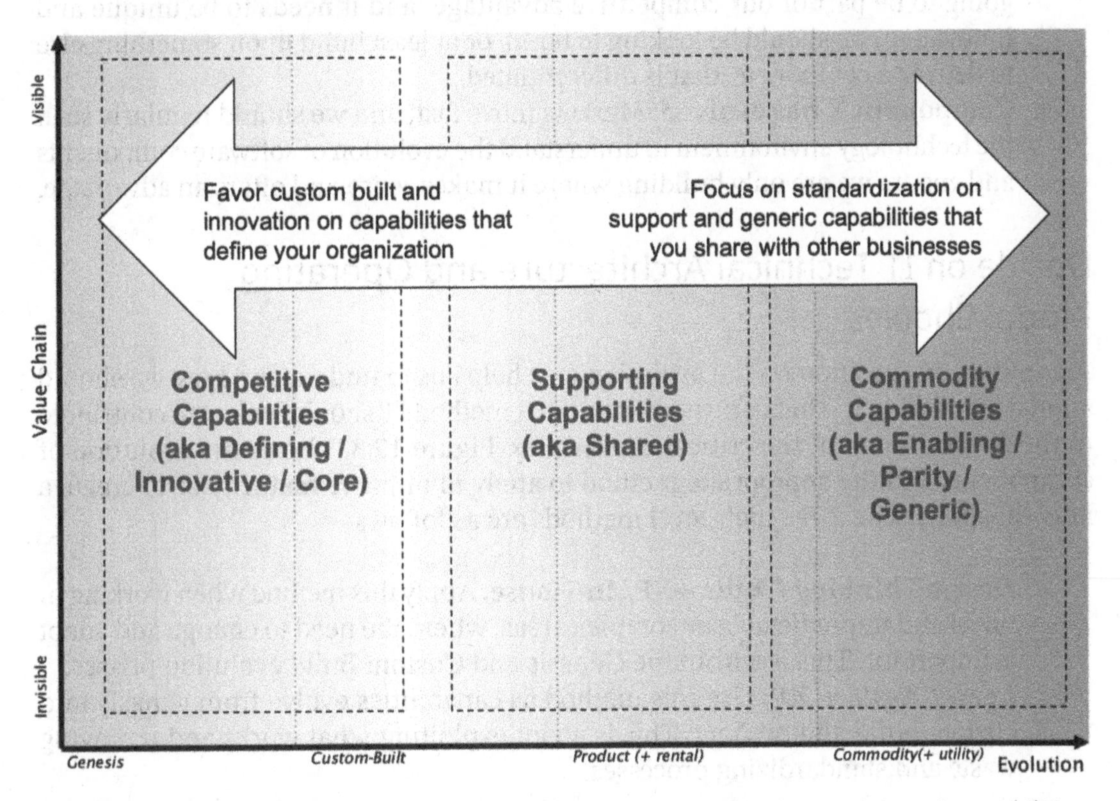

Figure 12.31: Mapping where components related to Core, Supporting, and Generic capabilities should all ideally reside on the Wardley map

- Are we treating a competitive component by outsourcing it? If so, we are unlikely to show the uniqueness that our organization brings.

To illustrate the point, consider Figure 12.32, where three components have been highlighted:

- **Duplication of Component B:** Different maps have similar needs; for example, one map requires a CRM capability for B2C customers, another has a CRM component for restaurant partners, and another has a component for delivery drivers. This potentially represents duplication that we can remove and thus consolidate and reduce cost. However, we need to ensure that each area doesn't have a specific and unique need for a different solution. This is where we need to attack the bias within each business unit to ensure there is no need for different solutions for the same problem.
- **Treating Component Y as a Product and not a Genesis:** If a component is going to be part of our competitive advantage, and it needs to be unique and innovative, we should be looking to build, or at least build upon something else to deliver a component that is differentiated.
- **Component X has evolved:** Markets move fast, and we should regularly scan the technology environment to understand the evolution of software components and ensure we are only building where it makes sense and offers an advantage.

Decide on IT Technical Architecture and Operating Model Choices

A component's position on the evolution axis helps us to understand how we should approach enabling it. That in turn informs what method IT should look at to contribute to the improvement of the critical capabilities. Figure 12.33 shows the evolution of components and the appropriate method to apply to improve maturity or to create a new capability. The three high-level methods are as follows:

- **Design Thinking / Agile +XP/ In-House:** Apply this method when working in novel and unpredictable or complex areas, where the need to change and adapt is important. This is within the Genesis and Custom Build evolution phases.
- **Lean / Agile + XP:** Use this method as capabilities evolve from genesis into custom build and product. This is about exploiting what works and removing waste and standardizing processes.

- **Six Sigma / Outsource:** Adopt this way of working when the capability is a commodity and is highly evolved. Six Sigma is an approach with a focus on process improvement and operational excellence. This is a prime candidate to outsource to specialists, otherwise we may fall into the trap of reinventing the wheel.

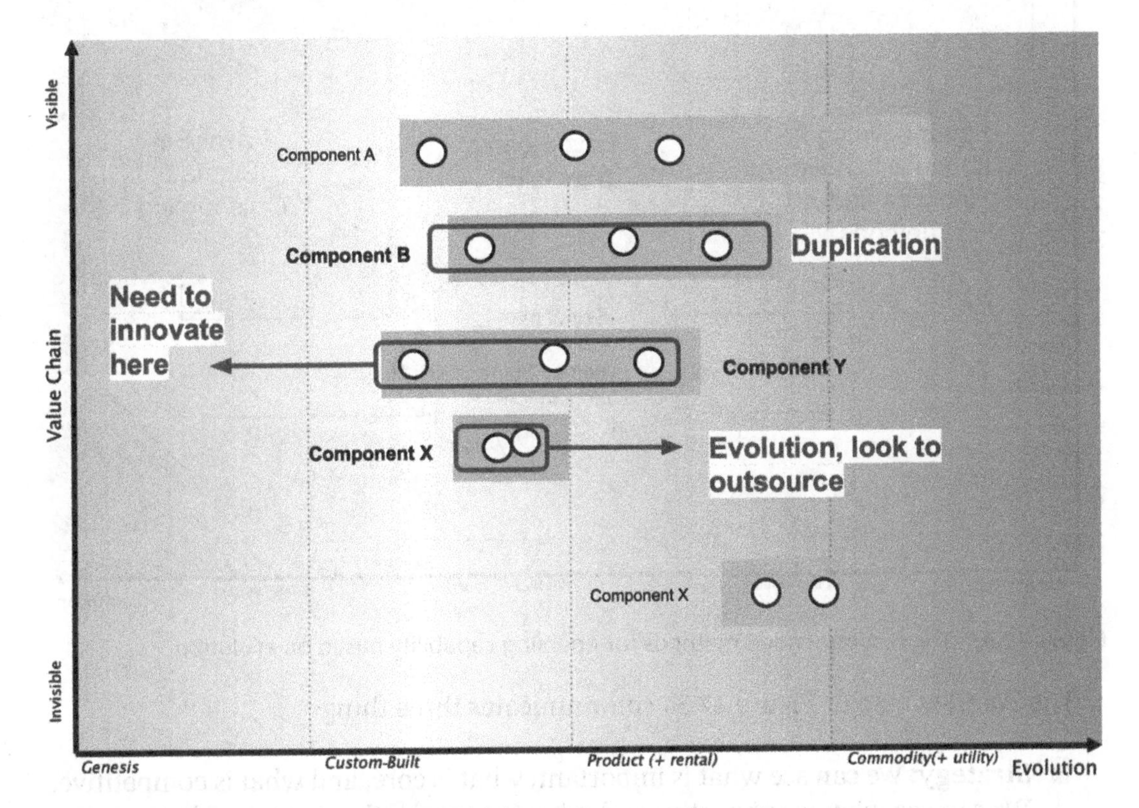

Figure 12.32: Identifying weaknesses in the maps

As approaches and techniques change depending on a component's stage of evolution, so too must the aptitudes of teams. For example, the mindset of people who are using agile and XP practices to experiment to drive innovation will be very different than the mindset of those that are focused on integrating COTS and reducing the variation in process. This recommendation on appropriate methods and culture that Wardley maps help to visualize, along with business architecture, is a critical input into how we organize teams. Figure 12.34 shows an example of how we can organize teams based on components, methods, and cultures.

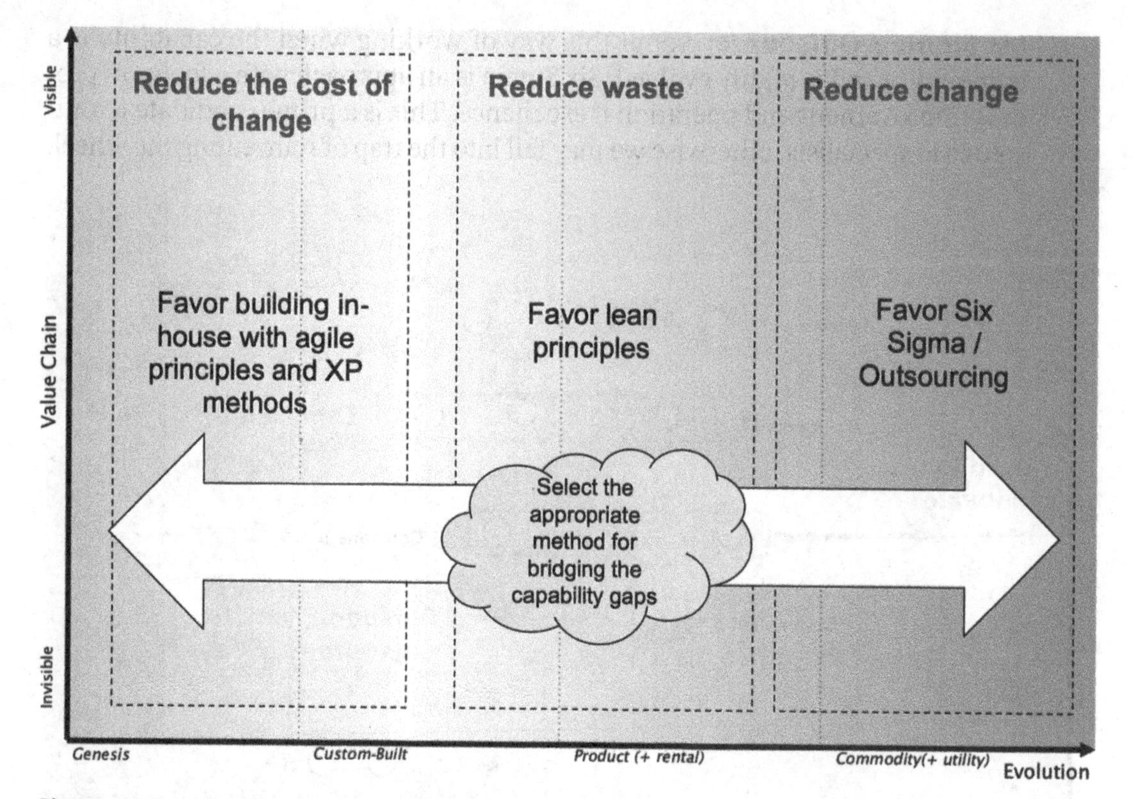

Figure 12.33: The recommended methods for enabling capability based on evolution

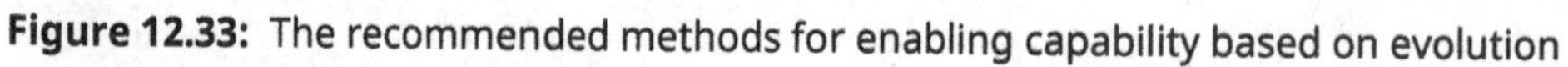

The Wardley map in Figure 12.34 communicates three things:

- **Strategy:** We can see what is important, what is core, and what is competitive. We can see what is supportive and what is generic, thus we are able to understand what is key to supporting business success.
- **Architecture:** We can start to understand dependencies and therefore components that logically belong together.
- **Team topology:** We can use the knowledge of the business architecture as well as chosen methods to inform our organization design.

As with the capability model and application portfolios to determine the priority of focus, you can still map to the matrices as presented before, shown in Figure 12.35. Therefore, we should still perform the method discussed earlier to map the portfolio of applications to the dependency chain to understand how each component is supported as well as understand the application's life cycle status, but this time informed by the position of the related capability on the map.

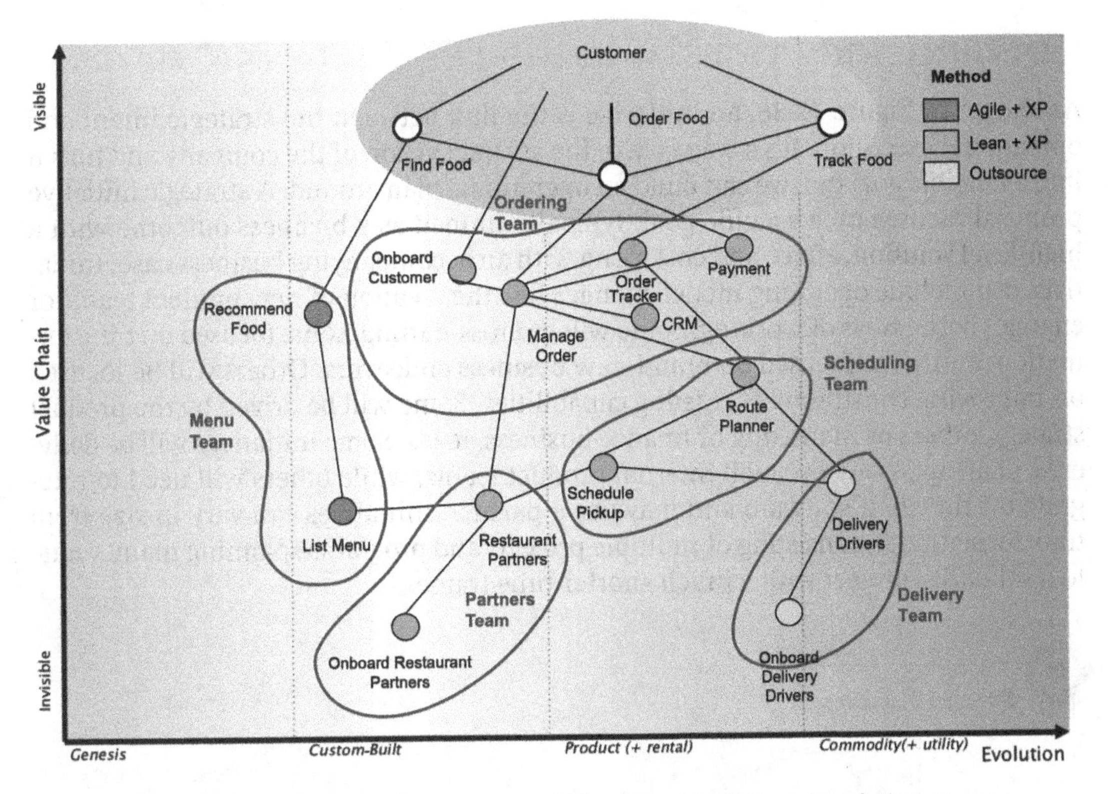

Figure 12.34: Organizing teams based on capability dependencies and evolution

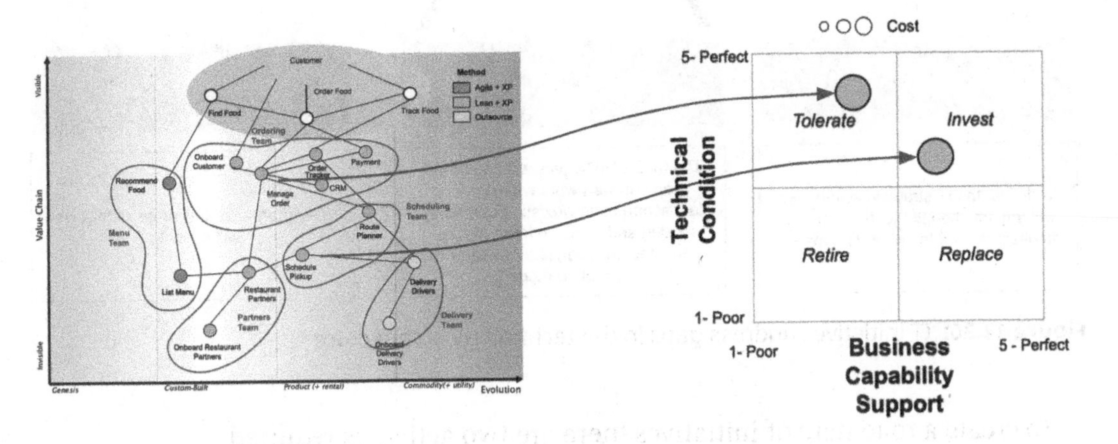

Figure 12.35: Complementing Wardley maps with a Boston matrix to determine priority of focus

Defining and Prioritizing the IT Initiatives

As shown in Figure 12.36, an initiative is the link between the strategic intent and operational execution. It's a way to take the abstract vision of the company and turn it into a tangible goal that we can build an operational plan around. A strategic initiative proposal focuses on a specific goal, typically framed as a business outcome with a high-level solution, effort, and cost along with an accompanying business case. Initiatives can include operating model changes like the creation of new product teams or changes in the ways of working. Some will be cross-cutting, some focused on transformation, and some focused on brand-new business endeavors. Others will be focused on improving constraints of existing capabilities. Some will be driven by the product strategy, others by marketing or finance business needs. Some initiatives will be delivered in an agile way by small internal product teams, while others will need to integrate off-the-shelf software and may use a partner. Initiatives can vary in size from transformational, consisting of multiple projects and programs spanning many years, to a solidarity project with a much shorter time frame.

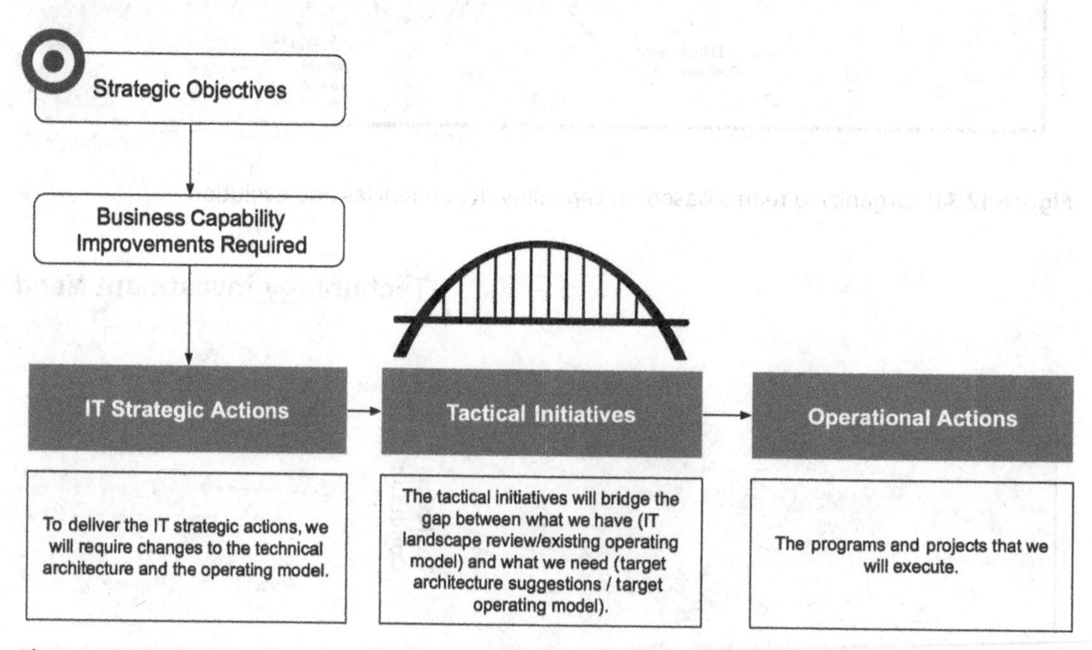

Figure 12.36: IT initiatives address gaps in the technology architecture

To create a road map of initiatives there are two activities required:

- **Defining the IT initiatives required to address capability gaps.** In previous steps, we have defined the areas in need of investment—namely, the business capability heat map and IT consolidation and rationalization suggestions. The next step is to turn this strategic intent into something more concrete—namely, a tactical initiative. Both IT and business unit leaders work together to define specific business outcome goals, high-level requirements, and solution options. The output is a document that describes a specific business need with a measure, impact, or benefit; alignment to a strategic objective; and a rough idea of solution, effort, and cost to address it. It's important to stress that a tactical initiative isn't actionable, but it is the link between the high-level strategic intent and the detailed operational execution solution designs. Working at this level we don't expect the costs and estimates to be accurate. It's more a tool to determine feasibility, identify dependencies, and understand the relative prioritization before we commit to something.

- **Prioritizing a road map of initiatives guided by investment targets.** The IT and business leadership teams can now assess the relative value of the portfolio of initiatives based on the rough estimates of their time, cost, value, and strategic fit. In addition to value, the prioritization process will be influenced by the exec team's investment strategy based on how they want to allocate funds to reach the business goals. Three other factors that also need to be considered are the operational essential needs, dependencies, and any IT architectural initiatives to address. The consistency in describing initiatives ensures that discussions on priority and what gets funded can be taken objectively.

Determine the IT Initiatives Required to Address Capability Gaps

With an understanding of the business needs, and the technical gaps that need to be closed, we can begin to form the IT initiatives by looking at solution options, benefits, measures, costs, complexities, and effort as well as deciding on concrete business outcomes. IT initiatives are mostly derived from business needs, but there are occasions where IT initiatives are purely focused on purely IT optimization. To determine the relative priority and build a tactical road map, we must ensure that all initiative proposals are documented in a consistent manner.

Exploring IT Initiatives to Address Business Needs

To determine the IT initiatives, IT leaders need to work with the exec, senior leaders, and product managers to better articulate their needs into focused business outcomes in order that solutions can be designed to address them. We also need to understand at a more granular level the opportunity or constraint. Are there other factors we need to consider? Does a solution need to be in place by a certain date to avoid a fine or to capitalize on an opportunity. What is the cost of delay? As well as a technical change, what is the impact on the people and process?

Essentially, we are taking the broad vision of improving or creating a business capability into something more concrete and tangible. By working closely with each and all the business units and departments, we can also identify commonalities and themes in terms of needs and gaps to maximize the impact of each initiative. This is like how Wardley maps allowed us to orientate and determine focus and direction, ensuring we provide consolidation solutions and deliver maximum impact.

The analysis from the application portfolio and the resultant target state suggestions to rationalize or optimize the technical landscape guide the development of the IT initiative with appropriate technologies and implementation approaches. As highlighted in the examples in Table 12.6, the business initiative is defined, the gap assessment is conducted, and the relevant target state suggestions or standards are reviewed before a solution is defined. In addition, any changes to the operating model are also taken into consideration, which themselves could be independent initiatives. The result of this process converts high-level strategic intent into a more concrete and tangible initiative with solution recommendations to address business needs.

Table 12.6: Using landscape analysis to identify solutions for IT initiatives

Strategic Action or Dept. (e.g., Marketing, Finance or Product) Operational Essential need	Target Architecture / Operating Suggestion from IT Landscape and Capability Assessment	What Are the IT Initiatives That Will Contribute to Meeting the Business Need?
1.1 Provide full caterer's information on the web for customers.	[Caterer Schedule] ■ Replace caterer schedule spreadsheet with COTs solution [List Menu] ■ Content Management System [Sales Channel] ■ Invest in website	■ Enable suppliers to provide full catering availability ■ Enable suppliers to provide full detail on menu options ■ Expose supplier information online for customer discovery
1.2 Enable caterers to self service quotes and customers to confirm bookings online.	[Customer Engagement] ■ Buy CRM [Manage Quotes] ■ Build	■ Provide ability for suppliers to service customer requests direct ■ Provide ability for customer to accept quote and book online
2.3 Automate the sign-up and approval process.	[Customer Registration] ■ Replace with RPA solution	■ Automate supplier sign up process ■ Automate supplier approval process
Payroll urgent issue - won't scale.	[Payroll] ■ Consolidate with Netsuite	■ Enable payroll to scale with employee growth
. . .	. . .	. . .

Exploring IT Initiatives to Address IT Capability Needs

Outside of initiatives to tackle business needs, there will be a set of initiatives focused on purely technical areas to address the IT capability optimization suggestions, covering the scores of applications, data, infrastructure, security, and the rest of the capabilities within the operating model. Table 12.7 shows some examples of architectural IT initiatives to manage risk, security, complexity, performance, and cost. There are two ways to include architectural initiatives in tactical plans:

- **Included as part of an IT initiative to address a business need.** Ideally, any IT capability initiative will be included in a corresponding business initiative. The approach in this case is that we incorporate the target architecture suggestion in the solution to address a business need if they occur in the same area.
- **Dedicated IT capability initiative.** Where there isn't an overlap in the target architecture suggestion and a business need, we must define a separate explicit architectural initiative.

Table 12.7: IT initiatives to address IT capability needs

IT Capability/Technology Need	IT Initiative
[Product Management] ■ Hire an experienced product manager. ■ Change topology of team to form product team around customer online journey. ■ Invest in product management skills and expertise.	■ Introduce product team for customer digital user journey for iterative improvements to conversion
[Application Development] ■ As this area is a supporting capability, partner with a team to provide an augment team who has expertise in RPA so we can free up in-house devs to focus on areas of differentiation. ■ Hire a product manager to manage this area.	■ Create new automation product team and partner with third party for expertise
[Security Management] ■ Partner with virtual CISO. ■ Align to a security framework to quantify security posture. ■ Aim to achieve compliance in all areas.	■ Partner with virtual CISO to gain compliance in an industry standard security framework
[Service Desk] ■ Outsource services desk in the US.	■ Outsource service desk support for US .
[Infrastructure Management] ■ Partner with expertise to move to the cloud and leverage managed services. ■ Invest in training and upskill team in cloud infrastructure.	■ Move to cloud-based native and managed infrastructure to support growth
...	...

Documenting Initiatives as Business Outcomes

IT initiatives should be described in a manner that is understandable to business leaders, clearly showing cost, effort, a solution overview, and the expected business impact. In the CSVLOD model, these are known as outlines. This lightweight initiative document can be based around an A3 report as introduced in Chapter 7, "How We Govern." An outline, as shown in a Figure 12.37, answers the following questions on a proposed initiative:

- What opportunity or constraint (business capability gap) is the initiative addressing?
- What is the target state? What capability improvements will there be?
- What is the business value or impact (e.g., cost saving / avoidance, margin improvement, risk reduction)?
- What does the solution (or the solution options) look like?
- How will we measure success?
- What is the cost, including capex/opex and impact to P&L?
- What are the timescales?
- Are there any dependencies?

The outline itself forms part of the overall business case used during the process of prioritizing and road map building. By describing outcomes in a consistent manner, we can make informed and objective decisions regarding priorities. To support business cases, larger multi-project initiatives can be broken down into small initiatives.

Phasing Approaches

The initial drafting of a road map helps to reveal the feasibility of the business strategy given the current investment strategy and capacity of the organization. The process of deciding how to phase the initiatives can be rather subjective. Gaining agreement on the phasing approach is an iterative and often political process. It isn't simply a case of picking the highest impact and lowest effort initiatives first. While we should use the priority approaches as laid out in Chapter 7, we need to consider a number of factors:

- How much money can the business afford to invest in each year?
- Outside of the strategic needs, what are the operational essential needs as well as architectural needs that require investment?
- Do initiatives have dependencies? Do we need operational changes to be made first?
- How much change can the organization manage? How much capacity do we have?

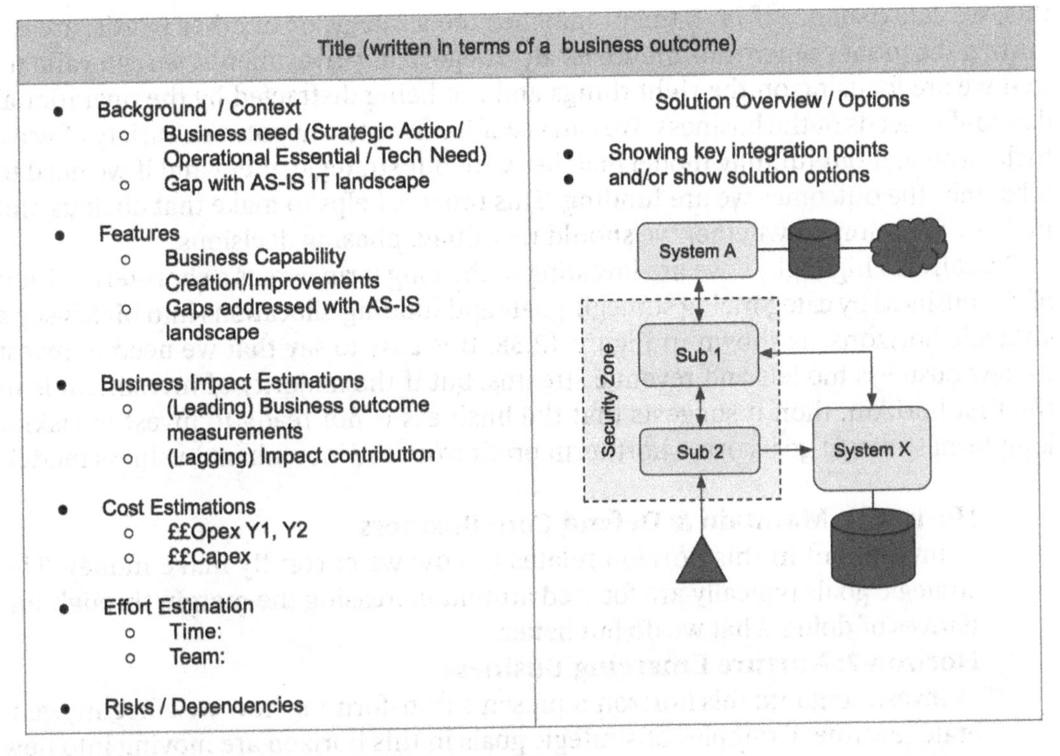

Figure 12.37: An example of an initiative overview

- Are we looking for a payback this financial year, or can we afford to invest in the long term?
- Are there business events that will have an impact on plans?
- How confident are we with delivering big change? Should we focus on lower effort albeit lower impact initiatives to prove we have the capability to manage change?
- What are the business goals? If in year 1 we need to launch into Europe, there is no point in having a localization initiative at the end of our plans.
- Think about TOC. Where can we make the greatest leverage? Where are the biggest constraints?
- Even if we can afford it, should we do big band, or should we do a phased approach?
- Have we considered financial covenants such as capital expenditure restrictions?

In systems thinking, to determine the purpose of a system, all we need to do is look at what it does rather than what it says it does. What objectives and outcomes is the system funding and are these linked to the strategic goals on how we win? To answer this, we can visualize if investment matches the strategy, or in other words, are we putting the money where our mouth is? By categorizing investments, we can validate that we are focusing on the right things and not being distracted by the operational day-to-day needs of the business. We can visualize investment pots in a variety of ways to determine if our funding model matches with our strategic intent and if we need to rebalance the outcomes we are funding. This process helps to make that obvious and invites discussion on whether we should reevaluate phasing decisions.

We can also highlight if we are investing in the long term versus a short-term vision of the business by categorizing strategic goals and funding allocations into McKinsey's strategic horizons, as shown in Figure 12.38. It is easy to say that we need to invest in new business models and revenue streams, but if the majority of investment is in the first horizon, then it suggests that the business is not ready to invest in riskier long-term strategic goals over short-term profit from the incumbent business model.

- **Horizon 1: Maintain & Defend Core Business**
  Investment in this horizon relates to how we currently make money. The strategic goals typically are focused around increasing the margin through initiatives of doing what we do but better.
- **Horizon 2: Nurture Emerging Business**
  Investment into this horizon represents transforming how we currently generate revenue. Examples of strategic goals in this horizon are moving into new geographies, tackling different customer segments, or launching new product lines. While not as safe in terms of profits compared to horizon 1 goals, this is building upon known knowns so payback should be fairly quick.
- **Horizon 3: Create Genuinely New Business**
  Investing in the third horizon represents a focus on completely new business models to generate money. Strategic goals will focus on R&D activities, perhaps even through the cannibalization of the main business. These strategic goals will be riskier and we should not expect them all to succeed. However, without looking to the future, we increase the likelihood of being disrupted.

There are a host of other methods we can use to analyze funding allocation to reveal how much investment is being put into strategic areas versus operational needs, such as splitting by channel (e.g., off- or online, mobile or desktop) or focusing on geography (e.g., global expansion vs. home location). All these methods and more can help us to visualize how money is invested to validate if we are focusing on the areas that will have the biggest impact to the business or being sidetracked by operational day-to-day activities.

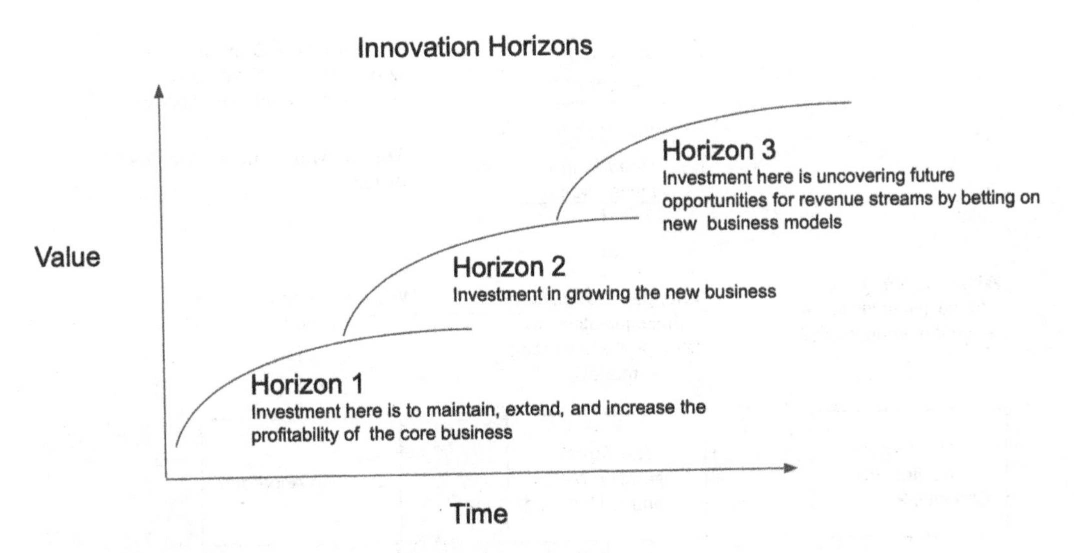

Figure 12.38: Plotting initiatives to determine priority

Communicating the IT Tactical Investment Road Map

To create the deck to communicate the IT tactical plan, we will again employ the Pyramid Principle as introduced in Chapter 11. Figure 12.39 shows the flow of the plan designed to build up clarity on why we are proposing the investments as well as ensuring clear alignment with the business strategy. The introduction is based on the SCQA framework, which stands for Situation Complication Question and Answer. The Situation, Complication, Question section will summarize the business strategy and IT strategic actions. The IT strategy executive summary from Chapter 11 is a good example for this as it clearly shows the red line from business goals to IT actions. The answer to how we will execute the actions is the IT tactical investment road map. This is the phasing of the IT initiatives designed to deliver the IT strategic actions.

The next layer of the pyramid summarizes why we have chosen the initiatives and the recommended approach to phasing. In this layer we answer that question with the target architecture suggestions, the target operating model changes, and the reasons for the phasing approach. The next layer supports these arguments with data to ensure your summaries are based on indisputable facts and are objectively verifiable to support your argument and opening answer. We can include the summary on the IT landscape, a review of the incumbent IT operating model. Last, we can include the portfolio of IT initiatives along with further supporting material such as the application inventories, reference architectures, vendor information, and more detailed financials.

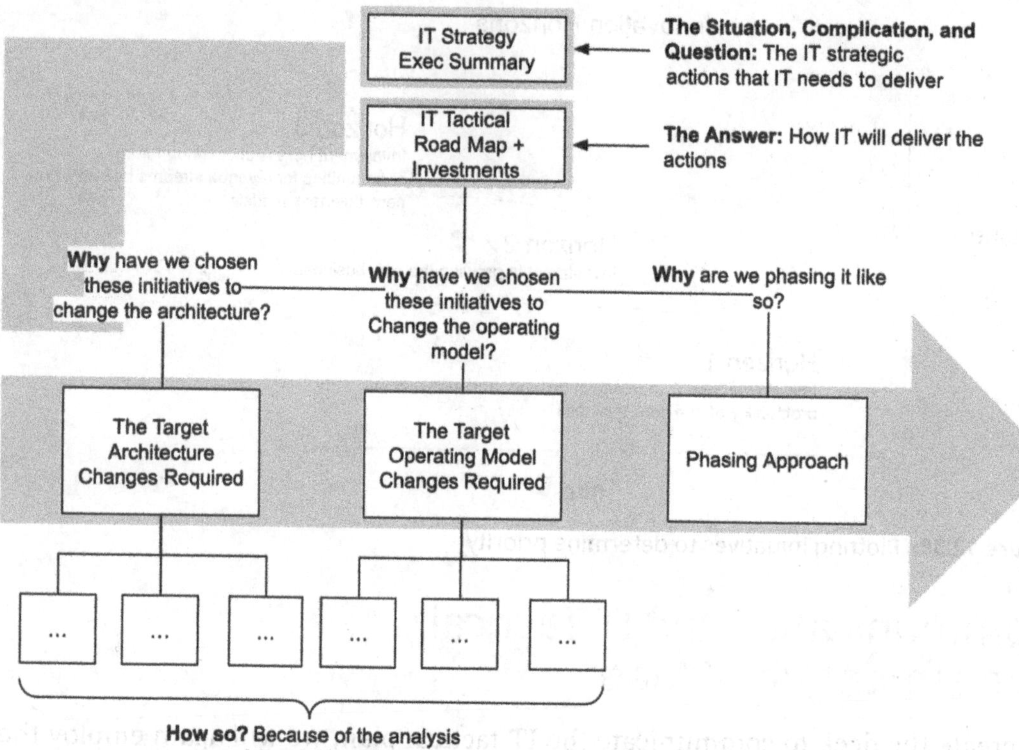

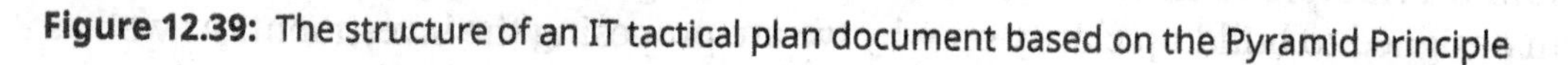

Figure 12.39: The structure of an IT tactical plan document based on the Pyramid Principle

The Introduction, Situation, Complication, and Question: The Business and IT Strategy

The first part of the document is a recap of the IT strategy so that the tactical plan can be understood in context. Figure 12.40 shows the strategic objectives required to achieve the business, and the IT actions that will contribute to enabling the business capabilities.

The Answer: Tactical Road Map

The tactical investment road map is the next step of taking strategic intent down to a tactical level. This is a method to understand the feasibility of business plans and aspirations and align and manage the expectations of the business on what is likely to change, and perhaps more important, what is not. Based on the exec groups' agreed prioritization, the IT tactical road map describes what IT is likely to focus on and invest in over the planning horizon.

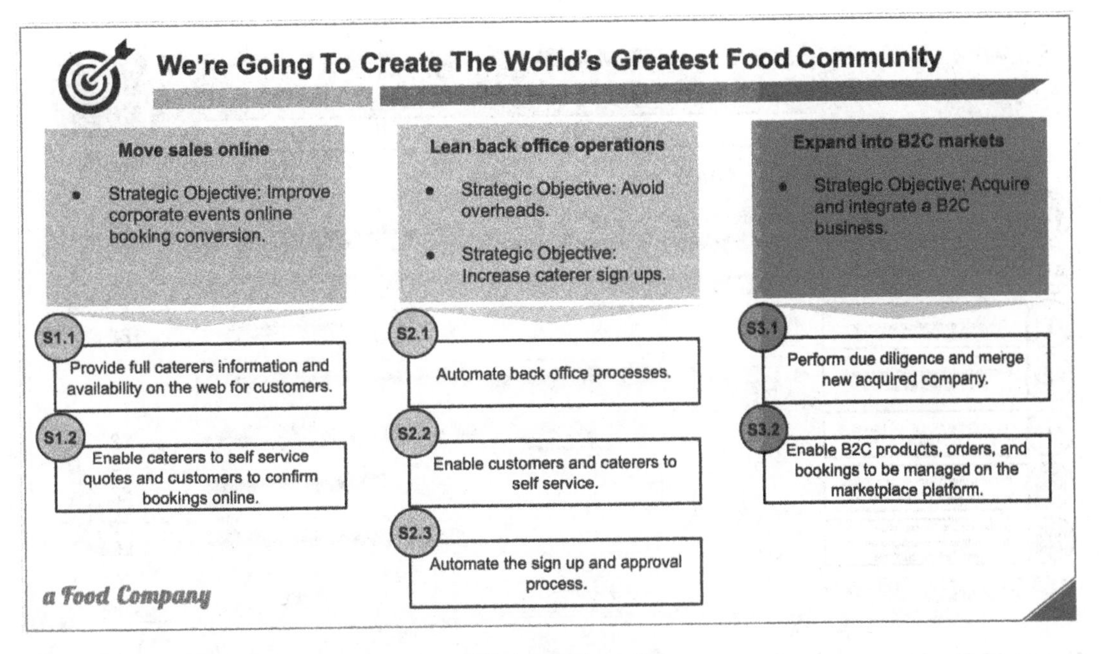

We're Going To Create The World's Greatest Food Community
Move sales online
Strategic Objective: Improve corporate events online booking conversion.
Lean back office operations
Strategic Objective: Avoid overheads.
Strategic Objective: Increase caterer sign ups.
Expand into B2C markets
Strategic Objective: Acquire and integrate a B2C business.
S1.1
Provide full caterers information and availability on the web for customers.
S1.2
Enable caterers to self service quotes and customers to confirm bookings online.
S2.1
Automate back office processes.
S2.2
Enable customers and caterers to self service.
S2.3
Automate the sign up and approval process.
S3.1
Perform due diligence and merge new acquired company.
S3.2
Enable B2C products, orders, and bookings to be managed on the marketplace platform.
a Food Company

Figure 12.40: The executive summary of IT strategic actions

There are several approaches to organize and visualize the tactical road map. Initiatives may be categorized in any number of ways, such as by strategic objective, product areas, business units, value streams, or customer journeys. I don't recommend you show by product team as this can complicate the road map because multiple teams may be involved in a single initiative. In terms of what you show on the road map, you must show the name of initiative in terms of business outcome so it's clear what value you will be delivering. You should show the estimated time. You can also show estimated cost and if the initiative has a dependency on anything else and the status of the initiative (e.g., in-flight, approved, and preapproval planning). If there are any major business events that will have a material bearing on what gets delivered, it is also a good idea to make that explicit. Figure 12.41 shows an example of a simple, and sometimes simple is best, tactical road map, organized by IT strategic actions.

Why? The Target States

The target state suggestions can range from showing large organizational transformations to smaller evolutionary changes. Figure 12.42 shows a major solution change at an e-commerce retailer. The target state is annotated with information on what is required, showing clearly how it relates to the strategic actions.

a Food Company — TACTICAL ROADMAP

	Year 1 Q1/Q2	Year 1 Q3/Q4	Year 2 Q1/Q2	
S1.1 Provide full caterers information and availability on the web for customers.	Enable suppliers to provide full catering availability; Enable suppliers to provide full detail on menu options	Enable suppliers to provide full catering availability; Enable suppliers to provide full detail on menu options	Expose supplier information online for customer discovery	££
S1.2 Enable caterers to self service quotes and customers to confirm bookings online.	Hire Product Team; Provide ability for suppliers to service customer requests direct	Provide ability for suppliers to service customer requests direct	Provide ability for customer to accept quote and book online	££
S2.1 Automate back office processes.			Hire Product Team; Automate Order Processing	££
S2.2 Enable customers and caterers to a self service for service	Hire Product Team; Enable customer self service for sales and service	Enable customer self service for sales and service	Enable customer self service for sales and service	££
S2.3 Automate the sign up and approval process.		Hire Product Team; Automate supplier sign up process	Automate supplier sign up process	££
S3.1 Perform due diligence and merge new acquired company.			Integrate new business	££
S3.2 Enable B2C products, orders, and bookings to be managed on the marketplace platform.		Enable B2C sales	Enable B2C sales	££
	££		££	

Figure 12.41: An example of a tactical road map

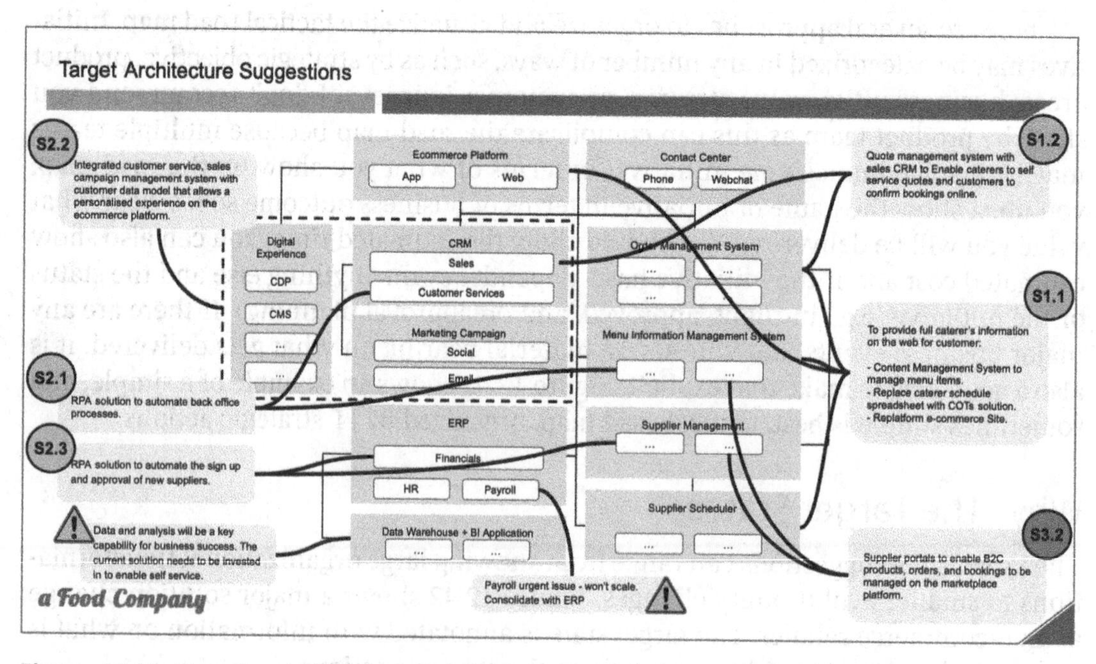

Figure 12.42: Transformational target architecture with annotated pain points

We can take transformational target architecture suggestions to the next stage and give a high-level idea of candidate solutions for the related solution spaces. Remember, at this level you only need to have an idea, not the exact answer. For example, Figure 12.43 shows solution options based on the target architecture of Figure 12.42. In this case we are comparing in-house build approach, against point solutions and a single vendor solution. Again, this is an abstraction; its purpose is to act as a high-level navigational aid on how you will likely fill capability gaps and the types of solution choices there are. As this is nontechnical visualization, it can be used to gain consensus and alignment on how you intend to evolve the landscape for both your technical teams and business peers.

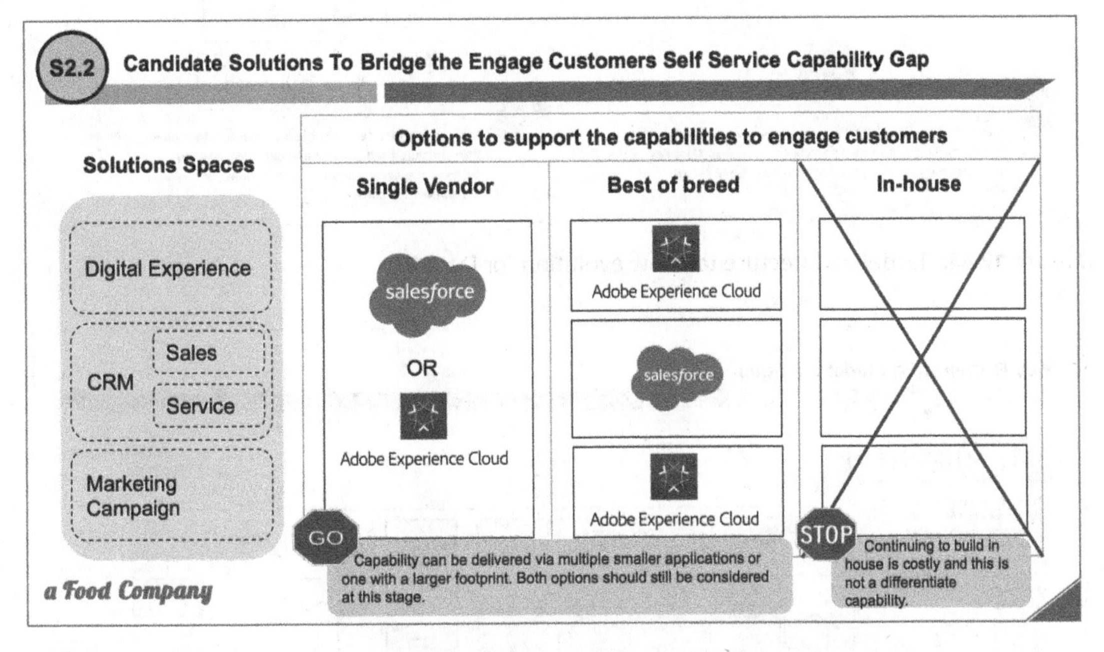

Figure 12.43: Transformational target architecture with annotated pain points

Target state suggestions can also be more evolutionary, focusing just on a single area. As shown in Figure 12.44, we are illustrating the need to change the architecture of the data and BI environment to address the business need of ingesting big data.

For the target operating model, we need to highlight the changes we require in terms of how we lead, how we govern, how we are organized, how we manage and source talent, and how we work to support the IT strategic actions. Figure 12.45 shows an example of how we can communicate this at a high level.

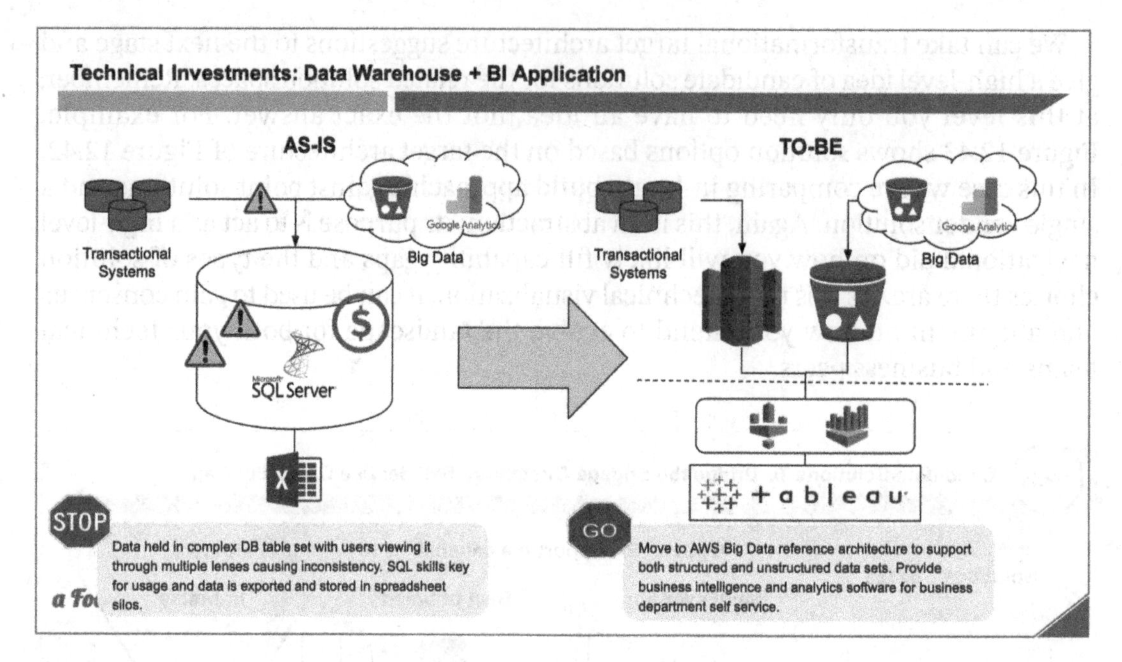

Figure 12.44: Target architecture to show evolution for Data BI

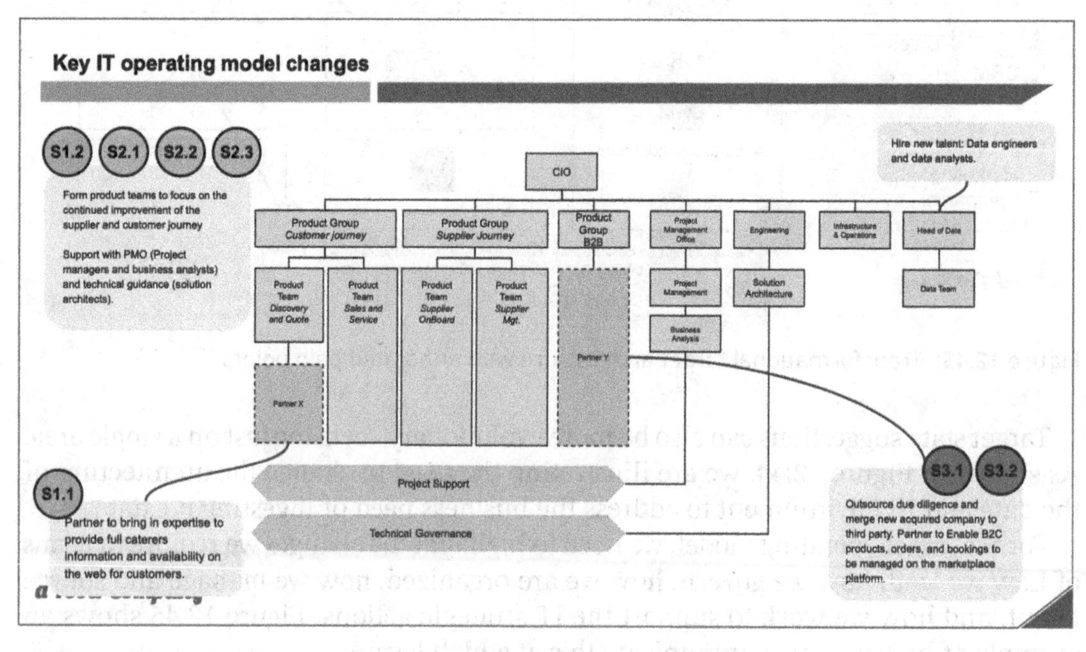

Figure 12.45: Changes required to the operating model

How So? The AS-IS States

Last, you need to support your arguments with data to ensure your summaries are based on indisputable facts and are objectively verifiable to support your argument and opening answer. Figure 12.46 shows an IT landscape annotated to show the challenges with how the technology will not support the IT strategic actions.

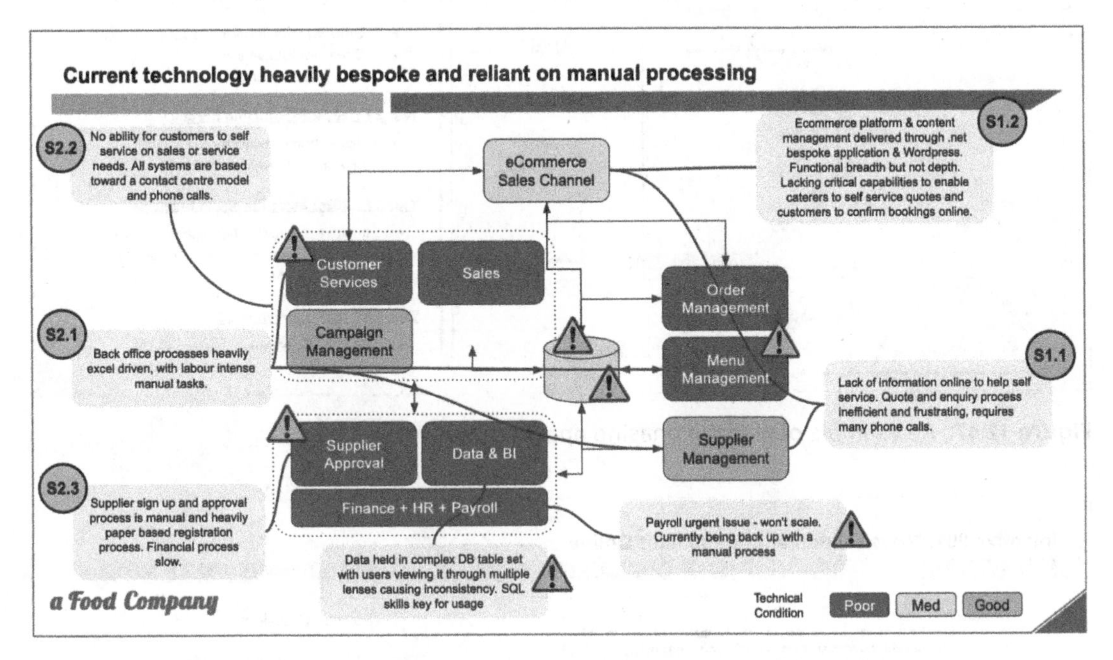

Figure 12.46: Highlighting the challenges with the IT landscape

How So? The Recommended Approach

The recommended phasing approach may have been agreed upon with the exec, but it is important to show the reasons behind those decisions. Figure 12.47 shows the reasons behind the recommended phasing approach.

Appendix: Portfolio of Initiatives and Supporting Documentation

The appendix of the IT tactical plan should contain the portfolio of initiatives. An example of one is shown in Figure 12.48. In addition, you can include the detailed application inventories or catalogs, current org charts, key vendor references' architectures or technology reference models, and any other information that is relevant to the decisions in the main body of this document.

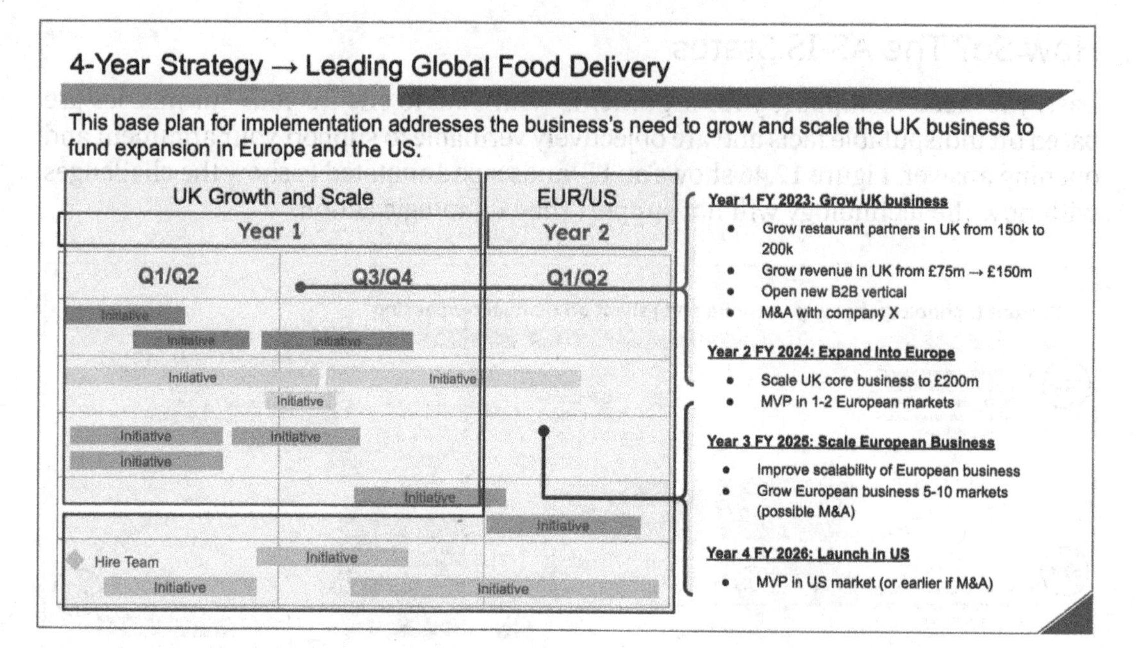

Figure 12.47: An example of why the phasing approach was taken

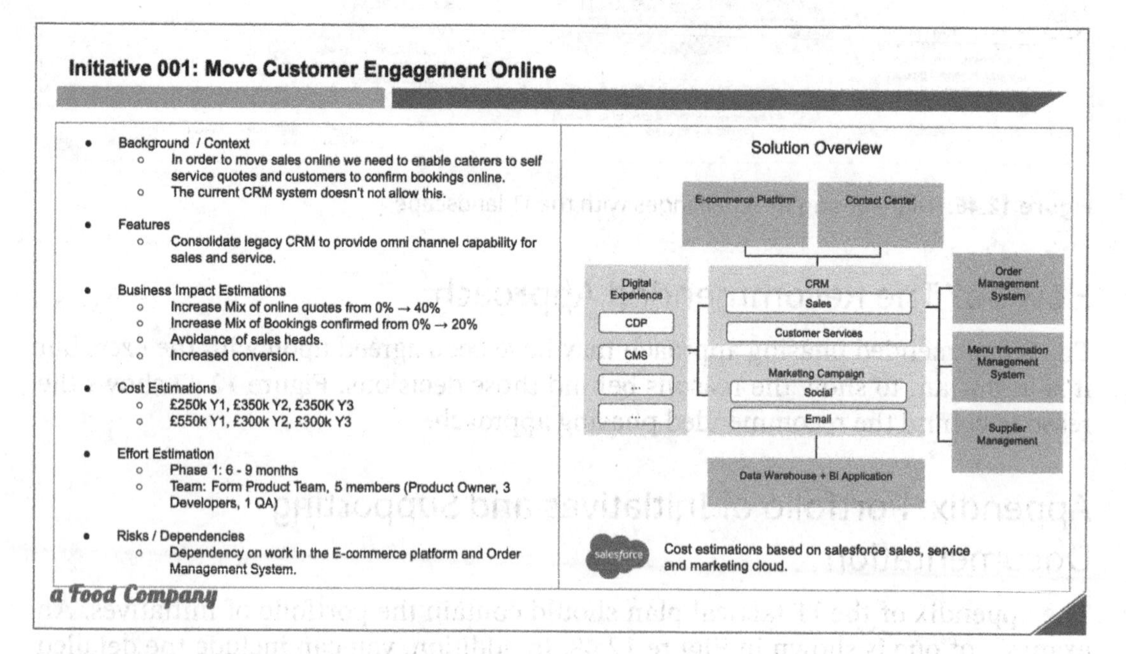

Figure 12.48: An example of an IT initiative.

Summary

We started the chapter by looking at three universal principles of tactical planning. The first was to create a technical vision to guide solution design. Without a clear future state for the IT architecture, we will inevitably end up with a big ball of mud. The second is that planning horizons depend on the context of your business. In some cases, where there is certainty, a longer planning time horizon makes sense; in other instances, where the environment is more dynamic, a shorter horizon is preferable. The third is that both planning and strategy are never set in stone. Both your strategy and planning efforts are based on the information you know now. If the context changes, then you should reevaluate if the strategy and tactical plans still make sense.

To provide a guide for strategy deployment and tactical planning, I introduced the Hoshin Kanri framework. In the simplest of terms, Hoshin Kanri is a framework for strategic alignment based on three important principles:

- Consensus and collaboration on targets and strategies at each level of the organization: strategic, tactical, and operational
- Autonomy to translate strategy into a team's own objectives and goals
- Honest and open feedback loops at all levels and the ability to change or adjust plans

The process of tactical planning itself is made up of three phases; however, these do not need to be run sequentially. They are as follows:

- **Clarifying the Business Needs:** Perhaps the most essential part of tactical planning is gaining consensus and alignment on the areas in need of investment. The techniques discussed in Chapter 11, and Chapter 6, can be used to build a business capability heat map that clearly visualizes the most critical needs for the business, both in delivering the strategy and in maintaining operational running.
- **Reviewing the Technology Landscape and the Operating Model:** Without due care and attention, the IT landscape can become a big ball of mud if we do not regularly review the evolving architecture and look for optimization and consolidation opportunities. Based on the capability needs identified in the previous section and the need to rationalize the IT estate to reduce complexity, we can produce a set of target architecture suggestions to influence and guide IT initiatives. As well as the IT architecture, we also need to review the IT operating model to identify any changes needed to support the technical investments to address capability gaps.

Wardley maps can help visualize both target architecture suggestions and operation model changes. Creating Wardley maps turns the process of creating a target arch and operating model into a map by providing movement through the plotting of capability evolution, which helps to provide context and guidance on the appropriate method of both technology and team organization to use to bridge capability gaps.

- **Defining and Prioritizing the IT Initiatives:** With clarity on the areas of the business that need investment and the suggestion of how to optimize the IT landscape, we then turned our attention to defining and prioritizing the IT initiatives.

The result is a tactical road map of initiatives describing the business outcomes we intend to deliver over, at least, the next 12 months. The plan is almost certainly likely to change and be altered by urgent initiatives or changes in the business context. However, based on the best of our knowledge, this represents how we will invest in technology to support the business strategy and operational needs of the organization. The tactical initiatives, or business outcomes, can then be assigned to product groups and product teams at the operational level to deliver, which is what we will look at in Chapter 13.

Operational Planning: Execution, Learning and Adapting

However beautiful the strategy, you should occasionally look at the results.
—Winston Churchill

When it is obvious that the goals cannot be reached, don't adjust the goals, adjust the action steps.
—Confucius

Plans are nothing; planning is everything.
—Dwight D. Eisenhower

In this final chapter we will explore the last step of strategy deployment—operational planning. Operational plans are the investments in programs, projects, or product teams that will be executed over the short term, typically the financial year, to deliver the prioritized tactical initiatives that will contribute to business success. We will explore how to create an operational plan using the approaches as covered in the operating model components of Chapter 6, "How We Work," and Chapter 7, "How We Govern."

However, as Mike Tyson once said, "Everyone has a plan until they get punched in the mouth." This is like another old saying, "No plan survives first contact with the enemy." In other words, we need to be able to adapt when things change at a strategic level, whether the change is in in market conditions, suppliers, or competitors. A change may also occur at a tactical level if our assumptions are proved invalid on which initiatives will contribute to achieving a goal, or it may be a change at the operational level based on feedback on how effective our execution is. To mitigate the fact that our plans will ultimately be wrong, we will look at how we build in feedback loops at the strategic, tactical, and operational levels, ensuring a regular rhythm of review and adjust.

Finally, we will end with a worked example to demonstrate the cascade and alignment of turning strategic intent into operational execution via tactical initiatives.

Operational Considerations

Making grand plans is one thing; ensuring the entire business, often composed of remote teams scattered across the country or countries, can deliver them is much harder. In his book *The Art of Action*, Nicholas Brealey Publishing; (2021), Stephen Bungay examines the friction that Clausewitz, a Prussian general and military theorist, observed when armies tried to execute strategy in volatile, uncertain, complex, and ambiguous situations. Friction occurs when we try to achieve a collective purpose in a complex and ever-changing environment with humans, who are also complex due to their free will and independent thought. It is these factors that inherently complicate even the simplest of tasks, or as Clausewitz states, they are "the concept that differentiates actual war from war on paper." This friction is just as applicable to executing IT or business strategy as it is with military strategy.

Bungay summarized that there were three gaps, shown in Figure 13.1, that cause friction and prevent strategies from achieving their goals. They are the knowledge gap (what we know versus what we want to know), the alignment gap (what we want people to do versus what they do) and the effects gap (what we want to happen versus what happens). As shown in Table 13.1, the typical response to encountering these gaps is tighter control and commanding with more granular instructions. However, the focus on command and control increases delays and reduces motivation due to a suppression of proactive and innovative action, which is precisely what is required in complex and unpredictable situations.

Table 13.1: Gaps between knowledge, alignment, and effectiveness

	Problem	Usual Reaction
Knowledge gap	Our plans will never be perfect due to the difference between what we would like to know and what we know.	Give more detailed information.
Alignment gap	People will never be perfectly aligned with a plan; therefore, what we want them to do will be different than what they do.	Give more detailed information.
Effects gap	Complex environments are unpredictable; therefore, there will be a difference between what we expect our actions to achieve and what they achieve.	Give more detailed and often tighter controls.

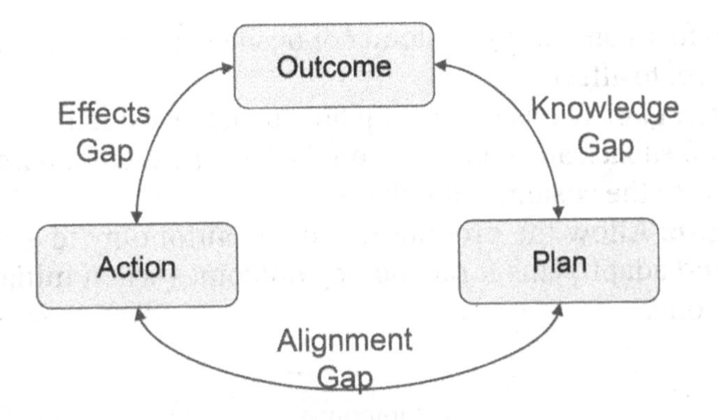

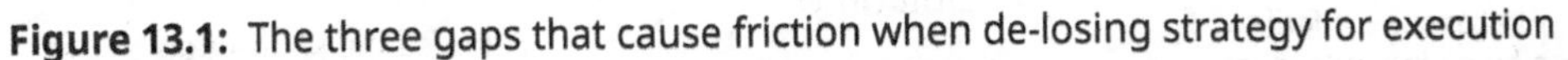

Figure 13.1: The three gaps that cause friction when de-losing strategy for execution

The way this manifests itself in the world of business and IT is giving teams explicitly defined scope and outputs to deliver, giving more and more granular instructions to people and suffocating teams with micromanagement. This is like asking people to read a book without the first three chapters, in that there is no context behind why these directions have been chosen and what business impacts we expect to generate from them. What we want is for them to deliver outcomes with business impacts. We want to give them problems focused on customers articulated in terms of a desired business outcome. What we need to do, as Bungay points out, is lead by intent.

Leading Through Intent

To mitigate the three gaps, the Prussian Army developed a doctrine called Auftragstaktik. Auftragstaktik was based on valuing independent thinking; in other words, in the fog of war commanders should use their initiative and improvisation to achieve the desired military outcome. It was Helmuth von Moltke the Elder who turned this principle into a way of operating, which achieved both high autonomy and high alignment with his commanders. As highlighted in Figure 13.2, to address the knowledge gap, he limited detailed direction in favor of expressing the essential intent. He allowed people to define what they could do to realize the intent to mitigate the alignment gap. And he gave people the autonomy to adapt their actions to better realize intent based on feedback to reduce the effects gap.

Applying this to IT and business strategy, we can address the friction and gaps in the following ways.

- **Knowledge gap:** Instead of asking teams to deliver output, we give teams business outcomes to deliver with meaningful constraints (technical guidance, standards, guardrails). The business outcomes cascade clearly from strategic objectives, ensuring alignment and a sense of purpose, but are at a level where teams can focus on a single customer or business problem and one that is within their control to affect.

- **Alignment gap:** Teams suggesting how to achieve outcomes is key as there is a need for feedback and consensus on the best the way forward, especially from those close to the customer problems.
- **Effects gap:** Allow the product teams the autonomy to explore alternative options and adapt plans for achieving outcomes when initial hypotheses are proved wrong.

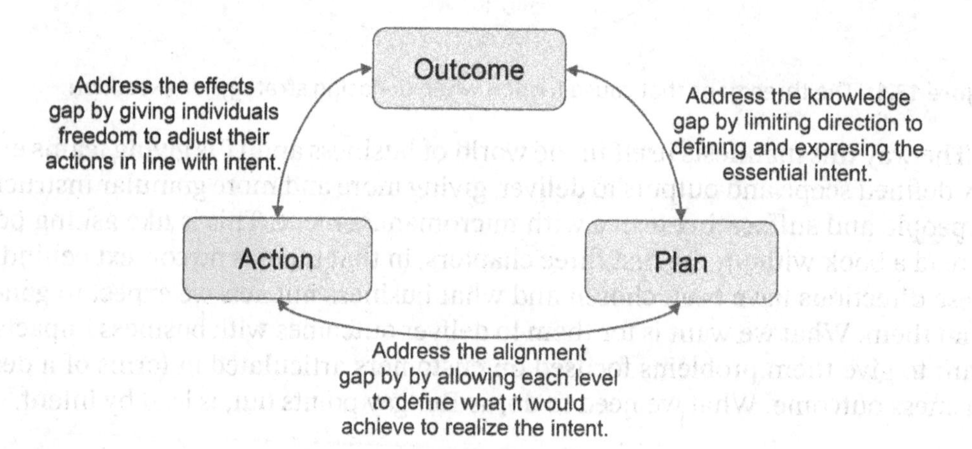
Outcome
Address the effects gap by giving individuals freedom to adjust their actions in line with intent.
Address the knowledge gap by limiting direction to defining and expresing the essential intent.
Action
Plan
Address the alignment gap by by allowing each level to define what it would achieve to realize the intent.

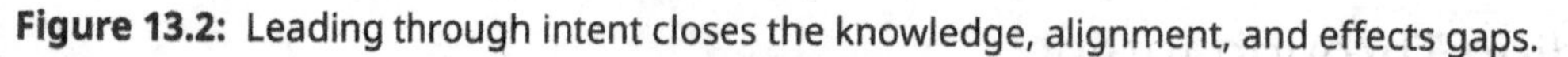
Figure 13.2: Leading through intent closes the knowledge, alignment, and effects gaps.

Objectives and Key Results

A framework that can be used to enable autonomy while ensuring alignment at the operational level is Objectives and Key Results (OKRs). OKRs is a lightweight goal setting and alignment framework that originated from Andy Grove at Intel but came to fame after it was later adopted by Google in 1999. OKRs is a tool for aligning effort and measuring progress toward organizational strategic objectives and goals. The premise around OKRs is that when people are responsible for deciding how to achieve a goal, they are more invested and motivated to see it through. In essence, OKRs can be distilled down to three simple principles:

- Tell people the outcome that needs to be achieved.
- Agree on measurable outcomes that can impact business strategies and organization goals, not outputs.
- Leave them to get on with it, supporting only through course correction and removing barriers.

The OKRs framework consists of the following elements:

- **An Objective** is a qualitative description of a future goal which gives clear direction and provides motivation and context behind the intent. An objective should be aspirational and ambitious and should stretch a team. That is not to say that objectives should be unrealistic, but they need to make a team feel uncomfortable. If teams are regularly hitting or exceeding goals, then they are clearly not ambitious enough.
- **Key results** are the measures of success, most often quantitative, which typically vary between two and five in number. Key results represent the specific targets that show how we are progressing toward the objective. Key results quantify the objective and articulate the target business outcome without stating how this will be achieved. This ensures we are focused on outcome over output and allows any number of initiatives to be trialed and executed to reach the outcome.

To achieve an objective, a set of tactics, aka operational actions is created. These are the operational tasks teams are going to perform to influence the metrics as defined in the key results. Key results describe the outcomes, not actions, and this by their very nature makes them flexible and adaptable to how the team will reach the outcome. Figure 13.3 shows an example of an objective with key results.

Reduce contacts into customer services

- Increase customer self-service from 40% to 80%.
- Increase one call resolution from 60% to 70%.
- Reduce contacts per order from 0.7 to 0.5.

Figure 13.3: An example of an OKR

As shown in Figure 13.4, OKRs can be cascaded; each level maps out its own objectives and key results. Because it's a top-down and bottom-up alignment like Hoshin Kanri, each lower-level team asks themselves what they can do to help progress toward the strategic goals. Consensus is reached on how lower levels can assist higher levels to ensure alignment, ownership, and commitment and ultimately act to promote personal investment in outcomes.

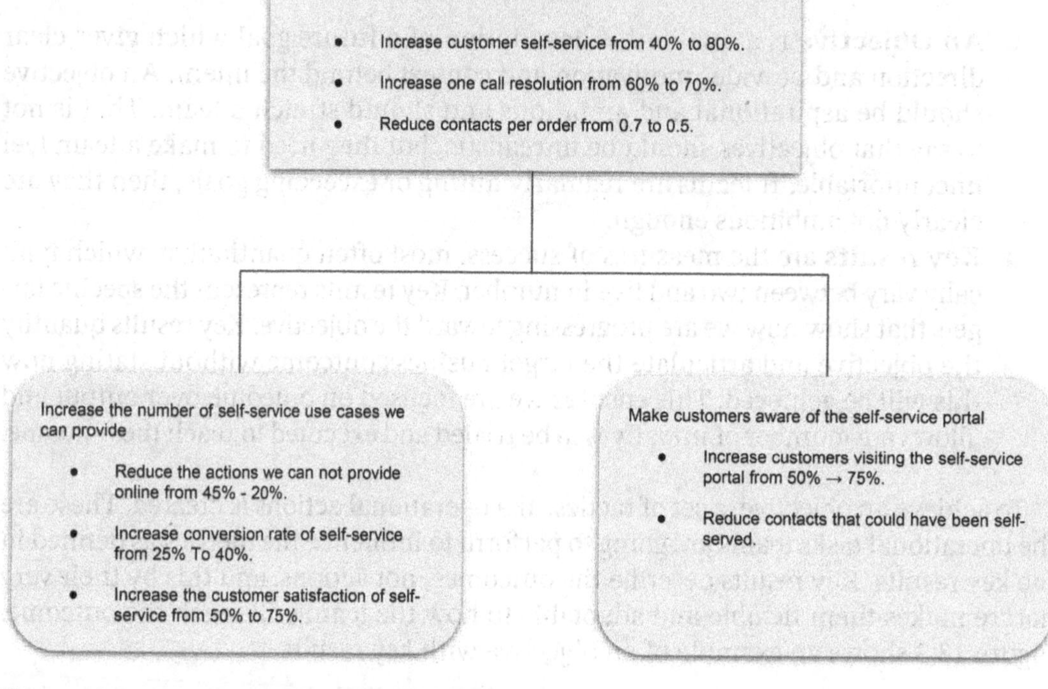

Reduce contacts into customer services
Increase customer self-service from 40% to 80%.
Increase one call resolution from 60% to 70%.
Reduce contacts per order from 0.7 to 0.5.
Increase the number of self-service use cases we can provide
Reduce the actions we can not provide online from 45% - 20%.
Increase conversion rate of self-service from 25% To 40%.
Increase the customer satisfaction of self-service from 50% to 75%.
Make customers aware of the self-service portal
Increase customers visiting the self-service portal from 50% → 75%.
Reduce contacts that could have been self-served.

Figure 13.4: An example of an OKR with child OKRs

As shown in Figure 13.5, OKRs can be used for deploying tactical initiatives aka business outcomes for complex problems to teams.

Operational Planning

The process of creating an operational plan, as illustrated in Figure 13.6, takes the tactical prioritized road map created in Chapter 12, "Tactical Planning: Deploying Strategy," along with any other urgent operational needs, available budgets, and inflight projects as an input. The actual creation of the operational plan is a three-step process:

1. The first step is to distill any of the larger tactical initiatives into separate investments.
2. Second, we need to determine the operational prioritization. This will take the priority of the tactical initiatives along with the other sources of business need, using the approach covered in Chapter 7.

3. The last step is to determine the solution design and team plans to implement all the approved investments. This investment is then added to the run budget of IT to give a full capex and opex budget.

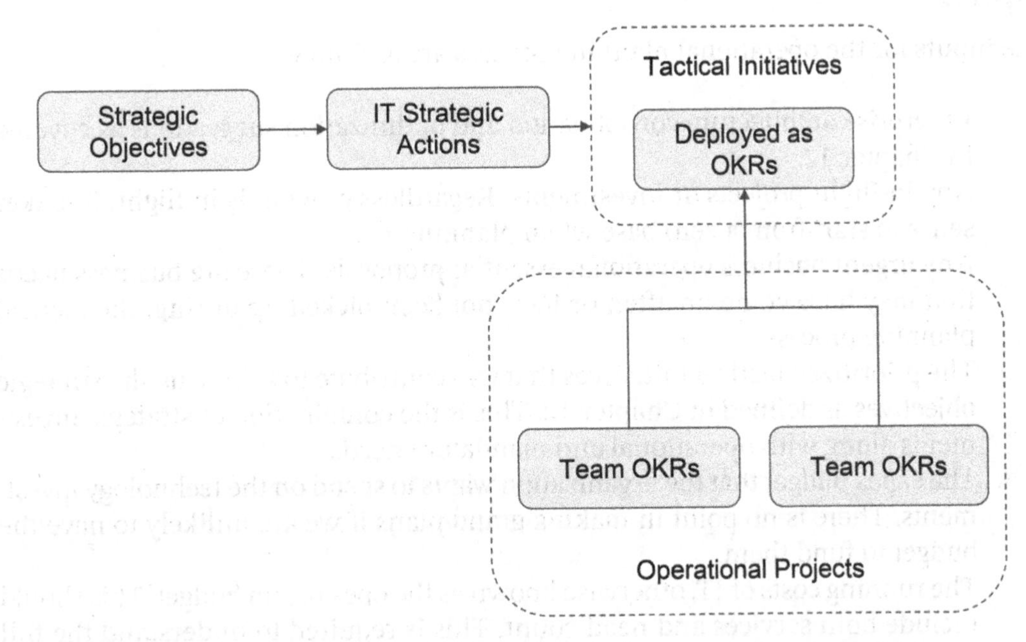

Figure 13.5: Using OKRs to deploy tactical initiatives aka business outcomes to teams

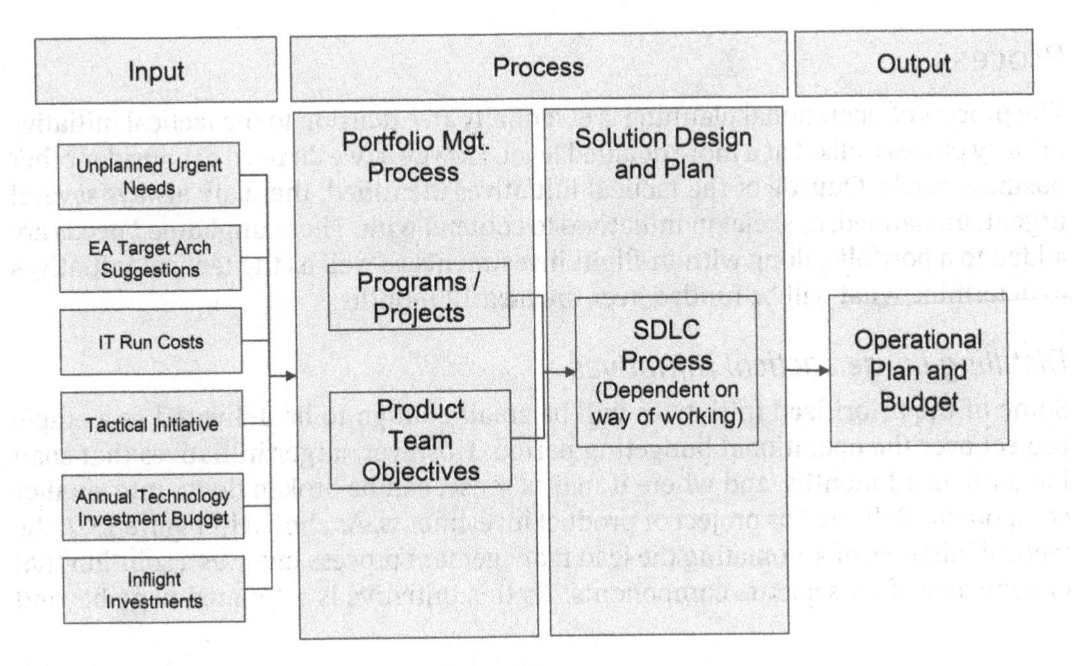

Figure 13.6: The process of operational planning

The output of this process is an operational plan for the year with accompanying budget.

Inputs

The inputs for the operational planning process are as follows:

- Enterprise architecture consolidation and optimization suggestions as covered in Chapter 12.
- Any in-flight projects or investments. Regardless of what is in flight, it makes sense to start from a zero base when planning.
- Any urgent business operational essential proposals. These are business needs that may have come up after, or have not been picked up during, the tactical planning process.
- The prioritized tactical initiatives that will contribute to achieving the strategic objectives as defined in Chapter 12. This is the combination of strategic investments along with operational and mandatory needs.
- The capex budget that the organization wants to spend on the technology investments. There is no point in making grand plans if we are unlikely to have the budget to fund them.
- The running costs of IT, otherwise known as the opex or run budget. This should include both services and head count. This is required to understand the full scope of investment for the financial year.

Process

The process of operational planning is essentially an extension to the tactical initiative priority process, albeit at a more detailed level. However, we do need to consider other business needs. Outside of the tactical initiatives identified, there are always several urgent, unplanned, or skeleton initiatives to contend with. These unplanned needs are added to a portfolio along with in-flight investments as well as the tactical initiatives to determine what will be funded over the next 12 months.

Distilling Large Tactical Initiatives

Some of the prioritized initiatives will be small enough to be delivered as a single project over the operational budgeting period. However, larger initiatives that span longer than 12 months, and where it makes sense, can be broken down into smaller components delivered as project or product investments. As shown in Figure 13.7, the tactical initiative of automating the lead management process involves a solution that encompasses four separate components. As this initiative is estimated to go beyond the financial year, it can be broken down into four operational projects that can be prioritized and potentially have their own business case.

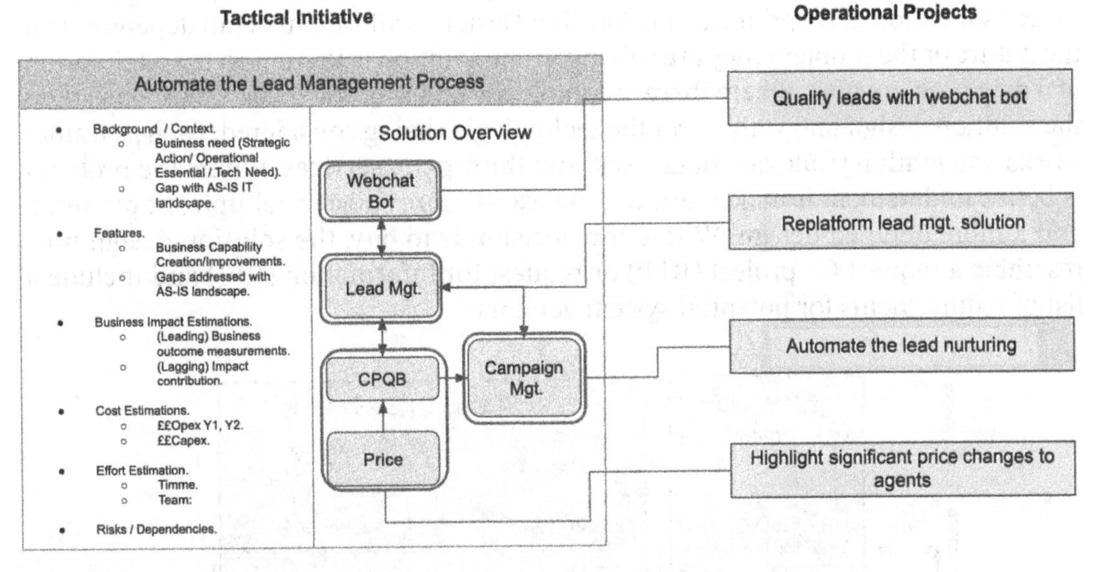

Figure 13.7: Distilling large initiatives into separate investments

Project Portfolio Management

As the operational planning process immediately follows tactical planning, the investment and priority decisions have already been made based on the high-level initiative proposals as covered in Chapter 12. What we need to do at the operational level is follow the portfolio management and prioritization process, as covered in Chapter 7, to be explicit on what we are going to invest in over the next 12 months. I say "be explicit" because the tactical plans are a level of abstraction higher than what is needed at an operational level. The range-based estimates at the tactical level are taken to the next level to improve accuracy. We also need to take into consideration any other urgent or unplanned projects and a more detailed look at dependencies between teams and/or projects.

Solution Designs

Solution Designs, the last step of the CSVLOD model as shown in Figure 13.8 and covered in Chapters 11, "IT Strategic Contribution," and 12, extend the high-level tactical initiatives by providing a more granular level of technical detail to show how teams will contribute to achieving the tactical initiative. Tactical initiatives focus on the longer term, so by design they are not particularly detailed. Whereas the role of a solution design is focused on the short term, explicitly defining how a solution will be delivered and how it will encompass the enterprise architectures target suggestions for consolidation and optimization. Solution Designs will vary in detail depending on the nature of the problem they are solving. If the solution is to support the exploration of a new problem area, where there is complexity and unknownable unknowns, then the solution design may only cover the technologies being considered, such as frameworks, integration strategies, databases, and third-party services. Where the problem is better understood, it makes sense to invest in more substantial upfront planning and a more detailed design. Where the decision is to buy, the solution design may resemble a request for project (RFP) or request for information (RFI) and include a list of requirements for potential system vendors.

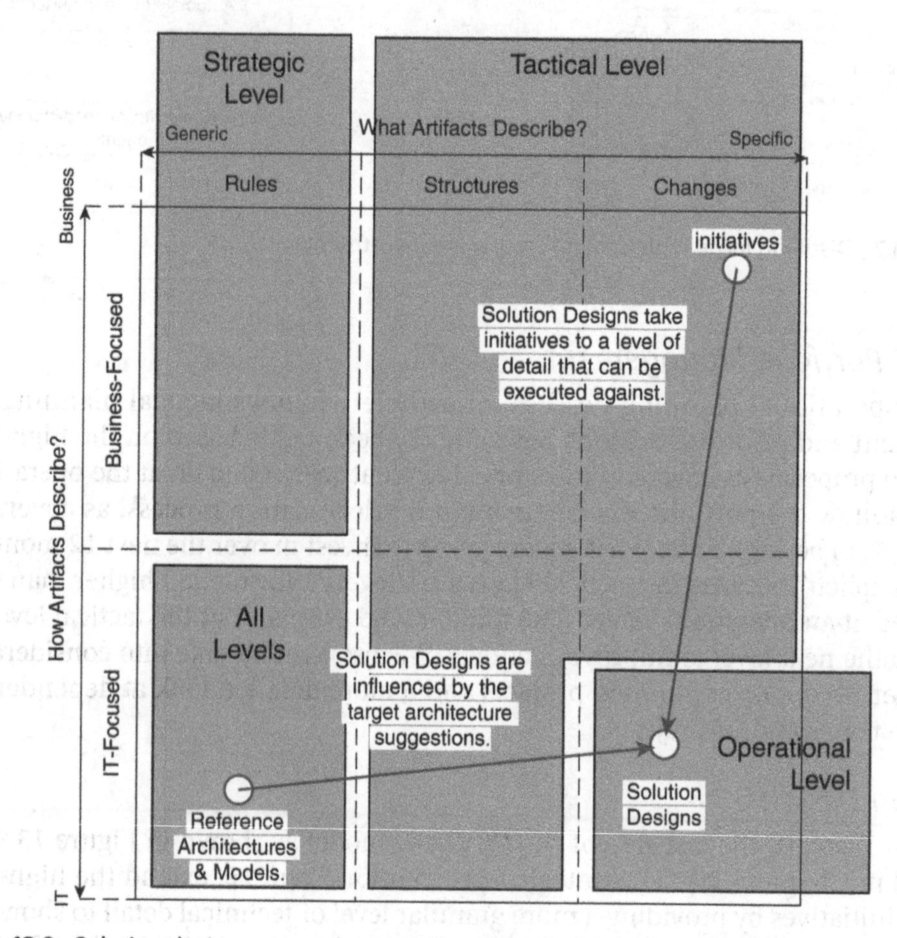

Figure 13.8: Solution designs are at the operational level of the CSVLOD model.

Regardless of whether we are building or buying, exploring or exploiting, taking an iterative approach or investing time in an upfront design, there are common aspects that all solution designs must cover. In his book *The Practice of Enterprise Architecture: A Modern Approach to Business and IT Alignment*, SK Publishing; 15 Jan. 2021, Svyatoslav Kotusev suggests three critical aspects:

- **Technologies:** The programming languages, frameworks, hosting designs, and database technologies to be used. This should be aligned with the enterprise architecture guidelines and/or policies.
- **Interactions:** How the solution design will integrate and interact with the existing IT landscape. This should cover data flows as well as changes to data models.
- **Nonfunctional requirements:** Security considerations, availability, recoverability, compliance, performance SLAs.

Even when taking an agile approach to a project there is still a need for technical governance to avoid overcomplicating the IT landscape or overlooking critical nonfunctional security or compliance requirements.

Implementation Plans

As well as the solution designs, we need the plans of how teams intend to implement them. Some solutions will be complemented with detailed project plans covering the tasks and milestones, like a traditional Gantt chart. For solutions that are more complex, exploratory, or focused on continuous investment, we can use product road maps that reflect more of the uncertainty and unpredictability of what is required to achieve the outcome.

The product road maps are made of the features that the team believes will contribute to delivering the desired tactical initiative outcome. As illustrated in Figure 13.9, teams look for opportunities that will result in desired outcomes. From these opportunities or sub-outcomes, teams identify the solutions and features they need to build. Just as a strategic objective has many tactical initiatives that can contribute to achieving it, so too an initiative can have many options on how to achieve it. Some opportunities will have obvious solutions with a clear correlation between cause and effect, some will be more of a hunch or a hypothesis that requires experimentation and testing.

These operational product-level road maps, as highlighted in Figure 13.10, become progressively less detailed as they plan features into the future. There is no need to spend time on deep discovery of features for outcomes on a road map that are over six months into the future and that are more than likely to change. For the current quarter, or the now, product teams are in deep discovery, experimenting based on the hypotheses that they believe will achieve the current outcomes. For the next quarter, those hypotheses are more like guesses on where the team thinks it will spend its time. In three and four quarters ahead, the near and far future, the teams may have a high-level idea of the themes and areas where they could investigate but no detail on exactly what they will build. As outcomes come into focus and teams discover more about the problems, they will begin to work on a detailed of solution designs and feature ideas.

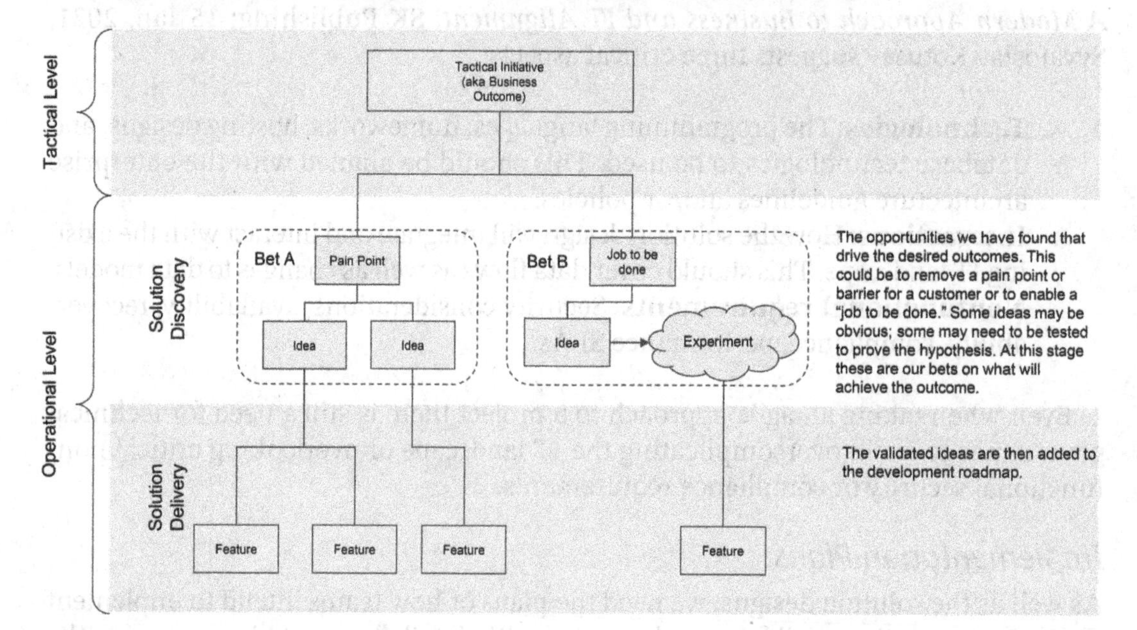

Figure 13.9: Product teams perform solution discovery to determine how best to deliver a business outcome.

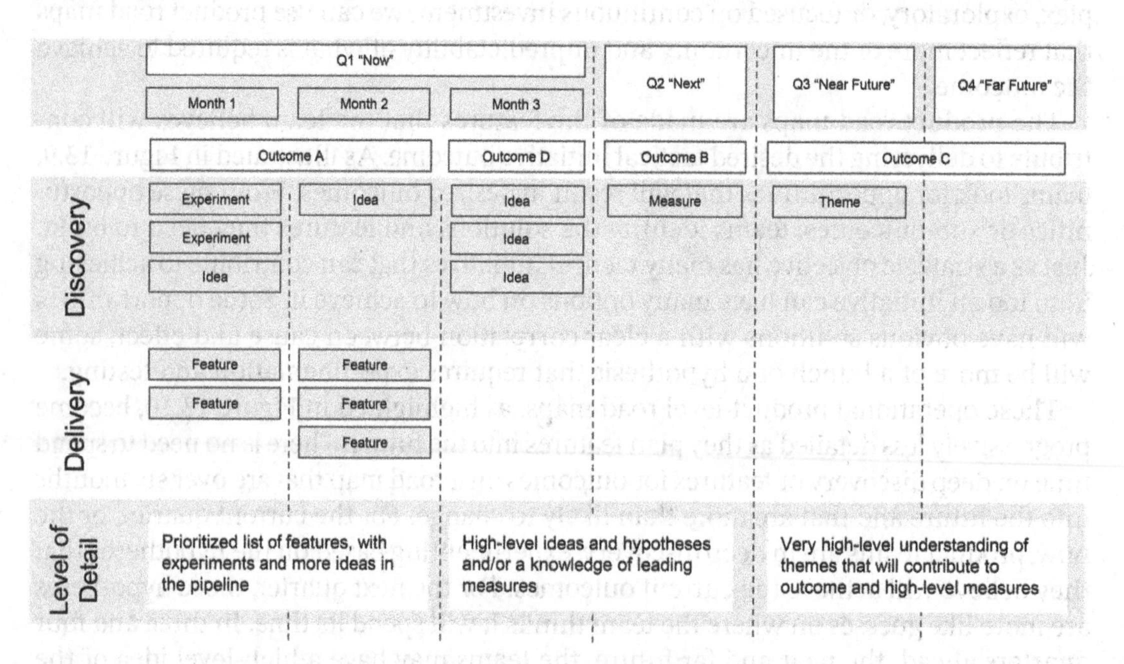

Figure 13.10: Product team road maps get more abstract the further out they plan.

Output

The output of the operational planning process is a road map of work with an accompanying budget for the financial year. As the audience for this is the board, we don't need to go into detail about what team is doing what. However, we need to show the business outcomes IT intends to deliver or contribute to (in other words, what the board is getting for its money). The full budget will also include the costs to run the business, typically the opex costs' lines along with the capex work that the initiatives are focused on. Figure 13.11 shows an example of how we can present the operational plan.

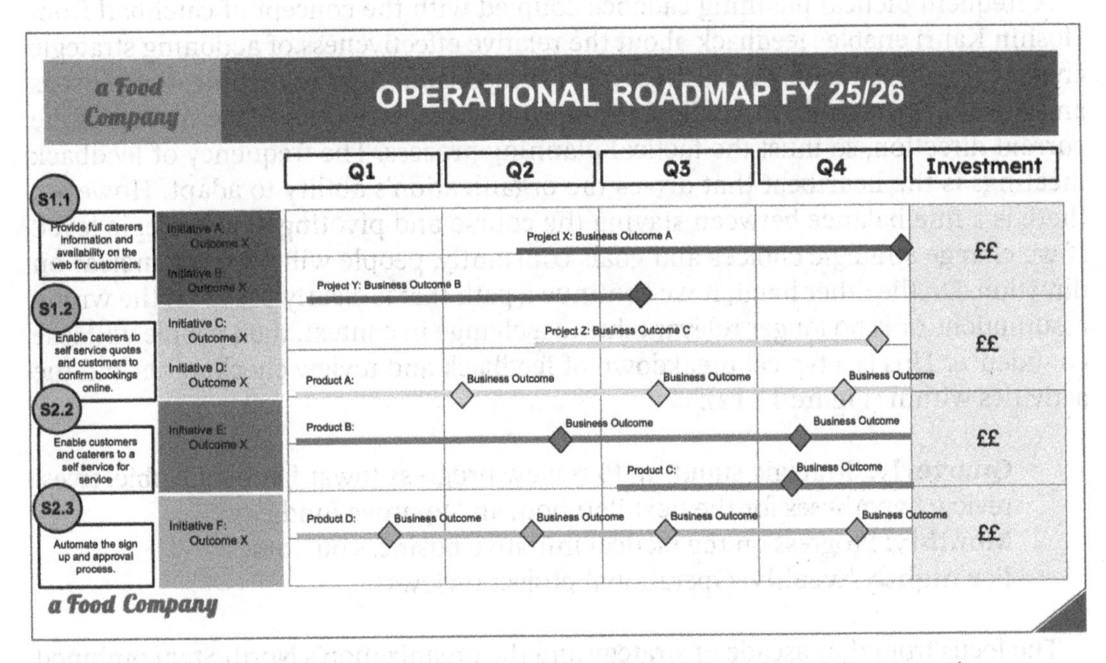

Figure 13.11: A high-level operational plan covering what IT intends to deliver over the next 12 months

Feedback, Learning, and Adapting

Change is the only thing that you can rely on. Strategies, tactics, and operational plans need to be constantly reevaluated, adaptable, and fluid, factoring in new learnings and changes in context on route toward delivery of a vision. Jeff Bezos, CEO of Amazon, once said, "We are stubborn on vision. We are flexible on details." Jeff Bezos knows that the path toward success is not set in stone. New competitors emerge, constraints appear, and technology advances open new opportunities. A business may need to change its proposition in response to customer demand or to respond to a new entrant into the marketplace. As Arie de Geus, a Dutch business theorist, commenting on an ever-changing business context puts it, "The ability to learn faster than competitors may be the only sustainable competitive advantage." We can only achieve this by accepting that change is the norm and learning is key. We need to accept the reality of working in complex problem domains with an unpredictable future and respond to change as opposed to religiously following a plan based on assumptions that may no longer be true. We achieve this by reviewing and adjusting, not only at an operational level but at a tactical and strategic level as well.

A frequent tactical planning cadence coupled with the concept of catchball from Hoshin Kanri enables feedback about the relative effectiveness of actioning strategic choices and investments across the layers of responsibility in an organization. Just as an agile project might have a daily stand-up and retrospective about the success of the current direction, so must the tactical planning process. The frequency of feedback meetings is the heartbeat that drives the organization's ability to adapt. However, there is a fine balance between staying the course and pivoting at a strategic level. If we change strategic choices and goals constantly, people will become unclear on direction. On the other hand, if we continue a path that is clearly based on the wrong assumptions or is no longer relevant due to a change in context, then people will lose confidence. This is a typical breakdown of feedback and review checkpoints and the activities within (Figure 13.12):

- **Quarterly:** Strategic stand-up to review progress toward strategic objectives, review hypotheses for the next iteration, and approve funds
- **Monthly:** Progress on the tactical initiative business outcomes
- **For nightly/weekly:** Operational project reviews

The focus from the cascade of strategy and the organization's North Star combined with the beat of the drum from the regular rhythm of planning iterations and feedback cycles drives the organization and keeps it in sync with strategic objectives. Conversations and analysis at regular intervals ensure the flow of ideas, the catchball process, up and down the organizational layers. This continuous flow of dialogue helps to cement the link between the strategic objectives and the operational day-to-day execution. The cadence of planning is designed to enable us to better cope with more rapid changes to the business context. This is important for a learning organization dealing with problems in the complex context as knowledge on how to proceed comes from experience, or put another way, we learn by doing. Experimentation and feedback are more valuable than trying to plan too far into the unpredictable future. It's only when we experiment that we generate data on our hypothesis; it's only when we stop to study the data that we can then make judgments on how to act on it—stay on the same path or pivot in a new direction. This combination of planning, doing, checking, and acting accelerates our learning and ensures that we embed a mental model of continuous improvement into the system. Governance comes through regular feedback, enabled by a frequent planning process, instead of through conformance to a plan.

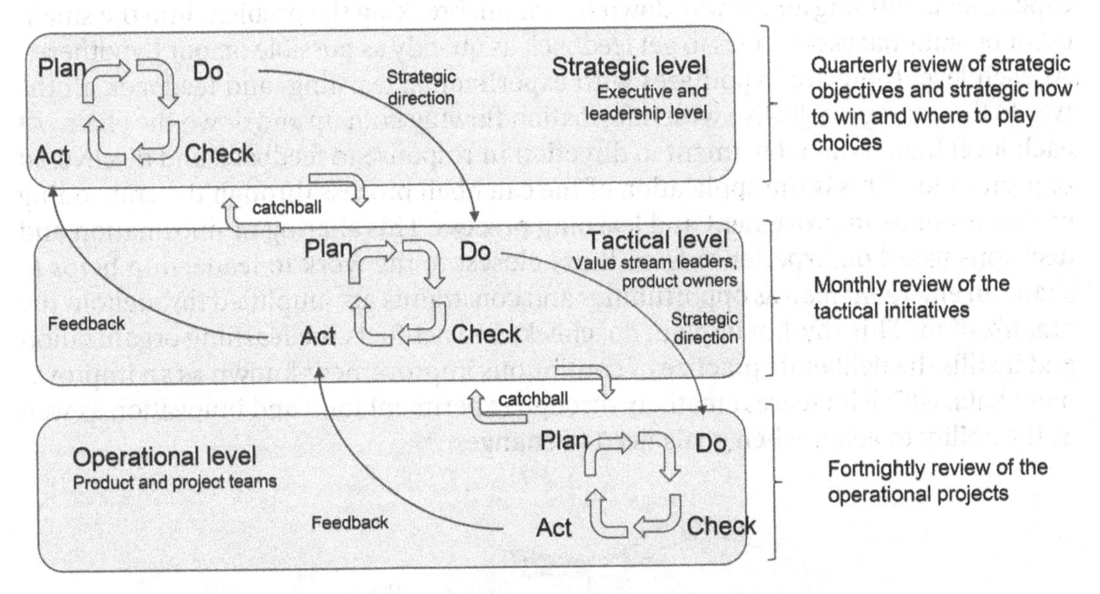

Figure 13.12: Feedback and review cadence based around Hoshin Kanri

Tactical and Operational Review: Use Nested PDCA Loops to Deliver Strategic Goals

As shown in Figure 13.13, we have nested PDCA feedback loops that integrate the operational plans with the tactical and strategic objectives:

- **The strategic objective loop:** Once a quarter the feedback of the tactical initiatives is studied at a strategic level to judge how well they have contributed to the strategic objective and, based on the learnings, how best to act on further investments. At the start, a strategic objective will have a rough outline of how it should be achieved, but as we constantly gather new information along the way, this plan will adapt and evolve using the PDCA loop.
- **The tactical initiative loop:** The plan part of the strategic objective loop is a portfolio of countermeasures or tactical initiatives. The "do" part of the strategic objective loop is the start of the tactical initiative loop.
- **The operational loop:** The plan part of the tactical initiative loop is composed of programs, projects, or continuous product investments. The "do" part of the tactical initiative is the start of the operational execution loop.

In complex problem domains, the tactical initiatives are the hypotheses on how to achieve the strategic objective, and the operational investments are the hypotheses on how to achieve the tactical initiative. In other words, we have experiments within experiments running all the way down the chain, breaking the problem into the smallest of operational executables to get feedback as quickly as possible on our hypotheses. At each level there are hypotheses with experiments, learning, and feedback all the way to the strategic objective with information flowing both up and down the chain. At each level there is an adjustment to direction in response to feedback and discovered opportunities. This is the application of the catchball process through the embedding of a continuous improvement and learning process. This sharing of information and decisions based on experience from those closest to the work to leadership helps to shape an emergent plan as opportunities and constraints are amplified throughout the organization. This rhythm of plan, do, check, and act fuels the learning organization and instills the deliberate practice of continuous improvement known as an improvement kata, which increases creativity through experimentation and innovation as well as the ability to adapt when goals need to change.

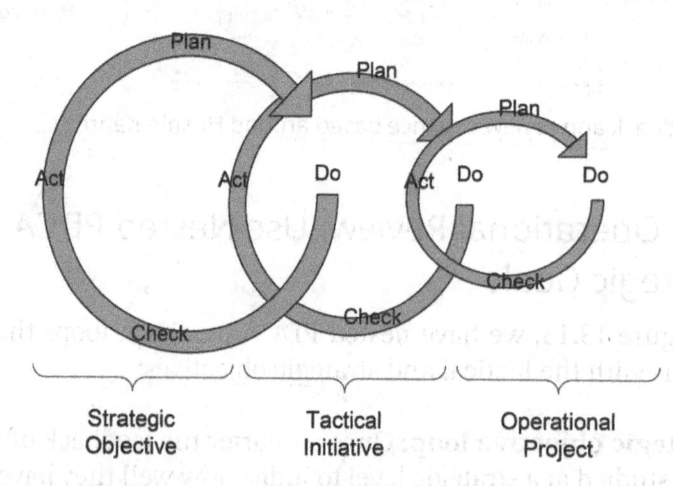

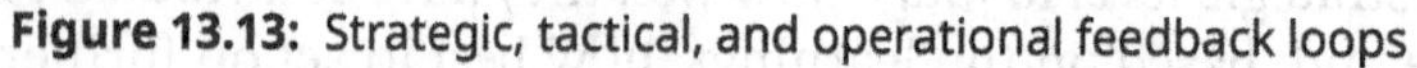

Figure 13.13: Strategic, tactical, and operational feedback loops

However, as covered in Chapter 6 and illustrated in the Wardley Map in Figure 13.14, the feedback loops will not always be based on a plan-do-check-act loop, which is best when optimizing an existing service or product. For problems focused on exploration, we may opt for the build-measure-learn loop based on the lean startup framework. Alternatively, for problems focused on high levels of quality and consistency, we may want to look at the Six Sigma define-measure-analyze-improve-control loop.

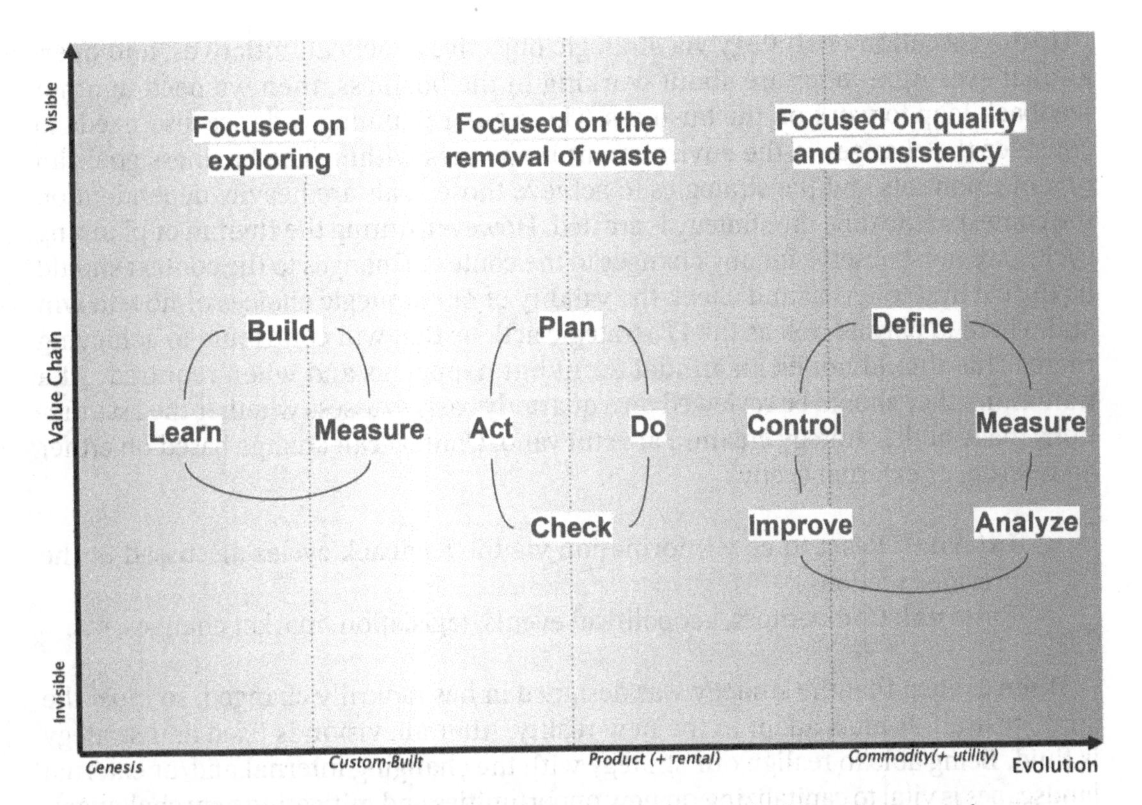

Figure 13.14: Actual feedback loops depend on the project methodology being used.

Strategic Review: Using the OODA Loop and Wardley's Strategic Cycle to React and Adapt to Contextual Changes

A strategy is not something you should do once every three years; it should evolve as and when the context changes. Internal feedback and insight from actions you take along with external impacts to changes in the business environment should trigger a review of the strategic choices, enabling the business to adapt to the analysis of new information. We must be willing to challenge and test our strategic assumptions if the context that they were based upon was to change. We would expect the high-level strategy to change and respond to major impacts to the environment that it was based on. Forecasting over years is hard; however, it is good to have a vision and an overall direction, but we must iteratively progress and adapt on the way, including how to continuously monitor the environment your organization operates in as well as the feedback from your actions.

If the execution of strategy via strategic objectives, tactical initiatives, and operational execution loops are about working in the business, then we need another feedback loop to work on the business. The regular planning cadence also needs to consider the changes to the environment it operates within. The business goals an organization sets and the strategies to achieve those goals are heavily dependent on the context at the time the strategy is created. However, during the rhythm of planning cycles, we must observe for any changes to the context. Changes to the context should be the catalyst to revisit and check the validity of our strategic choices of how to win and where to play as well as the IT strategic actions that will contribute to achieving them. This should not be an annual event but happen as and when required. At a minimum, they should be reviewed on a quarterly basis to assess whether the assumptions made at the strategy creation are still valid. Context can change based on either an internal or external event:

- **Internal:** React to new information via the feedback cycles discussed in the previous section.
- **External:** Competitors, geopolitical events, legislation, market changes, etc.

If the context that the strategy was designed in has radically changed, so must the strategy itself. It must adapt to the new reality. After all, vision is fixed and strategy is fluid. Being able to realign our strategy with the changing internal and/or external landscapes is vital to capitalizing on new opportunities and mitigating new challenges. To ascertain if our strategic choices are still relevant, we must constantly observe, orient, and decide and act, the four activities of the OODA (observe, orient, decide, act) loop.

The OODA loop is a decision-making model that focuses on distilling the available, and often incomplete, contextual information, understanding how it impacts you, and rapidly deciding on how to react. It was created by John Boyd, a US Air Force colonel, and designed so that fighter pilots could adjust strategies quickly in combat situations. When applied to business or IT strategy, it can be used to react to events in your business environment faster than your competitors and thus take a strategic advantage. Figure 13.15 shows how the OODA loop can be applied to organizational strategic thinking.

1. **Observe:** We need to observe and be aware of the environment we are working within. We should constantly gather data on external and internal events to gain an understanding of a threat of opportunity.
   a. **External:** Are there new competitors; has a change in technology made a competitive capability now a commodity? Is there talk of new legislation or compliance needs? Have security issues been exposed? Is there a chance of an acquisition or merger? Are there changes in the market?

   b. **Internal:** What information have we gathered from initiatives under way? Has a team made a discovery that has made a hypothesis invalid? Has an experiment unearthed a new opportunity or revealed a new constraint? Has internal feedback resulted in a change of strategic goal?

2. **Orient:** When we have gathered data and observed the event, we can analyze and reflect on how it impacts us. Is our business model still valid? Are our strategies no longer correct? Have we under- or over-invested in an area? What is different in the context now compared to when we set the strategic direction? At this stage, we also refer to our strategic principles, doctrine, and measures to guide how we observe and react to new changes in the context and how we act on new decisions.
3. **Decide:** After we have digested how the observed event affects us, we can decide on the best course of action. What should we stop or continue to do? Do we need to create a new goal, or should we evolve an existing one? Should we increase investment? Should we switch from an agile team to off-the-shelf software? Should we change the tech stack?
4. **Act:** Once we have decided, we need to act on it in a manner that enables fast feedback. We need to validate our decisions as they may have been made with limited and incomplete data. If we are unsure, we need to get clarity before we fully invest and drop into the PDCA loop. We should carry out experiments, use AB tests or pilots to gain clarity, and study how effective our response to the event will be.

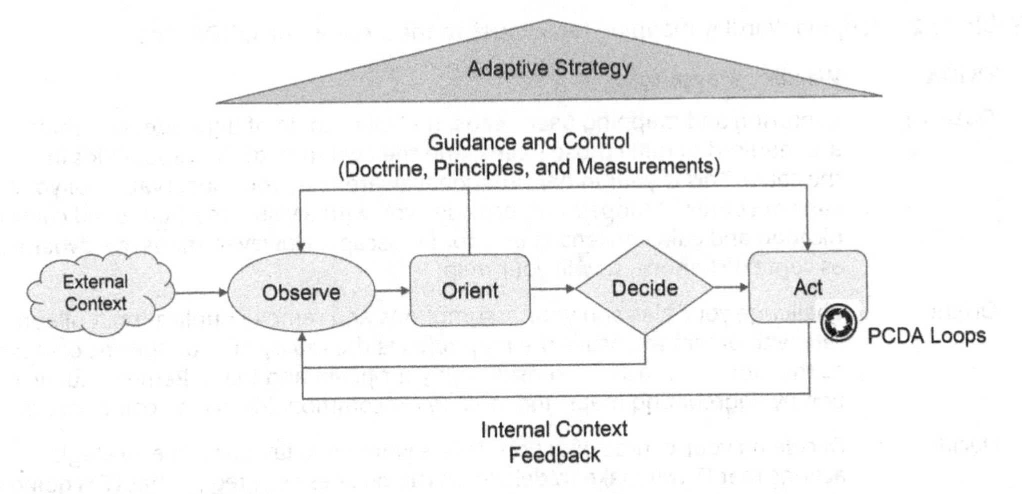

Figure 13.15: The OODA loop

The Wardley mapping methodology, a strategic thinking approach, is based on movement and situational awareness, which will change based on factors inside and outside the control of the organization. Therefore, the Wardley mapping process is introduced in the context of the OODA loop as shown in Figure 13.16. Table 13.2 shows the techniques of Wardley mapping adjacent to the related steps within the OODA loop.

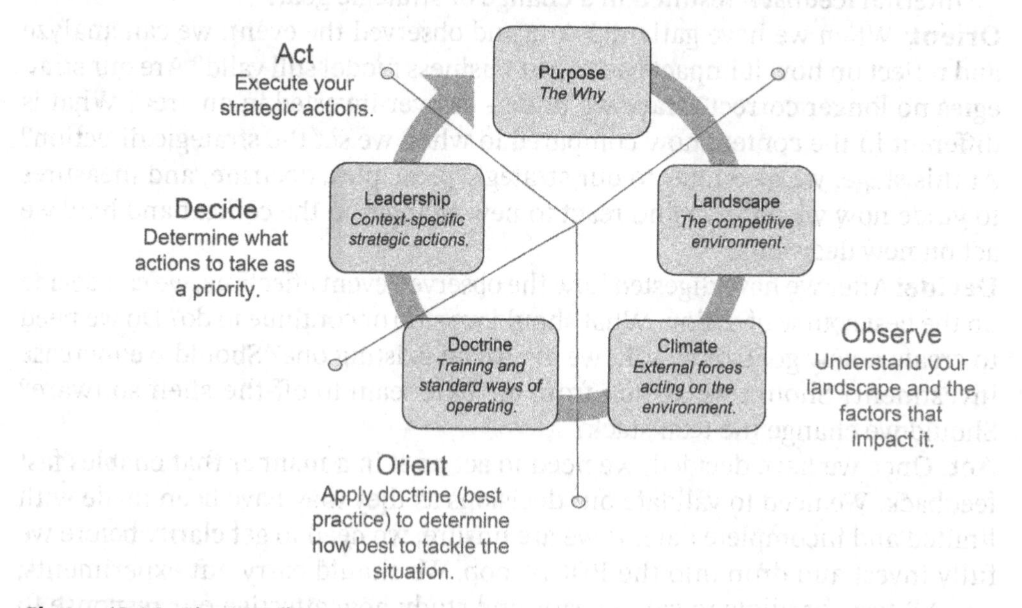

Figure 13.16: The Wardley strategic cycle

Table 13.2: Mapping Wardley mapping techniques to the steps in the OODA loop

OODA	Wardley Mapping
Observe	Capturing and mapping user needs, the value chain of dependencies that are required to fulfill those needs, and the evolution of the capabilities in the chain. This is your initial situational awareness, your observation of your current context. Mapped out, provides you with a visual method to aid communication and gain consensus on your landscape. However, maps are dynamic; as context changes, so will your map.
Orient	Challenge your bias and your assumptions and remove duplication. Collaborate with others to ensure the map reflects the reality of your specific context so that you can reach consensus with your peers and team. Remove duplication by aggregating maps and looking for commonality across components.
Decide	Decide on your course of action. This is where you lay down the strategic actions that IT will make to deliver on the business strategy. What IT is going to do to contribute to business success.
Act	The last action is to execute your strategy. However, through feedback from execution and or changes in your business environment, you will need to constantly iterate all the steps to ensure your strategic outlook is still relevant and effective. Mapping is not the end game; it is, however, an effective method to visualize and aid communication.

Integrated Feedback Loops

The OODA and PDCA loops may appear to differ only by the terminology used, but as shown in Figure 13.17, they are applicable to different levels of organizational planning. The PDCA loop is useful to follow when we are trying to solve a problem; the OODA loop will help us identify the most worthwhile and strategic problems to solve. The organization's strategy is concerned with setting intent and making explicit how and where an organization will win. Because of this, the factors that can influence the strategic outlook are chiefly, but not exclusively, external in nature. Strategic choices and intent are reevaluated based on a change to the environment that an organization operates within.

The OODA loop allows us to react to changes in the context, understand how they impact us, and decide, usually on insufficient data, how to act. This loop may result in just trying things to sense how we can deal with market changes, new opportunities, or challenges. However, as our experience of the new reality increases, we can move to the PDCA loop to help the organization execute a strategic objective, a tactical initiative, or an operational project.

The PDCA loop is applicable when the organization has more clarity and a strong hypothesis on a way forward to respond to changes in context, whereas events that question the strategic direction are managed by the OODA loop. The flow between the two loops is not one way. Feedback from operational execution can rise and influence the strategic outlook just as an external event can. The catchball method of Hoshin Kanri is about the flow of information and data between the strategic and tactical layers of an organization so that there is a synergy between strategic thinking and operational execution.

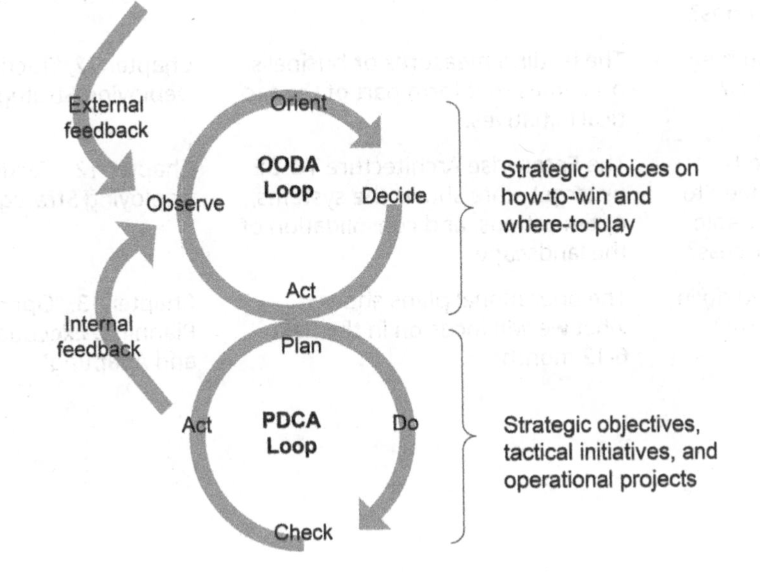

Figure 13.17: How the OODA and PDCA loops interrelate

Creating a Clear Line of Sight from Strategy to Execution

This chapter completes the strategic deployment process. By following the advice in the chapters of Part 3, "Strategy to Execution," the organization should be aligned and able to drive in the same direction to achieve business success. All employees should be able to answer the questions as detailed in Table 13.3, which will ensure alignment.

Table 13.3: Questions to verify strategic alignment

Question	Answer	Chapter Reference
Do you know where the organization is headed?	North Star: The company vision and mission.	Chapter 10, "Understand your Business"
How will you know when you get there?	3- to 5-year business goals will have been achieved.	Chapter 10, "Understand your Business"
Do you know how you will get there?	The business strategy articulates where we will play and how we will win.	Chapter 10, "Understand your Business"
Do you know what you will do to contribute?	The strategic actions as detailed in the IT strategy.	Chapter 11, "IT Strategic Contribution"
Do you know how you will deliver the strategic actions?	Deliver the initiatives as detailed in the tactical plan.	Chapter 12, "Tactical Planning: Deploying Strategy"
How will you measure progress?	The leading measures or business outcomes that form part of the tactical initiatives.	Chapter 12, "Tactical Planning: Deploying Strategy"
How will the IT landscape need to change to enable business success?	The Enterprise Architecture TO-BE or target state shows the systems, optimizations, and consolidation of the landscape	Chapter 12, "Tactical Planning: Deploying Strategy"
What will you do in the short term?	The operational plans show what we will focus on in the next 6-12 months.	Chapter 13, "Operational Planning: Execution, Learning, and Adapting"

Question	Answer	Chapter Reference
What if our assumptions on how to achieve business success are wrong?	We can use operational and tactical feedback loops to check and adjust our actions.	Chapter 13, "Operational Planning: Execution, Learning and Adapting"
What if things change and our plans become invalid?	We can use the OODA feedback loop at the strategic level to ensure business goals and strategic objectives are still valid and adjust plans if they are not.	Chapter 13, "Operational Planning: Execution, Learning and Adapting"

A Worked Example: From Strategy to Tactics to Operational Execution

To help cement the process of strategic deployment, we will work through a fictitious example. We will follow an e-commerce retail company and one of its strategic objectives that concerns the need to improve its customer experience to increase customer loyalty. This example has been kept deliberately simple to show strategic deployment end to end. Business problems are often much more complex.

The Strategic Objective

The first step is to understand the strategic objective and the context behind it—the why or the purpose. Too many times we rush to look for a solution without having a fundamental understanding of the problem we are trying to solve. This is an essential step for alignment. Even though we might think everyone knows why we are doing something, we must be explicit and even overcommunicate it. This is the principle of mission.

The business context is as follows:

> *Reputation for a business is vital to its success, and the effect of online reviews and customers recommendations can greatly impact upon this. We cannot afford to ignore what our customers are saying if we hope to retain their loyalty and turn them into brand ambassadors.*
>
> *We have been losing market share over the last 18 months. During this time, we have seen that the two leading competitors in our market space have made improvements observed through an increase in positive online review*

scores. During this time, we have also seen the sentiment of our customers' NPS steadily decreasing and Trustpilot reviews increasingly left with a negative score.

We need to improve NPS as that is a leading indicator for customer loyalty, and we need to turn the tide of negative reviews as that is contributing to reduction in new customers, put off by previous customer negative experiences.

Trustpilot is a website that focuses on business reviews. It allows customers to see honest reviews about customer service for different companies. A TrustScore is calculated on a scale from 1 to 5. The organization's Trustpilot review score over the last 12 months is 3.2. Our competitors' Trustpilot scores over the last 12 months are as follows:

- Competitors A - 4.0
- Competitors B - 4.1
- Competitors C - 3.6

Net promoter score (NPS) is a benchmarking tool that gives insights about customer loyalty by measuring customers' willingness to recommend a business. Customers rate a business on a 0–10 scale of how likely you would be to recommend it to a friend: 0–6 are detractors, 7–8 are passives, and 9–10 are promoters. The NPS score is determined by adding the count of detractors (-1), passives (0), and promoters (+1), dividing by the total number of ratings, and then multiplying the number by 100. The organization's NPS score is -21.

Our strategic objective is as follows:

- Regain the loyalty of our customers.
    - Increase NPS from -21 to 30.
    - Improve the average TP score from 3.2 to 4.2.

As this objective is about optimizing an existing product or service, we will apply lean thinking and follow the PDCA process. Figure 13.18 shows the start of an A3 document (covered in Chapter 7) to solve the strategic objective of regaining the loyalty of our customers.

Defining the IT Strategic Actions

As our objective is focused on the customer, the most appropriate set of tools to use is a customer journey map and customer research practices. To determine our strategic actions, we needed to discover the pain points, constraints, or barriers that prevent customers from achieving their jobs and therefore having a poor experience reflect in the NPS and Trustpilot scores. We can also use the service blueprint to trace the problems that customers have experienced to backend root causes in business capability and process.

Theme: Regain the loyalty of our customers

Background

Reputation for a business is vital to its success and the effect of online reviews and customers' recommendations can greatly impact upon this. We cannot afford to ignore what our customers are saying if we hope to retain their loyalty and turn them into brand ambassadors.

We have been losing market share over the last 18 months. During this time we have seen that the two leading competitors in our market space have made improvements observed through an increase in positive online review scores. During this time we have also seen the sentiment of our customers' NPS steadily decreasing and Trustpilot reviews increasing left with a negative score.

We need to improve NPS as that is a leading indicator for customer loyalty and we need to turn the tide of negative reviews as that is contributing to reduction in new customers, put off by previous customer negative experiences.

Current Condition

Our Trustpilot review scores

Competitors' Trustpilot score last 12 months:

- Competitor A - 4.0
- Competitor B - 4.1
- Competitor C - 3.6

Our NPS trend over time

Goal Statement

- Increase NPS from -21 to 30
- Improve TP average to beat competitor move from 3.2 to 4.2

Figure 13.18: Capturing information about the problem using an A3 template

Consider the customer journey map in Figure 13.19. This is a result of customer research, journey mapping, and connecting pain points to a business service blueprint and, finally, identifying waste in the related business process via value stream mapping and time and motion studies. The discovery process has identified the pain points and the related business metric evidence listed in Table 13.4. The pain points are also shown as a cause-and-effect diagram in Figure 13.20.

Table 13.4: Customer and business pain points impacting NPS and Trustpilot scores

Customer/ Business Pain Points	Capability
Slow and frustrating returns process ■ Customers are having to wait up to 30 days to receive a refund. ■ Customers have to phone customer services (CS) to arrange a collection. ■ A time and motion study has found the waste in organizing a collection and refund.	Order Fulfilment and Delivery
What I want is not in stock ■ Only 60% of products were in stock and available to order for delivery that week. ■ Customers are contacting CS to ask when stock is expected.	Allocation and Replenishment
Delivery not on time or in full (OTIF) ■ Only 85% of orders were OTIF. ■ Mapping the fulfilment value stream, we discovered many orders requiring manual intervention. ■ We also discovered that CS have had to refund customers as acts of goodwill, and therefore this is diluting our profits. It is also taking up lots of CS time to resolve.	Order Fulfilment and Delivery Stock Management

■ Faults with products
■ 5% of orders with faults
■ 10% of orders returned.
■ Top reasons for post-booking contacts:
  ■ Not as described.
  ■ Instructions not that great.
	Quality Management

Not happy with the level of customer service ■ Limited self-service capability. ■ Only 30% of customers were satisfied with resolution from CS. ■ CS first contact resolution is at 50%.	Customer Services
Not all customers are providing a Trustpilot review ■ 7% of customers completing an NPS score. ■ Only 3% of customers are completing a Trustpilot review and the majority of these are from customers who have had a negative experience.	Campaign Management
Site is so slooooow ■ Page speed averages at 7 seconds, which is frustrating for customers and will likely damage our organic SEO listing.	e-commerce Sales Channel

With an understanding of the pain, barriers, and obstacles both the business and customers are facing, we can focus on actions that IT can make to contribute to achieving the business goal. Table 13.5 shows the strategic contribution that IT will make in the form of strategic actions, along with measures, to help the business succeed.

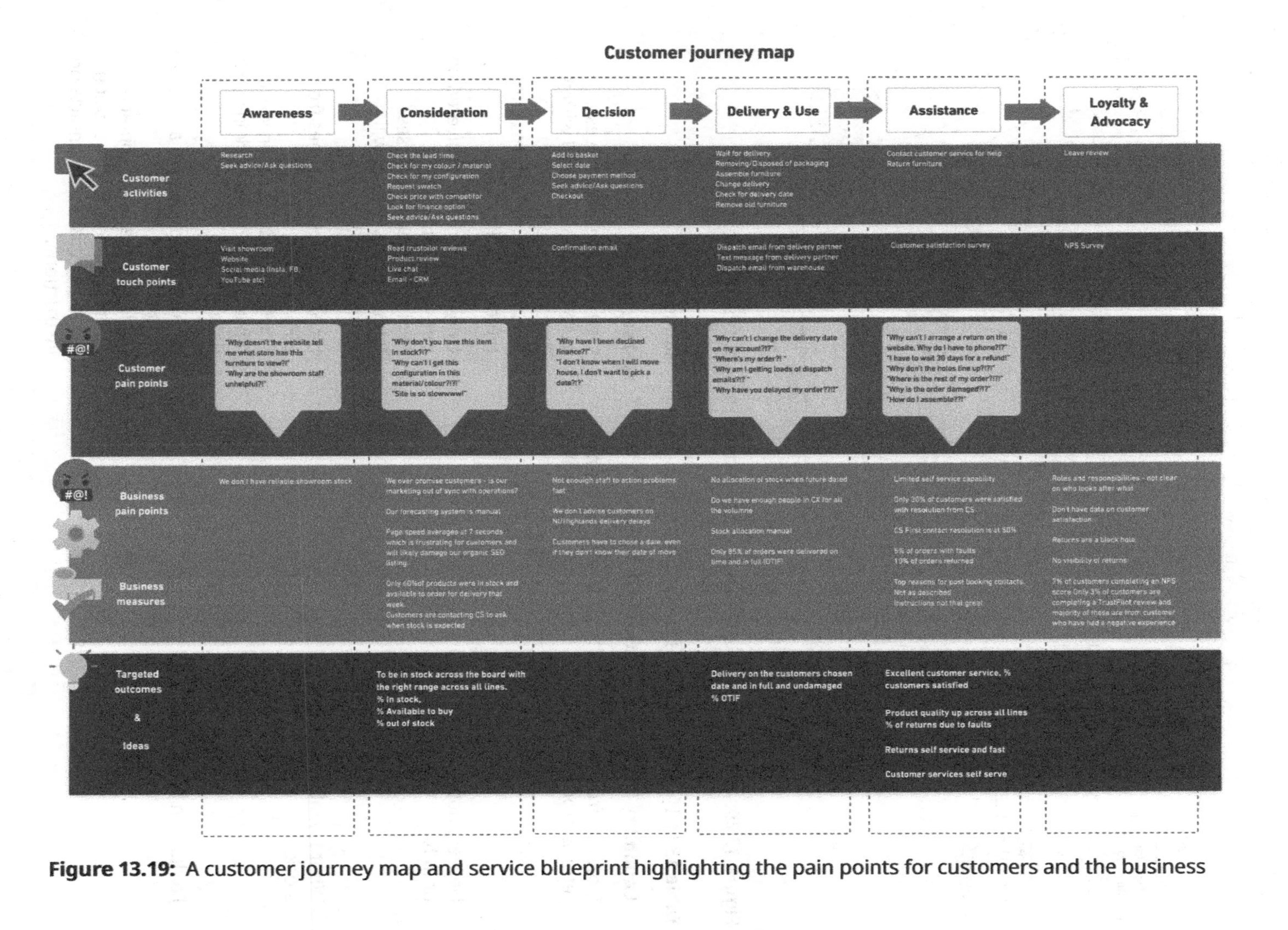

Figure 13.19: A customer journey map and service blueprint highlighting the pain points for customers and the business

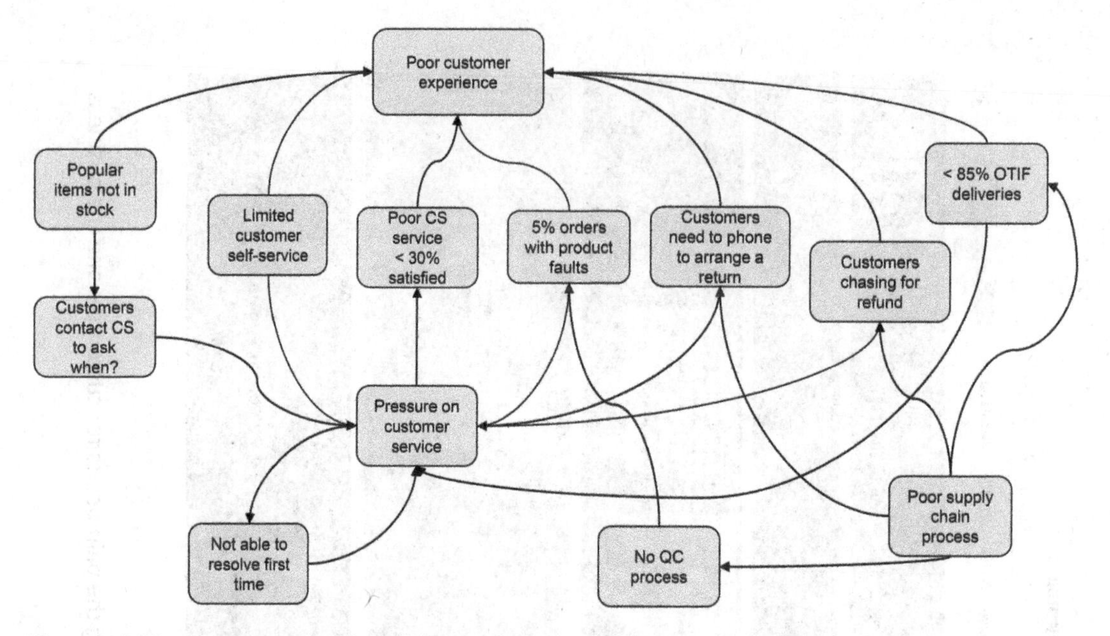

Figure 13.20: A cause-and-effect diagram showing the root causes of issues

Table 13.5: IT strategic actions to contribute to improving customer NPS

Capability Pain Point	Strategic Action	Measure
Slow and frustrating returns process.	S1 Automate the supply chain management.	■ Customers refund process automated and settled within 3 working days. ■ 90% of customer returns completed via self-service.
Delivery not on time or in full.		■ Only 97% of orders delivered on time and in full (OTIF). ■ Avoid overhead increases. ■ Reduced acts of goodwill refunds by 50%.
What I want is not in stock.		■ Increase stock of all lines from 60% to 90%. ■ 100% of products not in stock have an estimated back in stock date shown.
Faults with products.		■ Reduce returns due to a fault to 1%. ■ Avoid any customer contacts based on incorrect description. ■ Avoid any customer contacts on instructions.

Capability Pain Point	Strategic Action	Measure
Not happy with the level of customer service.	S2 Provide a customer services platform to enable customer self-service and agent effectiveness.	■ 75% of customers use a self-service option. ■ Increase first contact resolution to 50% → 80%. ■ Increase customer service satisfaction from 30% to 80%.
Not all customers are providing a Trustpilot review.	S3 Provide data on customers to power the campaign management system.	■ Increase NPS completion from 67% to 80%. ■ Increase customers adding a Trustpilot review from 43% to 60%.
Site is so sloooooow.	S4 Provide an e-commerce platform.	■ Average full-page load under 2 seconds.

Enterprise Architecture Suggestions

We can now turn our attention to reviewing how well the IT landscape of existing systems and technology are supporting the current capabilities of the business in the form of an AS-IS diagram. Where there is a technology gap, or where the technology is in poor condition we will highlight and annotate to show how it impacts achieving the strategic actions. In the TO-BE architecture we can highlight what the target state needs to look like in order to deliver the strategic actions as well as the opportunities to consolidate or optimize the enterprise architecture.

AS-IS

The IT landscape is shown in Figure 13.21 and the systems that are directly related to the IT strategic action have been annotated as to the technical condition and how well they currently support the business capabilities. Table 13.6 lists the review of the suitability of each system.

Table 13-6: The state of the AS-IS Landscape

Strategic Action	System	Comment
S4 Provide an e-commerce platform.	e-commerce sales channel.	■ The e-commerce platform and content management systems are delivered through a bespoke application with major performance and latency issues. ■ Needs development resource to update key product information and product augmentation e.g., instructions.

Continues

Table 13-6: The state of the AS-IS Landscape (*Continued*)

Strategic Action	System	Comment
S2 Provide a Customer Services platform to enable customer self-service and agent effectiveness.	Customer Services.	■ The customer self-service capability is managed via the corporate email system. ■ There is very little self-service post order functionality, all contact is directed via email.
S3 Provide data on customers to power the campaign management system.	Campaign Management.	■ No Customer data on orders is fed into the marketing campaign system to request product reviews.
S1 Automate the Supply Chain Management.	OMS.	■ Complex bespoke application poor capability for returns. ■ Stock availability based on manual spreadsheets.
	MMS.	■ Forecasting and planning is manual and heavily spreadsheet-based.
	WMS.	■ Enterprise-level system but missing critical quality management module. This capability has no clear process or supporting systems. ■ Errors with mis picks due to the paper-based process.
	Delivery Management.	■ Manual integration with a single delivery partner with poor visibility on delivery status and outgoing manifests. ■ Manual and inconsistent process for returns arrangements.

TO-BE

With a review of the AS-IS landscape and the knowledge of the strategic actions, we can design the future state of the architecture. This is a combination of optimization and consolidation to simplify the landscape at the same time as providing systems that can enable key business capabilities. Figure 13.22 shows a diagram of the target architecture suggestions with a reference to the strategic actions that the changes support. Table 13.7 lists the target architecture suggestions.

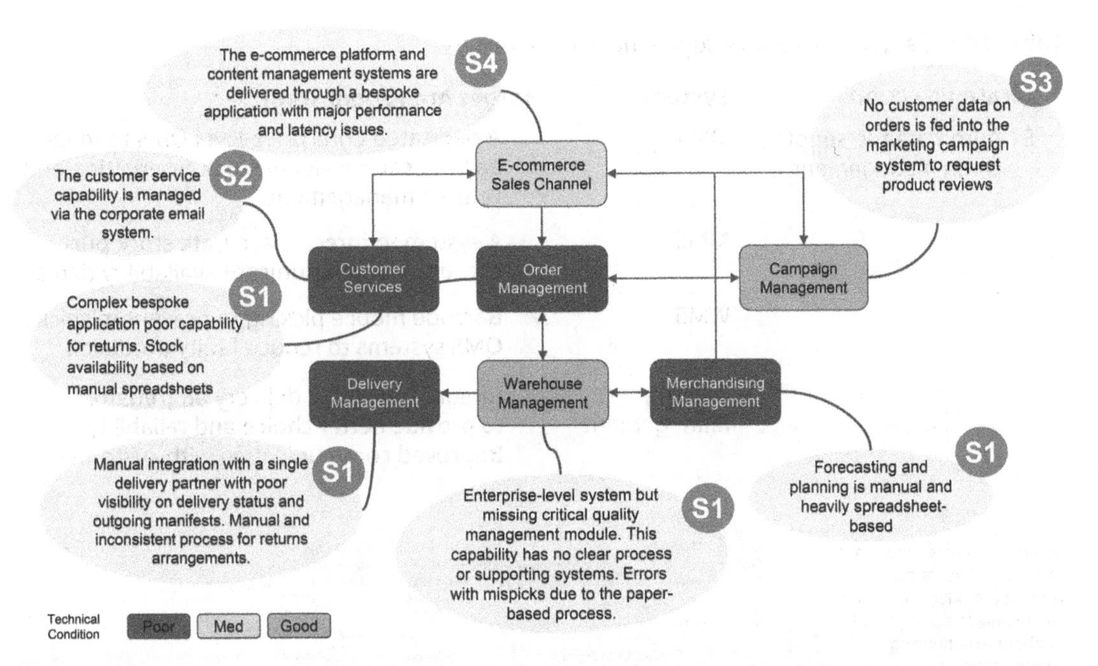

Figure 13.21: A diagram highlighting the state of the IT landscape in relation to supporting IT strategic actions

Table 13.7: Target architecture suggestions

Strategic Action	System	Target Arch Suggestion
S4 Provide an e-commerce platform.	e-commerce sales channel	■ An e-commerce platform where end users can manage product information and content that can scale without experiencing performance issues.
S2 Provide a customer services platform to enable customer self-service and agent effectiveness. S3 Provide data on customers to power the campaign management system.	CRM	■ Integrated customer service and campaign management system with data model augmented with web behavior and order data.

Continues

Table 13.7: Target architecture suggestions (*Continued*)

Strategic Action	System	Target Arch Suggestion
S1 Automate the supply chain management.	OMS	■ A dedicated enterprise-level OMS to manage the complexity of stock reservation and returns management.
	MMS	■ A system to forecast accurate stock purchases and communicate availability dates.
	WMS	■ Barcode mobile picking to prevent mispicks. ■ QMS systems to reduce faulty products.
	Delivery Management	■ Integration with a delivery aggregator to provide better choice and reliability; Improved communication with customers.

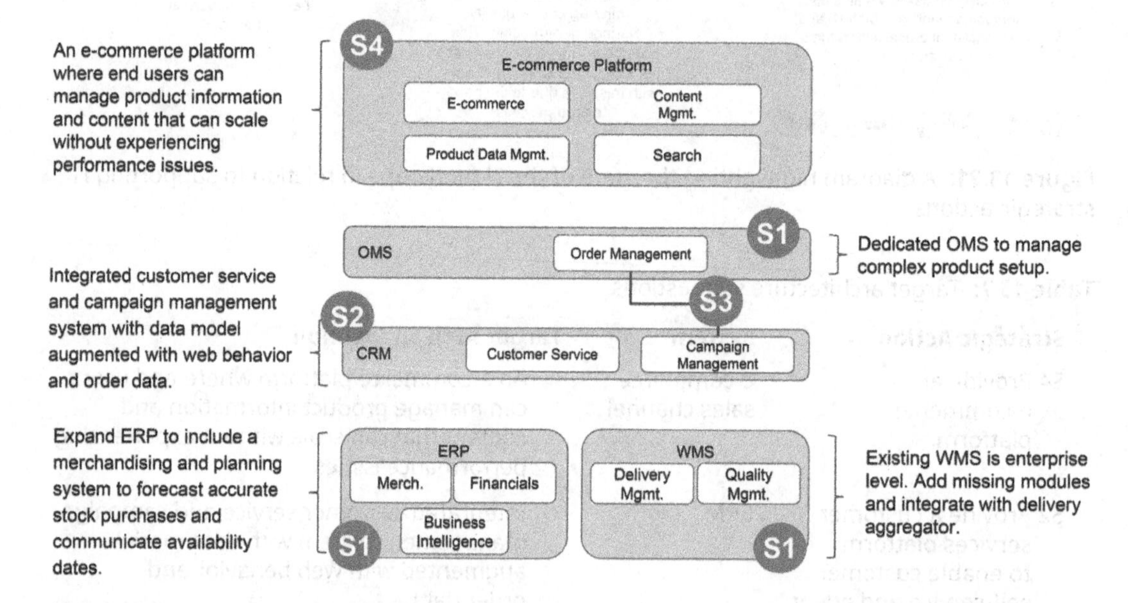

Figure 13.22: A diagram highlighting the target architecture suggestions to support the IT strategic actions

Operating Model Suggestions

To deliver on the strategic actions, we need to change the structure of the IT organization. As shown in Figure 13.23, there will be four teams based around the value streams and business capabilities that directly relate to the strategic actions:

- Discovery Team
    - Organized around the Visit to Order value stream
    - Internal team with continuous investment focused on conversion and customer engagement

- Order Team
  - Organized around the Order to Delivery value stream
  - Internal team with a focus on the continuous investment in optimizing the checkout and order fulfilment process
- Service Team
  - Organized around the Request to Response value stream
  - Internal team with a focus on self-service for post order needs
- Supply Chain Team
  - Organized around the Warehouse and MMS business capabilities
  - Outsourced to partner for deep expertise

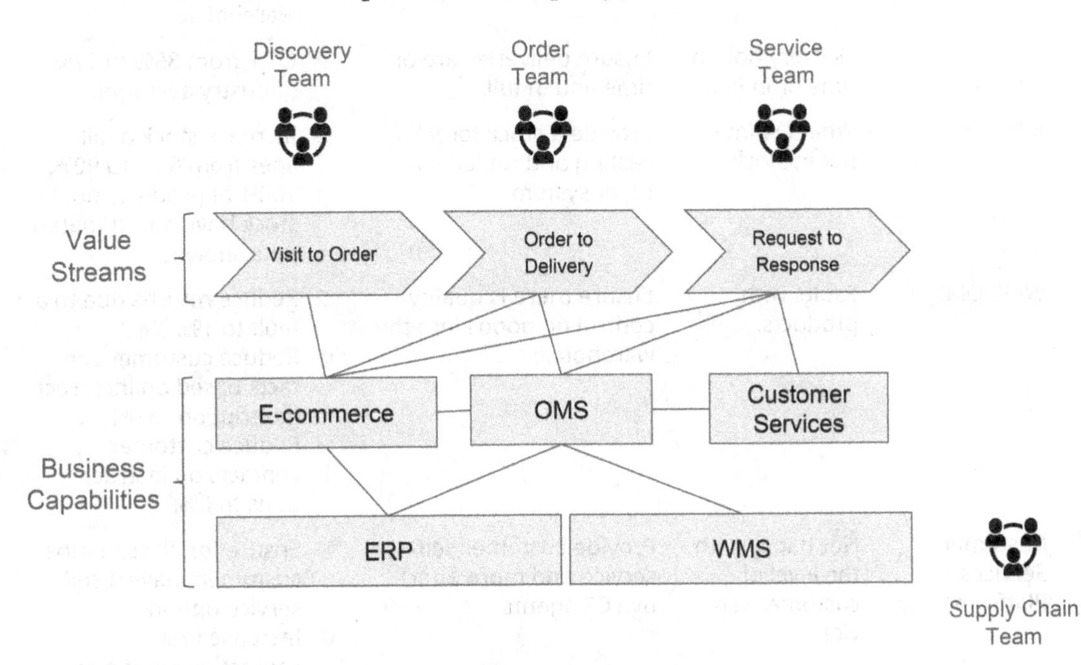

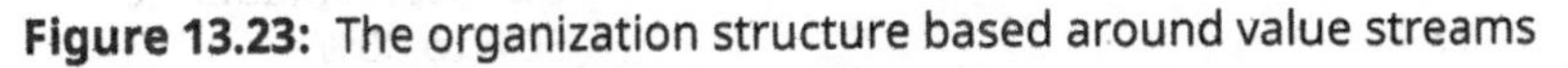
Figure 13.23: The organization structure based around value streams

Defining the Tactical Initiatives

With the strategic actions and the target architecture suggestions defined, we can determine the tactical initiatives that will bridge the gap from the AS-IS landscape to the TO-BE target vision. As listed in Table 13.8, the tactical initiatives are shown with the business outcome they will deliver and related measures.

As shown in Figure 13.24, we can update the A3 template for the strategic objective with the proposed countermeasures (tactical initiatives) that will contribute to achieving the goal. The result of the discovery and definition stages of problem exploration is a portfolio of tactical initiatives that can be prioritized for delivery. In addition, we have a deeper understanding of both the problem and the customer context.

Table 13.8: Tactical initiatives to move to the target state

TO-BE	AS-IS	Tactical Initiative	
Target state	**Gap**	**Business Outcome**	**Measure**
OMS	Slow and frustrating returns process.	Streamline the returns process.	■ 80% of customers to arrange returns via self-service channel. ■ Return inspection and refund within 5 hours of the product being received back into the warehouse.
	Delivery not on time or in full.	Ensure deliveries are on time and in full.	■ OTIF from 85% to 97% (industry average).
MMS	What I want is not in stock.	Provide a stock forecasting and replenishment system.	■ Increase stock of all lines from 60% to 90%. ■ 100% of products not in stock have an estimated date shown.
WMS QMS	Faults with products.	Ensure there is quality control on goods into the warehouse.	■ Reduce returns due to a fault to 1%. ■ Reduce customer contacts based on incorrect description to 0%. ■ Reduce customer contacts on instructions to 0%.
Customer Services Platform	Not happy with the level of customer service.	Provide customer self-service and more effective CS agents.	■ Ensure for all scenarios customers have a self-service option. ■ Increase first contact resolution to 50% to 80%. ■ Increase customer service satisfaction from 30% to 80%.
Campaign Management	Not all customers are providing a Trustpilot review.	Expose data so that marketing can create a campaign to improve NPS and Trustpilot reviews.	■ Increase NPS completion from 67% to 80%. ■ Increase customers adding a Trustpilot review from 43% to 60%.
E-commerce Platform	Bespoke website.	Stabilize the e-commerce sales channel.	■ Average full-page load under 2 seconds.

Theme: Regain the loyalty of our customers

Background

Reputation for a business is vital to its success and the effect of online reviews and customers' recommendations can greatly impact upon this. We cannot afford to ignore what our customers are saying if we hope to retain their loyalty and turn them into brand ambassadors.

We have been losing market share over the last 18 months. During this time we have seen that the two leading competitors in our market space have made improvements observed through an increase in positive online review scores. During this time we have also seen the sentiment of our customers' NPS steadily decreasing and Trust Pilot reviews increasing left with a negative score.

We need to improve NPS as that is a leading indicator for customer loyalty and we need to turn the tide of negative reviews as that is contributing to reduction in new customers, put off by previous customer negative experiences.

Current condition

Our Trustpilot review scores

Competitors Trustpilot scores last 12 months:

- Competitor A - 4.0
- Competitor B - 4.1
- Competitor C - 3.6

Our NPS trend over time

Goal Statement

- Increase NPS from -21 to 30
- Improve TP average to beat competitor move from 3.2 to 4.2

Analysis

Countermeasures (outcomes to target)

Initiative	Outcome Measure
Streamline the returns process	• 80% of customers to arrange returns via self-service channel • Return inspection and refund within 5 hours of the product being received back into the warehouse.
Ensure deliveries are on time and in full	• OTIF from 85% → 97% (industry average)
Provide a stock forecasting and replenishment system	• Increase stock of all lines from 60% to 90%. • 100% of products not in stock have an estimated date shown.
Ensure all quality control on goods in	• Reduce returns due to a fault to 1% • Reduce customer contacts based on incorrect description to 0% • Reduce customer contacts on instructions to 0%
Provide customer self-service and more effective CS agents	• Ensure for all scenarios customers have a self-service option • Increase first contact resolution to 50% → 80% • Increase customer service satisfaction from 30% to 80%
Expose data so that marketing can create a campaign to improve NPS and TrustPilot reviews	• Increase NPS completion from 67% to 80% • Increase customers adding a TrustPilot Review from 43% to 60%
Stabilize the Ecommerce sales channel	• Average full page load under 2 seconds

Confirmation (results)

Follow up (actions)

Figure 13.24: Strategic objective A3 with tactical initiative countermeasures

Operational Plans and Solution Design

To demonstrate breaking an initiative down into operational projects, we will use the tactical initiative of "Ensure deliveries are on time and in full," as highlighted in Figure 13.25.

Teams can leverage the techniques discussed previously in Chapter 6, such as value stream mapping and event storming to discover more about the obstacles standing in the way of achieving this tactical initiative outcome. As shown in Figure 13.26, work has been done to understanding the root cause of why the OTIF is only at 85%.

Discovery in the solution space is about generating multiple options for delivering a solution. It's about teams obsessing on the desired outcome rather than being wedded to a particular output. This is the essence of outcome over output, the need to focus on generating business outcomes rather than the successful delivery of a particular feature. Figure 13.27 shows an impact map based on the root cause analysis captured in Figure 13.26 to determine what projects to focus on. The target outcomes, aka projects, are also listed in Table 13.9, and finally, we have an A3 template, shown in Figure 13.28, for the tactical initiative with a portfolio of operational project countermeasures. In turn each team will have an A3 template to manage and review each project at an operational level.

Table 13.9: Operational projects distilled from the tactical initiative of improving OTIF

Tactical Initiative	**Operational Projects**		
	Project and Measure	**Outcomes to Target**	
		Actor	**Impact**
Delivery on time and in full. Measure: Increase OTIF from 85% to 97%.	Integrate delivery partner and address finder API.	Customer	Only select delivery dates with capacity to their post-code.
			No deliveries overscheduled.
		Delivery Partner	Notify customers when there is a delay in delivery.
			Allow customers to track deliveries.
	Configure OMS stock allocation system to avoid stock being oversold.	Operations	Reduce stock over allocation to zero.
	Integrate barcode mobile picking to avoid mispicks.		Reduce mispicks in the warehouse to zero.
	Auto upload manifest to delivery partner.		Reduce missing orders from the delivery partner manifest to zero.

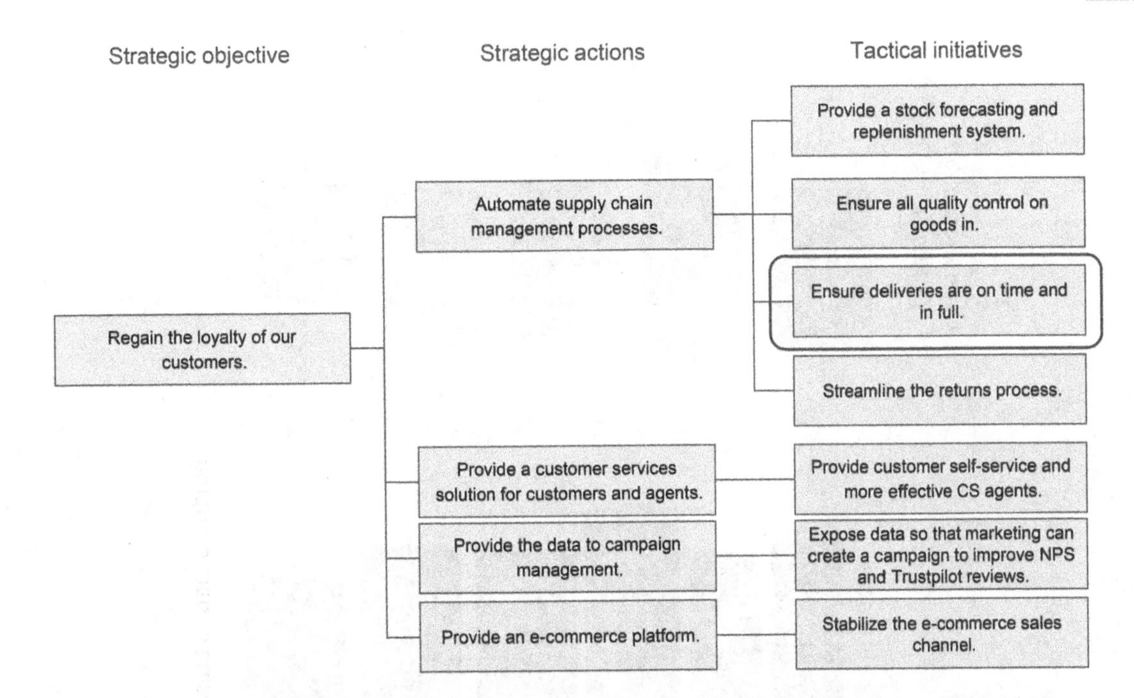

Figure 13.25: To demonstrate the operational level, we will focus on a single tactical initiative.

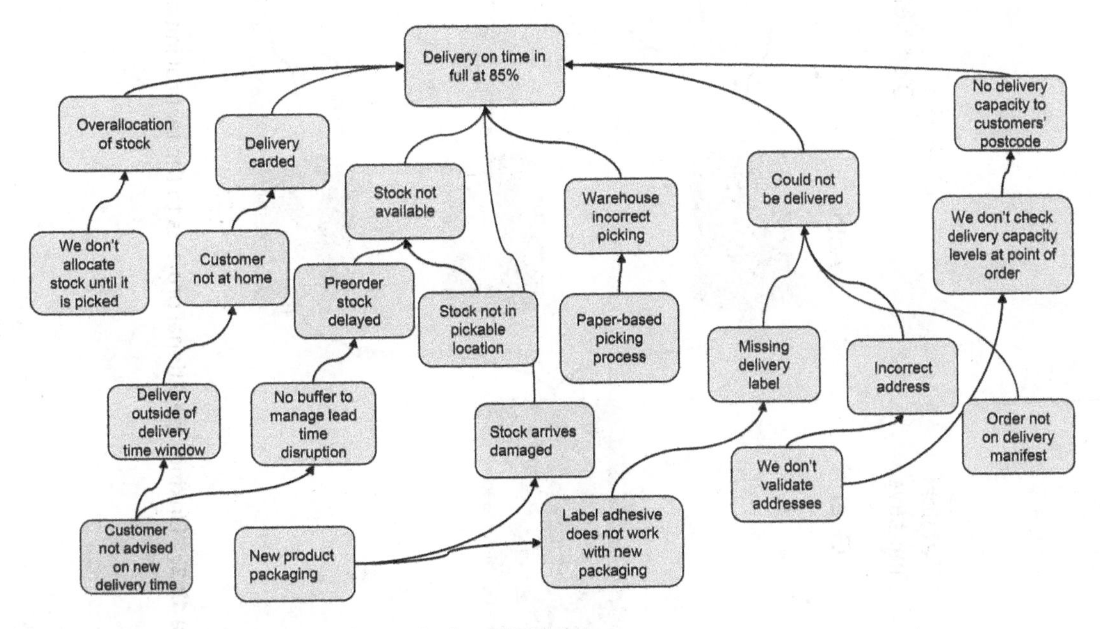

Figure 13.26: Root cause analysis of why OTIF is low

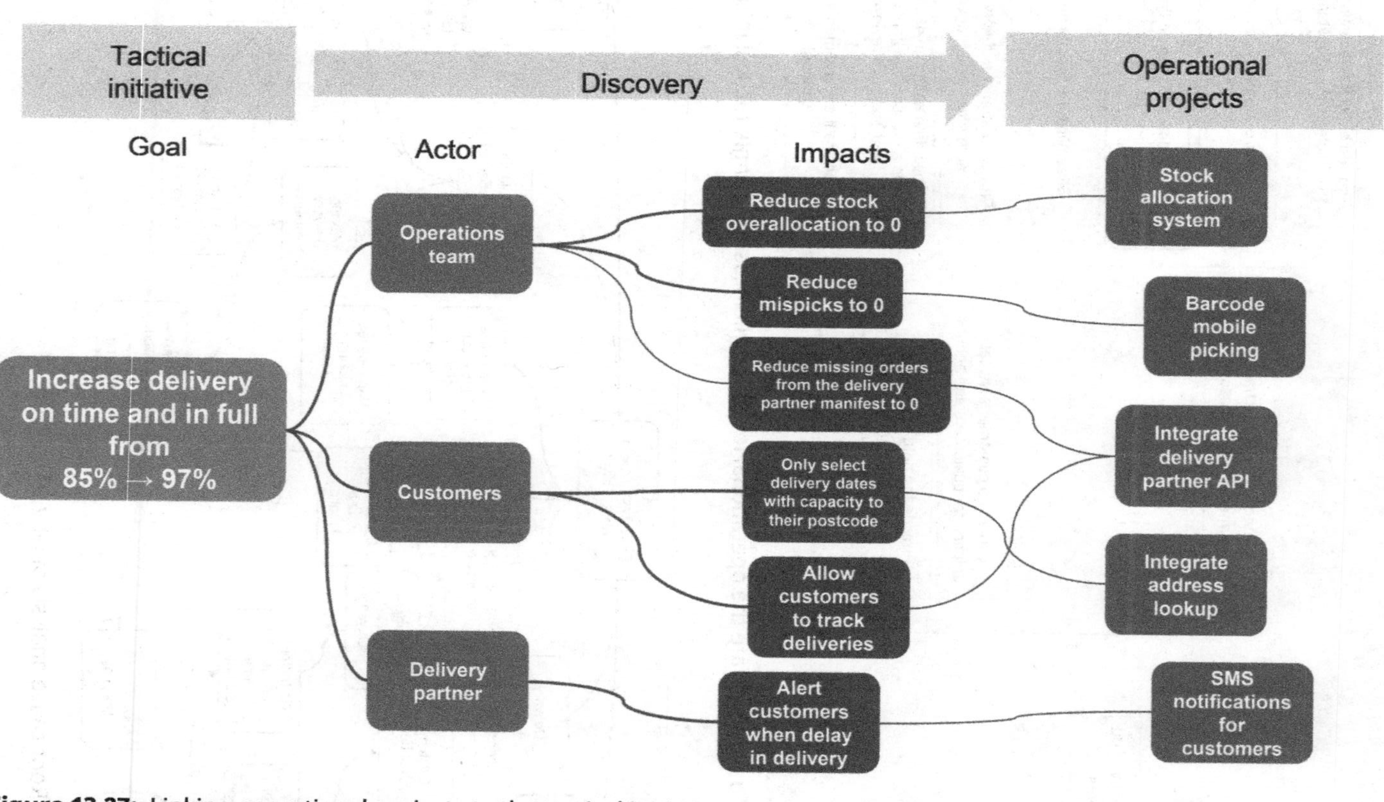

Figure 13.27: Linking operational projects to the tactical initiatives using impact mapping

Theme: Delivery on time and in full

Background

A key reason for customer frustration is the high amount of deliveries that are not delivered on time or in full. We are well below industry average and the highest single complaint relates to this issue.

Current condition

- OTIF 85%.
- Customer feedback 70% complain about late deliveries.

Goal Statement

- Increase OTIF from 85% → 97%.

Analysis

Delivery On Time in Full at 85%.
Over allocation of stock.
Delivery Carded.
Stock not available.
Warehouse incorrect picking.
Could not be delivered.
No delivery capacity to customers postcode.
We don't allocated stock until it is picked.
Customer not at home.
Preorder stock delayed.
Stock not in pickable location.
Paper based picking process.
Missing Delivery Label.
Incorrect address.
We don't check delivery capacity levels at point of order.
Delivery outside of delivery time window.
No buffer to manage lead time disruption.
Stock arrives damaged.
We don't validate addresses.
Order not on delivery manifest.
Customer not advised on new delivery time.
New product packaging.
Label adhesive does not work with new packaging.

Countermeasures (outcomes to target)

Action	Measure
Integrate delivery partner and configure SMS notification.	• Only select delivery dates with capacity to their postcode. • Missing orders from the delivery partner manifest to 0. • No deliveries overscheduled.
Configure OMS Stock allocation system to avoid stock being oversold.	• Reduce stock overallocation to 0.
Integrate Barcode Mobile Picking to Avoid Mispicks.	• Reduce mispicks to 0.
Integrate address finder API.	• Reduce delivery failures due to Incorrect address.

Confirmation (results)

Follow-up (actions)

Figure 13.28: Tactical initiative A3 with operational project countermeasures

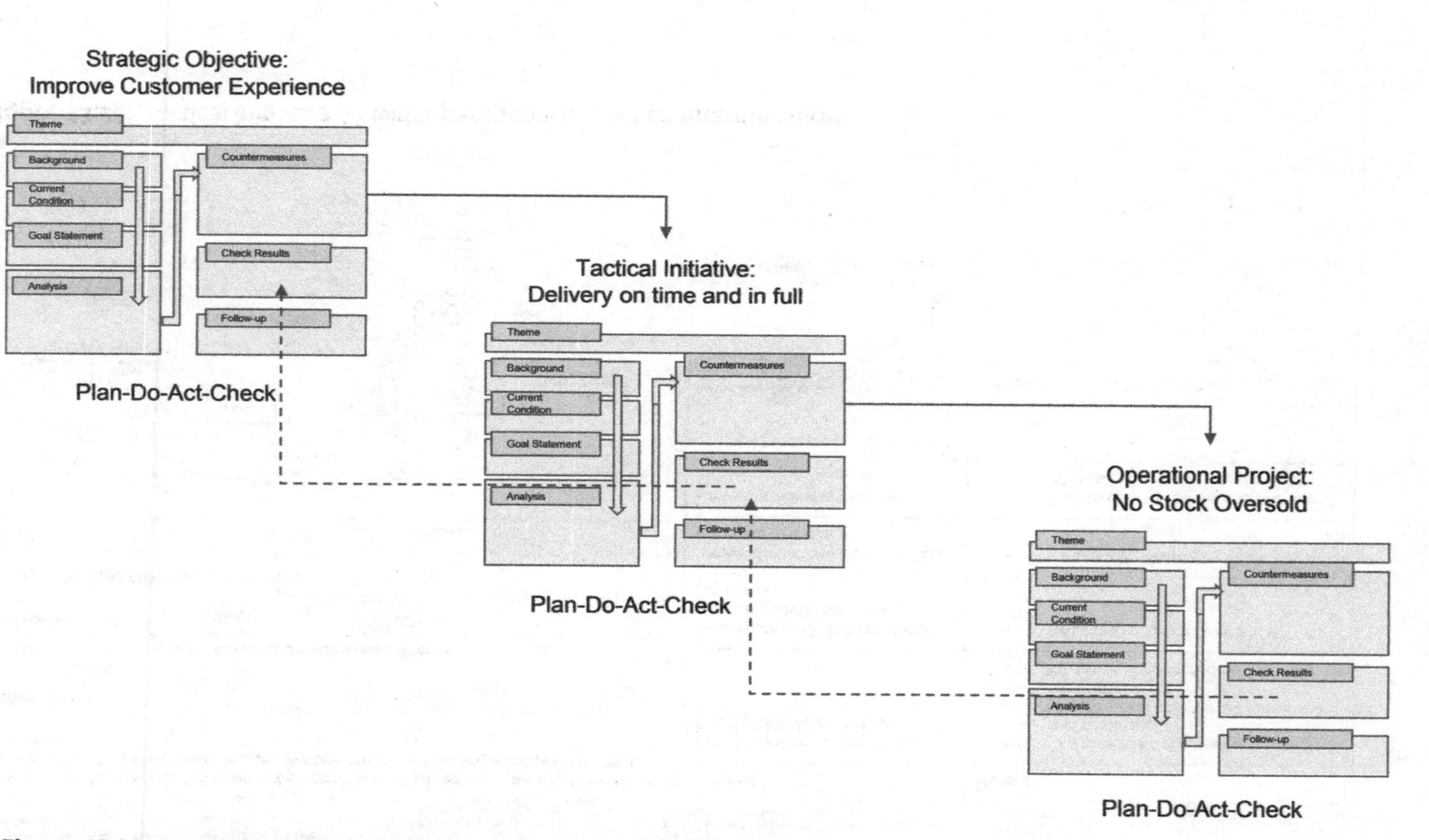

Figure 13.29: Integrated strategic, tactical, and operational PDCA loops

Feedback

As shown in Figure 13.29, the cascade of intent and measurement is illustrated using the A3 templates at each organizational planning level. This ensures the teams don't lose the purpose behind why we need to solve the problem. This offers a systematic and structured approach to problem solving, in this example a focus on optimizing or exploiting a current product as opposed to exploring a new opportunity.

Summary

Operational planning is the final part of strategy deployment. However, even the best laid plans are hard to deploy, especially in complex environments as observed by military theorist Clausewitz. To address the alignment, effects, and knowledge gaps, we looked at using OKRs to delegate business outcomes for teams to solve rather than output, ensuring teams understood intent.

The planning process itself is straightforward. The hard work has been done at the tactical level to determine where we will focus technology investments and in what order. At the operational level, it is all about getting to the next level of detail based on the target architecture suggestions and the appropriate way of working (Chapter 6) and ensuring we are taking all business needs into consideration (Chapter 7), and that we have the right people and structure to deliver it (Chapter 5, "How We Are Organised").

There is only one certainty when trying to plan beyond the next 6 months, and that is that we will be wrong. However, planning at a high level for the long term is not in vain. The act of planning looks at the feasibility of the organization's aspirations against the cost in both time and money of trying to achieve them. It also ensures that we are making decisions on what we invest in with the holistic picture or target vision of what needs to happen for our business to succeed. Of course, we will be wrong, our environment will change, our competitors will have their own strategies that we may need to react to, or our suppliers or political systems will make changes that have a material impact on our business. Technology will continue to evolve, turning the impossible and unfeasible into something that is achievable and affordable. Even within our business, things will change. Our assumptions or hypotheses on how we will deliver our strategic vision may be proved to be incorrect. We may have massively underestimated the complexity of delivering a solution to a problem. Our focus and investment on running the business could drastically outweigh the funding of changing it.

Because change is the only constant, our plans need to be adaptable. At a strategic objective, tactical initiative, and operational project level, we should employ feedback loops to ensure regular reviews to check assumptions and allow for course correction. The purpose of a frequent review cadence is to adapt plans based on feedback. Plans are useful, but often the design process is emergent in nature: we only learn about the problem when we interact with it; the more data and experience we gain allows us to make more informed planning decisions. By leveraging the principles and practices of lean and agile philosophies and applying them to planning and portfolio management, we combine the stable process of traditional planning with a more frequent cadence enabling us to review performance and check assumptions. This new planning process reduces risk by enabling the business to dynamically adapt to opportunities and constraints on an ongoing basis.

Creating a strategy is more of an art than a science. As well as plans being adaptable, we also need to ensure our strategy is kept relevant by allowing it to adapt based on external and internal feedback. From an IT perspective, our strategic actions can be impacted by external factors, more obviously by the advances in technology. This is why it is important for IT leaders to keep an eye on how technology evolves as it may change what we do, or at the least how we do it. Both internal and external contextual information will evolve the strategic outlook and challenge assumptions and bias. We looked at employing the OODA loop to check overall strategy against changes in context and to adapt quickly and outmaneuver competitors.

The act of defining a strategy and plan should not be considered a once and done activity. In the new reality, there is a need for a continuous evolving and adaptable process to allow strategy and our plans to emerge and guide us toward an organization's vision. Our strategy and plans will over time evolve based on choices from feedback made by independent teams across the organization in addition to events in the environment we operate in that question the assumptions our strategy was based upon. The actual path you take to deliver business success will differ from what you had when you set off, but this is optimal, as it shows your organization is learning and adapting. Remember, the value is in the planning, not the plan. In the end agility at an organizational level is the balance between a stable vision and fluid iterations of work toward that vision. Or as Simon Wardley puts it, "Crossing the river by feeling the stones."

Index

A

B

C

E

F

I

J

K

L

M

N

O

Q

R

S

T

U

V